Quantum and Non-Commutative Analysis

MATHEMATICAL PHYSICS STUDIES

Series Editor:

M. FLATO, *Université de Bourgogne, Dijon, France*

VOLUME 16

Quantum and Non-Commutative Analysis

Past, Present and Future Perspectives

edited by

Huzihiro Araki

Research Institute for Mathematical Sciences,
Kyoto University,
Kyoto, Japan

Keiichi R. Ito

Department of Mathematics and Information,
College of Liberal Arts,
Kyoto Unversity,
Kyoto, Japan

Akitaka Kishimoto

Department of Mathematics,
Hokkaido University,
Sapporo, Japan

and

Izumi Ojima

Research Institute for Mathematical Sciences,
Kyoto University,
Kyoto, Japan

KLUWER ACADEMIC PUBLISHERS

DORDRECHT / BOSTON / LONDON

Library of Congress Cataloging-in-Publication Data

Quantum and non-commutative analysis : past, present, and future
perspectives / edited by Huzihiro Araki ... [et al.].
p. cm. -- (Mathematical physics studies ; v. 16)
ISBN 0-7923-2532-X (alk. paper)
1. Mathematical physics--Congresses. 2. Quantum field theory--Congresses. 3. Solid state physics--Congresses. 4. Statistical physics--Congresses. 5. Quantum groups--Congresses. I. Araki, Huzihiro, 1932- . II. Series.
QC19.2.Q36 1993
530.1'43--dc20 93-31537

ISBN 0-7923-2532-X

Published by Kluwer Academic Publishers,
P.O. Box 17, 3300 AA Dordrecht, The Netherlands.

Kluwer Academic Publishers incorporates
the publishing programmes of
D. Reidel, Martinus Nijhoff, Dr W. Junk and MTP Press.

Sold and distributed in the U.S.A. and Canada
by Kluwer Academic Publishers,
101 Philip Drive, Norwell, MA 02061, U.S.A.

In all other countries, sold and distributed
by Kluwer Academic Publishers Group,
P.O. Box 322, 3300 AH Dordrecht, The Netherlands.

Printed on acid-free paper

Printed in the Netherlands

CONTENTS

Foreword

In the past decade, there has been a sudden and vigorous development in a number of research areas in mathematics and mathematical physics, such as theory of operator algebras, knot theory, theory of manifolds, infinite dimensional Lie algebras and quantum groups (as a new topics), etc. on the side of mathematics, quantum field theory and statistical mechanics on the side of mathematical physics. The new development is characterized by very strong relations and interactions between different research areas which were hitherto considered as remotely related.

Focussing on these new developments in mathematical physics and theory of operator algebras, the International Oji Seminar on Quantum Analysis was held at the Kansai Seminar House, Kyoto, JAPAN during June 25–29, 1992 by a generous sponsorship of the Japan Society for the Promotion of Science and the Fujihara Foundation of Science, as a workshop of relatively small number of (about 50) invited participants. This was followed by an open Symposium at RIMS, described below by its organizer, A. Kishimoto.

The Oji Seminar began with two key-note addresses, one by V.F.R. Jones on Spin Models in Knot Theory and von Neumann Algebras and by A. Jaffe on Where Quantum Field Theory Has Led. Subsequently topics such as Subfactors and Sector Theory, Solvable Models of Statistical Mechanics, Quantum Field Theory, Quantum Groups, and Renormalization Group Approach, are discussed. Towards the end, a panel discussion on Where Should Quantum Analysis Go? was held.

A half of this volume consists of written version of these lectures. The other half is from the subsequent RIMS Symposium described below. We hope that these lecture notes provide readers a glimpse at the present status of the fast developing subject of quantum analysis.

As the organizer, I would like to express my gratitude to many people who worked for the organization of this Oji Seminar for their devoted help, to all participants for their enthusiastic participation, to all speakers for their excellent talk and for writing up of their lectures for this volume and to the Japan Society for the Promotion of Science and the Fujihara Foundation for their generous support.

January, 1993 Huzihiro ARAKI

Preface

THE EDITORS

January, 1993

The symposium on Non-commutative Analysis held in Kyoto, June 29 – July 2, 1992, was supported for overseas participants by Commemorative Association for the Japan World Exposition (1970), The Great Britain Sasakawa Foundation, Japan Society for the Promotion of Science, Inoue Foundation for Science, and The Kajima Foundation. I would like to express my sincere thanks to them on behalf of the organizers.

I was advised and assisted by Professor Y. Nakagami among others during the course of the financial arrangements of the symposium and was helped very much by Ms. M. Ohmura, Hokkaido University, and Reiko for secretarial works. I am grateful to them for their generous help.

Akitaka Kishimoto
Department of Mathematics, Hokkaido University
Sapporo, Japan

Two international symposia, Oji International Seminar on Quantum Analysis and International Symposium on Non-commutative Analysis were held in Kyoto in early summer of 1992, with many distinguished participants from overseas. The Oji seminar was held at the Kansai Seminar House located 2.5 km north of Kyoto University. The International Symposium on Non-commutative Analysis was subsequently held at the Research Institute for Mathematical Sciences (RIMS).

The symposia were held on the occasion of the 60th birthday of Professor Huzihiro Araki, and the lectures were dedicated to him. There were more than 40 lectures, and most of them are described in these proceedings. We would like to thank the speakers for presenting their recent work.

Professor Araki had many students who are now working in many fields of mathematical sciences, and the two symposia were planned by these former students and materialized with many advices from Professor Araki himself and with enthusiatic help from secretaries in the RIMS, to whom we would like to express our sincere acknowledgements.

We list the names, addresses and e-mail addresses of the speakers in the final part of the proceedings. We here again would like to say "Arigatou gozai mashita (Thank you very much)" not only for their excellent talks but also for being an excellent audience, on behalf of all the organizers and editors.

Keiichi R. Ito

Department of Mathematics, College of Liberal Arts
Kyoto University, Kyoto 606, Japan, and
Department of Mathematics and Information
Konan College of Women, Konan 483, Japan

Izumi Ojima

Research Institute for Mathematical Sciences
Kyoto University, Kyoto 606-01, Japan

1. Quantum Field Theory

Local Quantum Physics and Beyond

RUDOLF HAAG
II. Institut für Theoretische Physik, Universität Hamburg

Abstract. Local quantum physics concerns the synthesis of the principle of locality on the level of special relativity with quantum theory. The frame of the theory, including its physical interpretation, may be characterized by a net of algebras of local observables satisfying a few general principles. This offers a concise language and an efficient tool for the study of the anatomy of the theory and its qualitative physical consequences. The most pertinent results for physics are

(i) particle structure and collision theory [1]–[5].
(ii) the theory of charge structure, exchange symmetry, global gauge groups [6]–[11]
(iii) characterization of thermal equilibrium, its relation to the mathematical theory of modular automorphisms, to stability and passivity and the rôle of the chemical potentials [12]–[18]
(iv) spontaneous breaking of symmetries [19].

There are many results on the mathematical structure and internal consistency, some of which will be mentioned. Some outstanding open problems in each of the areas (i) to (iv) will be pointed out. For a future theory – which includes the lessons from General Relativity – rather radical changes of the frame are needed. We shall discuss a few hints in this context.

1.

The title of my contribution is rather presumptuous for a talk of 45 minutes but there are two reasons for this choice. The first is the occasion for this seminar, celebrating the 60th birthday of Huzihiro Araki. When I first met Huzihiro, about 35 years ago at Princeton University he was almost at the beginning of his scientific career, though not quite. I heard rumours that he had already written some difficult papers. Anyway, it was extremely fortunate that in Princeton he was attracted to Local Quantum Physics and devoted much of his energy and talent in subsequent years to this subject which without him would be significantly poorer in results.

The other reason is that during the past four years I was essentially occupied by writing a book on this subject [25] which forced me to survey critically the whole area: what has been achieved, what has not been adequately understood, where are misfits which indicate that major changes are needed for the development of future theory.

I shall organize my talk in the following manner.

1) What is local Quantum Physics; aims, conceptual and mathematical

H. Araki et al. (eds.), Quantum and Non-Commutative Analysis, 3–10.

structure.

2) Principal results of physical interest
   (i) the theory of charge quantum numbers and exchange symmetry,
   (ii) particles and collision theory
   (iii) thermal equilibrium states.
3) Criticism and some suggestions concerning future tasks.

The aim of Local Quantum Physics is the synthesis of the *principle of locality* which was so immensely fruitful in classical physics with the principles of quantum theory. The traditional vehicle for this has been Quantum Field Theory and Local Quantum Physics evolved from the formalism of Quantum Field Theory with the aim of separating the essential ingredients from peripheral and possibly doubtful ones. In classical physics the principle of locality started from the ideas of Faraday; it led Maxwell to the formulation of his equations of electrodynamics, was sharpened by Einstein in the special theory of relativity and led ultimately to the general theory of relativity and the search for a unified field theory. It is, however, only on the level of special relativity that we have achieved a reasonable synthesis with quantum physics so far.

So, in local quantum physics we accept space-time as a given, classical arena, equipped with the causal and metric structure claimed by special relativity. It is the Minkowski space $\mathcal{M}$ with its symmetry group $\mathcal{P}$, the Poincaré group generated by translations and Lorentz transformations. In the first stage of the development we take over from quantum theory that we relate the mathematical structure to a Hilbert space $\mathcal{H}$ whose rays correspond to pure physical states, that we have a representation of the Poincaré group by unitary operators $U(g)$ acting on $\mathcal{H}$

$$g \in \mathcal{P} \rightarrow \mathcal{U}(g). \tag{1}$$

Further, that observables are represented by self-adjoint operators acting on $\mathcal{H}$.

In order to implement the principle of locality we need an ordering of the observables according to the regions in space-time where they can be measured. Thus, to every region $\mathcal{O} \subset \mathcal{M}$ there is an operator algebra $\mathcal{A}(\mathcal{O})$ whose elements are interpreted as physical operations performable within the region $\mathcal{O}$. In particular $\mathcal{A}(\mathcal{O})$ should contain all the observables which can be measured in $\mathcal{O}$.

This innocent observation has some important consequences. First, we may regard quantum fields as a tool for constructing the "net of algebras" $\{\mathcal{A}(\mathcal{O})\}$; alternatively put, we may regard a system of quantum fields as a coordinatization of the net, while the intrinsic information of the theory resides in the net, not in the way it is constructed from fields. In fact one must envisage the possibility that the net cannot be constructed from point-like objects like fields. Secondly, the basic physical interpretation of the theory is placed on local observables, not on particles. The central requirement of the principle of locality is Einstein's causality: observables referring to regions which lie space-like to each other are compatible and hence they must commute. Historically quantum field theory has strongly relied on a field-particle duality i.e. the assumption that to every species of "elementary" particle there should be a corresponding field and vice versa. This hinged the physical interpretation of the theory to an "asymptotic condition" relating the limits of field operators for times $t \to \pm\infty$ to the creation operators of incoming or outgoing particles. Basing instead the interpretation on local observables one is lead in a very simple and natural way to a collision theory, avoiding any distinction between elementary and composite particles and showing that the "asymptotic condition" is a consequence, not an additional input [20],[21].

Now, in traditional quantum field theory we use also fields which anti-commute at space-like distances. These "Fermi fields" cannot correspond to observables; their commutation relations mirror the antisymmetry of wave functions for particles obeying the Pauli principle. The use of non-observable quantities in the theory has the consequence that there are "superselection rules" [22]: the superposition principle cannot be valid between arbitrary vectors in $\mathcal{H}$. Rather $\mathcal{H}$ is broken up into a direct sum of coherent subspaces $\mathcal{H}_\kappa$ or "superselection sectors". In each of them we have the unrestricted superposition principle whereas phase relation between vectors in different sectors cannot be observable. Examples of superselection rules are integer spin vs. half integer spin and states with different charge. The observables transform each superselection sector into itself.

Obviously we cannot lose any physical information if we restrict attention to the algebras generated by the observables alone. For these algebras we have the Einstein causality principle in the simple form: observables belonging to space-like separated regions commute. The existence of different superselection sectors suggests then that we should not consider the observable algebras as operator algebras acting on a Hilbert space but as abstract algebras defined by their internal algebraic relations. Then we may expect that these abstract algebras have various inequivalent representations by operator algebras and this fact can explain the occurrence of superselection rules. This means in a way a return to the earliest days of quantum mechanics as understood by Heisenberg and Dirac who talked about "q-numbers" rather than operators on a Hilbert space. For quantum mechanics this point of view was put on a mathematically rigorous level by I. E. Segal who showed that the relevant mathematical object is (what we now call) an abstract C^*-algebra. For the quantum mechanics of a system of finitely many particles moving in $\mathbf{R}^3$ it was irrelevant whether we take the abstract algebra or a Hilbert space representation (i.e. the Heisenberg or the Schrödinger approach) because of von Neumann's uniqueness theorem, showing that there is essentially only one equivalence class of representations. It had early been noted by von Neumann and it was noticed again by various authors in the early 50s that in the case of infinitely many degrees of freedom there are many inequivalent representations of the canonical commutation relations. So, in quantum field theory the difference between the abstract and the concrete version becomes important. Segal championed the abstract version but could not make it work for physics because he was focusing on a single algebra rather than on a local net and was motivated by the inequivalent representations of canonical commutation relations instead of the distinction between observables and non-observables.

These remarks lead us to the second stage of the development of the theory. We characterize the theory by a net of abstract C^*-algebras generated by local observables. Let us restrict the use of the symbol $\mathcal{O}$ to denote a finitely extended, contractible, open region in Minkowski space and let $\mathfrak{A}(\mathcal{O})$ denote the abstract C^*-algebra generated by the bounded observables of this region. We may define the algebra of all "quasilocal observables" $\mathfrak{A}$ as the C^*-inductive limit of this net, i.e.

$$\mathfrak{A} = \overline{\mathfrak{A}(\mathcal{O})}.$$

We have then no longer the operators $\mathcal{U}(g)$ representing the Poincaré transformations nor can we consider these unitaries as elements of the algebra $\mathfrak{A}$ since they correspond to global, not quasilocal objects. Instead the Poincaré symmetry is represented by an automorphism group of $\mathfrak{A}$

$$g \in \mathcal{P} \to \alpha_g \in \operatorname{Aut} \mathfrak{A}$$

with the property

$$\alpha_g \mathfrak{A}(\mathcal{O}) = \mathfrak{A}(g\mathcal{O}).$$

The fact that the generators of translations are not associated to $\mathfrak{A}$ but defined as derivations on $\mathfrak{A}$ implies that we cannot define energy and momentum in every representation of $\mathfrak{A}$. This means in turn that it is more complicated to formulate the very important requirement of stability of the theory which, in the Hilbert space version is simply stated as the requirement that the energy spectrum has a lower bound or, more elaborately, that the joint spectrum of energy-momentum is confined to the forward cone in p^μ-space.

The automorphisms α_g allow us, however, to control the energy-momentum-transfer affected by elements of $\mathfrak{A}$. Specializing $g \in \mathcal{P}$ to translations x we can consider for, every element $t \in \mathfrak{A}$ and any smooth, absolutely integrable test function $f(x)$ the element

$$A(f) = \int f(x)(\alpha_x A) d^4x$$

and note that if the Fourier transform $\tilde{f}$ of f has its support within a bounded region Δ of 4-dimensional momentum space (p-space) then in a representation in which energy-momentum can be defined $A(f)$ changes it by 4-vectors lying in Δ. This was used by Doplicher for an algebraic formulation of the stability condition. Let I_Δ denote the norm closure of the linear span of the set $BA(f)$ with B and A running through $\mathfrak{A}$ and f through all test functions with supp $\tilde{f} \subset \Delta$. This is a left ideal in $\mathfrak{A}$. Call it the spectral ideal I_Δ. It might be all of $\mathfrak{A}$. The Doplicher ideal I_D is the spectral ideal for $\Delta = \overline{V^+}^c$ (the complement of the closed forward cone in p-space). The requirement that I_D is a proper ideal (not the whole algebra) is tantamount to the existence of representations in which energy-momentum can be defined such that there is a ground state, the vacuum with zero energy-momentum (possibly non-unique). Note that a "state" in the abstract algebraic setting is defined as an expectation functional, mathematically as a normalized, positive linear form on $\mathfrak{A}$. To sum up: if I_D is a proper ideal then we have one class of representations of $\mathfrak{A}$ in which energy-momentum is defined and has its spectrum confined to the forward cone in p-space. Let us denote this class by Rep_{Ep} because these are the representations used in elementary particle physics. There is another class of representations in which the energy is undefined (infinite) and this class is also interesting because it describes states with matter distribution extending to infinity. Such states appear for instance as hydrodynamic states if we do not introduce any artificial limitations by enclosing the system in a box. We denote such representations by Rep_H. The simplest ones among them contain the thermal equilibrium states and are labelled by the values of temperature and chemical potential.

Let us now come to the results. One can classify the set of representations in Rep_{Ep}. This is the theory of superselection sectors about which we heard much in this seminar. So I do not have to go into this in detail. It starts from the observation that the existence of different charge quantum numbers is encoded in the observable algebra by the existence of outer localized morphisms. From this one finds a composition law of charges, a charge conjugation and an exchange symmetry (statistics) for identical charges. To each basic charge one has a "statistics parameter" λ. In the case of 4-dimensional space-time (which we only considered so far) Einstein causality allows only the values $\lambda = \pm d^{-1}$ where d is a natural number. The sign corresponds to the Bose-Fermi distinction and the integer describes the order of the "parastatistics" in the sense of H. S. Green [24] i.e. it limits the type of Young-tableaux which can occur in the representation of the permutation group describing the exchange symmetry. Furthermore the whole structure can be related to the representation theory of a compact group, the global gauge group. Recently much interest has been devoted to theories in lower dimensional space-times. There the situation is more complicated. Instead of the permutation group the braid group appears and instead of the global gauge group one has presumably a quantum group or Hopf algebra structure. Still the analysis is not complete. For instance the case where the theory has no mass gap has not been analyzed.

The next physically important circle of questions about the states of Rep_{Ep} concerns the particle picture. Does the theory describe particles, how do we determine their collision cross sections, do we get a complete description of states in terms of asymptotic particle configurations? Approaching these questions from the point of view of local observables we are forced to translate the experimental procedures into the language of the net $\{\mathfrak{A}(\mathcal{O})\}$. This is a healthy exercise for a mathematical physicist and has its rewards in drawing attention to further important structural properties of the theory. The experimentalist uses detectors and coincidence arrangement of detectors. A single detector centered at the space-time point x can be modeled in our scheme by a positive element $C(x) = \alpha_x C \in \mathfrak{A}(\mathcal{O}_x)$ where $\mathcal{O}_x$ is some region around x, say with radius r. Furthermore the expectation value of $C(x)$ in the vacuum state should be below some tolerance level ε, insuring that the detector registers a local deviation from the vacuum. A coincidence arrangement of n detectors at time t can be modeled by the product

$$C_1(\mathbf{x}_1, t) \cdots C_n(\mathbf{x}_n, t)$$

with large spatial separations: $|\mathbf{x}_i - \mathbf{x}_K| > R \gg r$.

A state which can trigger some n-fold coincidence at a time t but no $(n+1)$-fold coincidence is called n-fold localized at the time. A single particle state is a state which is singly localized at all times. This gives an operational definition of the concept of particle. The asymptotic localization number for $t \to \infty$ becomes the number of outgoing particles. This approach, suggested in [1] and [2] has been developed in the last decade especially by Buchholz [3]–[5] to a remarkable degree so that the treatment of collision theory in the presence of zero mass particles and the discussion of completeness of the particle picture becomes possible.

One very important conclusion from this discussion is that the theory must have some "compactness" or "nuclearity" property. One version of this is the following. Let $\mathcal{S}_E$ denote the set of states with energy below E and $\mathcal{S}_E(\mathcal{O})$ their restriction to the region $\mathcal{O}$ (as positive linear forms over $\mathfrak{A}(\mathcal{O})$.) Then, to characterize a state in $\mathcal{S}_E(\mathcal{O})$ up to precision ε in the norm topology we need only a finite number $N_{(\varepsilon)}$ of parameters. We expect that the exploitation of this property can provide a key to the problem of classifying theories within the frame described.

The third class of physically interesting results concerns statistical mechanics. The formulation of the theory in terms of abstract local algebras gives us the benefit of treating states with non-vanishing matter density extending to infinity and if we wish we can do so in the relativistic setting. These are states in Rep_H and among them are the equilibrium states in the thermodynamics limit. Time is too short now to engage in this vast area. So I shall only say that the description of Gibbs ensembles given by Kubo and by Martin and Schwinger assumes a very natural form in the algebraic setting. It establishes contact with the powerful mathematical theory of modular automorphisms by Tomita and Takesaki, allows an operational characterization of equilibrium, and thermodynamic phases, relates the chemical potentials to the superselection structure of Rep_{Ep}.

Let us finally look at the shortcomings of the frame developed thus far and at open problems. The outstanding questions are of course: how can we classify the theories in local quantum physics? Can one construct a theory without recourse to a quantum field theoretical model? Can we incorporate the local gauge principle in this setting? Instead of addressing such general questions let me focus attention on seemingly minor points which suggest that some change in the emphasis on basic concepts is indicated.

1) It seems, empirically, that no superselection rules exist in physics which concern only a single, finite region. This is remarkable. It is swept under the rug in the superselection analysis by considering only representations which in restriction to an algebra $\mathfrak{A}(\mathcal{O})$ are equivalent to the vacuum representation. But if $\mathfrak{A}(\mathcal{O})$ is an abstract C^*-algebra we have to explain why only this one class is physically admitted.

2) The operational origin of the product in the observable algebras has remained in the dark.

3) The notion of observables in extremely small regions becomes questionable.

All these points seem to suggest to me that we should pay more attention to the structure of state space. The families $\mathcal{S}(\mathcal{O})$, $\mathcal{S}_E(\mathcal{O})$ (and their linear hulls) may be more basic than the algebras $\mathfrak{A}(\mathcal{O})$ (which may be regarded as their dual spaces). Instead of a net of algebras we have a (pre)-sheaf of states. Instead of assuming an algebraic structure on the side of $\mathfrak{A}$ we have the structure of faces in state space; instead of observables in very small regions we might talk about highly localized events. This change in emphasis, combined with a sharper formulation of the compactness or nuclearity requirement for $\mathcal{S}_E(\mathcal{O})$ in the limit of very small $\mathcal{O}$ may be helpful in addressing the larger questions mentioned above.

References

[1] Araki, H. and Haag, R., Collision Cross Section in Terms of Local Observables, *Commun. Math. Phys.*, **4**, 77 (1967).

[2] Haag, R. and Swieca, J. A., When Does a Quantum Field Theory Describe Particles? *Commun. Math. Phys.*, **1**, 308 (1965).

[3] Buchholz, D., The Physical State Space of Quantum Electrodynamics, *Commun. Math. Phys.*, **85**, 49 (1982).

[4] Buchholz, D., On Particles, Infraparticles and the Problem of Asymptotic Completeness, VIII th International Congress on Mathematical Physics, ed. M. Mebkhout and

R. Sénéor, World Scientific Publ. 1987.

[5] Buchholz, D. and Porrmann, M., Single Particle Weights, In preparation.

[6] Borchers, H. J., Local Rings and the Connections of Spin with Statistics, *Commun. Math. Phys.*, **1**, 291 (1965).

[7] Doplicher, S., Haag, R. and Roberts, J. E., Fields, Observables and Gauge Transformations II, *Commun. Math. Phys.*, **15**, 773 (1969); Local Observables and Particle Statistics I and II, *Commun. Math. Phys.*, **23**, 199 (1971), **35**, 49 (1974).

[8] Buchholz, D. and Fredenhagen, K., Locality and the Structure of Particle States, *Commun. Math. Phys.*, **84**, 1(1982).

[9] Doplicher, D. and Roberts, J. E., Endomorphisms of C^*–algebras, Cross Products and Duality for Compact Groups, *Ann. of Math.*, **130**, 75 (1989).

[10] Fredenhagen, K., Rehren, H. and Schroer, B., Superselection Sectors with Braid Group Statistics and Exchange Algebras, I General Theory, *Commun. Math. Phys.*, **125**, 201 (1989).

[11] Fröhlich, J. and Gabbiani, F., Braid Statistics in Local Quantum Theory, *Rev. Math. Phys.*, **2**, 251 (1990).

[12] Haag, R., Hugenholtz, N. M. and Winnink, M., On the Equilibrium States in Quantum Statistical Mechanics, *Commun. Math. Phys.*, **5**, 215 (1967).

[13] Takesaki, M., Tomita's Theory of Modular Hilbert Algebras and its Applications, Lecture Notes in Mathematics **128**, Springer 1970.

[14] Haag, R., Kastler, D. and Trych Pohlmeyer, E. B., Stability and Equilibrium States, *Commun. Math. Phys.*, **38**, 193 (1974).

[15] Pusz, W. and Woronowicz, S. L., Passive States and KMS-States for General Quantum Systems, *Commun. Math. Phys.*, **58**, 273 (1978).

[16] Araki, H., Haag, R., Kastler, D. and Takesaki, M., Extension of KMS-States and Chemical Potential, *Commun. Math. Phys.*, **53**, 97 (1977).

[17] Kastler, D., Equilibrium States of Matter and Operator Algebras, *Sympos. Math.*, **20**, 49 (1979).

[18] Buchholz, D. and Wichmann, E., Causal Independences and the Energy-level Density of States in Local Quantum Field Theory, *Commun. Math. Phys.*, **106**, 321 (1986).

[19] Bratteli, O. and Robinson, D. W., Operator Algebras and Quantum Statistical Mechanics, Vol. I and II. Springer 1979, 1981.

[20] Haag, R., Quantum Field Theories with Composite Particles and Asymptotic Conditions, *Phys. Rev.*, **112**, 669 (1958).

[21] Ruelle, D., On the Asymptotic Condition in Quantum Field Theory, *Helv. Phys. Acta*, **35**, 144 (1962)

[22] Wick, J. C., Wightman, A.S. and Wigner, E.P., The Intrinsic Parity of Elementary Particles, *Phys. Rev.*, **88**, 101 (1952).

[23] Doplicher, S., An Algebraic Spectrum Condition, *Commun. Math. Phys.*, **1**, 1 (1965).

[24] Green, H.S., A Generalized Method of Field Quantization, *Phys. Rev.*, **90**, 270 (1953).

[25] Haag, R., Local Quantum Physics, Springer 1992.

A Non-commuting Realization of Minkowski Space

H.J. BORCHERS
Institut für Theoretische Physik der Universität Göttingen
Bunsenstrasse 9, D 3400 Göttingen

Abstract. In this paper we construct an orderpreserving map from the open sets of the Minkowski space into a family of non-commuting projections. This map is naturally associated to quantum field theory in the sense of Araki, Haag, and Kastler. The image contains some of the informations encoded in the manner how the algebra of the smaller domain is imbedded in the algebra of the bigger domain. As an illustrative example we will treat the classification problem of chiral field theories.

Key words: non-commutative analysis, quantum field theory

1. Introduction

This paper is dedicated to Huzihiro Araki on the occasion of his 60th birthday.

The appearence of non-commutativity in physics has a long tradition. The first example seems to be wave-vector **k** and the position-vector **x** appearing in optics. Here the non-commutativity is a consequence of the Fourier transformation, but only in the middle of the twenties it has been realized that this is a conseqence of general principles in physics which are known as quantum physics. It took another sixty years until non-commutativity has been realized also as a principle of geometry [8,9].This formulation of geometry was the break-through of a long series of results containing non-commutative versions of standard methods in physics and in geometry. In physics one has to mention the Yang-Mills equation in connection with non-commutative gauge groups and mainly the theory of supersymmetry. The idea of fermionic charges appeared first in the two dimensional dual model. The attempt to make these ideas applicable to four dimensional theories appeared in a paper of Volkov and Akulov [23] and a systematic theory by Wess and Zumino [24].Since Segal [18] the theory of normal linear functionals of a von Neumann algebra is considered as a non-commutative generalization of integration theory. In this spirit non-commutative geometry has been invented by A. Connes [8,9].

There have been several attempts to find applications for physics. First they went in the direction of making the space of internal symmetries to a non-commutative one or unifying the spaces of configuration and internal

H. Araki et al. (eds.), Quantum and Non-Commutative Analysis, 11–30.

symmetry to a non-commutative one. In this field one should mention the papers of Dubois-Violette, Kerner, and Madore [12], Coqueraux, Esposito-Farèse, and Scheck [11] and mainly that of Connes and Lott [10]. These investigations have the aim to get a better understanding first of the Yang-Mills theory and second of the unified theory of electro-weak interactions.

In the examples discussed so far the starting point was the attempt to apply non-commutative geometry. Up to now there is missing a convincing argument for using this or any other ansatz. In this paper I want to present an approach to non-commutative geometry which is naturally associated to quantum field theory in the sense of Araki, Haag and Kastler [1],[13]. Using the nuclearity condition of Buchholz and Wichmann [6] one knows from the investigation of Buchholz, D'Antoni, and Fredenhagen [5] that all the local von Neumann algebras, if they are factors, are all the same, namely hyperfinite type III_1 factors. This shows that the algebraic structure of the local von Neumann algebras can contain only physical informations of a general nature. All informations specific for the model under consideration must, therfore, be encoded in the manner how the algebra of the smaller domain is imbedded in the algebra of the bigger domain. What we will do is closely related to the investigation of H. Araki [2] concerning the free field. However, our discussion suggests to look for possible descriptions for this imbedding. We will treat one aspect of the problem which uses concepts introduced in the Tomita-Takesaki theory.

We will look at the operator $S(\mathcal{O})$ which is associated to every local algebra $\mathcal{M}(\mathcal{O})$ and which is defined by the relation $SA\Omega = A^*\Omega$. This setting has the property: $\mathcal{O}_1 \subset \mathcal{O}_2$ implies that the operator $S(\mathcal{O}_2)$ is an extension of $S(\mathcal{O}_1)$. This property is a kind of isotony, which is best be expressed by using the graphs of these operators. M. Stone [19] has introduced the projection onto the graph of a closed operator as a suitable object describing its properties. These projections are the objects we will deal with. However, we first have to take care of the fact that the operators $S(\mathcal{O})$ are antilinear. This can be done in such a way that we do not lose the correct behaviour under the action of the translations.

In the next section we will, for the convenience of the reader, recollect the properties of the Stone matrices. In the third section we will make use of these objects in order to define a map Φ from the sets in Minkowski space into a family of projections. $\Phi : \quad \mathcal{O} \longrightarrow E(\mathcal{O})$. There we shall also derive some of the properties of this map and in particular we show that it preserves isotony and maps into a non-commuting family.

Together with this map there exists a projection $\Pi : \mathcal{M}(\mathcal{O}_2) \to \mathcal{M}(\mathcal{O}_1)$. It will be constructed and investigated in the fourth section. In particular we will show that these projections cannot be orderpreserving and normal at the same time.

In the fifth section we apply our results to chiral quantum field theory.

In this situation the structure of the von Neumann algebra generated by the projections $E(\mathcal{O})$ can be explicitely computed.

In the last section we will discuss our approach to non-commutative geometry and we will mention some of the problens to be solved in order to make our theory into a powerful tool for quantum field theory.

2. Assumptions and preliminaries

In this note we will work with the standard setting of the theory of local observables in the vacuum-sector. This theory is defined by the following

II.1 Assumptions:
1. To every bounded open set $\mathcal{O}$ in Minkowski space is associated a von Neumann algebra $\mathcal{M}(\mathcal{O})$ acting on a Hilbert space $\mathcal{H}$.
2. For any open region $\mathcal{R}$ the algebra $\mathcal{M}(\mathcal{R})$ denotes the von Neumann algebra generated by $\cup\{\mathcal{M}(\mathcal{O});\ \mathcal{O} \subset \mathcal{R}\}$.
3. We assume that we have a continuous unitary representation $U(a)$ of the translation group of the Minkowski space $\mathbf{R}^d$. This representation fulfills the spectrum condition. Moreover, we assume that there exists a unique invariant state Ω, i.e. $U(a)\Omega = \Omega$ for all $a \in \mathbf{R}^d$.
4. $U(a)$ induces a group of automorphisms α_a on $\mathcal{M}(\mathbf{R}^d)$ fulfilling the equation $\alpha_a \mathcal{M}(\mathcal{O}) = \mathcal{M}(\mathcal{O} + a)$.
5. This family of von Neumann algebras shall fulfill isotony, i.e $\mathcal{O}_1 \subset \mathcal{O}_2$ implies $\mathcal{M}(\mathcal{O}_1) \subset \mathcal{M}(\mathcal{O}_2)$, and weak additivity i.e. for every set $\mathcal{O}$ the family of operators $\cup\{\mathcal{M}(\mathcal{O}+a);\ a \in \mathbf{R}^d\}$ generates $\mathcal{M}(\mathbf{R}^d)$.
6. If the theory is invariant under Lorentz-transformations then we denote the unitary representation of this group by $U(\Lambda)$. From the assumptions above it follows that Lorentz covariance implies the invariance of Ω under $U(\Lambda)$.
7. If we are dealing with a CPT invariant theory then we write Θ_0 for the CPT operator. Again the vacuum vector is invariant under this operator.

Note: These conditions imply that Ω is a vacuum vector and that $\mathcal{M}(\mathbf{R}^d)$ is the von Neumann algebra of all bounded operators on $\mathcal{H}$. In addition the vector Ω is cyclic and separating for all the von Neumann algebras $\mathcal{M}(\mathcal{O})$.

In the following we will have to deal extensively with unbounded operators. One way of doing this is the method introduced by M. Stone. We will give a short description of this technique.

II.2 The Stone matrix: Let X be a closed linear operator on the Hilbert space $\mathcal{H}$. Its graph $G(X)$ is a closed linear subspace of $\mathcal{H} \oplus \mathcal{H}$. M. Stone [19] introduced the projection onto $G(X)$ which can be represented as a 2×2 matrix $(P_{i,j}(X))$ of bounded operators. This matrix is called the *Stone*-or *characteristic matrix* of the operator X. It has the following properties (see also [17]):

1) A matrix $P_{i,j}$ defines a projection on $\mathcal{H} \oplus \mathcal{H}$ if the following holds:

$$P_{i,j}^* = P_{j,i}, \qquad \sum_k P_{i,k} P_{k,j} = P_{i,j}. \tag{1}$$

2) Let $(P_{j,i})$ represent a projection, then it projects onto a graph iff

$$N(1 - P_{2,2}) = \{0\}. \tag{2}$$

3) If $(P_{i,j})$ projects onto a graph then the operator X is defined as follows:

$$D(X) = \sum_j P_{1,j}\xi_j, \qquad X : \sum_j P_{1,j}\xi_j \longrightarrow \sum_j P_{2,j}\xi_j. \quad \forall \xi_j \in \mathcal{H}.$$

This implies in particular $P_{2,j} = XP_{1,j}, \qquad j = 1, 2.$
4) If $(P_{i,j})$ projects onto a graph then the operator X is densely defined iff, with $N(A)$ the nullspace of A, one has

$$N(P_{1,1}) = \{0\}. \tag{3}$$

Notation: A closed densely defined operator will be called regular.
5) If X is a regular operator, then the same holds for X^* . The two graphs $G(X)$ and $G(X^*)$ are related by the unitary operator V:

$$V = \begin{pmatrix} 0 & 1 \\ -1 & 0 \end{pmatrix}. \tag{4}$$

It is $G(X)^\perp = VG(X^*)V$. Hence if $Q_{i,j}$ denotes the characteristic matrix of X^* then one finds

$$\begin{aligned} \begin{pmatrix} Q_{1,1} & Q_{1,2} \\ Q_{2,1} & Q_{2,2} \end{pmatrix} &= \begin{pmatrix} 1 - P_{2,2} & P_{2,1} \\ P_{1,2} & 1 - P_{1,1} \end{pmatrix}, \\ Q &= V^*(1 - P)V. \end{aligned} \tag{5}$$

6) From 3),4), and 5) one derives the following representation for the operators $P_{i,j}$:

$$\begin{aligned} P_{1,1} &= (1 + X^*X)^{-1}, & P_{1,2} &= X^*(1 + XX^*)^{-1}, \\ P_{2,1} &= X(1 + X^*X)^{-1}, & P_{2,2} &= XX^*(1 + XX^*)^{-1}. \end{aligned}$$

7) Let $(P_{i,k})$ be the characteristic matrix of a regular operator X. Then X has a densely defined inverse iff

$$N(P_{2,2}) = \{0\}, \qquad N(1 - P_{1,1}) = \{0\}. \tag{6}$$

If $(Q_{i,k})$ denotes the characteristic matrix of X^{-1} then one has with

$$W = \begin{pmatrix} 0 & 1 \\ 1 & 0 \end{pmatrix} \tag{7}$$

the relation

$$\begin{pmatrix} Q_{1,1} & Q_{1,2} \\ Q_{2,1} & Q_{2,2} \end{pmatrix} = \begin{pmatrix} P_{2,2} & P_{2,1} \\ P_{1,2} & P_{1,1} \end{pmatrix}, \qquad Q = WPW. \tag{8}$$

8) The closed operator Y is an extension of the closed operator X iff $G(X) \subset G(Y)$.

For later use we want to fix some notations. Some of them are often used in the Tomita-Takesaki theory [21,20].

II.3 Notations:
a. By $\Delta(\mathcal{O})$ and $J(\mathcal{O})$ we denote the modular operator and the modular conjugation of the pair $\{\mathcal{M}(\mathcal{O}), \Omega\}$. The operator $J(\mathcal{O})\Delta(\mathcal{O})^{1/2}$ will be denoted by $S(\mathcal{O})$. We also use the operators $J(\mathcal{R})$ etc. provided $\mathcal{R}$ is a domain with spacelike complement.
b. For any open set $\mathcal{R}$ we denote by $\mathcal{R}'$ the spacelike complement of the closure of $\mathcal{R}$.
c. We say a pair of sets $\mathcal{R}, \mathcal{R}$‘ fulfills the duality condition if the relation

$$\mathcal{M}(\mathcal{R})' = \mathcal{M}(\mathcal{R}')$$

holds.
d. A theory of local observables fulfills the *duality condition* if there exists a family $\mathcal{D}$ of sets such that every pair $(\mathcal{R}, \mathcal{R}') \in \mathcal{D}$ fulfills the duality condition. For $\mathcal{D}$ we require:
(α) $\mathcal{R} \in \mathcal{D}$ implies $\mathcal{R}' \in \mathcal{D}$.
(β) $\mathcal{D}$ is translationally invariant i.e. $\mathcal{R} \in \mathcal{D} \longrightarrow \mathcal{R} + a \in \mathcal{D}$.
(γ) $\mathcal{D}$ contains all wedges and all double cones.
e. By Θ we will denote a conjugation commuting with the generators of the translations. Therefore one has $\Theta U(a) = U(-a)\Theta$. Such an operator exists since $\{U(a)\}$ is a commutative group. If we are dealing with a theory having a TCP operator Θ_0, then we take this operator for Θ.
f. By $\mathcal{H}^{\perp}$ we denote the subspace of vectors perpendicular to the vacuum vector Ω.
g. If E_1, E_2 are two projections then $E_1 < E_2$ means $E_1 E_2 = E_2 E_1 = E_1$.

3. The canonical map

Let us denote by $\widehat{\mathcal{H}}$ the Hilbert space $\mathcal{H} \oplus \mathcal{H}$. If X is a closed linear operator then the elements of its graph are written as $\psi \oplus X\psi$, where ψ is in the domain of X. Let $\mathcal{R}$ be an open set in Minkowski space with $\mathcal{R}' \neq \emptyset$. In this situation the vector Ω is cyclic and separating for the von Neumann algebra $\mathcal{M}(\mathcal{R})$ and the operator $S(\mathcal{R})$, as defined in the last section, is a closed antilinear operator. Therefore, the product $\Theta S(\mathcal{R})$ is a closed linear operator.

If $\mathcal{N}$ is a von Neumann algebra, not necessarily of the form $\mathcal{M}(\mathcal{R})$ which has Ω as cyclic and seperating vector, then we denote its operator S by $S(\mathcal{N})$.

III.1 Definition:
(*i*) Let $E(\mathcal{R}) \in \mathcal{B}(\widehat{\mathcal{H}})$ be the projection onto the graph of the operator $\Theta S(\mathcal{R})$. The map

$$\mathcal{R} \longrightarrow E(\mathcal{R}) \in \mathcal{B}(\widehat{\mathcal{H}}) \tag{9}$$

will be denoted by Φ i.e. we write $E(\mathcal{R}) = \Phi(\mathcal{R})$. All these objects are only defined if $\mathcal{R}' \neq \emptyset$.

Occasionally we deal with von Neumann algebras $\mathcal{N}$ not belonging to a set in $\mathbf{R}^n$. In this situation we write $E(\mathcal{N})$ if we need the projection onto the graph of $\Theta S(\mathcal{N})$.
(*ii*) On $\widehat{\mathcal{H}}$ we define a unitary representation of the translationgroup by the equation

$$\widehat{U}(a) = U(a) \oplus U(-a). \tag{10}$$

(*iii*) If we are dealing with a Lorentz covariant theory then we define $\widehat{U}(\Lambda) = U(\Lambda) \oplus U(\Lambda)$. The operators $\widehat{U}(a)$ and $\widehat{U}(\Lambda)$ together define again a representation of the Poincaré group.
(*iv*) If the theory admits a CPT operator then we put $\widehat{\Theta}_0 = \mathrm{diag}(\Theta_0, \Theta_0)$.

In the last section II.2.5 and 7 we have introduced on $\widehat{\mathcal{H}}$ the two unitary operators V and W. These operators together with $\widehat{U}(a)$ fulfill the following relations:

III.2 Lemma:*Amongst the operators $\widehat{U}(a), V$ and W we have the following relations:*

$$\begin{aligned} W\widehat{U}(a) &= \widehat{U}(-a)W, & V\widehat{U}(a) &= \widehat{U}(-a)V, \\ W &= W^*, & W^2 &= 1 = WW^*. \\ V &= -V^*, & -V^2 &= 1 = VV^*, \\ & & WV &= -VW. \end{aligned} \tag{11}$$

If the theory is also Lorentz- or CPT invariant then we have the additional relations

$$\begin{aligned} V\widehat{U}(\Lambda) &= \widehat{U}(\Lambda)V, & W\widehat{U}(\Lambda) &= \widehat{U}(\Lambda)W, \\ V\widehat{\Theta}_0 &= \widehat{\Theta}_0 V, & W\widehat{\Theta}_0 &= \widehat{\Theta}_0 W. \end{aligned} \tag{11'}$$

These relations are easily checked. The details are left to the reader.

Next we are looking at some of the properties of the projections $E(\mathcal{R})$ which are only defined if $\mathcal{R}$ has a spacelike complement.

III.3 Proposition: *The map Φ and the projections $E(\mathcal{R})$ have the following properties:*
(1) *The map Φ is isotonic i.e.*

$$\mathcal{R}_1 \subset \mathcal{R}_2 \longrightarrow E(\mathcal{R}_1) < E(\mathcal{R}_2). \tag{12}$$

(2) *The map Φ has the following transformation properties:*

$$\widehat{U}(a)E(\mathcal{R})\widehat{U}(-a) = E(\mathcal{R}+a). \tag{13}$$

If the theory is Lorentz- and CPT invariant then one has the relations

$$\widehat{U}(\Lambda)E(\mathcal{R})\widehat{U}(\Lambda^{-1}) = E(\mathcal{R}_\Lambda), \quad \widehat{\Theta}_0 E(\mathcal{R})\widehat{\Theta}_0 = E(-\mathcal{R}).$$

(3) *Assume we are dealing with a CPT invariant theory then $VE(\mathcal{R})V^*$ and $E(-\mathcal{R}')$ are perpendicular projections.*

$$VE(\mathcal{R})V^*E(-\mathcal{R}') = E(-\mathcal{R}')VE(\mathcal{R})V^* = 0. \tag{14}$$

The pair of sets $\mathcal{R}$ and $\mathcal{R}'$ fulfills duality iff one has the stronger result

$$VE(\mathcal{R})V^* + E(-\mathcal{R}') = 1. \tag{14'}$$

(4) *If the theory is CPT invariant then one has*

$$WE(\mathcal{R})W^* = E(-\mathcal{R}). \tag{14''}$$

If we do not have this invariance then the projection $WE(\mathcal{R})W^*$ is associated to the von Neumann algebra $\Theta\mathcal{M}(\mathcal{R})\Theta$ which is not necessarily associated with a domain in $\mathbf{R}^d$.

Proof: If $\mathcal{N}$ is a sub von Neumann algebra of $\mathcal{M}$ with a common cyclic and separating vector then $S(\mathcal{M})$ is an extension of $S(\mathcal{N})$. This implies (1) by the definition of $E(\mathcal{R})$ and II.2.8. For proving the first part of (2) we have to remember that $E(\mathcal{R})$ is the projection onto the graph of $\Theta S(\mathcal{R})$. Since by the covariance of the local net we have $U(a)S(\mathcal{R})U(-a) = S(\mathcal{R}+a)$ it follows that the projection $E(\mathcal{R})$ transforms with $\widehat{U}(a)$. The results of equation (14') are obtained in the same manner by remembering that $U(\Lambda)$ and Θ_0 commute and that $E(\mathcal{R})$ is the projection onto the graph of $\Theta_0 S(\mathcal{R})$. For proving (3) notice that $V(1-E(\mathcal{R}))V^*$ is the projection onto the graph of $(\Theta S(\mathcal{R}))^* = S^*(\mathcal{R})\Theta$. If Θ is the CPT operator Θ_0 then one has $S^*(\mathcal{R})\Theta_0 = \Theta_0 S^*(-\mathcal{R})$. From locality we know that $S^*(-\mathcal{R})$ is an extension of $S(-\mathcal{R}')$. Hence we obtain the first statement. If we have duality for the pair $\mathcal{R}, \mathcal{R}'$ then we also have duality for the pair $-\mathcal{R}, -\mathcal{R}'$

by CPT invariance. Duality means that $\mathcal{M}(\mathcal{R}')$ is the commutant of $\mathcal{M}(\mathcal{R})$ or equivalently $S^*(\mathcal{R}) = S(\mathcal{R}')$. Putting everything together we obtain the second part of (3). Since $S(\mathcal{R})$ is its own inverse we have $(\Theta_0 S(\mathcal{R}))^{-1} = S(\mathcal{R})\Theta_0 = \Theta_0 S(-\mathcal{R})$. From this we obtain equation (14″) by remembering the meaning of W. (See II.2.7). □

III.4 Remarks:
(1) The operator $S(\mathcal{R})$ is antilinear. Therefore, we have introduced ΘS in order to obtain a linear operator, so that we can use the standard theory of Stone. A different possibility is to work with the space $\mathcal{H}\oplus\overline{\mathcal{H}}$,where $\overline{\mathcal{H}}$ is the conjugate linear image of $\mathcal{H}$. This would have the advantage that the concept is formally independent of Θ. However, since it is awkward to work with a vector space which is not complex linear I prefer to treat the operator ΘS. In reality the operator Θ appears also in the theory on $\mathcal{H}\oplus\overline{\mathcal{H}}$.It is hidden in the isometry between $\mathcal{H}$ and $\overline{\mathcal{H}}$.
(2) If we are starting with a non CPT invariant theory then neither the operator Θ nor the map Φ is unique. If we have CPT invariance and if we take Θ_0 for Θ then ad W defines a symmetry of the image of Φ. If in addition one deals with a theory enjoying the duality condition for a family of sets then the image of this set is also invariant under ad V.
(3) The two elements V, W are important. The relations of equation (11) imply that $\frac{1}{2}(1+VW)$ and $\frac{1}{2}(1+WV)$ are two orthogonal projections which commute and which add up to 1. Both these projections commute with $\widehat{U}(a)$ and $\widehat{U}(\Lambda)$. They project onto the two copies of the original Hilbert space. If V^+ denotes the closed forward lightcone then one has:

$$\begin{aligned}&\text{spectrum of } \widehat{U}(a)\ \tfrac{1}{2}(1+VW) \subset V^+,\\ &\text{spectrum of } \widehat{U}(a)\ \tfrac{1}{2}(1+WV) \subset -V^+.\end{aligned} \tag{15}$$

Moreover, the two representations $\widehat{U}(a)\frac{1}{2}(1+VW)$ and $\widehat{U}(-a)\frac{1}{2}(1+WV)$ are isomorphic. The operator W interchanges the two Hilbert spaces. The operator $\frac{1}{2}(W-V)$ is an isometry mapping the first Hilbert space onto the second and $\frac{1}{2}(W+V)$ maps the second onto the first one.
(4) Using these two projections one can for every operator $\widehat{A} \in \mathcal{B}(\widehat{\mathcal{H}})$ construct its matrix $A_{i,k}$. If $P_{i,k}$ is the matrix of $E(\mathcal{R})$ then by II.2.2),4) and 7) the diagonal elements have the properties:

$$N(P_{1,1}) = N(1-P_{1,1}) = N(P_{2,2}) = N(1-P_{2,2}) = 0. \tag{16}$$

(5) Since E is the projection onto the graph of ΘS it follows that $\Omega\oplus\Omega$ is an eigenvector of every $E(\mathcal{R})$. On the other hand by the representation given in II.2.6 one has $P_{1,1} = (1+\Delta)^{-1}$, $P_{2,1} = \Theta S(1+\Delta)^{-1}$. This shows:

$$\begin{aligned}E(\Omega\oplus\Omega) &= \Omega\oplus\Omega, & E(\Omega\oplus 0) &= \tfrac{1}{2}\Omega\oplus\Omega,\\ E(0\oplus\Omega) &= \tfrac{1}{2}\Omega\oplus\Omega, & E(\Omega\oplus(-\Omega)) &= 0.\end{aligned} \tag{17}$$

(7) For later calculations it is useful to have the exact form of the Stone matrix of $E(\mathcal{R})$. By using II.2.6 we obtain:

$$E(\mathcal{R}) = \begin{pmatrix} (1+\Delta)^{-1} & \Delta^{1/2}(1+\Delta)^{-1}J\Theta \\ \Theta J\Delta^{1/2}(1+\Delta)^{-1} & \Theta(1+\Delta)^{-1}\Theta \end{pmatrix} \tag{18}$$

where Δ and J are the modular operators and conjugation of the pair $(\mathcal{M}(\mathcal{R}),\Omega)$.

By isotony we find a large number of decreasing respectively increasing sequences of projections in the image of Φ. With regard to the limit of such sequences we have the following result:

III.5 Lemma:
(a) Let $\mathcal{R}_n$ be a decreasing sequence of sets such that
α. $\cap_n\mathcal{R}_n$ is one point, or
β. $\cup_n\mathcal{R}_n'$ contains all of $\mathbf{R}^n$
then we have

$$\lim_n E(\mathcal{R}_n) = E_0 \tag{19_0}$$

where E_0 is the projection onto $\mathbf{C}(\Omega\oplus\Omega)$.
(b) Let $\mathcal{R}_n$ be an increasing sequence of sets such that $\cup_n\mathcal{R}_n$ contains all of $\mathbf{R}^n$ then one has

$$\lim_n E(\mathcal{R}_n) = E_\infty \tag{19_∞}$$

with

$$E_\infty\mathcal{H} = \hat{\mathcal{H}} \ominus \mathbf{C}(\Omega\oplus(-\Omega)) = (\mathcal{H}^\perp\oplus\mathcal{H}^\perp)\oplus\mathbf{C}(\Omega\oplus\Omega).$$

Proof: By Remark III.4.(5) we know that $\Omega\oplus\Omega$ is an eigenvector of every $E(\mathcal{R})$ and that these projections anihilate the vector $\Omega\oplus(-\Omega)$. Hence it is sufficient to look at the vectors in $\mathcal{H}^\perp\oplus\mathcal{H}^\perp$. First we show part b). Assume $\varphi_1\oplus\varphi_2 \in \mathcal{H}^\perp\oplus\mathcal{H}^\perp$ with $||\varphi_i|| \leq 1$ and define $\psi_2 = \Theta\varphi_2$. These vectors are perpendicular to Ω, hence there exists an operator $B \in \mathcal{B}(\mathcal{H})$ with $||B|| \leq 1$ and $\varphi_1 = B\Omega$, $\psi_2 = B^*\Omega$. Since $\cup\mathcal{M}(\mathcal{R}_n)$ is strong-*-dense in $\mathcal{B}(\mathcal{H})$ exists for every $\epsilon > 0$ a number n and an operator $A \in \mathcal{M}(\mathcal{R}_n)$ with $||(B-A)\Omega|| \leq \epsilon/2$ and $||(B^*-A^*)\Omega|| \leq \epsilon/2$. Since we have $S(\mathcal{R}_n)A\Omega = A^*\Omega$ it follows that $A\Omega\oplus\Theta A^*\Omega$ is an eigenvector of $E(\mathcal{R}_n)$. This implies that $\varphi_1\oplus\varphi_2$ is an eigenvector of E_∞.

For proving a) notice first that $\cap\mathcal{M}(\mathcal{R}_n) = \mathbf{C}1$, so that $\cup\mathcal{M}(\mathcal{R}_n)'$ is strong-*-dense in $\mathcal{B}(\mathcal{H})$. This holds also for $\cup\Theta\mathcal{M}(\mathcal{R})'\Theta$. Notice the relation $\Theta S(\Theta\mathcal{M}(\mathcal{R})'\Theta) = S^*(\mathcal{R})\Theta$. Consequently the corresponding projections converge to E_∞. Hence we obtain from II.2.5: $\cap E(\mathcal{R}_n) = \cap V^*(1 - E(\Theta\mathcal{M}(\mathcal{R})'\Theta))V = 1 - V\{\cup E(\Theta\mathcal{M}(\mathcal{R})'\Theta)\}V^* = 1 - V^*E_\infty V$. But this projection is nothing else but E_0. □

Next we want to show that the image of Φ is, indeed, a non-commutative family of projections. We will restrict this investigation to the situation where we have CPT-invariance and duality, because then the arguments are much simpler. We first derive some properties of the von Neumann algebra generated by the image of Φ, which we denote by $\mathcal{P}$.

III.6 Lemma: *Suppose we are dealing with a CPT-invariant theory of local observables fulfilling duality. Then the von Neumann algebra $\mathcal{P}$ has the following properties:*
(1) *If $\widehat{A} \in \mathcal{P}$ and if $\widehat{A}$ is the matrix $\begin{pmatrix} A_{1,1} & A_{1,2} \\ A_{2,1} & A_{2,2} \end{pmatrix}$ then also*

$$\widehat{A}_D = \begin{pmatrix} A_{1,1} & 0 \\ 0 & A_{2,2} \end{pmatrix} \quad \text{and} \quad \widehat{A}_N = \begin{pmatrix} 0 & A_{1,2} \\ A_{2,1} & 0 \end{pmatrix} \tag{20}$$

belong to $\mathcal{P}$.
(2) *If we put $J = J(\mathcal{R})$ then the matrices*

$$\begin{pmatrix} 0 & J\Theta_0 \\ \Theta_0 J & 0 \end{pmatrix}, \quad \begin{pmatrix} 0 & \Theta_0 J \\ J\Theta_0 & 0 \end{pmatrix} \quad \text{and} \quad \begin{pmatrix} J_1 J_2 & 0 \\ 0 & \Theta_0 J_1 J_2 \Theta_0 \end{pmatrix} \tag{21}$$

belong to $\mathcal{P}$.
(3) *For every $J = J(\mathcal{R})$ one can show that* ad $\begin{pmatrix} J & 0 \\ 0 & \Theta_0 J \Theta_0 \end{pmatrix}$ *defines an automorphism of $\mathcal{P}$.*

Proof: From Proposition III.3.(3) and (4) we know that $\mathcal{P}$ is invariant under the action of ad V, ad W and ad $\widehat{\Theta}_0$. Consequently it is also invariant under ad VW. Let $\widehat{A}$ be in $\mathcal{P}$ then ad VW changes the sign of the off-diagonal elements. Consequently $\widehat{A}_D$ and $\widehat{A}_N$ belong to $\mathcal{P}$. This shows (1). Taking now one of the projections $E(\mathcal{R})$ then we obtain from eqn. (18):

$$E_D = \begin{pmatrix} (1+\Delta)^{-1} & 0 \\ 0 & \Theta_0(1+\Delta)^{-1}\Theta_0 \end{pmatrix}, \; E_N = \begin{pmatrix} 0 & \frac{\Delta^{1/2}}{(1+\Delta)} J\Theta_0 \\ \Theta_0 J \frac{\Delta^{1/2}}{(1+\Delta)} & 0 \end{pmatrix}.$$

If we now multiply E_N from the left by

$$(E_D^{-1} - 1)^{-1/2} E_D^{-1} = \begin{pmatrix} \Delta^{-1/2}(1+\Delta) & 0 \\ 0 & \Theta_0 \Delta^{-1/2}(1+\Delta)\Theta_0 \end{pmatrix}$$

then we obtain the first matrix. Taking ad $\widehat{\Theta}_0$ of this matrix we obtain the second one. The third one is the product of the other two. Multiplying the first matrix on the right by $\widehat{\Theta}_0 W$ we obtain statement (3). □

From this result we easily can deduce that the set of projections $\{E(\mathcal{R})\}$ is not commutative.

III.7 Corollary: *If a theory of local observables fulfills the assumptions* II.1, *duality, and admits a CPT-operator then the algebra $\mathcal{P}$ is not commutative except all $\mathcal{M}(\mathcal{O})$ are only the multiples of the identity.*

Proof: By contradiction. If $\mathcal{P}$ is commutative we know by the proceeding lemma that the matrix E_D commutes with the first matrix of eqn. (21). The (1,2) component of this relation gives

$$(1+\Delta(\mathcal{R}_1))^{-1}J(\mathcal{R}_2)\Theta_0 = J(\mathcal{R}_2)(1+\Delta(\mathcal{R}_1))^{-1}\Theta_0$$

where $\Delta = \Delta(\mathcal{R}_1)$ and $J = J(\mathcal{R}_2)$ for any two domains. Choosing $\mathcal{R}_1 = \mathcal{R}_2$ we find that Δ and J commute. This implies that $\Delta = 1$ and hence Ω is a tracevector for every algebra $\mathcal{M}(\mathcal{O})$. Consequently the vacuum expectation-values of any commutator vanish and the representation of the observables is an abelian algebra. Since we only have one vacuum this algebra is onedimensional. □

III.8 Remark: The map Φ as in Definition III.1 depends on the choice of the conjugation Θ. If we replace Θ by Θ' then we obtain $\Phi_{\Theta'} = V\Phi_\Theta V^*$, where V is the unitary operator $V = \mathrm{diag}(1, \Theta'\Theta)$.

4. A projection of von Neumann algebras

The problems we are handling in this section are concerned with two von Neumann algebras $\mathcal{N} \subset \mathcal{M}$ having a common cyclic and separating vector Ω. Therefore, we will not use the notation of the theory of local observables.

Starting from two von Neumann algebras $\mathcal{N} \subset \mathcal{M}$ we saw in the last section that $E(\mathcal{N})$ is a subprojektion of $E(\mathcal{M})$. Now we want to lift the relation $E(\mathcal{N})E(\mathcal{M}) = E(\mathcal{N})$ to a projection map

$$\Pi(\mathcal{N},\mathcal{M}) : \mathcal{M} \longrightarrow \mathcal{N}. \tag{22}$$

To this end recall (see e.g. Bratteli and Robinson [4] proof of Proposition 2.5.9) that to every vector ψ in the domain of $S(\mathcal{M})$ there exist two operators Q and $Q^\dagger$ affiliated with $\mathcal{M}$ and defined on $\mathcal{M}'\Omega$. They have the property

$$Q\Omega = \psi, \quad \text{and} \quad Q^\dagger\Omega = S(\mathcal{M})\psi.$$

IV.1 Definition: The set of operators described above will be denoted by $\widetilde{\mathcal{M}}$.

Note that we have a one to one correspondence between the elements of $\widetilde{\mathcal{M}}$ and the domain of definition of $S(\mathcal{M})$. The operators Q and $Q^\dagger$ are closable operators. The map $Q \to Q^\dagger$ is an involution. $Q^\dagger$ is the restriction of Q^* to the domain $\mathcal{M}'\Omega$. These properties can now be used to define the map Π.

IV.2 Definition:
(i) We denote by $D(S(\mathcal{M}))$ the domain of definition of $S(\mathcal{M})$.
(ii) We denote by $\alpha(\mathcal{M})$ the map $\widetilde{\mathcal{M}} \to D(S(\mathcal{M}))$ just described and by $\alpha^{-1}(\mathcal{M})$ its inverse.
(iii) We denote by $\beta(\mathcal{M})$ the map $\psi \in D(S(\mathcal{M})) \to \psi \oplus \Theta S(\mathcal{M})\psi \in G(\Theta S(\mathcal{M}))$ and by $\beta^{-1}(\mathcal{M})$ the inverse map, i.e. the restriction onto the first component.
(iv) Let $\mathcal{N} \subset \mathcal{M}$ then we define the map

$$\Pi(\mathcal{N},\mathcal{M}) = \alpha^{-1}(\mathcal{N})\beta^{-1}(\mathcal{N})E(\mathcal{N})\beta(\mathcal{M})\alpha(\mathcal{M}). \tag{22'}$$

$\Pi(\mathcal{N},\mathcal{M})$ maps $\widetilde{\mathcal{M}}$ onto $\widetilde{\mathcal{N}}$.

It is the aim of this section to investigate the map $\Pi(\mathcal{N},\mathcal{M})$. We start with the trivial part.

IV.3 Lemma: *Assume $\mathcal{N} \subset \mathcal{M} \subset \mathcal{N}_1$ are three von Neumann algebras.*
(i) $\Pi(\mathcal{N},\mathcal{M})$ is a linear map commuting with the involution $\dagger$.
The map $\Pi(\mathcal{N},\mathcal{M})$ is independent of the choice of the operator Θ.
(ii) The map $\Pi(\mathcal{N},\mathcal{M})$ is a projection i.e. if $Q\,\eta\,\mathcal{N}$ then one has $\Pi(\mathcal{N},\mathcal{M})Q = Q$.
(iii) Furthermore one has

$$\Pi(\mathcal{N},\mathcal{M})\Pi(\mathcal{M},\mathcal{N}_1) = \Pi(\mathcal{N},\mathcal{N}_1). \tag{23}$$

Proof: From the definition of the maps α and β one sees that they are linear. Since the projection on a Hilbert space is a linear operator we get that $\Pi(\mathcal{N},\mathcal{M})$ is linear. From the definition of the operators Q and $Q^\dagger$ we see that map α intertwines between the involution on $\widetilde{\mathcal{M}}$ and the multiplication by S on $D(S)$. The map β intertwines between the multiplication by S on $D(S)$ and the multiplication by $\begin{pmatrix} S & 0 \\ 0 & \Theta S(\mathcal{M})\Theta \end{pmatrix}$ on the graph. Therefore, we have to show the relation

$$\begin{pmatrix} S(\mathcal{N}) & 0 \\ 0 & \Theta S(\mathcal{N})\Theta \end{pmatrix} E(\mathcal{N}) = E(\mathcal{N}) \begin{pmatrix} S(\mathcal{M}) & 0 \\ 0 & \Theta S(\mathcal{M})\Theta \end{pmatrix}.$$

If $\psi \in D(S(\mathcal{M}))$ with $S(\mathcal{M})\psi = \psi$ then $\beta\psi = \psi \oplus \Theta\psi$. Multiplying this by the projection $E(\mathcal{N})$, where it is given by eqn. (18), we obtain with $\Delta = \Delta(\mathcal{N})$ and $J = J(\mathcal{N})$:

$$\{(1+\Delta)^{-1}\psi + \Delta^{1/2}(1+\Delta)^{-1}J\psi\} \oplus \{\Theta J\Delta^{1/2}(1+\Delta)^{-1}\psi + \Theta(1+\Delta)^{-1}\psi\}.$$

Using the equation $J\Delta^{1/2}(1+\Delta)^{-1} = \Delta^{1/2}(1+\Delta)^{-1}J$ we obtain

$$E(\mathcal{N})\psi \oplus \Theta\psi = \chi \oplus \Theta\chi$$

with $\chi = (1+\Delta)^{-1}\psi + \Delta^{1/2}(1+\Delta)^{-1}J\psi$. From this representation we see that $S(\mathcal{N})\chi = \chi$ holds and Π commutes with the involution. This representation of χ also shows that the map Π is independent of Θ, so (i) is proved. If $Q\ \eta\, \mathcal{N}$ then we have $S(\mathcal{M})Q\Omega = S(\mathcal{N})Q\Omega$. Hence $Q\Omega \oplus \Theta S(\mathcal{M})Q\Omega$ belongs to the graph of $\Theta S(\mathcal{N})$ and (ii) follows. (iii) is a simple conseqence of properties of subprojections. □

IV.4 Discussion:

(1) We have seen that the map $\Pi(\mathcal{N},\mathcal{M})$ is a projection. Hence we know the action of Π on $\mathcal{N}$. There is another subset of $\mathcal{M}$ where the action of $\Pi(\mathcal{N},\mathcal{M})$ can easily be computed. From the Tomita-Takesaki-theory one knows that to every $A' \in \mathcal{M}'$ exists an operator $A \in \mathcal{M}$ with $(1+\Delta)^{-1}A'\Omega = A\Omega$. For these operators we obtain:

$$\Pi(\mathcal{N},\mathcal{M})A\Omega = \{(1+\Delta(\mathcal{N}))^{-1}(1+\Delta(\mathcal{M}))^{-1} + \Delta(\mathcal{N}))^{1/2} \\ (1+\Delta(\mathcal{N}))^{-1}\ J(\mathcal{N})J(\mathcal{M})(1+\Delta(\mathcal{M}))^{-1}\}A'\Omega = (1+\Delta(\mathcal{N}))^{-1}A'\Omega.$$

For the last transformation we have used the (1,1) component of the equation $E(\mathcal{N})E(\mathcal{M}) = E(\mathcal{N})$. The right-hand side defines an operator in $\mathcal{N}$ since $A' \in \mathcal{M}' \subset \mathcal{N}'$.

(2) The map $A' \to A$ is given by the formula

$$A = \int dt \frac{\sigma^t(j(A'))}{e^{\pi t/2} + e^{-\pi t/2}}$$

which is an orderpreserving map. The converse map is only defined on a dense subset of $\mathcal{M}$ and is $A' = j(\sigma_{\imath/2}(A) + \sigma_{-\imath/2}(A))$. This map is in general not orderpreserving. However, it is not immediately clear how the combination

$$\Pi(\mathcal{N},\mathcal{M})(A) = \int dt \frac{\sigma_t^{\mathcal{N}}}{e^{\pi t/2} + e^{-\pi t/2}} j^{\mathcal{N}}(j^{\mathcal{M}}(\sigma_{\imath/2}^{\mathcal{M}}(A) + \sigma_{-\imath/2}^{\mathcal{M}}(A)))$$

behaves.

IV.5 Lemma: *The projection $\Pi(\mathcal{N},\mathcal{M})$ does not be both normal and order-preserving if $\mathcal{N}$ is a proper subalgebra of $\mathcal{M}$.*

Proof: Assume the contrary then $(\Omega, \Pi(\mathcal{N},\mathcal{M})(A)\Omega)$ defines a normal positive linear functional on $\mathcal{M}$. Hence exists a uniqe vector $\psi \in \mathcal{H}^+(\mathcal{M},\Omega)$ with

$$(\psi, A\psi) = (\Omega, \Pi(\mathcal{N},\mathcal{M})(A)\Omega).$$

On the other hand using the representation of $\Pi(\mathcal{N},\mathcal{M})(A)$ given in IV.4.(2) we see

$$(\Omega, \Pi(\mathcal{N},\mathcal{M})(A)\Omega) = (\Omega, A\Omega).$$

But this is a contradiction if $\mathcal{N} \neq \mathcal{M}$. □

This result corresponds with a result of Tomiyama [22] telling that projections of norm one are often not normal.

We have constructed a projection map $\Pi(\mathcal{N}, \mathcal{M})$. This map turned out not to be normal and positive at the same time. It is not known whether or not it has one of these properties. Therefore it is not possible to build something like an index-theory in the sense of Jones [14], Kosaki [15], and Longo [16]. The standard index is infinite in the situation of physics because of the big relative commutant.

5. Example: Chiral quantum field theory

In this section we want to discuss the case of only one variable. This means we are dealing with a twodimensional theory where the action along one of the lightlike directions is trivial. We choose the situation where the translations along the lightcone coordinate a^+ are not trivial. The translations along this axis will be denoted by $U(a)$. This is a representation of the one dimensional translation group with invariant vector Ω and with non-negative generator. $\mathcal{M}$ is a von Neumann algebra having Ω as cyclic and separating vector. The modular operator and modular conjugation will be denoted by Δ and J. We will assume the following:
(i) $U(a)\mathcal{M}U^*(a) \subset \mathcal{M}$ for $a \geq 0$.
(ii) Ω is the only vector invariant under $U(a)$ and Δ.
(iii) $\{\cup\{U(a)\mathcal{M}U^*(a); a \in \mathbf{R}\}\}'' = \mathcal{B}(\mathcal{H})$.
(iv) $\{\cup\{U(a)\mathcal{M}'U^*(a); a \in \mathbf{R}\}\}'' = \mathcal{B}(\mathcal{H})$.

V.1 Notation and results: (i) Let Δ be the modular operator and J the modular conjugation of the pair $(\mathcal{M}, \Omega)$ then we put $T(t) = \Delta^{it/2\pi}$.
(ii) We denote $\mathcal{M}_a = U(a)\mathcal{M}U^*(a)$, and $\mathcal{M}'_a = U(a)\mathcal{M}'U^*(a)$.
(iii) Using the result obtained by the author [3] there are the following relations between the operators $T(t), U(a), J$:

$$T(t)U(a)T^*(t) = U(e^{-t}a), \qquad JU(a)J = U(-a). \tag{24}$$

This implies

$$J\mathcal{M}_a J = \mathcal{M}'_{-a}, \qquad T(t)\mathcal{M}_a T^*(t) = \mathcal{M}_{e^{-t}a}. \tag{25}$$

The first of the last equations means in particular that J is the CPT operator of this theory.
(iv) The projections introduced in Definition III.1 which are associated with the algebras $\mathcal{M}_a$ will be denoted by E_a and those associated with the algebras $\mathcal{M}'_a$ by E'_a.
(v) Recall that E_0 is the projection onto $\mathbf{C}(\Omega \oplus \Omega)$ and that this projection is contained in every E_a and E'_a. We define $F_a = E_a - E_0$ and $F'_a = E'_a - E_0$.

Using the result of Lemma III.5 we find that the two families of projections $\{F_a\}$ and $\{F'_a\}$ are two resolutions of the identity in $\mathcal{H}^\perp \oplus \mathcal{H}^\perp$. F_a tends to zero for $a \to \infty$ and to one for $a \to -\infty$. Contrary to this F'_a tends to zero for $a \to -\infty$ and to one for $a \to \infty$. This leads to the introduction of the following two operators:

V.2 Definition: With the two resolutions of the identity $\{F_a\}_{+\infty}^{-\infty}$ and $\{F'_a\}_{-\infty}^{+\infty}$ put:

$$\begin{aligned} \widehat{X} &= \int_{+\infty}^{-\infty} -a\, dF_a = \int_{-\infty}^{+\infty} \alpha\, dF_\alpha, \quad (\alpha = -a) \\ \widehat{X}' &= \int_{-\infty}^{+\infty} a\, dF'_a. \end{aligned} \tag{26}$$

Note that the two operators are defined in $\mathcal{H}^\perp \oplus \mathcal{H}^\perp$.

These two operators contain all our knowledge about the projections E_a and E'_a. The relations of the two operators with the others are described in the following (We use the same notation $\widehat{U}(a)$ etc. also for the restriction of these operators to $\mathcal{H}^\perp \oplus \mathcal{H}^\perp$.):

V.3 Lemma: *Writing for the translations* $\widehat{U}(a) = \exp\{\imath \widehat{P} a\}$ *and using equation* (13) *we obtain:*

$$\widehat{U}(a)\widehat{X}\widehat{U}^*(a) = \widehat{X} + a, \quad \widehat{U}(a)\widehat{X}'\widehat{U}^*(a) = \widehat{X}' - a.$$

Differentiating these equations we get:

$$[\widehat{P}, \widehat{X}] = -\imath, \qquad [\widehat{P}, \widehat{X}'] = \imath. \tag{27}$$

The equations (25) *lead to the relations*

$$\widehat{T}(t)\widehat{X}\widehat{T}^*(t) = e^t \widehat{X}, \qquad \widehat{T}(t)\widehat{X}'\widehat{T}^*(t) = e^t \widehat{X}', \tag{28}$$

and from the equations (14) $\div$ (14„) *we obtain*

$$\begin{aligned} V\widehat{X}V^* &= -\widehat{X}, \qquad V\widehat{X}'V^* = -\widehat{X}', \\ W\widehat{X}W &= \widehat{J}\widehat{X}\widehat{J} = \widehat{X}'. \end{aligned} \tag{29}$$

Proof: We recollect from Proposition III.3 the properties of the projections F_a and F'_a and add to this the consequence of equation (28):

$$\begin{aligned} \widehat{U}(b)F_a\widehat{U}^*(b) &= F_{a+b}, & \widehat{U}(b)F'_a\widehat{U}^*(b) &= F'_{a+b}, \\ VF_aV^* &= 1 - F_{-a}, & VF'_aV^* &= 1 - F'_{-a}, \\ WF_aW = JF_aJ &= F'_{-a}, & WF'_aW = JF'_aJ &= F_{-a}, \\ T(t)F_aT^*(t) &= F_{e^{-t}a}, & T(t)F'_aT^*(t) &= F'_{e^{-t}a}. \end{aligned}$$

With this we obtain from equation (26)

$$\widehat{U}(a)\widehat{X}\widehat{U}^*(a) = \int_{-\infty}^{+\infty} \alpha\, dF_{\alpha-a} = \int_{-\infty}^{+\infty} (\alpha + a)\, dF_\alpha = \widehat{X} + a,$$

and

$$\widehat{U}(a)\widehat{X}'\widehat{U}^*(a) = \int_{-\infty}^{+\infty} -b\, d\widehat{F}'_{b+a} = \widehat{X}' - a.$$

Differentiating these equations we obtain (27).

Using equation (25) we get:

$$\widehat{T}(t)\widehat{X}\widehat{T}^*(t) = \int_{+\infty}^{-\infty} -b\, d\widehat{F}_{e^{-t}b} = e^t \int_{+\infty}^{-\infty} -b\, d\widehat{F}_b,$$

and the same for $\widehat{X}'$. This leads to equation (28).

Since by reflection of the real line the interval $(-\infty, a]$ is mapped onto the interval $[-a, \infty)$ we see that $\{1 - F_{-a}\}$ is the spectral family of $-X$ if $\{F_a\}$ is that of X, so we obtain:

$$V\widehat{X}V^* = -\widehat{X},$$

and a similar result for $\widehat{X}'$. This gives the first two lines of (29). Finally we find:

$$W\Big(\int_{+\infty}^{-\infty} -b\, d\widehat{F}_b\Big)W = \int_{-\infty}^{+\infty} b\, d\widehat{F}'_b,$$

and the same for $\widehat{J}$. This implies the last line of (29). □

From the equations (27) and (28) we learn that we have two representations of the canonical commutation relations covariant under dilatations. Since the translations appear in both representations we decompose the Hilbert space $\mathcal{H}^\perp \oplus \mathcal{H}^\perp$ with respect to the Abelian algebra generated by $\widehat{P}$. The covariance under dilatation and the symmetry induced by V (equation (11')) tells us that we have homogenious multiplicity, i.e. we have:

$$\mathcal{H}^\perp \oplus \mathcal{H}^\perp \cong \mathcal{L}^2(\mathbf{R}) \otimes \mathcal{K} \tag{30}$$

where $\mathcal{K}$ is some Hilbert space which has to be determined.

In this situation we get:

$$\begin{aligned} \widehat{P} &= P \otimes 1_{\mathcal{K}}, & \widehat{U}(a) &= e^{\imath Pa} \otimes 1_{\mathcal{K}}, \\ \widehat{T}(t) &= T_1(t) \otimes T_2(t), & (T_1(t)\phi)(p) &= e^{-t/2}\phi(e^{-t}p), \end{aligned} \tag{31}$$

which gives the relation

$$T_1(t)PT_1^*(t) = e^{-t}P.$$

This equation is in accordance with equation (24). With the notation $Q = \imath\frac{d}{dp}$, the solution of (27) becomes:

$$\widehat{X} = Q \otimes 1_{\mathcal{K}} + X_2, \qquad \widehat{X}' = -Q \otimes 1_{\mathcal{K}} + X_2' \tag{32}$$

where X_1 and X_2' are operators commuting with P and hence they belong to the algebra $\mathcal{M}(P)\otimes\mathcal{B}(\mathcal{K})$. Here $\mathcal{M}(P)$ denotes the Abelian algebra generated by P.

$$X_2;\ X_2' \in \mathcal{M}(P) \otimes \mathcal{B}(\mathcal{K}). \tag{33}$$

Next let us look at the operators V and W. We know from the equation (15) that the operators $\frac{1}{2}(1 + VW)$ and $\frac{1}{2}(1 - \widehat{V}W)$ commute with $\widehat{U}(a)$ and $\widehat{T}(t)$ and project onto the subrepresentation of $\widehat{U}(a)$ with positive and negative spectrum, respectively. This implies the following relations

$$\begin{aligned} \tfrac{1}{2}(1+VW)\mathcal{H}^{\perp} \oplus \mathcal{H}^{\perp} &= \mathcal{L}^2(\mathbf{R}^+) \otimes \mathcal{K}, \\ \tfrac{1}{2}(1-VW)\mathcal{H}^{\perp} \oplus \mathcal{H}^{\perp} &= \mathcal{L}^2(\mathbf{R}^-) \otimes \mathcal{K} \end{aligned} \tag{34}$$

with $\mathcal{L}^2(\mathbf{R}) = \mathcal{L}^2(\mathbf{R}^+) \oplus \mathcal{L}^2(\mathbf{R}^-)$. The operators V and W must have the following structure:

$$V = \begin{pmatrix} 0 & r \\ -r & 0 \end{pmatrix} \otimes V_2, \qquad W = \begin{pmatrix} 0 & r \\ r & 0 \end{pmatrix} \otimes V_2 \tag{35}$$

where r is the reflection $(r\phi)(p) = \phi(-p)$,

$$r : \mathcal{L}^2(\mathbf{R}^+) \longrightarrow \mathcal{L}^2(\mathbf{R}^-), \qquad r : \mathcal{L}^2(\mathbf{R}^-) \longrightarrow \mathcal{L}^2(\mathbf{R}^+),$$

and V_2 is a unitary operator in $\mathcal{B}(\mathcal{K})$ with the property $V_2^* = V_2$, $V_2^2 = 1$.

We want to use these equations for a structure analysis of the two operators X and X'. From the equations (28) and (29) we obtain:

$$\begin{aligned} \widehat{T}(t)\begin{Bmatrix} X_2 \\ X_2' \end{Bmatrix}\widehat{T}^*(t) &= e^t \begin{Bmatrix} X_2 \\ X_2' \end{Bmatrix}, \\ V\begin{Bmatrix} X_2 \\ X_2' \end{Bmatrix}V^* &= -\begin{Bmatrix} X_2 \\ X_2' \end{Bmatrix}, \qquad W\begin{Bmatrix} X_2 \\ X_2' \end{Bmatrix}W = \begin{Bmatrix} X_2' \\ X_2 \end{Bmatrix}. \end{aligned} \tag{36}$$

At this point it should be remarked that the operator VW contains the function $\epsilon(p)$ in momentumspace. Therefore the domain of definition of the operators $\widehat{X}$ and $\widehat{X}'$ contains only function $f(p)$ such that $f(p)$ and $\epsilon(p)f(p)$ are both differentiable.

If we now put all informations together then we obtain:

V.4 Lemma: *Let us write*

$$\mathcal{L}^2(\mathbf{R}) \otimes \mathcal{K} = \int_{-\infty}^{\infty} \mathcal{H}_p \, dp$$

with $\mathcal{H}_p = \mathcal{K}$ *and*

$$X_2 = \int_{\oplus} X_2(p)\, dp, \qquad X_2' = \int_{\oplus} X_2'(p)\, dp$$

then the structure of $\widehat{T}(t)$ *given in* (31) *and the structure of* V *and* W *given in* (35) *leads to:*

$$\begin{aligned} T_2(t)\left\{ \begin{matrix} X_2(e^{-t}p) \\ X_2'(e^{-t}p) \end{matrix} \right\} T_2^*(t) &= e^t \left\{ \begin{matrix} X_2(p) \\ X_2'(p) \end{matrix} \right\}, \\ V_2 \left\{ \begin{matrix} X_2(p) \\ X_2'(-p) \end{matrix} \right\} V_2 &= \left\{ \begin{matrix} X_2(-p) \\ X_2'(p) \end{matrix} \right\}, \\ V_2 X_2(p) V_2 &= X_2'(-p). \end{aligned} \tag{37}$$

We learn from these equations that one can choose freely only one of the $X_2(p)$. *The others are determined by the above equations. We obtain the following solution:*

$$\begin{aligned} X_2(p) &= \begin{cases} \frac{1}{p} T_2(\log|p|) X T_2^*(\log|p|) & p > 0, \\ \frac{1}{|p|} T_2(\log|p|) V_2 X V_2 T_2^*(\log|p|) & p < 0, \end{cases} \\ X_2' &= X_2, \qquad X \in \mathcal{B}(\mathcal{K}). \end{aligned} \tag{38}$$

Proof:
Let $\psi(p)$ be in $\mathcal{K}$ then we have $(V\psi)(p) = \epsilon(p)V_2\psi(-p)$, This amounts to $(VX_2V^*\psi) = \epsilon(p)V_2(X_2V^*\psi)(-p) = \epsilon(p)V_2X_2(-p)(V^*\psi)(-p) = \epsilon(p)\epsilon(-p)V_2X_2(-p)V_2\psi(p)$. From this we obtain the second line of (37). Comparing the second line of (37) with the last line of (37) we obtain $X_2' = X_2$. Inserting into the first line of (37) $X_2(1) = X$ and $e^{-t}p = 1$ we obtain for $p > 0$ the stated result. For $p < 0$ we note that we have $p = -|p|$ and $X_2(-1) = VXV$. This amounts to the second part of (38). Therfore, (38) is reduced to (37) and it remains to show these equations. The first line of (36) together with (31) gives

$$\begin{aligned} e^t X_2(p)\psi(p) &= \{\widehat{T}(t) X_2(p) \widehat{T}^*(t)\psi\}(p) \\ = e^{-t/2} T_2(t) X_2(e^{-t}p)\{\widehat{T}^*(t)\psi\}(e^{-t}p) &= T_2(t) X_2(e^{-t}p) T_2^*(t)\psi(p). \end{aligned}$$

This is the first equation of (37). The others follow in the same manner together with the structure of V and W given in equation (35). □

Let us remark that in an integral decomposition the Hilbert spaces are only fixed up to a unitary transformation which can be chosen once and for all. In our case this means we can write without loss of generality $X(p) = \frac{1}{|p|}X$ or

$$\begin{aligned} \widehat{X} &= Q \otimes 1_{\mathcal{K}} + \tfrac{1}{|p|} \otimes X, \\ \widehat{X}' &= -Q \otimes 1_{\mathcal{K}} + \otimes X, \quad X \in \mathcal{B}(\mathcal{K}). \end{aligned} \tag{39}$$

We finally have to look at the action of J. From (24) we see that J acts on $\mathcal{L}^2(\mathbf{R})$ as the operator which goes to the conjugate complex. Therefore, it remains to look at the action on $\mathcal{K}$. From (29) and (39) we obtain

$$J 1 \otimes X J = 1 \otimes X. \tag{40}$$

This shows us that the self-adjoint operator X is real with respect to the conjugation induced by J on the Hilbert space $\mathcal{K}$.

This example confirms the result of Corollary III.7 namely, that the algebra generated by the projection $\{F_a, F'_b\}$ is not commutative. We obtain

$$[\widehat{X}, \widehat{X}'] = -2\imath \frac{p}{|p|^3} \otimes X.$$

With this result we have obtained an algebraic invariant of the theories described in the beginning of this section. The algebraic invariant is the unitary equivalence class of the operator X, which means the type of the spectrum of X and the multiplicities.

If we know the operator X then we can construct Δ. For the full characterization of the theory we need the operator J and whatever is needed to construct the algebra $\mathcal{M}$ if J and Δ is known.

6. Conclusions and problems

We have constructed a map $\mathcal{M}(\mathcal{O}) \to E(\mathcal{O})$ but nothing is known of the converse map. In particular what is the meaning of the invariance-group of one element $E(\mathcal{O})$ which respects also the decomposition of the Hilbert space $\widehat{\mathcal{H}}$? These operators have the form $\widehat{U} = \mathrm{diag}(U, \Theta J U J \Theta)$ where the unitary operator U commutes with the modular operator $\Delta(\mathcal{O})$. But is this all? There are two problems. First we do not know wether or not the operator U also commutes with J. If we assume this, then we also need to know wether U maps the positive cone $\mathrm{cl}\{\mathcal{M}^+(\mathcal{O})\Omega\}$ onto itself in order to obtain an automorphism of $\mathcal{M}(\mathcal{O})$ [7].

The difficulty in solving these problems is the fact that the map $A\Omega \to A^*\Omega$ is defined on every selfadjoint subspace of $\mathcal{M}(\mathcal{O})$. A closer look seems to indicate that again the concept of positivity is missing in our setting.

References

[1] Araki,H., Einführung in die axiomatische Quantenfeldtheorie, *Vorlesungen an der E.T.H. Zürich*, I. W.S. 1961/62 ausgearbeitet von K. Hepp u. F. Riahi, II. S.S. 1962 ausgearbeitet von W. Wyss

[2] Araki,H., A Lattice of von Neumann Algebras Associated with the Quantum Theory of a Free Bose Field, *J. Math. Phys.*, **4**, 1343-1362 (1963)

[3] Borchers,H.-J., The CPT-Theorem in Two-dimensional Theories of Local Observables, *Commun. Math. Phys.*, **143**, 315-332 (1992)

[4] Bratteli,O. and Robinson,D.W., Operator Algebras and Quantum Statistical Mechanics I, *Springer-Verlag*, New York, Heidelberg, Berlin, (1979)

[5] Buchholz,D., D'Antoni,C., and Fredenhagen,K., The Universal Structure of Local Algebras, *Commun. Math. Phys.*, **111**, 123-135 (1987)

[6] Buchholz,D, Wichmann,E.H., Causal Independence and the Energy-Level Density of States in Local Quantum Field Theory, *Commun. Math. Phys.*, **106**, 321 (1986)

[7] Connes,A., Caractérisation des algèbres de von Neumann comme espaces vectoriels ordonnés, *Ann. Inst. Fourier*, **26**, 121-155. (1974)

[8] Connes,A., C*-algèbres et géométrie différentielle, *C.R. Acad. Sci. Paris*, **290**, Série A, 599 (1980)

[9] Connes,A., Noncommutative differential geometry, *Publ. Math. IHES*, **62** (1985)

[10] Connes,A., and Lott,J., Particle models and noncommutative geometry, *Nucl. Phys.*, (proc. Suppl.) **B 18**, 29 (1990)

[11] Coquereaux,R., Esposito-Farèse,G., and Scheck,F., The Theory of Electro-weak Interactions Described by $SU(2|1)$, Algebraic Superconnections, *preprint Marseille*, CPT-90/P.E.2464 (1991)

[12] Dubois-Violette,M., Kerner,M., and Madore,J., Non-Commutative Differential Geometry and a new Model of Gauge Theory, *J. Math. Phys.*, **13**, 323 (1989)

[13] Haag,R., and Kastler,D., An algebraic approach to quantum field theory, *J. Math. Phys.*, **5**, 848 (1964)

[14] Jones,V.F.R., Index for subfactors, *Invent. Math.*, **71**, 1-25, (1983)

[15] Kosaki,H., Extension of Jones' theory of index to arbitrary factors, *J. Funct. Anal.*, **66**, 123-149, (1986)

[16] Longo,R., Index of Subfactors and Statistics of Quantum Fields, *Commun. Math. Phys.*, **126**, 217-249 (1989)

[17] Nussbaum,A.E., Reduction Theory for Unbounded Closed Operators in Hilbert Space, *Duke Math. J.*, **31**, 33-44 (1964)

[18] Segal,I.E., A non-commutative extension of abstract integration, *Ann. Math.*, **57**, 401-457 (1953); Correction: Ann. Math. **57**, 595-596 (1953)

[19] Stone,M.H., On unbounded operators in Hilbert space, *Jour. Ind. Math. Soc.*, **15**, 155-192 (1951)

[20] Takesaki,M., Tomita's Theory of Modular Hilbert Algebras and its Applications, Lecture Notes in Math. Vol. **128**, (1970) *Springer-Verlag*, Heidelberg

[21] Tomita,M., Standard Forms of von Neumann Algebras, Fifth Functional Analysis Symposium of Math. Soc. of Japan, *Sendai*, (1967)

[22] Tomiyama,J., On the Projections of Norm One in W^*-Algebras, III, *Tôhoku Math. J.*, **11**, 125-129 (1959)

[23] VolkovD.V., and Akulov,V., Is the Neutrino a Goldstone Particle?, *Phys. Lett.*, **46B** 109-110 (1973)

[24] WessJ., and Zumino,B., Supergauge Transformations in Four Dimensional Theories, *Nucl. Phys. B*, **70** 39-50 (1974)

Goldstone's Theorem Revisited

JOHN E. ROBERTS
Dipartimento di Matematica
Università di Roma "Tor Vergata"
via della Ricerca Scientifica
I–00133 Roma, Italy

My talk reports on joint work with D. Buchholz, S. Doplicher and R. Longo [1]. We discuss spontaneously broken symmetries and prove theorems of Goldstone type without assuming that the symmetry in question is generated by a conserved current.

The setting is the following: we consider a 1–parameter group $\lambda \mapsto \gamma_\lambda$ of automorphisms of a field net $\mathcal{O} \to \mathfrak{F}(\mathcal{O})$ of von Neumann algebras defined over the set $\mathcal{K}$ of double cones. These automorphisms are to be local, to commute with spacetime translations and to be weak operator continuous, i.e.

$$\gamma_\lambda(\mathfrak{F}(\mathcal{O})) = \mathfrak{F}(\mathcal{O}); \quad \mathcal{O} \in \mathcal{K} ,$$
$$\gamma_\lambda \alpha_a = \alpha_a \gamma_\lambda , \quad \lambda \in \mathbb{R} , \quad a \in \mathbb{R}^{s+1} ,$$
$$\lambda \mapsto \gamma_\lambda(\mathbf{F}) \quad \text{is continuous in the weak operator topology, } \mathbf{F} \in \mathfrak{F} .$$

Here s denotes the number of space dimensions, α_a denotes the spacetime translation through a and the weak operator topology refers to the representation defined by the vacuum state ω_0.

The 1–parameter group of symmetries is said to be *spontaneously broken* if the vacuum state is not invariant, i.e. if for some $\lambda \in \mathbb{R}$, $\omega_0 \circ \gamma_\lambda \neq \omega_0$. The notion of spontaneous symmetry breaking therefore depends on the choice of some (physically important) state. The notion is, for example, often discussed in connection with equilibrium states but I confine myself here to the case of a given fixed vacuum state ω_0.

Now let δ denote the derivation which is the infinitesimal generator of the 1–parameter group and let $\mathcal{D}$ denote the domain of δ, i.e.

$$\mathcal{D} = \{\mathbf{F} \in \mathfrak{F} : \lim \frac{1}{\lambda}(\gamma_\lambda(\mathbf{F}) - \mathbf{F}) \text{ exists in norm}\} .$$

If $\mathbf{F} \in \mathcal{D}$, then $\delta(\mathbf{F})$ is, of course, defined to be the above limit. The set

$$\mathcal{D}_0 := \bigcup_{\mathcal{O} \in \mathcal{K}} \mathfrak{F}(\mathcal{O}) \cap \mathcal{D}$$

H. Araki et al. (eds.), Quantum and Non-Commutative Analysis, 31–39.

is a core for δ. The concept of spontaneously broken symmetry can equally well be formulated in terms of the generating derivation δ:

$$\omega_0 \circ \gamma_\lambda = \omega_0 \ , \quad \lambda \in \mathbb{R} \quad \Leftrightarrow \quad \omega_0 \circ \delta = 0 \ .$$

Our aim is to derive conditions under which spontaneous symmetry breaking, $\omega_0 \circ \delta \neq 0$, is possible and what matters is the behaviour of the functional $\omega_0 \circ \delta$ on $\mathfrak{F}(\mathcal{O}) \cap \mathcal{D}$ for large $\mathcal{O}$. Let $\mathcal{O}_\mathbf{R}$ denote the double cone with base $\{\mathbf{x} : |\mathbf{x}| \leq \mathbf{R}\}$ then, since $\mathfrak{F}(\mathcal{O}_\mathbf{R})$ is in standard form in the vacuum representation, there is a skew–adjoint operator $J_\mathbf{R}$ implementing δ on $\mathfrak{F}(\mathcal{O}_\mathbf{R})$, i.e.

$$\delta(\mathbf{F}) = [J_\mathbf{R}, \mathbf{F}] \ , \quad \mathbf{F} \in \mathfrak{F}(\mathcal{O}_\mathbf{R}) \cap \mathcal{D} \ .$$

If the vacuum vector Ω, $\omega_0(\mathbf{F}) = (\Omega, \mathbf{F}\Omega)$, $\mathbf{F} \in \mathfrak{F}$, lies in $\mathcal{D}(J_\mathbf{R})$, the domain of $J_\mathbf{R}$, we have the following simple estimate:

$$|\omega_0 \circ \delta(\mathbf{F})| \leq c_\mathbf{R}(\|\mathbf{F}\Omega\| + \|\mathbf{F}^*\Omega\|) \ , \quad \mathbf{F} \in \mathfrak{F}(\mathcal{O}_\mathbf{R}) \cap \mathcal{D} \ ,$$

with $c_\mathbf{R} \leq \|J_\mathbf{R}\Omega\|$. However, even if $\Omega \notin \mathcal{D}(J_\mathbf{R})$, then simple manipulations involving the resolvent of $J_\mathbf{R}$ show that given any $\varepsilon > 0$, we still have the following estimate:

$$|\omega_0 \circ \delta(\mathbf{F})| \leq c_{\mathbf{R},\varepsilon}(\|\mathbf{F}\Omega\| + \|\mathbf{F}^*\Omega\|) + \varepsilon\|\delta(\mathbf{F})\| \ , \quad \mathbf{F} \in \mathfrak{F}(\mathcal{O}_\mathbf{R}) \cap \mathcal{D} \ ,$$

which is equally effective.

Our results of Goldstone type are expressed in terms of these a priori estimates and they therefore do not require any explicit assumptions that the symmetry in question is generated by some conserved current. Instead we simply exploit the fact that the functional $\omega_0 \circ \delta$ is invariant under space and time translations. Set

$$\alpha_f(\mathbf{F}) := \int dt f(t) \alpha_t(\mathbf{F}) \ , \quad f \in L^1(\mathbb{R}) \ ,$$

where α_t denotes a time translation. When $\int f = 1$, then

$$\omega_0 \circ \delta \alpha_f(\mathbf{F}) = \omega_0 \circ \alpha_f \delta(\mathbf{F}) = \omega_0 \circ \delta(\mathbf{F}) \ .$$

Hence the l.h.s. of our estimate is unchanged if we replace $\mathbf{F}$ by $\alpha_f(\mathbf{F})$, however, the r.h.s. changes in general and we can attempt to minimize it. This leads to the following result:

Proposition. *In a theory with mass gap $m > 0$ there is a model–independent constant $c > 0$ such that if*

$$|\omega_0 \circ \delta(\mathbf{F})| \leq c_\varepsilon e^{\mu \mathbf{R}}(\|\mathbf{F}\Omega\| + \|\mathbf{F}^*\Omega\|) + \varepsilon\|\delta(\mathbf{F})\| \ , \quad \mathbf{F} \in \mathfrak{F}(\mathcal{O}_\mathbf{R})$$

for some $0 \le \mu < cm$, *then* $\omega_0 \circ \delta = 0$.

The best possible value of c is not known. However, if we are in the case where we can take $\varepsilon = 0$ then our result holds with $c = 1$ and this result is best possible since there is a model with spontaneous symmetry breaking satisfying the estimate for all $\mu > m$.

The above result covers one facet of Goldstone's Theorem since it shows how a mass gap can inhibit spontaneous symmetry breaking. The other facet of Goldstone's Theorem is the relation between spontaneous symmetry breaking and particles of zero mass. At this point, we need to make use of the Bose–Fermi commutation relations in the field net: there is a $\mathbb{Z}_2$–grading of the field net such that if $\mathcal{O}_1$ and $\mathcal{O}_2$ are spacelike separated double cones then

$$\mathbf{F}_+\mathbf{F}'_+ = \mathbf{F}'_+\mathbf{F}_+ \,, \quad \mathbf{F}_+\mathbf{F}'_- = \mathbf{F}'_-\mathbf{F}_+ \,, \quad \mathbf{F}_-\mathbf{F}'_- = -\mathbf{F}'_-\mathbf{F}_- \,,$$
$$\mathbf{F}_\pm \in \mathfrak{F}(\mathcal{O}_1) \,, \quad \mathbf{F}'_\pm \in \mathfrak{F}(\mathcal{O}_2) \,.$$

To derive our results, we now make use of space rather than time translations and apply the techniques of the Jost–Lehmann–Dyson representation guided by the discussion in [2]. The key formula is the following:

$$\lim_{\mathbf{R}\to\infty} \mathbf{R}^{s-1}\|\alpha_{f_\mathbf{R}}(\mathbf{F})\Omega\|^2 = (2\pi)^{-s}\int \frac{d\mathbf{p}}{|\mathbf{p}|}|\tilde{f}(\mathbf{p})|^2 \int d\mathbf{a}(\Omega, \alpha_\mathbf{a}(\mathbf{F})E_1 P^0 \mathbf{F}\Omega) \,.$$

It tells us what happens to $\|\alpha_f(\mathbf{F})\Omega\|^2$ if we dilate the function f,

$$f_\mathbf{R}(\mathbf{a}) := \mathbf{R}^{-s} f(\mathbf{R}^{-1}\mathbf{a}) \,,$$

and pass to the limit. Here, $\mathbf{a}$ denotes a spatial translation, $\tilde{f}$ denotes the Fourier transform of f and $\mathbf{F}$ must be smooth, local and self–adjoint with vanishing vacuum expectation value. P^0 is the zero component of the momentum operator and E_1 is the spectral projection of the momentum corresponding to the Borel set $\partial V_+ \backslash \{0\}$, the boundary of the forward light cone with the apex removed. E_1 is therefore the projection onto the subspace associated with particles of mass zero in the vacuum representation of $\mathfrak{F}$.

Proposition. *Suppose* $s > 1$ *and for each* $\varepsilon > 0$ *we have*

$$|\omega_0 \circ \delta(\mathbf{F})| \le c_{\mathbf{R},\varepsilon}(\|\mathbf{F}\Omega\| + \|\mathbf{F}^*\Omega\|) + \varepsilon\|\delta(\mathbf{F})\| \,, \quad \mathbf{F} \in \mathfrak{F}(\mathcal{O}_\mathbf{R}) \cap \mathcal{D} \,.$$

Now if

a) $\liminf_{\mathbf{R}\to\infty} c_{\mathbf{R},\varepsilon}\mathbf{R}^{-\frac{1}{2}(s-1)} = 0$, *then* $\omega_0 \circ \delta = 0$;

b) $\liminf_{\mathbf{R}\to\infty} c_{\mathbf{R},\varepsilon}\mathbf{R}^{-\frac{1}{2}(s-1)} < +\infty$ *and* $\omega_0 \circ \delta \ne 0$, *then there are massless Bosons in the theory;*

c) $c_{\mathbf{R},\varepsilon}$ *is polynomially bounded and* $\omega_0 \circ \delta \neq 0$ *then the spectrum of spacetime translations coincides with the forward light cone but there may be no massless particles in the theory.*

Remarks. a) The proof that massless Goldstone particles exist in theories with a spontaneously broken symmetry depends on the specific a priori bound $c_{\mathbf{R},\varepsilon} \leq c_\varepsilon \mathbf{R}^{\frac{1}{2}(s-1)}$. One way of getting such a bound is first to require that the 1–parameter group γ of symmetries is generated by a conserved current j_μ. This means that the derivation δ is implemented on $\mathfrak{F}(\mathcal{O}_\mathbf{R})$ by

$$J_\mathbf{R} = i \int \mathrm{d}x f(x^0) g\left(\frac{\mathbf{x}}{\mathbf{R}}\right) j_0(x) \,,$$

where $f \in \mathcal{D}(\mathbb{R})$ with $\int \mathrm{d}x^0 f(x^0) = 1$ and $g \in \mathcal{D}(\mathbb{R}^s)$ with $g(\mathbf{x}) = 1$ for $|\mathbf{x}| \leq 1$. If the current j_μ also satisfies the Wightman axioms and is hence in particular translation covariant then it follows that

$$\|J_\mathbf{R}\Omega\| \leq \text{ const.}\, \mathbf{R}^{\frac{1}{2}(s-1)}$$

so we are in case b) and recover the traditional version of Goldstone's Theorem. However, our models provide examples of spontaneously broken symmetries where these hypotheses are violated. In particular, there is a spontaneously broken symmetry associated with the free electromagnetic field derived from a conserved current which is not translation covariant. This last example is interesting for another reason because it shows that we cannot in general conclude that the massless Goldstone Bosons are scalars.

b) It is easy to understand why the massless particles in Goldstone's Theorem must be Bosons. Any $\mathbf{F} \in \mathfrak{F}$ can be written as a sum of a Bose and a Fermi part: $\mathbf{F} = \mathbf{F}_+ + \mathbf{F}_-$. Since γ_λ acts locally and commutes with translations, $\gamma_\lambda(\mathbf{F})_\pm = \gamma_\lambda(\mathbf{F}_\pm)$ giving $\delta(\mathbf{F})_\pm = \delta(\mathbf{F}_\pm)$. Now $\omega_0(\mathbf{F}_-) = 0$, so $\omega_0 \circ \delta$ vanishes identically on the Fermi part of $\mathcal{D} \subset \mathfrak{F}$ and any spontaneous breaking must be due to the Bose part, i.e. there must be an $\mathbf{F}_+$ with $\omega_0(\delta(\mathbf{F}_+)) \neq 0$ so the key formula gives $E_1\mathbf{F}_+\Omega \neq 0$.

c) If γ_λ commutes with rotations, the Goldstone Boson has helicity zero for if we average over the rotation group setting

$$\overline{\mathbf{F}}_+ := \int \mathrm{d}\mu(\mathbf{R})\alpha_\mathbf{R}(\mathbf{F}_+)$$

and $0 \neq \omega_0 \circ \delta(\mathbf{F}_+) = \omega_0 \circ \delta(\overline{\mathbf{F}}_+)$ then, from the key formula, $E_1\overline{\mathbf{F}}_+\Omega \neq 0$ and is a vector invariant under rotations. Thus the subspace of massless particles contains states of helicity zero and the Goldstone Boson has helicity zero. However, as we shall discuss below, the free electromagnetic field admits a group of spontaneously broken symmetries shifting the zero point of the field and the only massless particle has helicity one.

d) Geometrical symmetries can also be treated. Let $\lambda \to D_\lambda$ be a 1–parameter group of diffeomorphisms of spacetime (e.g. translations or Lorentz transformations) and suppose

$$\gamma_\lambda(\mathfrak{F}(\mathcal{O})) = \mathfrak{F}(D_\lambda \mathcal{O}) , \quad \mathcal{O} \in \mathcal{K} ,$$

$$\gamma_\lambda \alpha_a = \alpha_{D_\lambda a} \gamma_\lambda , \quad \lambda \in \mathbb{R} , \quad a \in \mathbb{R}^{s+1} ,$$

then the generating derivation δ still has the property that $\omega_0 \circ \delta \alpha_a = \omega_0 \circ \delta$, $a \in \mathbb{R}^{s+1}$ and the results extend.

e) Supersymmetries are defined by densely defined linear maps $\overline{\delta}$ on $\mathfrak{F}$ satisfying the graded Leibnitz rule

$$\overline{\delta}(\mathbf{F}_\sigma \mathbf{G}_\tau) = \overline{\delta}(\mathbf{F}_\sigma)\mathbf{G}_\tau + \sigma \mathbf{F}_\sigma \overline{\delta}(\mathbf{G}_\tau) , \quad \sigma, \tau \in \{+, -\}$$

with respect to the Bose–Fermi grading. $\overline{\delta}$ is local and commutes with translations so

$$\omega_0 \circ \overline{\delta} \alpha_{\mathbf{a}} = \omega_0 \circ \overline{\delta} , \quad \mathbf{a} \in \mathbb{R}^{s+1}$$

and the results extend. In fact the power $\mathbf{R}^{-\frac{1}{2}(s-1)}$ in the proposition can be replaced by $\mathbf{R}^{-\frac{1}{2}s}$. The corresponding Goldstone particles are now massless Fermions hence having half–integral helicity.

Examples. It has also been our aim to complement our results on spontaneous symmetry breaking by providing a few simple examples which show that the scope for proving stronger results is strictly limited. We begin with the simplest of all examples which illustrates the traditional setting for Goldstone's Theorem.

a) Consider the free massless scalar field φ in $s+1$–dimensions. We work as usual in terms of the Weyl operators,

$$W(f) := e^{i\varphi(f)} , \quad f \in \mathcal{D}(\mathbb{R}^{s+1}) .$$

The vacuum functional is given by

$$\omega_0(W(f)) = e^{-\frac{1}{2}\|f\|^2} , \quad \|f\|^2 := \int \frac{d\mathbf{p}}{2p_0} |\tilde{f}(p_0, \mathbf{p})|^2 , \quad p_0 = |\mathbf{p}| .$$

If we consider automorphisms γ_t of the form

$$\gamma_t(W(f)) = e^{it(f)} W(f) ,$$

where t is a linear functional on the 1–particle space then we certainly get local automorphisms of the net of Weyl algebras but these automorphisms do not necessarily extend to automorphisms of the net $\mathcal{O} \to \mathfrak{F}(\mathcal{O})$ of von Neumann algebras defined through the vacuum representation. A necessary and sufficient condition for this to be the case is that t should be continuous in restriction to the corresponding local subspaces of the 1–particle space. Since, we furthermore require γ_t to commute with spacetime translations, we must have

$$t(f) = c \int dx f(x) = c(2\pi)^{\frac{1}{2}s+1} \tilde{f}(0) .$$

We have therefore exhibited a 1–parameter group of spontaneously broken symmetries. In this case, the constants $c_{\mathbf{R}}$ in our bounds can be chosen to be proportional to $\mathbf{R}^{\frac{1}{2}(s-1)}$. We have an associated conserved covariant current $j_\mu = \partial_\mu \varphi$ and of course we have a Boson with mass zero and helicity zero in this case.

b) Next consider the free electromagnetic field with Weyl operators

$$W(f) := e^{i\mathbf{F}_{\mu\nu}(f^{\mu\nu})} ,$$

where the test functions f are antisymmetric in the spacetime indices μ, ν and a summation over the indices is understood. The vacuum functional is given by

$$\omega_0(W(f)) = e^{-\frac{1}{2}\|f\|^2} , \quad \|f\|^2 := \int \frac{d\mathbf{p}}{2p_0} \overline{\tilde{f}^{\mu\nu}(p)} p_\mu g_{\nu\rho} p_\sigma \tilde{f}^{\rho\sigma}(p) , \quad p_0 = |\mathbf{p}| .$$

We again consider automorphisms of the Weyl algebra given by a linear functional t on the 1–particle space and if these are to yield automorphisms of the net $\mathfrak{F}$ of von Neumann algebras defined through the vacuum representation and which furthermore commute with the spacetime translations then t must have the form:

$$t(f) = \lambda_{\mu\nu} \int dx f^{\mu\nu}(x) , \quad \lambda_{\mu\nu} = -\lambda_{\nu\mu} .$$

The corresponding group $t \mapsto \gamma_t$ of automorphisms of $\mathfrak{F}$ is isomorphic to $\mathbb{R}^{\frac{1}{2}s(s+1)}$. In particular, we have 1–parameter groups $\lambda \mapsto \gamma_{\lambda t}$, $t \neq 0$, of spontaneously broken symmetries. The constants $c_{\mathbf{R}}$ can now be chosen proportional to $\mathbf{R}^{\frac{1}{2}(s+1)}$ and this is the best possible choice. Hence these 1–parameter groups cannot be generated by conserved currents satisfying the Wightman axioms. Nevertheless they are generated by conserved currents of the form

$$j_\mu(x) = d_{\mu\nu\rho\sigma} x^\nu \mathbf{F}^{\rho\sigma}(x) ,$$

which are not translation covariant. The relation between the coefficients λ determining t and d determining j_μ is not given here since it is relatively involved.

c) Next consider the generalized free fields. We have Weyl operators $W(f)$ labelled by $f \in \mathcal{D}(\mathbb{R}^{s+1})$ as in the case of the free field but the norm on the 1–particle spaces is given

$$\|f\|^2 = \int d\mu(m) \int \frac{d\mathbf{p}}{2p_0} |\tilde{f}(p_0, \mathbf{p})|^2 , \quad p_0 = (\mathbf{p}^2 + m^2)^{\frac{1}{2}} ,$$

and therefore involves an additional integration over the mass m. For our purposes, we will assume that the measure μ is absolutely continuous with respect to Lebesgue measure on $\mathbb{R}_+$,

$$d\mu(m) = \rho(m)dm \ .$$

The key observation is that we now have a 1–parameter group of automorphisms $\lambda \mapsto \gamma_\lambda$ of the Weyl algebra given by the same formula,

$$\gamma_\lambda(W(f)) = e^{i\lambda \tilde{f}(0)} W(f) \ , \quad \lambda \in \mathbb{R} \ ,$$

as for the free scalar field since $\tilde{f}$ is entire analytic so that $\|f\| = 0$ implies $\tilde{f} = 0$ and hence $\tilde{f}(0) = 0$. However, we still need to know whether these automorphisms extend to the net $\mathfrak{F}$ of von Neumann algebras defined by the vacuum representation, i.e. whether the linear functional in question is continuous in restriction to the local 1–particle spaces.

Lemma. *Suppose that for some $M > 0$,*

$$k(M)^2 := \int_M^\infty dm \frac{1}{m\rho(m)} < +\infty \ ,$$

then for any real test function f with support in $\mathcal{O}_{\mathrm{R}}$,

$$|\tilde{f}(0)| \leq \ \text{const.}\, k(\mathbf{M})\mathbf{R}^{\frac{1}{2}s} \left(1 + \frac{1}{\mathbf{MR}}\right)^{\frac{1}{2}} e^{\mathbf{MR}} \|f\| \ .$$

If furthermore $k(0) < +\infty$, then

$$|\tilde{f}(0)| \leq \ \text{const.}\, \mathbf{R}^{\frac{1}{2}s} \|f\| \ .$$

We conclude that if $k(\mathbf{M}) < +\infty$ for some $\mathbf{M} \geq 0$, then the above formula does indeed define a 1–parameter group of local automorphisms of $\mathfrak{F}$ commuting with spacetime translations.

If $\rho(m) = 0$ for $m \leq \mathbf{M}$ and $k(\mathbf{M}) < +\infty$ then we have a theory with a mass gap $\mathbf{M}$ and a spontaneously broken symmetry where we can take

$$c_{\mathrm{R}} = \ \text{const.}\, \mathbf{R}^{\frac{1}{2}s} e^{\mathbf{MR}} \ .$$

The growth here is only slightly stronger than in our first proposition and since the bounds in that proposition cannot hold, this symmetry cannot be derived from a tempered conserved current.

On the other hand, if we have no mass gap and $k(0) < +\infty$, we have a theory with spontaneously broken symmetry where we can take

$$c_{\mathrm{R}} = \ \text{const.}\, \mathbf{R}^{\frac{1}{2}s} \ .$$

The fact that we now have no mass gap is consistent with our second proposition however we also have no Goldstone particle so we only need a slightly faster growth to violate the conclusions of Goldstone's Theorem.

This really concludes my discussion of Goldstone's Theorem and I hope I have convinced you that by avoiding any hypotheses on the existence of an underlying conserved current, the proof of Goldstone's Theorem has been simplified and its scope widened. The fact that one could derive results starting with the operators $J_{\mathbf{R}}$ implementing δ rather than the conserved current j_μ had been recognized by H. Reeh [3] but this line of research was apparently not pursued further. At the same time, the examples show that, whenever Goldstone's Theorem is invoked, certain symmetries are being explicitly, or implicitly, excluded.

There is, however, still one important point of principle concerning spontaneously broken symmetries which I would like to discuss briefly even if it has no direct bearing on Goldstone's Theorem. In the above, we have simply treated symmetries as automorphisms of a field net $\mathfrak{F}$. It should, however, be possible to define any physically relevant concept in terms of the observables. We should therefore ask how spontaneously broken symmetries are to be defined if we start from the observable net $\mathcal{O} \to \mathfrak{A}(\mathcal{O})$, $\mathcal{O} \in \mathcal{K}$, which is a subnet of the field net $\mathfrak{F}$. It seems to be an empirical fact that this net has no local symmetries commuting with spacetime translations, spontaneously broken or not. Hence any such symmetries must be gauge symmetries, i.e. they must act trivially on the subnet $\mathfrak{A}$. This means, incidentally, that the vacuum state of $\mathfrak{A}$ is invariant under such symmetries so that we need not give up the idea of there being a unique physical vacuum state if there are spontaneously broken symmetries. However, to have an intrinsic concept of spontaneously broken symmetry it is necessary to be able to make a canonical choice of $\mathfrak{F}$. The possibility of making such a choice was demonstrated in [4]: there is a canonical field net $\mathfrak{F}$ with normal Bose–Fermi commutation relations which is acted on by a compact gauge group $\mathbf{G}$. The fixed–point net is just the dual net

$$\mathfrak{F}^{\mathbf{G}} = \mathfrak{A}^d , \quad \mathfrak{A}^d(\mathcal{O}) = \mathfrak{A}(\mathcal{O}')' , \quad \mathcal{O} \in \mathcal{K} ,$$

where the commutant refers to the vacuum representation of $\mathfrak{A}$. Furthermore,

$$\mathbf{G} = \{g \in \mathrm{Aut}\mathfrak{F} : g(A) = A , \quad A \in \mathfrak{A}^d\} .$$

We now define the full gauge group of the theory Γ by

$$\Gamma := \{\gamma \in \mathrm{Aut}\mathfrak{F} : \gamma(A) = A , \quad A \in \mathfrak{A}\} .$$

Proposition.

i) $\gamma(\mathfrak{F}(\mathcal{O})) = \mathfrak{F}(\mathcal{O})$, $\mathcal{O} \in \mathcal{K}$, $\gamma \in \Gamma$.

ii) *Each* $\gamma \in \Gamma$ *is locally normal, i.e. normal in restriction to each* $\mathfrak{F}(\mathcal{O})$.

iii) *If* $\gamma \in \Gamma$ *then* $\omega_0 \circ \gamma = \omega_0$ *if and only if* $\gamma \in G$.

We therefore see that the elements of Γ turn out to be *local* automorphisms of the net $\mathfrak{F}$ and that G turns out to be precisely the subgroup of Γ consisting of automorphisms which are not spontaneously broken. Conditions ensuring that the elements of Γ commute with spacetime translations are presented in [1]. Hence we may regard the set of spontaneously broken symmetries in the theory as being given by $\{\gamma \in \Gamma : \gamma \notin G\}$.

It has proved possible to compute Γ in two simple models. If we turn back to the example of the free massless scalar field and take the observable net $\mathcal{O} \rightarrow \mathfrak{A}(\mathcal{O})$ to be generated in the vacuum representation by those Weyl operators $W(f)$ for which $\int f = 0$. Then the field net $\mathfrak{F}$ turns out to be the canonical field net and $\Gamma \simeq \mathbb{R}$ is just the 1–parameter group of automorphisms defined before.

Similarly if we go back to the example of the free electromagnetic field and take the observable net to be generated in the vacuum representation by those Weyl operators $W(f)$ for which $\int f^{\mu\nu} = 0$ then the field net $\mathfrak{F}$ again turns out to be the canonical field net and $\Gamma \simeq \mathbb{R}^{\frac{1}{2}s(s+1)}$ turns out to be just the group of automorphisms discussed in the example.

References

[1] D. Buchholz, S. Doplicher, R. Longo, J.E. Roberts, A New Look at Goldstone's Theorem. *Rev. Math. Phys.*, to appear.

[2] H. Araki, K. Hepp, D. Ruelle, On the Asymptotic Behaviour of Wightman Functions in Space–like Directions. *Helv. Phys. Acta*, **35**, 164–174 (1962).

[3] H. Reeh, One–Parameter Symmetry Transformations. *Nuovo Cimento*, **51A**, 638–645 (1967).

[4] S. Doplicher, J.E. Roberts, Why there is a Field Algebra with a Compact Gauge Group Describing the Superselection Structure in Particle Physics. *Commun. Math. Phys.*, **131**, 51–107 (1990).

Global Observables in Local Quantum Physics

KLAUS FREDENHAGEN
II. Institut für Theoretische Physik
Universität Hamburg

Abstract. An algebraic construction of global observables which characterize superselection sectors of chiral conformal field theories is described.

Key words: Quantum field theory, operator algebras, superselection sectors, conformal field theory

Local quantum physics [1], better known under its traditional name algebraic quantum field theory, was developed by Araki, Haag, Kastler and others [2-7] in the beginning sixties as a conceptual framework in which the principles of special relativity are incorporated into quantum physics, thereby formalizing the essential physical content of quantum field theory. The basic feature of this approach is the emphasis on local observables. Only local observables are considered to be fundamental, and the physical interpretation is based solely on the knowledge of the system of von Neumann algebras $\mathcal{A} = (\mathcal{A}(\mathcal{O}))_{\mathcal{O}\in\mathcal{K}}$ where $\mathcal{K}$ is the set of open double cones in Minkowski space $\mathcal{M}$ and $\mathcal{A}(\mathcal{O})$ is the algebra of observables measurable within $\mathcal{O}$. $\mathcal{A}$ is supposed to satisfy the requirements of locality and isotony. The global algebra $\mathcal{A}(\mathcal{M})$ is defined as the C*-inductive limit

$$\mathcal{A}(\mathcal{M}) = \overline{\bigcup_{\mathcal{O}\in\mathcal{K}} \mathcal{A}(\mathcal{O})}, \tag{1}$$

and algebras of arbitrary regions $\mathcal{S}$ in Minkowski space are defined as the smallest C*-algebras $\mathcal{A}(\mathcal{S}) \subset \mathcal{A}(\mathcal{M})$ such that $\mathcal{A}(\mathcal{O}) \subset \mathcal{A}(\mathcal{S})$ for all $\mathcal{O} \in \mathcal{K}$ with $\mathcal{O} \subset \mathcal{S}$.

Global observables, such as energy, charge, univalence etc. are not elements of $\mathcal{A}(\mathcal{M})$. They only arise as weak limit points of local observables in certain representations of $\mathcal{A}(\mathcal{M})$. It was therefore very important to develop a representation theory for $\mathcal{A}(\mathcal{M})$, a project which was carried through in several steps in [7-17], see [1] and [18] for further references and more details. A typical feature is that $\mathcal{A}(\mathcal{M})$ is simple, so all representations are faithful, and the state space of one single representation is weakly dense in the whole state space [19] (all sectors are "physically equivalent" in the sense of [7]).

H. Araki et al. (eds.), Quantum and Non-Commutative Analysis, 41–51.

In the DHR theory [10] one starts from a distinguished representation π_0 (interpreted as the vacuum representation) which satisfies Haag duality

$$\pi_0(\mathcal{A}(\mathcal{O})) = \pi_0(\mathcal{A}(\mathcal{O}'))' \quad , \mathcal{O} \in \mathcal{K}. \tag{2}$$

That this property is satisfied for the vacuum representation of the free scalar field was verified by Araki [4, 6]. It is now known to hold in more general cases, but there are also exceptions, typically, but not always associated with spontaneous symmetry breakdown [20, 22, 23]. One then selects representations π which are local excitations of π_0 in the sense that π and π_0 are unitarily equivalent at infinity, i.e.

$$\pi|_{\mathcal{A}(\mathcal{O}')} \simeq \pi_0|_{\mathcal{A}(\mathcal{O}')} \quad , \mathcal{O} \in \mathcal{K} \tag{3}$$

(DHR criterion). These representations violate Haag duality in general, and the degree of violation can be measured in terms of a number $d(\pi) \geq 1$, called the statistical dimension, whose square was shown by [14] to coincide with the Jones index [21] of the inclusion

$$\pi(\mathcal{A}(\mathcal{O})) \subset \pi(\mathcal{A}(\mathcal{O}'))' \tag{4}$$

which is actually independent of $\mathcal{O}$.

In $d \geq 3$ space time dimensions the structure of all DHR representations with finite statistical dimension was completely clarified by Doplicher and Roberts [12]. There is a uniquely determined compact group G, a $\gamma \in G$ with $\gamma^2 = 1$ which changes the sign of Fermi fields, a net of von Neumann algebras $(\mathcal{F}(\mathcal{O}))_{\mathcal{O}\in\mathcal{K}}$, an action $g \mapsto \alpha_g$ of G by automorphisms of $\mathcal{F}(\mathcal{M})$ such that each local algebra $\mathcal{F}(\mathcal{O})$ is stable under the action of G and that $\mathcal{A}(\mathcal{O})$ is the set of G-invariant elements of $\mathcal{F}(\mathcal{O})$, and $\mathcal{F}(\mathcal{O})$ satisfies graded locality with respect to the grading induced by γ. Moreover, there is a faithful irreducible representation $\hat{\pi}_0$ of $\mathcal{F}(\mathcal{O})$ such that

$$\hat{\pi}_0|_{\mathcal{A}(\mathcal{M})} \simeq \bigoplus_{\xi \in \hat{G}} d(\xi)\pi_\xi \tag{5}$$

with a bijective mapping $\xi \mapsto [\pi_\xi]$ of equivalence classes ξ of irreducible representations of G (with dimension $d(\xi)$) to equivalence classes of irreducible representations of $\mathcal{A}(\mathcal{M})$ satisfying the DHR criterion and having finite statistical dimension $d(\pi_\xi) = d(\xi)$. Furthermore, there is a strongly continuous unitary representation U of G in $\mathcal{H}_{\hat{\pi}_0}$ which implements the automorphisms α_g and whose restriction to $\mathcal{H}_{\pi_\xi}$ is equivalent to a multiple of ξ.

At the very end, one finds global charge operators in the weak closure $\hat{\pi}_0(\mathcal{A})''$ as conjugacy class means,

$$Q([g]) = \int_G dh U(hgh^{-1}) \quad , g \in G, \tag{6}$$

and the character table of G arises from

$$\chi_\xi(g)1_{\mathcal{H}_{\pi_\xi}} = Q([g])|_{\mathcal{H}_{\pi_\xi}} \tag{7}$$

In $d = 2$ space time dimensions (as well as for nonlocalizable charges [11] in $d = 3$) the statistical dimensions can be nonintegers, so a simple correspondence to group representations is no longer possible. But the smallest possible values have the form of "quantum dimensions" $[d]_q = (q^d - q^{-d})/(q - q^{-1})$ for $q = e^{i\pi n}$, $n \in \mathbb{Z}$ where d is the dimension of a representation of the quantum group $\mathrm{sl}(2)_q$, so quantum groups are the natural candidates for a generalized internal symmetry. Actually, the situation is more complicated as may be seen from the analysis of the chiral Ising model [24].

A detailed analysis of DHR sectors in 2 dimensions [13, 14] led to the following results. Associated with each sector there is a unitary representation of the braid group characterizing the statistics of particles, and there are Markov traces given in terms of the left inverses of the DHR theory inducing link and ribbon invariants in the usual way and also invariants of 3-manifolds, both of the Reshetikin-Turaev [25] and of the Turaev-Viro [26] type [13, II].

Examples for such a structure are chiral conformal field theories in 2 dimensions where the structure of sectors ("primary fields") fits perfectly into the DHR scheme. There is, however, one important gap in the identification of sectors of conformal field theory with DHR sectors. Namely, as Buchholz and Schulz-Mirbach [23] showed, Haag duality is violated in typical cases (Virasoro algebras with central charge $c > 1$).

Let me briefly describe what a chiral theory is from the algebraic point of view. Let $(\mathcal{A}(\mathcal{O}))_{\mathcal{O}\in\mathcal{K}}$ be a translation covariant Haag-Kastler net on 2-dimensional Minkowski space. 2-dimensional Minkowski space may be written in terms of light cone coordinates $u = t - x$ and $v = t + x$. A chiral theory is characterized by the triviality of translations in one lightlike (say v-) direction. A double cone is a product of two open bounded intervals, and in a chiral theory the algebra of a double cone $I \times J$, where I and J are open bounded intervals on the real axis, does not change under shifts of J on the v-axis. Therefore one may introduce a net of algebras indexed by open bounded intervals on $\mathbb{R}$ by

$$\mathcal{A}(I) = \bigvee_J \mathcal{A}(I \times J) \quad . \tag{8}$$

Locality of the original net implies that algebras of disjoint intervals commute.

We now consider the case that not only the translations but the whole real Moebius group $\mathrm{M} = \mathrm{SL}(2,\mathbb{R})/\mathbb{Z}_2$ acts covariantly on this net,

$$\alpha_g(\mathcal{A}(I)) = \mathcal{A}(gI) \tag{9}$$

provided I and gI are bounded intervals. Let (π_0, U) be a vacuum representation of $(\mathcal{A}(\mathbb{R}), \alpha)$, i.e. a representation π_0 of $\mathcal{A}(\mathbb{R})$ in a Hilbert space $\mathcal{H}_0$ together with a positive energy representation U of M in $\mathcal{H}_0$ which implements α,

$$\mathrm{Ad}U(g) \circ \pi_0 = \pi_0 \circ \alpha_g \quad , g \in \mathrm{M} \tag{10}$$

and has a one dimensional subspace $\mathbb{C}\Omega$ of invariant vectors. It is known that under quite general assumptions the vacuum representation satisfies essential Haag duality,

$$\pi_0(\mathcal{A}(\mathbb{R}_+))' = \pi_0(\mathcal{A}(\mathbb{R}_-))'' \tag{11}$$

(see [27,23,28-31]). One may then extend the net of local algebras to $\mathrm{S}^1 = \mathbb{R} \cup \{\infty\}$ by defining

$$\pi_0(\mathcal{A}(I)) = U(g)\pi_0(\mathcal{A}(g^{-1}I))U(g)^{-1} \quad , g \in \mathrm{M} \tag{12}$$

where I is an open interval on S^1 with noncoinciding boundary points and $g^{-1}I$ is a bounded interval on $\mathbb{R} = \mathrm{S}^1 \setminus \{\infty\}$. By Moebius covariance, Haag duality on S^1 follows from essential duality on $\mathbb{R}$,

$$\pi_0(\mathcal{A}(I))' = \pi_0(\mathcal{A}(I')) \tag{13}$$

where I' is the interior of the complement of I in S^1. Haag duality on $\mathbb{R}$, however, might not hold since the complement of a bounded interval in $\mathbb{R}$ is a union of two disjoint intervals $I_>$ and $I_<$ having ∞ as a joint boundary point, and the algebra $\pi_0(\mathcal{A}(I_>)) \vee \pi_0(\mathcal{A}(I_<))$ might be smaller than $\pi_0(\mathcal{A}(I_> \cup \{\infty\} \cup I_<))$. Actually, this is the case for the algebra generated by the energy momentum tensor with central charge $c > 1$ [23], and a similar phenomenon was observed much earlier in a different context by Hislop and Longo (failure of timelike duality for a massless scalar field in 4 dimensions [32]).

A positive energy representation may be defined as a family (π^I) of representations of $\mathcal{A}(I)$ in a Hilbert space $\mathcal{H}_\pi$, $I \in \mathcal{K}$ where $\mathcal{K}$ is now the set of open intervals on S^1 with noncoinciding boundary points, with consistent restrictions

$$\pi^I|_{\mathcal{A}(J)} = \pi^J \quad \text{for} \quad I \supset J \tag{14}$$

together with a positive energy representation U of the covering group of the Moebius group such that

$$\mathrm{Ad}U(g) \circ \pi^I = \pi^{gI} \circ \alpha_g. \tag{15}$$

By a result of Buchholz, Mack and Todorov [33] which relies on a result of Takesaki and Winnink [34] the representations π^I are automatically normal.

Instead of dealing with families of representations of local algebras, it turns out to be useful to deal with representations of a global algebra associated to S^1. $\mathcal{A}(S^1)$ may be defined as the free C*-algebra generated by all local algebras $\mathcal{A}(I)$ modulo all relations within local algebras. $\mathcal{A}(S^1)$ can be characterized by the following universality condition: for each family of normal representations (π^I) in some Hilbert space $\mathcal{H}$ which satisfies the consistency condition (14) there is a unique representation π of $\mathcal{A}(S^1)$ in $\mathcal{H}$ such that $\pi|_{\mathcal{A}(I)} = \pi^I$. (See [35, 36] and [13, II].)

These global algebras are not as abstract as one might think. Actually, they have already been used in the algebraic treatment of models of conformal field theory. A simple example I want to mention is the theory of the U(1) current algebra. This model was analyzed in [33]. One considers the *-algebra generated by unitaries $W(f)$ with $f \in C^\infty(S^1, \mathbb{R})$ which satisfy the Weyl relation

$$W(f)W(g) = e^{-i\int f'g} W(f+g) \tag{16}$$

where $\int f'g = \int_{S^1} dz \frac{df}{dz} g$. The local algebras $\mathcal{A}_0(I)$ are defined as the subalgebras generated by $W(f)$ with $\mathrm{supp} f \subset I$. Using the Weyl relation one easily sees that each consistent family of local representations characterizes a unique representation of $\mathcal{A}_0(S^1)$. One now may look at the vacuum representation π_0 induced by the vacuum state

$$\omega_0(W(f)) = e^{-||f||^2} \tag{17}$$

where $||f||^2 = \sum_{n\in\mathbb{N}} n|f_n|^2$, f_n being the nth Fourier coefficient of f. The local algebra is defined as the completion of $\mathcal{A}_0(I)$ with respect to the family of seminorms $||\cdot||_T = |\mathrm{Tr}\pi_0(\cdot)T|$, with trace class operators T in $\mathcal{H}_{\pi_0}$. One then considers the set of all states ω on $\mathcal{A}_0(S^1)$ such that the induced representation π_ω has restrictions to the local algebras $\mathcal{A}_0(I)$ which extend to normal representations of $\mathcal{A}(I)$ and introduces the "universal locally normal" representation $\hat{\pi} = \bigoplus \pi_\omega$. The map $A \mapsto \hat{\pi}(A)$, $A \in \mathcal{A}_0(I)$, extends to an isomorphism of $\mathcal{A}(I)$ and $\hat{\pi}(\mathcal{A}_0(I))''$. $\mathcal{A}(S^1)$ can now be defined as the C*-algebra in $\mathcal{B}(\mathcal{H}_{\hat{\pi}})$ generated by $\hat{\pi}(\mathcal{A}_0(I))''$, $I \in \mathcal{K}$.

We finally have to check that $\mathcal{A}(S^1)$ satisfies the universality condition. So let (π^I) denote a consistent family of local representations of $\mathcal{A}(I)$. The family $(\pi^I|_{\mathcal{A}_0(I)})$ defines a unique representation $\pi^{(0)}$ of $\mathcal{A}_0(S^1)$. $\pi^{(0)}$ is a direct sum of cyclic representations π_i each of which is equivalent to a subrepresentation of $\hat{\pi}$. Let V_i be an isometric intertwiner from $\mathcal{H}_{\pi_i} \subset \mathcal{H}_{\pi^{(0)}}$ into $\mathcal{H}_{\hat{\pi}}$. Then the representation π of $\mathcal{A}(S^1)$ is given by

$$\pi(A) = \sum_i V_i^* A V_i \quad , A \in \mathcal{A}(S^1) \subset \mathcal{B}(\mathcal{H}_{\hat{\pi}}). \tag{18}$$

Let us return to the general discussion. We are interested in representations of the universal algebra $\mathcal{A}(S^1)$ satisfying the DHR criterion which assumes here the form

$$\pi \in (\mathrm{DHR}) \Longleftrightarrow \pi|_{\mathcal{A}(I)} \simeq \pi_0|_{\mathcal{A}(I)} \quad \forall I \in \mathcal{K} \quad . \tag{19}$$

According to [33] positive energy representations with a cyclic vector automatically satisfy this criterion. In analogy to the standard DHR theory, we look for endomorphisms ρ of $\mathcal{A}(S^1)$ with the property $\pi \simeq \pi_0 \circ \rho$. But since π_0 is not necessarily faithful, the construction of ρ is somewhat delicate. It has been described in [35, 36] and, in more detail, in [13, II]. One obtains endomorphisms which are localizable in the following sense:

Definition: *An endomorphism ρ of $\mathcal{A}(S^1)$ is called localizable within $I \in \mathcal{K}$ if for all $I_0 \subset I$, $I_0 \in \mathcal{K}$ there is a unitary $U_0 \in \mathcal{A}(I)$ such that*

$$\rho(A) = \mathrm{Ad}U_0(A) \quad , A \in \mathcal{A}(I_0'), \tag{20}$$

$$\mathrm{Ad}U_0^* \circ \rho(\mathcal{A}(I_1)) \subset \mathcal{A}(I_1) \quad \textit{for} \quad I_1 \supset I_0 \quad , I_1 \in \mathcal{K}. \tag{21}$$

We remark that an endomorphism which is localizable within I in the above sense acts trivially on $\mathcal{A}(I')$ (so it is localized in I in the usual sense) and restricts to an endomorphism of $\mathcal{A}(I_1)$ for $I_1 \supset I$. It has the further property that the endomorphism $\rho_0 = \mathrm{Ad}U_0^* \circ \rho$ used in the definition is localizable within I_0. Namely, let $I_{00} \subset I_0$, $I_{00} \in \mathcal{K}$. Since ρ is localizable within I and $I_{00} \subset I$ there is some unitary $U_{00} \in \mathcal{A}(I)$ satisfying (20) and (21) with I_0 being replaced by I_{00}. Then the unitary $U_0^*U_{00}$ satisfies (20) and (21) with ρ being replaced by ρ_0 and I_0 by I_{00}. It remains to show that $U_0^*U_{00} \in \mathcal{A}(I_0)$. From (20) and the fact that ρ_0 acts trivially on $\mathcal{A}(I_0')$ we conclude that $\pi_0(U_0^*U_{00}) \in \pi_0(\mathcal{A}(I_0'))'$ which coincides with $\pi_0(\mathcal{A}(I_0))$ by Haag duality. But $U_0^*U_{00} \in \mathcal{A}(I)$ and π_0 is faithful on $\mathcal{A}(I) \supset \mathcal{A}(I_0)$, hence $U_0^*U_{00} \in \mathcal{A}(I_0)$.

It is crucial for the whole construction that intertwiners between representations can be lifted to intertwiners between localizable endomorphisms:

Proposition: *Let ρ, ρ' be endomorphisms of $\mathcal{A}(S^1)$ which are localizable within $I \in \mathcal{K}$ (in the sense of the above definition), and let $S \in \mathcal{B}(\mathcal{H}_{\pi_0})$ be an intertwiner from $\pi_0 \circ \rho$ to $\pi_0 \circ \rho'$, i.e.*

$$S\pi_0 \circ \rho(A) = \pi_0 \circ \rho'(A)S \quad , A \in \mathcal{A}(S^1). \tag{22}$$

Then there exists a unique $S_0 \in \mathcal{A}(I)$ such that $\pi_0(S_0) = S$, and S_0 intertwines ρ with ρ', i.e.

$$S_0\rho(A) = \rho'(A)S_0 \quad , A \in \mathcal{A}(S^1). \tag{23}$$

Proof: ρ and ρ' act trivially on $\mathcal{A}(I')$, hence $S \in \pi(\mathcal{A}(I'))' = \pi_0(\mathcal{A}(I))$, and there is a unique $S_0 \in \mathcal{A}(I)$ with $\pi_0(S_0) = S$. We have to show that S_0 satisfies the intertwining relation (23).

Since $\mathcal{A}(S^1)$ is generated, as a norm closed algebra, by local operators, it is sufficient to check (23) for local algebras $\mathcal{A}(I_1)$. For a given $I_1 \in \mathcal{K}$ there are $I_2, I_0 \in \mathcal{K}$ such that $I_1 \subset I_2$ and $I_0 \subset I_2 \cap I$. We choose unitaries $U_0, U_0' \in \mathcal{A}(I)$ such that (20) and (21) hold for ρ and I_0, and ρ' and I_0, respectively. Let $\rho_0 = \mathrm{Ad}U_0^* \circ \rho$, $\rho_0' = \mathrm{Ad}U_0'^* \circ \rho'$ and $S_1 = U_0'^* S_0 U_0$. Then by (22)

$$\pi_0(S_1\rho_0(A)) = \pi_0(\rho_0'(A)S_1) \quad , A \in \mathcal{A}(S^1). \tag{24}$$

ρ_0 and ρ_0' act trivially on $\mathcal{A}(I_0')$, so $\pi_0(S_1) \in \pi_0(\mathcal{A}(I_0'))' = \pi_0(\mathcal{A}(I_0))$, and by the faithfulness of π_0 on $\mathcal{A}(I) \supset \mathcal{A}(I_0)$ we get $S_1 \in \mathcal{A}(I_0)$. Now let $A \in \mathcal{A}(I_1)$. ρ_0 and ρ_0' are localizable within I_0 and therefore induce endomorphisms of $\mathcal{A}(I_2)$. Hence (24) is a relation in the vacuum representation of the local algebra $\mathcal{A}(I_2)$ and thus remains valid in the universal algebra $\mathcal{A}(S^1)$, i.e.

$$S_1\rho_0(A) = \rho_0'(A)S_1 \quad , A \in \mathcal{A}(I_1). \tag{25}$$

But this implies the desired relation (23) on $\mathcal{A}(I_1)$. Since $I_1 \in \mathcal{K}$ was arbitrary the proof is complete. □

A representation $\pi \in$ (DHR) leads to localizable endomorphisms which are in addition transportable in the following sense:

Definition: *An endomorphism ρ of $\mathcal{A}(S^1)$ is called transportable if for each $I \in \mathcal{K}$ there exists some endomorphism ρ' of $\mathcal{A}(S^1)$ which is localizable within I and which is inner equivalent to ρ, i.e. there is some unitary $U \in \mathcal{A}(S^1)$ such that $\rho' = \mathrm{Ad}U \circ \rho$.*

We have a 1-1 correspondence between equivalence classes of DHR representations of $\mathcal{A}(S^1)$ and inner equivalence classes of transportable endomorphisms of $\mathcal{A}(S^1)$. We denote by Δ the set of all transportable endomorphisms and by $\Delta(I)$ the subset of endomorphisms which are localizable within $I \in \mathcal{K}$. Note that transportable endomorphisms which are localizable within some interval I are automatically also localizable in every larger interval. This property might not hold without the transportability assumption.

One now can proceed as in the DHR analysis, using as crucial input the correspondence between intertwiners of representations and local intertwiners of localizable endomorphisms established in the Proposition. There are, however, new features which are connected with the existence of nonlocal intertwiners. The following structure was found in [13, II].

Let $I, I_1 \in \mathcal{K}$ have disjoint closures, and let $J_\pm \in \mathcal{K}$ with $J_\pm \supset I \cup I_1$ and $J_+ \cup J_- = S^1$, where, within J_+, I_1 lies on the right hand side of I. Let $\rho \in \Delta(I), \rho_1 \in \Delta(I_1)$ and $V \in \mathcal{B}(\mathcal{H}_{\pi_0})$ unitary such that $\mathrm{Ad}V \circ \pi_0 \circ \rho = \pi_0 \circ \rho_1$. According to the Proposition there exist unique intertwiners $V_\pm \in \mathcal{A}(J_\pm)$ with $\pi_0(V_\pm) = V$ and $\mathrm{Ad}V_\pm \circ \rho = \rho_1$. The unitary

$$V_I(\rho) = V_+^{-1} V_- \tag{26}$$

is an intertwiner from ρ to ρ with $\pi_0(V_I(\rho)) = 1$. It depends only on ρ and, possibly, also on the choice of $I \in \mathcal{K}$ with $\rho \in \Delta(I)$. The latter dependence is compatible with inclusions, $V_I(\rho) = V_J(\rho)$ if $I \subset J$; actually, I am not aware of any example where $V_I(\rho)$ really depends on I. $V_I(\rho)$ transforms covariantly under local intertwiners; if $\rho, \sigma \in \Delta(I)$ and $S \in \mathcal{A}(I)$ intertwines ρ with σ we have

$$S V_I(\rho) = V_I(\sigma) S \quad . \tag{27}$$

It also has a simple behavior under products; if $\rho, \sigma \in \Delta(I)$ we find

$$V_I(\rho\sigma) = V_I(\rho)\rho(V_I(\sigma)) \quad . \tag{28}$$

The global selfintertwiners $V_I(\rho)$ have an interesting relation to the statistics operators of the DHR theory. Let $\rho, \sigma \in \Delta(I)$. Then, with the above notations,

$$\sigma(V_I(\rho)) = (\sigma(V_+^{-1})V_+) \cdot (V_+^{-1}V_-) \cdot (V_-^{-1}\sigma(V_-)) \tag{29}$$

$$= \varepsilon_I(\rho,\sigma) \cdot V_I(\rho) \cdot \varepsilon_I(\sigma,\rho) \tag{30}$$

where $\varepsilon_I(\cdot,\cdot)$ are the statistics operators of the DHR theory. They satisfy $\varepsilon_I = \varepsilon_J$ for $I \subset J$, but they depend, in general, on the choice of the interval I with $\rho, \sigma \in \Delta(I)$. Relation (30) together with the usual braid relations of the statistics operators means that the operators $\rho^{i-1}(V_I(\rho))$, $\rho^{i-1}(\varepsilon_I(\rho,\rho))$, $i \in \mathbb{N}$ generate a unitary representation of the braid group of the cylinder. Here $\rho^{i-1}(V_I(\rho))$ represents a braid where the i-th strand is wrapped once around the cylinder whereas all other strands are straight.

In the vacuum representation, (30) implies

$$\pi_0\sigma(V_I(\rho)) = \pi_0(\varepsilon_I(\rho,\sigma)\varepsilon_I(\sigma,\rho)) \tag{31}$$

where the operator on the right hand side is the so-called monodromy operator which is nontrivial for some pair ρ, σ if and only if the theory has nontrivial braid group statistics. We observe that, in general, π_0 is not faithful and $\mathcal{A}(S^1)$ has nontrivial ideals (cf. [12] for a related observation).

$\mathcal{A}(S^1)$ even has a nontrivial center if it has sectors with finite statistical dimension and braid group statistics. Let $I, I_1, J_\pm$ as before, let $\rho \in \Delta(I)$ and let $\bar{\rho} \in \Delta(I_1)$ be conjugate to ρ in the sense that $\pi_0 \circ \rho\bar{\rho}$ contains a subrepresentation equivalent to π_0. Let R be an isometric intertwiner from π_0 to $\pi_0 \circ \rho\bar{\rho}$, and let $R_\pm$ be its preimages under π_0 in $\mathcal{A}(J_\pm)$, respectively. Then

$$R_+^* R_- \in \mathrm{Z}(\mathcal{A}(\mathrm{S}^1)) \quad . \tag{32}$$

It turns out that for ρ irreducible (i.e. the representation $\pi_0 \circ \rho$ is irreducible) the central element $C_\rho = d(\rho) R_+^* R_-$ is uniquely determined by the equivalence class of ρ. One finds $C_\rho^* = C_{\bar{\rho}}$, the multiplication rule

$$C_\rho C_\sigma = \sum_{[\tau]} N_{\rho\sigma}^\tau C_\tau \tag{33}$$

where the nonnegative integers $N_{\cdot,\cdot}^{\cdot}$ are identical to the fusion coefficients occuring in the reduction of composed representations

$$\pi_0 \circ \rho\sigma = \bigoplus_{[\tau]} N_{\rho\sigma}^\tau \pi_0 \circ \tau \tag{34}$$

and the generalized character table

$$d(\sigma)\pi_0\sigma(C_\rho) = \sum_{[\tau]} N_{\rho\sigma}^\tau d(\tau) \frac{\kappa(\tau)}{\kappa(\rho)\kappa(\sigma)} =: Y_{\rho\sigma} \quad . \tag{35}$$

where $\kappa(\cdot)$ denotes the statistics phase. In case the number of DHR sectors is finite ("rational theories") and the vacuum sector is the only one with trivial monodromies with all other sectors, i.e. for all $\rho \in \Delta(I)$, $\rho \not\simeq \mathrm{id}$ there exists some $\sigma \in \Delta(I)$ such that $\varepsilon_I(\rho,\sigma)\varepsilon_I(\sigma,\rho) \neq 1$, the matrix $(Y_{\rho,\sigma})$ is invertible [37], hence the system of charge operators (C_ρ) distinguishes the sectors. The matrix Y is proportional to Verlinde's [38] matrix S [37, 17]. So in these cases the global charge operators are obtained by an algebraic construction from local observables, and also remainders of a generalized group structure occur (33) in analogy to the multiplication structure of conjugacy classes of finite groups. Note that this multiplicative structure is identical to that of the representation classes (34); this selfduality property is in the group case true only for finite abelian groups.

References

[1] Rudolf Haag, *Local Quantum Physics. Fields, Particles, Algebras* Springer-Verlag Berlin Heidelberg 1992

[2] R. Haag: Discussion des "axiomes" et des propriétés asymptotiques d'une théorie des champs locales avec particules composées. In *Les problèmes mathématiques de la théorie quantique des champs*, Lille 1957. Paris: Centre National de la Recherche Scientifique 1959

[3] R. Haag, B. Schroer: Postulates of Quantum Field Theory. J. Math. Phys. **3**, 248 (1962)

[4] H. Araki: A Lattice of von Neumann Algebras Associated with the Quantum Theory of a Free Bose Field. J. Math. Phys. **4**, 1343 (1963)

[5] H.J. Borchers, R. Haag, B. Schroer: The vacuum state in quantum field theory. Nuovo Cim. **24**, 214 (1963)
[6] H. Araki: Von Neumann Algebras of Local Observables for Free Scalar Field. J. Math. Phys. **5**, 1 (1964)
[7] R. Haag, D. Kastler: An Algebraic Approach to Field Theory. J. Math. Phys. **5**, 848 (1964)
[8] H.J. Borchers: Local Rings and the Connection of Spin with Statistics. Comm. Math. Phys. **1**, 281 (1965)
[9] S. Doplicher, R. Haag, J.E. Roberts: Fields, Observables and Gauge Transformations. Comm. Math. Phys. **13**, 1, (1969) and Comm. Math. Phys. **15**, 173 (1969)
[10] S. Doplicher, R. Haag, J.E. Roberts: Local Observables and Particle Statistics. Comm. Math. Phys. **23**, 199 (1971) and **35**, 49 (1974)
[11] D. Buchholz, K. Fredenhagen: Locality and the Structure of Particle States. Comm. Math. Phys. **84**, 1 (1982)
[12] S. Doplicher, J.E. Roberts: Why there is a Field Algebra with a Compact Gauge Group Describing the Superselection Structure in Particle Physics. Comm. Math. Phys. **131**, 51 (1990)
[13] K. Fredenhagen, K.-H. Rehren, B. Schroer: Superselection Sectors with Braid Group Statistics and Exchange Algebras I: General Theory. Comm. Math. Phys. **125**, 201 (1989) and II: Geometric Aspects and Conformal Covariance (to appear in Rev. Math. Phys.)
[14] R. Longo: Index of Subfactors and Statistics of Quantum Fields. Comm. Math. Phys. **126**, 217 (1989) and **130**, 285 (1990)
[15] J. Fröhlich, F. Gabbiani, P.-A. Marchetti: Superselection Structure and Statistics in Three-dimensional Local Quantum Theory. In *Knots, Topology, and Quantum Field Theories*, ed. G. Lusanna, p. 335 (World Scientific, 1989)
[16] J. Fröhlich, P.-A. Marchetti: Spin-Statistics Theorem and Scattering in Planar Quantum Field Theory with Braid Statistics. Nucl. Phys. **B356**, 533 (1991)
[17] J. Fröhlich, F. Gabbiani: Braid Statistics in Local Quantum Theory. Rev. Math. Phys. **2**, 251 (1991).
[18] D. Kastler (ed.): *The Algebraic Theory of Superselection Sectors. Introduction and Recent Results.* World Scientific, 1990
[19] J.M.G. Fell: The dual spaces of C*-algebras. Trans. Amer. Math. Soc. **94**, 365 (1960)
[20] J.E. Roberts: Spontaneously Broken Gauge Symmetries and Superselection Rules. Proceedings of the International School of Mathematical Physics, Camerino 1974, ed. G. Gallavotti, Universita di Camerino 1976
[21] V.F.R. Jones: Index for Subfactors. Invent. Math. **72**, 1 (1983)
[22] D. Buchholz, S. Doplicher, R. Longo, J. E. Roberts: A New Look at Goldstone's Theorem. Rev. Math. Phys. (to appear)
[23] D. Buchholz, H. Schulz-Mirbach: Haag duality in conformal quantum field theory. Rev. Math. Phys. **2**, 105 (1990)
[24] G. Mack, V. Schomerus: Conformal field algebras with quantum symmetry from the theory of superselection sectors. Comm. Math. Phys. **134**, 391 (1990)
[25] N.Yu. Reshetikhin, V.G. Turaev: Invariants of 3-manifolds via link polynomials and quantum groups. Invent. Math. **103**, 547 (1991).
[26] V.G. Turaev, O. Viro: State sum invariants of 3-manifolds and quantum $6j$-symbols. Topology **31**, 865 (1992)
M. Karowski, W. Müller, R. Schrader: State sum invariants of compact 3-manifolds with boundary and $6j$-symbols. Journ. Phys. A **25**, 4847 (1992)
[27] J.J. Bisognano, E.H. Wichmann: On the Duality Condition for a Hermitean Scalar Field. J. Math. Phys. **16**, 985 (1975)
[28] M. Jörß: Lokale Netze auf dem eindimensionalen Lichtkegel. Diploma thesis FU Berlin (1991).
M. Jörß: On the existence of pointlike localized fields in conformally invariant quantum physics. DESY 92-156

[29] H.J. Borchers: The CPT theorem in two-dimensional theories of local observables. Comm. Math. Phys. **143**, 315 (1992)

[30] F. Gabbiani, J. Fröhlich: Operator algebras and conformal field theory. ETH-TH/92-30 (preprint)

[31] R. Brunetti, D. Guido, R. Longo: Modular structure and duality in conformal quantum field theory. (To appear)

[32] P.D. Hislop, R. Longo: Modular structure of the local algebras associated with the free massless scalar field theory. Comm. Math. Phys. **84**, 71 (1982)

[33] D. Buchholz, G. Mack, I. Todorov: The Current Algebra on the Circle as a Germ for Local Field Theories. In *Conformal Field Theories and Related Topics*, eds. P. Binetruy et al., Nucl. Phys. B (Proc. Suppl.) **5B**, 20, (1988)

[34] H. Takesaki, W. Winnink: Local normality in quantum statistical mechanics. Comm. Math. Phys. **30**, 129 (1973)

[35] K. Fredenhagen: Generalizations of the Theory of Superselection Sectors. In [18], p. 379 (1990)

[36] D. Guido, R. Longo: Relativistic invariance and charge conjugation in quantum field theory. Comm. Math. Phys. **148**, 521 (1992)

[37] K.H. Rehren: Braid group statistics and their superselection rules. In [18] p. 333 (1990)

[38] E. Verlinde: Fusion rules and modular transformations in 2D conformal field theory. Nucl. Phys. **B300**, 360 (1988)

State of the art of Alain Connes' version of the standard model

D.KASTLER
CPT CNRS Luminy Case 907 F-13288 France

Abstract. Survey of the improved technique, based on the non-commutative DeRham complex, for deriving the standard model lagrangian in the non-commutative Yang-Mills scheme, under stressing of tensorial aspects.

Key words: Non-commutative geometry. Theory of elementary particles

Last year Alain Connes' version of the (classical level of the) standard model in non-commutative differential geometry was modernized in several ways: most importantly, the coupling of the electroweak and the chromodynamic sectors was recognized to pertain to (a novel concept of) *non-commutative Poincaré duality*; and the cumbersome adynamical fields parasiting the initial treatment were eliminated by replacing the formal differential forms by the genuine quantum forms (elements of the *non-commutative De Rham complex*) [1]. Accessorily, the theory was presented in a more transparent way by systematic use of the skew tensor-product structure, and of 2×2 matrices with 2×2 matrix-entries instead of 4×4 matrices [2] [4]. This presentation is sketched in what follows, after an introduction (sections [1] through [4]) to the non-commutative De Rham complex showing how the classical De Rham complex arises as a special case of its non-commutative generalization.

[1] *The classical De Rham complex via Dirac operator. Usual differential forms as a special case of quantum forms.* We begin by presenting the classical De Rham complex $(\Omega(\mathbf{M}), \mathbf{d})$ of a d-dimensional compact spin^c riemannian manifold $\mathbf{M}$ with spin bundle $\mathcal{S}_{\mathbf{M}}$[1] in a way which will generalize to the non-commutative frame. The algebra at hand $\mathcal{A} = C^\infty(\mathbf{M}, \mathcal{R})$ acts by point-wise multiplication on the Hilbert space $H = L^2(\mathcal{S}_{\mathbf{M}})$ of square-integrable spinor-fields, also enacted by the Dirac operator D. The "chirality operator" $\chi = \gamma^{2d+1}$ makes H a $\mathcal{Z}/2$-graded Hilbert space. For future reference we note the following structural features:

1. There is a $*$-representation: $\mathcal{A} \ni a \to \underline{a}$ on the $\mathcal{Z}/2$-graded Hilbert space

[1] $\Omega(\mathbf{M})$ is the set of Cartan forms; $\mathbf{d}$ is the exterior differentiation. With Λ the exterior product and with the $\mathcal{N}$-grading stemming from the order of forms, $(\Omega(\mathbf{M}), \Lambda, \mathbf{d})$ is an $\mathcal{N}$-graded differential algebra.

H. Araki et al. (eds.), Quantum and Non-Commutative Analysis, 53–71.

H by even bounded operators.
2. D is an odd self-adjoint operator.
3. All commutators $[D, \underline{a}], a \in \mathcal{A}$, are bounded.
4. D^{-1} is compact.[2]
5. One has $D^{-d} \in L^{+d}$.

We now address the unital differential envelope $\Omega\mathcal{A} = \oplus_{n\in\mathbf{N}}\Omega\mathcal{A}^n$ of $\mathcal{A} : \Omega\mathcal{A}^n$ consists of linear combinations of formal expressions $a_0\delta a_1...\delta a_n$, $a_0, .., a_n \in \mathcal{A}$, handled as follows: the symbols a_k behave in $\Omega\mathcal{A}$ as they do in $\mathcal{A}$, the δa_k, linear in a_k, fulfill a Leibnitz rule:

$$\delta(a_k a_1) = (\delta a_k)a_1 + a_k(\delta a_1), \tag{1}$$

we set $\delta 1 = 0$, and:

$$\delta(a_0\delta a_1...\delta a_n) = \delta a_0\delta a_1...\delta a_n, \quad a_0, a_1, ..., a_n \in \mathcal{A}, \tag{2}$$

these rules making $(\Omega\mathcal{A} = \oplus_{n\in\mathcal{N}}\Omega\mathcal{A}^n, \delta)$ an $\mathbf{N}$-graded differential algebra, which we equip with the $*$-operation extending that of $\mathcal{A}$ and skew-commuting with δ. A crucial item is now the $*$-representation π_D of the algebra (mind: *not* of the differential algebra!) $\Omega\mathcal{A}$ given as:[3]

$$\pi_D(a_0\delta a_1...\delta a_n) = (-i)^n\underline{a_0}[D, \underline{a_1}]...[D, \underline{a_n}], \quad a_0, a_1, ..., a_n \in \Omega\mathcal{A}. \tag{3}$$

Let $K = Ker\pi_D$: K is an algebraic ideal – neither a graded nor a differential ideal. *However,* $J = K^* + \delta K^*$ where $K^* = \oplus_{n\in\mathbf{N}}K^n, K^n = K \cap \Omega\mathcal{A}^n$, *is a graded differential ideal*: consequently, the $\mathbf{N}$-grading and the differential δ pass to the quotient $\Omega\mathcal{A}_D = \Omega\mathcal{A}/J$ thus becoming an $\mathbf{N}$-graded differential algebra $(\Omega\mathcal{A}_D, \delta)$. The astonishing fact is now that *one has an isomorphism of* $\mathbf{N}$*-graded differential algebras.* $(\Omega\mathcal{A}_D, \delta) \approx (\Omega(\mathbf{M}), \mathbf{d})$, specifically:

$$\Omega_D\mathcal{A} \ni P_2\omega \to \frac{1}{n}\omega_{cl} \quad \text{where} \quad \begin{cases} \omega_{cl} = a_0\mathbf{d}a_1 \wedge ... \wedge \mathbf{d}a_n \\ \text{for } \omega = a_0\delta a_1...\delta a_n, a_0, a_1, ..., a_n \in \mathcal{A}. \end{cases} \tag{4}$$

as $\mathbf{N}$*-graded differential algebras.* What we obtained is thus a construction of the classical De Rham complex $\mathcal{A} = C^\infty(\mathbf{M})$ in terms of the Dirac operator D. Here are two supplementary remarks on the specification of $\Omega\mathcal{A}_D$: first, one can easily see that one has:

$$\Omega\mathcal{A}_D^n = \pi_D(\Omega\mathcal{A}^n)/\pi_D(\delta K^{n-1}), \tag{5}$$

and, moreover, property (5) above makes the *Dixmier trace* Tr_ω available to define the following graded trace τ_D on $\Omega\mathcal{A}$:

$$\tau_D(\omega) = \mathrm{Tr}_\omega(D^{-d}\pi_D(\omega)), \quad \omega \in \Omega\mathcal{A}, \tag{6}$$

---

[2] This assumes that D has a trivial kernel. Otherwise take $(D^2 + m^2)^{1/2}$ instead.

[3] pi_D is multiplicative because $[D, \cdot]$ is a derivation, however, not of vanishing square since $D^2 \neq 0$.

with ensuing scalar product $(\cdot,\cdot)_D$. This allows to pick up a canonical representative in each class of $\pi_D(\Omega\mathcal{A}^n) mod \pi_D(\delta\Omega\mathcal{A}^{n-1})$, namely for any ω the element $P_n\omega$, where P_n is the projection in $\pi_D(\Omega\mathcal{A}^n)$ onto the subspace perpendicular to $\pi_D(\delta K^{n-1})$ for the scalar product $(\cdot,\cdot)_D$.

[2] *The non-commutative De Rham complex.* We now have everything at hand to define the non-commutative generalization of the De Rham complex. We shall define a *non-commutative d-dimensional metric space* as a couple of a $*$-algebra $\mathcal{A}$ and a d^+*-summable* K*-cycle* (H, D) *over* $\mathcal{A}$, namely a $\mathcal{Z}/2$-graded Hilbert space H carrying a bounded $*$-representation: $a \rightarrow \underline{a}$ of $\mathcal{A}$, and an unbounded operator D, assumed to fulfil the properties (1) through (5) above. Under these assumptions we can define as above the $\mathbf{N}$-graded differential algebra $(\Omega\mathcal{A}_D = \oplus_{n\in\mathbf{N}}\Omega\mathcal{A}_D^n, \delta)$ now yielding the notion of non-commutative De Rham complex.

[3] *Connections, curvatures and the non-commutative Yang-Mills action.* Let the couple $\mathcal{A}, (H, D)$ be a non-commutative d-dimensional metric space. We describe the notion of connection in the special case, sufficient for our purposes, where the "vector bundle over $\mathcal{A}$" is $\mathcal{A}$ itself, in other terms where the relevant finite projective module over $\mathcal{A}$ is $\mathcal{A}$ itself considered as a hermitean right $\mathcal{A}$-module with $\mathcal{A}$-valued scalar product $(a, b) = a^*b; a, b \in \mathcal{A}$. A *compatible connection* is then defined as an antihermitean grade-one δ-anti-derivation ∇ of $\Omega_D\mathcal{A}$ where the latter is taken as a right module over itself: specifically we have a linear antihermitean $\nabla : \Omega_D\mathcal{A} \rightarrow \Omega_D\mathcal{A}$ fulfilling

$$\nabla(\omega\omega') = (\nabla\omega)\omega' + (-1)^{\partial\omega}\omega\delta\omega', \tag{7}$$

$\omega, \omega' \in \Omega_D\mathcal{A}, \omega$ of grade $\partial\omega$. The differential δ is a compatible connection in this sense. Since differences of δ-anti-derivations are endomorphisms, it follows that each compatible connection ∇ is of the form $\nabla = \delta + i\rho$ with $\rho = \rho^* \in \Omega\mathcal{A}_D^1$ the corresponding *vector potential.* Iterating ∇ we then get the *curvature*

$$\nabla^2 = \theta, \tag{8}$$

where

$$\theta = \theta^* = \delta\rho + \rho^2 \in \Omega\mathcal{A}_D^2. \tag{9}$$

The *non-commutative Yang-Mills action* is then the following functional on vector potentials:

$$\theta = \theta^* = \delta YM(\rho) = (\theta, \theta)_D = \mathrm{Tr}_\omega(D^{-d}\pi_D(\theta^*\theta)). \tag{10}$$

The fact that (10) is the right non-commutative generalization of the usual Yang-Mills action follows from the fact that one falls back on the latter if applying this notion to the classical algebra $\mathcal{A} = C^\infty(\mathrm{M})$ equipped with the classical Dirac K-cycle $(L^2(\mathcal{S}_\mathrm{M}), D)$.

The computation of the non-commutative Yang-Mills action will require a description of the grade-one and grade-two parts of the non-commutative De Rham complex:

$$\begin{cases} \Omega\mathcal{A}^1_D = \pi_D(\Omega\mathcal{A}^1) \text{ in the usual case where } K^0 = 0 \\ \Omega\mathcal{A}^2_D = \pi_D(\Omega\mathcal{A}^2)/\pi_D(\delta K^1) \end{cases} \tag{11}$$

[4] *Poincaré duality. Non-commutative riemannian manifolds.* We refrained from calling manifolds the above d-dimensional metric spaces because they do not in general fulfil Poincaré duality. Poincaré duality in the classical case of $C^\infty(\mathrm{M})$ consists in a one-to-one map between homology and cohomology. Alain Connes found that the appropriate non-commutative generalization of Poincaré duality pertains to couples $(\mathcal{A}, \mathcal{B})$ of (non-commutative) algebras with a one-to-one map between homology of the one and cohomology of the other, this arising from the following situation: we say that we have a *d-dimensional non-commutative riemannian manifold* (abridged to NCRM) whenever we have a d-dimensional metric space consisting of a couple $\mathcal{A} \otimes \mathcal{B}, (H, D)$ of an algebraic tensor product of two $*$-algebras $\mathcal{A}, \mathcal{B}$ and a d^+-summable K-cycle (H, D) over $\mathcal{A} \otimes \mathcal{B}$, in such a way that one has[4]

$$[[D, a \otimes 1;], 1 \otimes b] = 0, \quad a \in \mathcal{A}, b \in \mathcal{B}, \tag{12}$$

and

$$\mathrm{Tr}_\omega\{\chi \underline{a}_0[D, \underline{a}_1]...[D, \underline{a}_k]|D|^{-d}\} = 0, \quad a_0, a_1, ..., a_d \in \mathcal{A}. \tag{13}$$

We shall define the vector potentials of the d-dimensional non-commutative riemannian manifold $\mathcal{A} \otimes \mathcal{B}, (H, D)$ as the $\rho \otimes 1 + 1 \otimes \sigma$, ρ and σ vector potentials of the respective non-commutative metric factor spaces $\mathcal{A}, (H, D)$ and $\mathcal{B}, (H, D)$.[5]

We now turn to Alain Connes' description of the standard model of elementary particles, as we computed it for ourselves, attempting to stress the tensor product and quaternionic aspects. Alain Connes interprets the traditional "inner degrees of freedom" of particle physics as inherent features of a non-commutative space, in fact a 4-dimensional non-commutative riemannian manifold in the sense above. This non-commutative space is obtained in a simple way as the product of a "finite space" carrying the inner degrees of freedom by ordinary space time. The next paragraph is devoted to the description of this "finite space" which is a (discrete) riemannian manifold in the sense of the last paragraph.[6]

---

[4] Condition (13) is mentioned for completeness. We shall not use it.

[5] This refers to the tensor-product structure discussed below.

[6] Poincaré duality for the product of finite space and ordinary space arises from the fact that a product of non-commutative riemannian manifolds is a non-commutative riemannian manifold.

[5] *The finite space (discrete riemannian manifold of inner degrees of freedom).* The algebra is the tensor product $\mathbf{A}_{ew} \otimes \mathbf{B}_{chr}$ of electro-weak and chromodynamic factors:

$$\begin{cases} \mathbf{A}_{ew} = \mathbf{C} \oplus \mathcal{H} = \{(p,q); p \in \mathbf{C}, q \in \mathcal{H}\} \\ \qquad = \mathcal{H}_{diag} \oplus \mathcal{H} = \{(p = \left(\begin{smallmatrix} \overline{p} & 0 \\ 0 & p \end{smallmatrix}\right), q); p \in \mathbf{C}, q \in \mathcal{H}\} \\ \mathbf{B}_{chr} = \mathbf{C} \oplus M_3(\mathbf{C}) = \{(p', m); p' \in \mathbf{C}, m \in M_3(\mathbf{C})\} \end{cases} \tag{14}$$

(here the set $\mathcal{H}$ of quaternions and the set $M_3(\mathbf{C})$ of complex 3×3 matrices are considered as $*$-algebras over the reals. We give a useful alternative description of $\mathbf{A}_{ew}$ obtained by reinterpreting $\mathbf{C}$ as the set $\mathcal{H}_{diag}$ of diagonal quaternions via the identification $p = \left(\begin{smallmatrix} \overline{p} & 0 \\ 0 & p \end{smallmatrix}\right), p \in \mathbf{C}$).

We now describe the K-cycle. The latter is the direct sum $(H_1, D_1, \chi_1) \oplus (H_q, D_q, \chi_q)$ of a leptonic K-cycle (H_1, D_1, χ_1) and a quark K-cycle (H_q, D_q, χ_q). We first[7] describe the latter: the Hilbert space is a tensor product

$$H_q = [\underbrace{\mathbf{C}^2_R \otimes \mathbf{C}^N}_{u_R d_R} \oplus \underbrace{\mathbf{C}^2_L \otimes \mathbf{C}^N}_{u_L d_L}] \otimes \mathbf{C}^3_{chr} \tag{15}$$

of an electro-weak and a chromodynamic factor, the first of which splits into a direct sum of a "right-handed" and a "left-handed" space, each of them tensored by a factor $\mathbf{C}^N$ corresponding to the N fermions families ($N = 3$ in physics). We indicated the corresponding type of particles: upper and lower right-handed quarks and upper and lower left-handed quarks. For the specification of D_q and of the representation π_q of $\mathbf{A}_{ew} \otimes 1_{chr}$ we shall handle the endomorphisms of $\mathbf{C}^2_R \otimes \mathbf{C}^N \oplus \mathbf{C}^2_L \otimes \mathbf{C}^N$ as 2×2 matrices with entries endomorphisms of $\mathbf{C}^2 \otimes \mathbf{C}^N$: with this notation we define:[8]

$$\begin{array}{c} \quad L \quad R \\ \begin{pmatrix} \mathbf{p} & 0 \\ 0 & \mathbf{q} \end{pmatrix} \begin{array}{c} L \\ R \end{array} \otimes 1, \end{array} \tag{16}$$

where

$$\begin{cases} \mathbf{p} = p \otimes \mathbf{1}_N \\ \mathbf{q} = q \otimes \mathbf{1}_N \end{cases} \tag{17}$$

(p the 2×2 matrix $\left(\begin{smallmatrix} \overline{p} & 0 \\ 0 & p \end{smallmatrix}\right)$, further

$$\pi_q(1_{ew} \otimes (p', m)) = \begin{array}{c} \quad L \quad R \\ \begin{pmatrix} 1 & 0 \\ 0 & 1 \end{pmatrix} \begin{array}{c} L \\ R \end{array} \otimes m, \end{array} \tag{18}$$

---

[7] The leptonic K-cycle is obtained from the quark K-cycle by the simple procedure of "leptonic reduction" – see below.

[8] $1_{ew}(1_{chr})$ denotes both the unit of $\mathbf{A}_{ew}(\mathbf{B}_{chr})$ and the identity operator of the electro-weak (chromodynamic) Hilbert space.

$$D_q = \begin{pmatrix} 0 & \mathbf{M}^* \\ \mathbf{M} & 0 \end{pmatrix} \begin{matrix} L \\ R \end{matrix} \otimes 1, \quad (\text{columns } L \ R) \tag{19}$$

where, with M_u and M_d the respective mass-matrices of the upper and lower quarks:

$$\begin{cases} \mathbf{M} = \begin{pmatrix} 1 & 0 \\ 0 & 0 \end{pmatrix} \otimes M_u + \begin{pmatrix} 0 & 0 \\ 0 & 1 \end{pmatrix} \otimes M_d \\ \quad = 1 \otimes \frac{1}{2}(M_u + M_d) - iK \otimes \frac{1}{2}(M_u - M_d) \\ \quad \text{with } K \text{ the quaternion } K = \begin{pmatrix} i & 0 \\ 0 & -i \end{pmatrix} \end{cases} \tag{20}$$

finally

$$\chi_q = \begin{pmatrix} 1 \otimes 1_N & 0 \\ 0 & -1 \otimes 1_N \end{pmatrix} \begin{matrix} L \\ R \end{matrix} \otimes 1. \quad (\text{columns } L \ R) \tag{21}$$

The leptonic K-cycle (H_1, D_1, χ_1) is as follows: the Hilbert space H_1 is:

$$H_1 = [\mathbf{C}_R^{1;} \otimes \mathbf{C}^N \oplus \mathbf{C}_L^{2;} \otimes \mathbf{C}^N] \otimes \mathbf{C}_{chr}^1, \tag{22}$$

$$e_R \qquad \nu_L \ e_L$$

with indicated particle-types right-handed electron, left-handed neutrino, left-handed electron.[9] The operators D_1 and $\pi_1((p,q) \otimes 1_{chr})$ are obtained from D_q and $\pi_q((p,q) \otimes 1_{chr})$ by means of the following procedure ("leptonic reduction"): write the 2×2 matrices (16), (19) and (21) as the 4×4 matrices:

$$\pi_q((p,q)) = \begin{pmatrix} \overline{p}1_N & 0 & 0 & 0 \\ 0 & p1_N & 0 & 0 \\ 0 & 0 & a1_N & b1_N \\ 0 & 0 & -\overline{b}1_N & \overline{a}1_N \end{pmatrix}, (p = \begin{pmatrix} \overline{p} & 0 \\ 0 & p \end{pmatrix}, q = \begin{pmatrix} a & b \\ -\overline{b} & \overline{a} \end{pmatrix} \in \mathbf{A}_{ew}, \tag{17a}$$

$$D_q = \begin{pmatrix} 0 & 0 & M_u^* & 0 \\ 0 & 0 & 0 & M_d^* \\ M_u & 0 & 0 & 0 \\ 0 & M_d & 0 & 0 \end{pmatrix}, \tag{19a}$$

$$\chi_q = \begin{pmatrix} 1_N & 0 & 0 & 0 \\ 0 & 1_N & 0 & 0 \\ 0 & 0 & -1_N & 0 \\ 0 & 0 & 0 & -1_N \end{pmatrix}, \tag{21a}$$

[9] There is no right-handed neutrino! The term electron should be taken generically for electron-muon-tau.

and then effect the changes $M_u \to 0, m_d \to M_e$ followed by restriction to the right-lower corner 3×3 matrix (M_e is the electron mass-matrix, the neutrinos have vanishing masses). We note that the above situation fulfils property (12) above required as a feature of generalized Poincaré duality.

As a next item, we describe $\Omega_{D_q}\mathbf{A}^1_{ew}$ and $\Omega_{D_q}\mathbf{A}^2_{ew}$. The one and two-quantum forms are then easily deduced using leptonic reduction.

The $\mathbf{A}_{ew}$-bimodule $i\Omega_{D_q}\mathbf{A}^1_{ew} = i\pi_q(\Omega(\mathbf{A}_{ew})^1)$ is isomorphic to $\mathcal{H} \times \mathcal{H}$ as a real vector space. With the above 2×2 matrix notation its elements

$$\begin{pmatrix} & Q' \\ Q & \end{pmatrix} = \begin{pmatrix} 0 & \mathbf{M}^*(Q' \otimes 1_N) \\ (Q \otimes 1_N)\mathbf{M} & 0 \end{pmatrix} \qquad , Q, Q' \in \mathcal{H} \tag{23}$$

(concretely for $\rho = \sum_{i=1,\ldots,k}(p_i, q_i)\delta(\pi_i, \chi) \in \Omega_{D_q}\mathbf{A}^1_{ew}, (p_i, q_i), (\pi_i, \chi_i) \in \mathbf{A}_{ew}$:

$$i\pi_q(\rho) = \begin{pmatrix} & Q' \\ Q & \end{pmatrix} \quad \text{with} \quad \begin{cases} Q = \sum_{i=1,\ldots,k} q_i(\pi_i - \chi_i); \\ Q' = \sum_{i=1,\ldots,k} p_i(\chi_i - \pi_i) \end{cases}, \quad) \tag{23a}$$

give rise to the following rules:

– for the *-operation

$$\begin{pmatrix} & Q' \\ Q & \end{pmatrix}^* = \left({}_{Q'^*,}{}^{Q^*}\right), \quad \begin{pmatrix} & Q' \\ Q & \end{pmatrix} \in \Omega_{D_q}\mathbf{A}^1_{ew}, \tag{24}$$

– for the bimodule multiplication[10]

$$(p_1, q_1)\begin{pmatrix} & Q' \\ Q & \end{pmatrix}(p_2, q_2) = \left({}_{q_1 Q p_2,}{}^{p_1 Q' q_2}\right), \begin{cases} (p_1, q_1), (p_2, q_2) \in \mathbf{A}_{ew} \\ \begin{pmatrix} & Q' \\ Q & \end{pmatrix} \in \Omega_{D_q}\mathbf{A}^1_{ew} \end{cases}, \tag{25}$$

– for the differential in grade zero

$$i \cdot \delta(p, q) = \left({}_{p-q,}{}^{q-p}\right), \quad (p, q) \in \mathbf{A}_{ew}, \tag{26}$$

– for the scalar product[11]

$$\begin{cases} \left(\left({}_{Q_1,}{}^{Q_1'}\right), \left({}_{Q_2,}{}^{Q_2'}\right)\right)_{D_q} = Tr\Sigma[(Q_1, Q_2) + (Q_1', Q_2')] \\ \left({}_{Q_1,}{}^{Q_1'}\right), \left({}_{Q_2,}{}^{Q_2'}\right) \in \Omega_{D_q}\mathbf{A}^1_{ew} \\ \text{with } \Sigma = \frac{1}{2}(\mu_u + \mu_d), \mu_u = M_u M_u^\star, \mu_d = +M_d M_d^\star. \end{cases} \tag{27}$$

[10] This rule as well as the rule (31) below is easily memorized by writing $(p, q) \in \mathbf{A}_{ew}$ as $\left({}_{p,}{}^{q}\right)$.

[11] We recall that we defined the (real euclidean) scalar product of quaternions as $(q, q') = (\frac{1}{2})\mathrm{Tr}(q^* q'), q, q' \in \mathcal{H}$. For the complex structure of $\mathcal{H}$ defined by K, (q, q') becomes the real part of a hermitean scalar product with purely imaginary part $i(q, Kq')$.

The $\mathbf{A}_{ew}$-bimodule $\Omega_{D_q}\mathbf{A}^2_{ew} = \pi_q(\Omega(\mathbf{A}_{ew})^2)/\pi_q(\delta K^1) \simeq P_2\pi_q(\Omega\mathbf{A}_{ew})^2$ is isomorphic to $\mathcal{H}\times\mathcal{H}$ as a real vector space. Its elements

$$({}^{Q}{}_{Q'}) = \begin{pmatrix} \mathbf{M}^*(Q\otimes 1_N)\mathbf{M} & 0 \\ 0 & Q'\otimes\Sigma \end{pmatrix}, \tag{28}$$

are obtained as the components $P_2({}^{Q}{}_{Q',Q''}) = ({}^{Q}{}_{Q',0})$ orthogonal to $\pi_q(\delta K^1)$ of general elements of $\pi_q(\Omega(\mathbf{A}_{ew})^2 \simeq \mathcal{H}\times\mathcal{H}\times\mathcal{H}$:

$$({}^{Q}{}_{Q',Q''}) = \begin{pmatrix} \mathbf{M}^*(Q\otimes 1_N)\mathbf{M} & 0 \\ 0 & Q'\otimes\Sigma + iQ''\otimes\Delta \end{pmatrix} \tag{29}$$

(concretely, for $\theta = \sum_{i=1,\dots,k}(s_i,r_i)\delta(p_i,q_i)\delta(\pi_i,\chi_i), (s_i,r_i),(p_i,q_i),(\pi_i,\chi_i) \in \mathbf{A}_{ew}$:

$$\pi_q(\theta) = ({}^{Q}{}_{Q',Q''}) \quad\text{with}\quad \begin{cases} Q = \sum_{i=1\dots k} s_i(p_i-q_i)(\pi_i-\chi_i, \\ Q' = \sum_{i=1\dots k} r_i(p_i-q_i)(\pi_i-\chi_i), \\ Q'' = \sum_{i=1\dots k} r_i(p_i-q_i)K(\pi_i-\chi_i) \end{cases} \tag{29a)}$$

One has the following rules:

– for the *-operation

$$({}^{Q}{}_{Q'})^* = ({}^{Q^*}{}_{Q'^*}), \quad ({}^{Q}{}_{Q'}) \in \Omega_{D_q}\mathbf{A}^2_{ew}, \tag{30}$$

– for the bimodule multiplication

$$(p_1,q_1)({}^{Q}{}_{Q'})(p_2,q_2) = ({}^{p_1Qp_2}{}_{q_1Q'q_2}), \quad \begin{cases} (p_1,q_1),(p_2,q_2) \in \mathbf{A}_{ew} \\ ({}^{Q}{}_{Q'}) \in \Omega_{D_q}\mathbf{A}^2_{ew} \end{cases}, \tag{31}$$

– for the multiplication rule $\Omega_{D_q}\mathbf{A}^1_{ew}\times\Omega_{D_q}\mathbf{A}^1_{ew} \to \Omega_{D_q}\mathbf{A}^2_{ew}$

$$({}_{Q_1}{}^{Q'_1})({}_{Q_2}{}^{Q'_2}) = ({}^{Q'_1Q_2}{}_{Q_1Q'_2}), \quad ({}_{Q_1}{}^{Q'_1}),({}_{Q_2}{}^{Q'_2}) \in \Omega_{D_q}\mathbf{A}^1_{ew} \tag{32}$$

– for the differential in grade one:[4]

$$i\cdot\delta({}^{Q}{}_{Q'}) = ({}^{Q+Q'}{}_{Q+Q'}), \quad ({}^{Q}{}_{Q'}) \in \Omega_{D_q}\mathbf{A}^1_{ew}, \tag{33}$$

– for the scalar product of the pair $({}^{Q_1}{}_{Q'_1,Q''_1}),({}^{Q_2}{}_{Q'_2,Q''_2}) \in \Omega_{D_q}\mathbf{A}^2_{ew}$ with $Q_1 = \begin{pmatrix} \alpha_1 & \beta_1 \\ -\bar\beta_1 & \bar\alpha_1 \end{pmatrix}, Q_2 = \begin{pmatrix} \alpha_2 & \beta_2 \\ -\bar\beta_2 & \bar\alpha_2 \end{pmatrix}$:

$$\begin{aligned} (({}^{Q_1}{}_{Q'_1,Q''_1}),({}^{Q_2}{}_{Q'_2,Q''_2}))_{D_q} &= 2\mathrm{Tr}(\Sigma^2)(Q'_1,Q'_2) + 2\mathrm{Tr}(\Delta^2)(Q''_1,Q''_2 \\ &+\tfrac{1}{2}\mathrm{Tr}(\mu_u^2+\mu_d^2)(\bar\alpha_1\alpha_2+\alpha_1\bar\alpha_2) + Tr(\mu_u\mu_d)(\beta_1\bar\beta_2+\bar\beta_1\beta_2). \end{aligned} \tag{34}$$

The Yang-Mills action of the quark sector of the finite space has no physical significance, is however instructive as offering a baby-model of Higgs fields and Higgs potentials. The vector potentials are of the form:

$$V = i\left({}_{Q}, {}^{Q^*}\right), \quad Q \in \mathcal{H}, \tag{35}$$

with the corresponding curvature:

$$\theta = -[|Q+1|^2 - 1]^2\left({}^{\mathbf{1}}, {}_{\mathbf{1}}\right) = -[|Q+1| - 1]\begin{pmatrix} \mathrm{M^*M} & 0 \\ 0 & 1 \otimes \Sigma \end{pmatrix}, \tag{36}$$

leading to the Yang-Mills action

$$YM(V) = \mathrm{Tr}(\theta^2) = 6\mathrm{Tr}[\mu_u^2 + \mu_d^2 + \mu_u\mu_d][|Q+1|^2 - 1]^2. \tag{37}$$

The reason for studying separately the finite space of inner degrees of freedom is that the information thus gathered can be used to investigate the full space obtained by tensorization of the finite space by usual space-time[12]. We thus interpose a paragraph devoted to the tensor-product structure.

[6] *The tensor-product structure.* (i): General facts. Let $\mathcal{A}_1, \mathcal{A}_2$ be unital *-algebras equipped with respective K-cycles (H_1, D_1, χ_1) and (H_2, D_2, χ_2).
(a) The *tensor product of the K-cycles* (H_1, D_1, χ_1) and $H_2, D_2, \chi_2)$ is by definition the K-cycle $(H = H_1 \otimes H_2, D, \chi)$ over $\mathcal{A} = \mathcal{A}_1 \otimes \mathcal{A}_2$, where

$$\begin{cases} \underline{a_1 \otimes a_2} = \underline{a}_1 \otimes \underline{a}_2, & a_1 \in \mathcal{A}_1, a_2 \in \mathcal{A}_2, \\ D = D_1 \otimes \mathbf{1}_2 + \chi_1 \otimes D_2 & \\ \chi = \chi_1 \otimes \chi_2. & \end{cases} \tag{38}$$

In the case where (H_1, D_1, χ_1) is a d^+-summable, d even, $d > 1$, and where $dim\mathcal{A}_2 = p < \infty, dim\mathcal{H}_2 = q < \infty$, one has that (H, D, χ) is d^+-summable, and that $\tau_D = \tau_{D_1} \otimes \mathrm{Tr}_{H_2}$.
(b) The differential envelope of the tensor product projects homomorphically onto the skew tensor product of the differential envelopes:

$$\Omega\mathcal{A} \to \Omega\mathcal{A}_1 \hat{\otimes} \Omega\mathcal{A}_2. \tag{39}$$

(c) The representation π_D is the skew tensor product of the representations π_{D_1} and π_{D_2} in the following sense: one has

$$\pi_D(\omega_1 \hat{\otimes} \omega_2) = \pi_1(\omega_1) \hat{\otimes} \pi_2(\omega_2) = \pi_1(\omega_1)\chi_2^q \otimes \pi_2(\omega_2), \; \omega_1 \in \Omega\mathcal{A}_1^p, \omega_2 \in \Omega\mathcal{A}_2^q, \tag{40}$$

where $\otimes$ denotes ordinary, and $\hat{\otimes}$ skew tensor products. In other terms, one has a homomorphism $\pi : \Omega\mathcal{A}' \hat{\otimes} \Omega\mathcal{A}'' \to EndH' \hat{\otimes} EndH''$ of skew tensor products of $\mathcal{Z}/2$-graded algebras.

[12] meaning euclidean space-time. This theory, based upon elliptic operators, is within the euclidean frame.

(ii) Application to the case at hand[13]: We consider the full electroweak algebra[14]

$$\begin{aligned} A &= C^\infty(\mathbf{M},\mathbf{R}) \otimes \mathbf{A}_{ew} = C^\infty(\mathbf{M},\mathbf{C}) \oplus C^\infty(\mathbf{M},\mathcal{H}) \\ &= \{(f,q); f \in C^\infty(\mathbf{M},\mathbf{C}), q \in C^\infty(\mathbf{M},\mathcal{H})\} \end{aligned} \tag{41}$$

which we equip with the K-cycle $(\mathbf{H}_q, \mathbf{D}_q, \chi_q)$ tensor product of the K-cycles $(L^2(\mathcal{S}_{\mathbf{M}}), D, \gamma^5)$ and (H_q, D_q, χ_q): we thus have (with the shorthands $\mathcal{A} = C^\infty(\mathbf{M},\mathbf{R})$ and π_q for $\pi_{\mathbf{D}_q}$ and π_{D_1}):

$$\begin{cases} \mathbf{H}_q = L^2(\mathcal{S}_{\mathbf{M}}) \otimes H_q = \mathbf{H}_{qR} \oplus \mathbf{H}_{qL}, \\ \pi_q(f \otimes (p,q)) = \underline{f} \otimes \pi_q((p,q)), f \in \mathcal{A}, (p,q) \in \mathbf{A}_{ew}, \\ \mathbf{D}_q = D\hat{\otimes}\mathbf{1}_q + \mathbf{1}_H\hat{\otimes}D_q = D \otimes \mathbf{1}_q + \gamma^5 \otimes D_q, \\ \chi_q = \gamma^5 \otimes \chi_q. \end{cases} \tag{42}$$

Writing, according to the direct sum decomposition in the first line, the endomorphisms of $\mathbf{H}_q$ as 2×2 matrices with entries endomorphisms of $L^2(\mathcal{S}_{\mathbf{M}} \otimes [\mathbf{C}^2 \otimes \mathbf{C}^N]$, the latter prescriptions read:

$$\pi_q(f \otimes (p,q)) = \begin{pmatrix} \underline{f} \otimes \mathbf{p} & 0 \\ 0 & \underline{f} \otimes \mathbf{q} \end{pmatrix} \begin{matrix} L \\ R \end{matrix}, \quad \text{(columns } L, R\text{)} \tag{43}$$

with $\begin{cases} \mathbf{p} = p \otimes \mathbf{1}_N \\ \mathbf{q} = q \otimes \mathbf{1}_N \end{cases}, \begin{cases} f \in \mathcal{A} \\ (p,q) \in \mathbf{A}_{ew} \end{cases},$

$$\mathbf{D}_q = \begin{pmatrix} D \otimes \mathbf{1} & \gamma^5 \otimes \mathbf{M}^* \\ \gamma^5 \otimes \mathbf{M} & D \otimes \mathbf{1} \end{pmatrix} \begin{matrix} L \\ R \end{matrix} \quad \text{(columns } L, R\text{)} \tag{44}$$

with $\mathbf{M} = \begin{pmatrix} 1 & 0 \\ 0 & 0 \end{pmatrix} \otimes M_u + \begin{pmatrix} 0 & 0 \\ 0 & 1 \end{pmatrix} \otimes M_d$ (M_e, M_d, M_u are the $N \times N$ complex mass matrices of the respective electron, lower quark, and upper quark families, assumed to be invertible); and:

$$\chi_q = \begin{pmatrix} \gamma^5 \otimes \mathbf{1} & 0 \\ 0 & -\gamma^5 \otimes \mathbf{1} \end{pmatrix} \begin{matrix} L \\ R \end{matrix}. \quad \text{(columns } L, R\text{)} \tag{45}$$

We then have the following tensor decompositions:

$$\pi_{\mathbf{D}_q}(\Omega A^1) = \pi_D(\Omega\mathcal{A}^1)\hat{\otimes}\pi_{D_q}(\mathbf{A}_{ew}) \oplus \pi_D(\mathcal{A})\hat{\otimes}\pi_{D_q}(\Omega\mathbf{A}^1_{ew}) \tag{46}$$

[13] For the moment we consider only the quark K-cycle, and the electroweak sector. Full consideration of the standard model will come later.

[14] tensor product over the reals. The gauge group $\mathbf{G}$ of A is by definition $\mathbf{G} = \{\mathbf{u} = (u,v) \in A; u^*u = 1_{C^\infty(M,C)} v^*v = 1_{C^\infty(M,H)}\} = C^\infty(M,T) \times C^\infty(M,$ unitary quaternions) with structural group $U(1) \times SU(2)$.

i.e.

$$\Omega_{\mathbf{D}_q} A^1 = \Omega_D \mathcal{A}^1 \hat{\otimes} \mathbf{A}_{ew} \oplus \mathcal{A} \hat{\otimes} \Omega_{D_q} \mathbf{A}^1_{ew} (= \Omega_{\mathbf{D}_q} A^1_{[1,0]} \oplus \Omega_{\mathbf{D}_q} A^1_{[0,1]}), \tag{46a}$$

$$\begin{aligned} \pi_{\mathbf{D}_q}(\Omega A^2) &= \pi_D(\Omega \mathcal{A}^2) \hat{\otimes} \pi_{D_q}(\mathbf{A}_{ew}) + \pi_D(\Omega \mathcal{A}^1) \hat{\otimes} \pi_{D_q}(\mathbf{A}^1_{ew}) \\ &\quad + \pi_D(\mathcal{A}) \hat{\otimes} \pi_{D_q}(\mathbf{A}^2_{ew}) \\ &(= \pi_{\mathbf{D}_q}(\Omega A^2_{[2,0]}) + \pi_{\mathbf{D}_q}(\Omega A^2_{[1,1]}) + \pi_{\mathbf{D}_q}(\Omega A^2_{[0,2]}),) \end{aligned} \tag{47}$$

and (key property):

$$\pi_q(\delta K^1_A) = \pi_D(\delta K^1_{\mathcal{A}}) \otimes \pi_q(\mathbf{A}_{ew}) + \pi_D(\mathcal{A}) \otimes \pi_q(\delta K^1_{\mathbf{A}_{ew}}). \tag{48}$$

We now describe the quantum one and two-forms, beginning with the $\mathbf{D}_q$-quantum forms and curvatures as a first step.

[7] *The sets* $\Omega_{\mathbf{D}_q}(A)^1 = \pi_{\mathbf{D}_q}(\Omega A^1), \pi_{\mathbf{D}_q}(\Omega A^2)$ *and the projection* P_2. As we have seen, the elements of $\pi_{D_q}(\mathbf{A}^0_{ew}), \pi_{D_q}(\mathbf{A}^1_{ew}), \pi_{D_q}(\mathbf{A}^2_{ew})$, and the D_q-quantum zero, one and two-forms are 2×2 matrices with entries endomorphisms of $\mathbf{C}^2 \otimes \mathbf{C}^N$ (diagonal for zero and two-forms, anti-diagonal for onw-forms). As for $\pi_D(\Omega\mathcal{A})$, resp. $\Omega_D\mathcal{A}$, according to

$$\pi_D(a_0 \delta a_1 ... \delta a_n) = \begin{cases} \text{multiplication by} a_0 \gamma^\mu \partial_\mu a_1 ... \gamma^\mu \partial_\mu a_n, \\ a_0, a_1, ..., a_n \in \mathcal{A}, \end{cases} \tag{49}$$

their elements belong to the Clifford bundle of $\mathbf{M}$ concretely acting on its spin bundle in a "fiber-preserving" way: namely of the type $\underline{f}$ with $f \in C^\infty(\mathbf{M}, \mathbf{R})$ in grade zero, resp. $\underline{\gamma}(a)$ with $a \in \Omega(\mathbf{M}, \mathbf{R})^1$ in grade one, and $\underline{\gamma}(\lambda)$ with $\lambda \in \Omega(\mathbf{M}, \mathbf{R}) \oplus C^\infty \mathbf{M}, \mathbf{R}$ in grade two.[15] These remarks are the basis of the following results.

The elements ρ of the A-bimodule $\pi_{\mathbf{D}_q}(\Omega A^1) = \Omega_{\mathbf{D}_q} A^1 = \Omega_{\mathbf{D}_q} A^1_{[1,0]} \oplus \Omega_{\mathbf{D}_q} A^1_{[0,1]}$ (multiplied by i)[16] are given as follows in terms of quadruples $i\rho = (\mathbf{a}, \mathbf{b}, \mathbf{H}, \mathbf{H}'), \mathbf{a} \in \Omega(\mathbf{M}, \mathbf{C})^1, \mathbf{b} \in (\Omega(\mathbf{M}, i\mathcal{H})^1, \mathbf{H}, \mathbf{H}' \in C^\infty(\mathbf{M}, \mathcal{H})$: one has for the bihomogeneous components the following 2×2 matrices with entries acting on the fibers of $L^2(\mathcal{S}_{\mathbf{M}}) \otimes \mathbf{C}^N$:

$$i\rho = (\mathbf{a}, \mathbf{b}, \mathbf{H}, \mathbf{H}') : \begin{cases} i\rho_{[1,0]} = \begin{pmatrix} \underline{\gamma}(\mathbf{a}^{\cdot}.) \otimes \mathbf{1}_N & 0 \\ 0 & \underline{\gamma}(\mathbf{b}^{\cdot}.) \otimes \mathbf{1}_N \end{pmatrix} \\ i\rho_{[0,1]} = \begin{pmatrix} 0 & \mathbf{M}^*(\gamma^5 \underline{\mathbf{H}}' \otimes \mathbf{1}_N) \\ (\gamma^5 \underline{\mathbf{H}} \otimes \mathbf{1}_N)\mathbf{M} & 0 \end{pmatrix} \end{cases} \tag{50}$$

where we used the shorthands ($\mathbf{a} \in \Omega(\mathbf{M}, \mathbf{C})^1$):

---

[15] with the projection P_2 consisting in suppressing the $C^\infty(\mathbf{M}, \mathbf{R})$-component.

[16] We are describing $i\Omega_{\mathbf{D}_q} A^1$ because its self-adjoint elements will correspond to compatible vector potentials multiplied by i (the physicists' vector potentials).

$$\begin{cases} \underline{\gamma}(\mathbf{a}^{\cdot}.) = \begin{pmatrix} -\underline{\gamma}(\bar{\mathbf{a}}) & 0 \\ 0 & \underline{\gamma}(\mathbf{a}) \end{pmatrix}, & \mathbf{a}^{\cdot}. = \begin{pmatrix} -\bar{\mathbf{a}} & 0 \\ 0 & \mathbf{a} \end{pmatrix} \in \Omega(\mathrm{M}, i\mathcal{H}_{diag})^1, \\ \underline{\gamma}(\mathbf{b}^{\cdot}.) = \begin{pmatrix} \underline{\gamma}(\mathbf{b}_1^1) & \underline{\gamma}(\mathbf{b}_2^1) \\ \underline{\gamma}(\mathbf{b}_1^2) & \underline{\gamma}(\mathbf{b}_2^2) \end{pmatrix}, & \mathbf{b}^{\cdot}. = \begin{pmatrix} \mathbf{b}_1^1 & \mathbf{b}_2^1 = \bar{\mathbf{b}}_1^2 \\ \mathbf{b}_1^2 & \mathbf{b}_2^2 = \bar{\mathbf{b}}_1^1 \end{pmatrix} \in \Omega(\mathrm{M}, i\mathcal{H})^1. \end{cases} \tag{51}$$

The elements of $\pi_q(\delta K_A^1)$ have the form:[17]

$$\begin{cases} \begin{pmatrix} (\underline{S}_k^i) \otimes 1_N & 0 \\ 0 & (\underline{T}_k^i) \otimes 1_N \end{pmatrix} + \begin{pmatrix} 0 & 0 \\ 0 & i\underline{\mathbf{R}} \otimes \Delta \end{pmatrix} \\ \text{where} \begin{cases} (S_k^i) \in C^\infty(\mathrm{M}, \mathcal{H}_{diag}) \\ (T_k^i) \in C^\infty(\mathrm{M}, \mathcal{H}) \\ \mathbf{R} \in C^\infty(\mathrm{M}, \mathcal{H}) \end{cases} \end{cases} \tag{52}$$

The vector potentials (antihermitean elements of $\Omega_{\mathbf{D}_q} A^1$, singled out as the antihermitean $V = \rho$ above are accordingly specified by triples $(\mathbf{a}, \mathbf{b}, \mathbf{H})$ of a hermitean $U(1)$-connection-one-form $\mathbf{a}$, a hermitean $SU(2)$-connection-one-form $\mathbf{b}^{\cdot}.$, and a doublet field $\mathbf{H} \in C^\infty(\mathrm{M}, \mathcal{H})$: one has

$$iV = (\mathbf{a}, \mathbf{b}, \mathbf{H}, \mathbf{H}^*): \begin{cases} iV_{[1,0]} = \begin{pmatrix} \underline{\gamma}(\mathbf{a}^{\cdot}.) \otimes \mathbf{1}_N & 0 \\ 0 & \underline{\gamma}(\mathbf{b}^{\cdot}.) \otimes \mathbf{1}_N \end{pmatrix} \\ iV_{[0,1]} = \begin{pmatrix} 0 & \mathbf{M}^*(\gamma^5 \underline{\mathbf{H}}^* \otimes \mathbf{1}_N) \\ (\gamma^5 \underline{\mathbf{H}} \otimes \mathbf{1}_N)\mathbf{M} & 0 \end{pmatrix} \end{cases} \tag{53}$$

$$\begin{cases} \mathbf{a} = \bar{\mathbf{a}} \\ \mathbf{b} = \mathbf{b}^* \text{ i.e. } \mathbf{b}_k^i = \bar{\mathbf{b}}_i^k, \ i, k = 1, 2 \end{cases}$$

(Note that a quaternion is antihermitean iff it is traceless. The field $\mathbf{H} \in C^\infty(\mathrm{M}, \mathcal{H})$ is identified with a doublet-field according to $\mathbf{H} = \begin{pmatrix} \bar{\mathbf{H}}^2 & \mathbf{H}^1 \\ -\bar{\mathbf{H}}^1 & \mathbf{H}^2 \end{pmatrix} \leftrightarrow \mathbf{H}^{\cdot} = (\mathbf{H}^1, \mathbf{H}^2) \leftrightarrow \mathbf{H}. = (\bar{\mathbf{H}}^1, \bar{\mathbf{H}}^2)$).

Under the gauge transformation $\mathbf{u} = (u, v = (v_k^i)) \in \mathbf{G}$ the $\mathbf{D}_q$-quantum-one-form $-i(\mathbf{a}, \mathbf{b}, \mathbf{H}, \mathbf{H}')$ becomes $-i({}^{\mathbf{u}}\mathbf{a}, {}^{\mathbf{u}}\mathbf{b}, {}^{\mathbf{u}}\mathbf{H}, {}^{\mathbf{u}}\mathbf{H}')$ with:

$$\begin{cases} {}^{\mathbf{u}}\mathbf{a} = \mathbf{a} + iu\mathbf{d}u^* \\ {}^{\mathbf{u}}\mathbf{b} = v\mathbf{b}v^* + iv\mathbf{d}v^* \\ {}^{\mathbf{u}}(\mathbf{H} + \mathbf{1}) = u^* v(\mathbf{H} + \mathbf{1}) \\ {}^{\mathbf{u}}(\mathbf{H}' + \mathbf{1}) = u^* v(\mathbf{H}' + \mathbf{1}) \end{cases} \tag{54}$$

In the particular case of vector potentials, $\mathbf{a}$ and $(\mathbf{a}_k^i)$ thus behave as the respective one-forms of a $U(1)$ and an $SU(2)$-connection, whereas $\phi. = H. + (0,1)$ and $\phi^{\cdot} = H^{\cdot} + (0,1)$ behave as respective covariant and contravariant $SU(2)$-doublets.

[17] This statement is synonymous with (48).

The elements of $\pi_{\mathbf{D}_q}(\Omega A^2)$ are of the type:[18]

$$\begin{cases} \eta_{[2,0]} = \begin{pmatrix} [(\gamma(\lambda^i_k)) - (\underline{X}^i_k)] \otimes 1_N & 0 \\ 0 & [(\gamma(\mu^i_k)) - (\underline{Y}^i_k)] \otimes 1_N \end{pmatrix} \\ \eta_{[1,1]} = \begin{pmatrix} 0 & \mathbf{M}^*[i\underline{\gamma}(\mathbf{q}')\gamma^5 \otimes 1_N] \\ [i\underline{\gamma}(\mathbf{q})\gamma^5 \otimes 1_N]\mathbf{M} & 0 \end{pmatrix} \\ \eta_{[0,2]} = \begin{pmatrix} \mathbf{M}^*(\mathbf{Q} \otimes 1_N)\mathbf{M} & 0 \\ 0 & \underline{\mathbf{Q}}' \otimes \Sigma + i\underline{\mathbf{Q}}'' \otimes \Delta \end{pmatrix} \end{cases} \tag{55}$$

where

$$\begin{cases} (\lambda^i_k) \in \Omega(\mathbf{M}, \mathcal{H}_{diag})^2 \\ (X^i_k) \in C^\infty(\mathbf{M}, \mathcal{H}_{diag}) \\ (\mu^i_k) \in \Omega(\mathbf{M}, \mathcal{H})^2 \\ (Y^i_k) \in C^\infty(\mathbf{M}, \mathcal{H}) \\ \mathbf{q}, \mathbf{q}' \in \Omega(\mathbf{M}, \mathcal{H})^1 \\ \mathbf{Q}, \mathbf{Q}', \mathbf{Q}'' \in C^\infty(\mathbf{M}, \mathcal{H}) \end{cases}$$

with the following action of P_2:

$$\begin{cases} (P_2\eta)_{[2,0]} = \begin{pmatrix} [(\gamma(\lambda^i_k)) - (2N)^{-1}\mathrm{Tr}(\mu_u + \mu_d)\mathbf{Q}_{diag}] \otimes 1_N & 0 \\ 0 & [(\gamma(\mu^i_k)) - (2N)^{-1}\mathrm{Tr}(\mu_u + \mu_d)\mathbf{Q}'] \otimes 1_N \end{pmatrix} \\ (P_2\eta)_{[1,1]} = \begin{pmatrix} 0 & \mathbf{M}^*[i\underline{\gamma}(\mathbf{q}')\gamma^5 \otimes 1_N] \\ [i\underline{\gamma}(\mathbf{q})\gamma^5 \otimes 1_N]\mathbf{M} & 0 \end{pmatrix} = \eta_{[1,1]} \\ (P_2\eta)_{[0,2]} = \begin{pmatrix} \mathbf{M}^*(\mathbf{Q} \otimes 1_N)\mathbf{M} & 0 \\ 0 & \mathbf{Q}' \otimes \Sigma \end{pmatrix} \end{cases} \tag{56}$$

[8] *The electroweak sector.* We now adapt the previous results to the full electroweak sector $(\mathbf{H}, \mathbf{D}, \chi) = (\mathbf{H}_q, \mathbf{D}_q, \chi_q) \oplus (\mathbf{H}_1, \mathbf{D}_1, \chi_1)$, with scalar product stemming from the graded trace $\alpha_1\tau_{\mathbf{D}_1} + \alpha_q\tau_{\mathbf{D}_q}$, convex combination of the leptonic and quark scalar products with the coefficients $\alpha_1 = (1+x)/2, \alpha_q = (1-x)/2$. The results for the leptonic K-cycle are deduced from those in the last paragraph by leptonic reduction. We have that $\Omega A^1 \cap Ker\pi_1 = \Omega A^1 \cap Ker\pi_q = K^1$. Correspondingly, the A-modules $i\pi_{\mathbf{D}}(A^1) = i\Omega_{\mathbf{D}}A^1$, resp. $\pi_{\mathbf{D}}(A^2)$, are one-to-one with $i\pi_{\mathbf{D}_q}(A^1) = i\Omega_{\mathbf{D}_q}A^1$, resp. $\pi_{\mathbf{D}_q}(A^2)$: their elements consist of direct sums of elements (50). resp. (55) plus their leptonic reductions. The elements of $\pi_{\mathbf{D}}(\delta K^1_A)$ are likewise direct sums of elements (52), plus their leptonic reductions. The projection P_2 is accordingly modified as follows: its action on elements (55) is now, with $L = \mathrm{Tr}[\alpha_1\mu_e + \alpha_q(\mu_u + \mu_d)], \mu_e = M_e M_e^*, \mu_u = M_u M_u^*, \mu_d = M_d M_d^*$:

[18] While $\rho_{[1,0]}$ and $\rho_{[0,1]}$ are linearly independent, this does not hold for $\eta_{[2,0]}, \eta_{[1,1]}$ and $\eta_{[0,2]}$ (in the subscript $[i,j]$ i refers to the γ^5 and j to the χ_q parity).

$$\begin{cases} (P_2\eta)_{[2,0]} = \begin{pmatrix} [(\gamma(\lambda^i_k)) - (\alpha_1+6\alpha_q)^{-1}L\mathbf{Q}_{diag}]\otimes 1_N & 0 \\ 0 & [(\gamma(\mu^i_k)) - (\alpha_1+3\alpha_q)^{-1}(2N)^{-1}L\mathbf{Q}']\otimes 1_N \end{pmatrix} \\ (P_2\eta)_{[1,1]} = \begin{pmatrix} 0 & \mathbf{M}^*[i\underline{\gamma}(\mathbf{q}')\gamma^5\otimes 1_N] \\ [i\underline{\gamma}(\mathbf{q})\gamma^5\otimes 1_N]\mathbf{M} & 0 \end{pmatrix} = \eta_{[1,1]} \\ (P_2\eta)_{[0,2]} = \begin{pmatrix} \mathbf{M}^*(\mathbf{Q}\otimes 1_N)\mathbf{M} & 0 \\ 0 & \mathbf{Q}'\otimes\Sigma \end{pmatrix} \end{cases} \tag{57}$$

the action of P_2 on the leptonic reduction of η yielding the leptonic reduction of $P_2\eta$.

We now describe the curvature corresponding to the vector potential specified in (53): it is specified as follows by its quark component θ_q:

$$\begin{cases} -\theta_{q[2.0]} = \begin{pmatrix} [\frac{1}{2}\gamma(\mathbf{f}^{\cdot}.) - (\alpha_1+6\alpha_q)^{-1}N^{-1}Lv_\phi\mathbf{1}]\otimes\mathbf{1}_N & 0 \\ 0 & [\frac{1}{2}\gamma(\mathbf{h}^{\cdot}.) - (2N)^{-1}\underline{L}v_\phi\mathbf{1}]\otimes\mathbf{1}_N \end{pmatrix} \\ -\theta_{q[1,1]} = \begin{pmatrix} 0 & \mathbf{M}^*[\underline{\gamma}(i\mathbf{D}\phi^*)\gamma^5\otimes\mathbf{1}_N] \\ [\underline{\gamma}(i\mathbf{D}\phi)\gamma^5\otimes\mathbf{1}_N]\mathbf{M} & 0 \end{pmatrix} \\ -\theta_{q[0,2]} = \begin{pmatrix} v_\phi\mathbf{M}^*\mathbf{M} & 0 \\ 0 & v_\phi\mathbf{1}\otimes\Sigma \end{pmatrix} \end{cases} \tag{58}$$

the leptonic component θ_1 resulting by leptonic reduction. Here we use the shorthands:[19]

$$\begin{cases} \underline{\gamma}(\mathbf{f}^{\cdot}.) = \begin{pmatrix} -\underline{\gamma}(\mathbf{f}) & 0 \\ 0 & \underline{\gamma}(\mathbf{f}) \end{pmatrix}, \quad \mathbf{f} = \mathbf{da}\in C^\infty(\mathbf{M},\mathbf{C}), \\ \underline{\gamma}(\mathbf{h}^{\cdot}.) = \begin{pmatrix} \underline{\gamma}(\mathbf{h}^1_1) & \underline{\gamma}(\mathbf{h}^1_2) \\ \underline{\gamma}(\mathbf{h}^2_1) & \underline{\gamma}(\mathbf{h}^2_2) \end{pmatrix}, \quad \mathbf{h}^j_k = \mathbf{db}^j_k - i\mathbf{b}^j_m\wedge\mathbf{b}^m_k, \\ \qquad\qquad (\text{i.e. } \mathbf{h}_{\mu\nu} = \partial_\mu\mathbf{b}_\nu - \partial_\nu\mathbf{b}_\mu - i[\mathbf{b}_\mu,\mathbf{b}_\nu]) \\ \phi = \mathbf{H}+\mathbf{1}, v_\phi = \phi_i\phi^i - 1, \quad \mathbf{H}\in C^\infty(\mathbf{M},\mathcal{H}), \\ \mathbf{D}\phi = \mathbf{d}\phi + i(\phi\mathbf{a}^{\cdot}. - \mathbf{b}^{\cdot}.\phi) \quad \mathbf{D}\phi^* = (\mathbf{D}\phi)^* = \mathbf{d}\phi^* - i(\mathbf{a}^{\cdot}.\phi^* - \phi^*\mathbf{b}^{\cdot}.). \end{cases} \tag{59}$$

Here, $\mathbf{f}, \mathbf{h}^{\cdot}. = (\mathbf{h}^i_j), i,j = 1,2$, are the respective curvatures of the $U(1)$-connection $\mathbf{a}$ and $SU(2)$-connection $\mathbf{b}^{\cdot}. = (\mathbf{b}^i_j) = \mathbf{b}^a\frac{\tau_a}{2}$, $\alpha = 1,2,3$:

$$\begin{cases} \mathbf{f} = \mathbf{da} = \frac{1}{2}\mathbf{f}_{\mu\nu}dx^\mu\wedge dx^\nu \\ \text{with } \mathbf{f}_{\mu\nu} = \partial_\mu\mathbf{a}_\nu - \partial_\nu\mathbf{a}_\mu \end{cases}, \tag{60}$$

$$\begin{cases} \mathbf{h}^{\cdot}. = \nabla\mathbf{b}^{\cdot}. = \frac{1}{2}\mathbf{h}_{\mu\nu}dx^\mu\wedge dx^\nu = (\mathbf{h}^j_k) = \mathbf{h}^a\frac{\tau_a}{2} \\ \text{with } \mathbf{h}^j_k = \mathbf{db}^j_k - i\mathbf{b}^j_m\wedge\mathbf{b}^m_k, \text{ i.e. } \mathbf{h}^j_{k\mu\nu} = \partial_\mu\mathbf{b}^j_{k\nu} - \partial_\nu\mathbf{b}^j_{k\mu} - i[\mathbf{b}_\mu,\mathbf{b}_\nu]^j_k, \\ \text{or } \mathbf{h}_{\mu\nu} = \partial_\mu\mathbf{b}_\nu - \partial_\nu\mathbf{b}_\mu - i[\mathbf{b}_\mu,\mathbf{b}_\nu], \text{ i.e. } \mathbf{g}^a_{\mu\nu} = \partial_\mu\mathbf{b}^a_\nu - \partial_\nu\mathbf{b}^a_\mu + \epsilon_{abc}\mathbf{b}^b_\mu\mathbf{b}^c_\nu \end{cases} \tag{61}$$

[19] We denoted covariant derivatives by the boldfaced letter $\mathbf{D}$ to avoid confusion with the Dirac operator. We recall that $\Sigma = \frac{1}{2}(\mu_u + \mu_d)$ with $\mu_u = M_u M_{u^*}$ and $\mu_d = M_d M_{d^*}$.

The usual combined $U(1)\times SU(2)$-covariant derivatives

$$\begin{cases}\mathbf{D}\phi^{\cdot}:\mathbf{D}\phi^{j}=\mathbf{d}\phi^{j}+i(\mathbf{a}\phi^{j}-\mathbf{b}^{j}_{k}\phi^{k}),\ \text{or}\ \mathbf{D}\phi^{\cdot}=\mathbf{d}\phi^{\cdot}+i(\mathbf{a}-\mathbf{b}^{a}\frac{\tau_i}{2})\phi^{\cdot}\\ \mathbf{D}\phi_{.}:\mathbf{D}\phi_{j}=\mathbf{d}\phi_{j}-i(\mathbf{a}\phi_{j}-\mathbf{b}^{k}_{j}\phi_{k}),\ \text{or}\ \mathbf{D}\phi_{.}=\mathbf{d}\phi_{.}-i\phi_{.}(\mathbf{a}-\mathbf{b}^{a}\frac{\tau_i}{2})\end{cases}\tag{62}$$

has the following relationshiop with the above quaternion covariant derivatives: for $\phi=\begin{pmatrix}\bar{\phi}^2 & \phi^1\\ -\bar{\phi}^1 & \phi^2\end{pmatrix}\leftrightarrow\phi^{\cdot}=(\phi^1,\phi^2)\leftrightarrow\phi_{.}=(\bar{\phi}^1,\bar{\phi}^2)$ one has:

$$\mathbf{D}\phi=\mathbf{d}\phi+i(\phi(\mathbf{a}^{\cdot}.)-(\mathbf{b}^{\cdot}.)\phi)=\begin{pmatrix}\mathbf{D}\phi_2 & \mathbf{D}\phi^1\\ -\mathbf{D}\phi_1 & \mathbf{D}\phi^2\end{pmatrix}.\tag{63}$$

We now describe the simpler chromodynamics sector.

[9] *The chromodynamics sector.* We now tensorize the Dirac K-cycle by the chromodynamics finite-space K-cycle; this yields the chromodynamics *-algebra (over the reals):

$$\begin{aligned}B&=C^{\infty}(\mathbf{M},\mathcal{R})\otimes\mathbf{B}_{chr}=C^{\infty}(\mathbf{M},\mathbf{C})\oplus C^{\infty}(\mathbf{M},M_3(\mathbf{C}))\\&=\{\mathbf{b}=(f',m);\ f'\in C^{\infty}(\mathbf{M},\mathbf{C}),m\in C^{\infty}(\mathbf{M},M_3(\mathbf{C}))\}.\end{aligned}\tag{64}$$

The operator $\mathbf{D}$ is that specified by its action $\mathbf{D}_q$ on $\mathbf{H}_q$ in (44), yielding by leptonic reduction the action $\mathbf{D}_l$ on $\mathbf{H}_l$. The action of B is as follows: on the quark space $\mathbf{H}_q$:

$$\pi_q((f',m)=(m^i_k))=\pi_q(0,m)=1_{ew}\otimes\underline{m}$$

$$=\begin{array}{c}\begin{matrix}u_R & d_R & u_L & d_L\end{matrix}\\ \begin{pmatrix}\underline{m}^j_k & 0 & 0 & 0\\ 0 & \underline{m}^j_k & 0 & 0\\ 0 & 0 & \underline{m}^j_k & 0\\ 0 & 0 & 0 & \underline{m}^j_k\end{pmatrix}\end{array}\otimes e^k_j\ \begin{matrix}u_R\\ d_R\\ u_L\\ d_L\end{matrix},\tag{65}$$

(here we denoted by 1_3, and e^k_j the unit, resp. the matrix units of $M_3(\mathbf{C})$); and on the leptonic space $\mathbf{H}_1$:

$$\pi_1((f',m))=\pi_1((f',0))=\begin{array}{c}\begin{matrix}e_R & \nu_L & e_L\end{matrix}\\ \begin{pmatrix}\underline{f}'\otimes 1_N & 0 & 0\\ 0 & \underline{f}'\otimes 1_N & 0\\ 0 & 0 & \underline{f}'\otimes 1_N\end{pmatrix}\end{array}\begin{matrix}e_R\\ \nu_L\\ e_L\end{matrix}.\tag{66}$$

Note the important fact that *the discrete (antidiagonal) part of* $\mathbf{D}$ *commutes with the action of* B, this implying that the gluon field is an $SU(3)$-gauge field; and also that the full standard model space is a 4-dimensional noncommutative Poincaré manifold.

We now describe the chromodynamics part of the curvature. We have the quark and leptonic components:

$$\theta_q = \mathbf{1}_{ew} \otimes i\underline{\gamma}(\mathbf{g}) = \begin{array}{c} \begin{matrix} u_R & d_R & u_L & d_L \end{matrix} \\ \begin{pmatrix} i\gamma(\mathbf{g}_k^j) \otimes \mathbf{1}_N & 0 & 0 & 0 \\ 0 & - & 0 & 0 \\ 0 & 0 & - & 0 \\ 0 & 0 & 0 & - \end{pmatrix} \otimes e_j^k \end{array} \begin{matrix} u_R \\ d_R \\ u_L \\ d_L \end{matrix}, \qquad (67)$$

$$\theta_1 = \mathbf{1}_{ew} \otimes i\underline{\gamma}(\mathbf{f}') = \begin{array}{c} \begin{matrix} e_R & \nu_L & e_L \end{matrix} \\ \begin{pmatrix} i\gamma(\mathbf{f}') & 0 & 0 \\ 0 & - & 0 \\ 0 & 0 & - \end{pmatrix} \end{array} \begin{matrix} e_R \\ \nu_L \\ e_L \end{matrix}. \qquad (68)$$

In order to isolate the $U(1)$-part of the group $U(3)$, we rewrite (65) and (67) according to the following decomposition of 3×3 matrices

$$m = m_k^j e_j^k = m^0 \mathbf{1} + m^a \frac{\lambda_a}{2} \quad \text{(summation over repeated indices)}, \quad m \in M_3(\mathbf{C}), \qquad (69)$$

where the $\lambda_a, a = 1, ..., 8$ are the usual Gell-Mann matrices. This yields:

$$\pi_q(f', m) = \begin{array}{c} \begin{matrix} u_R & d_R & u_L & d_L \end{matrix} \\ \begin{pmatrix} \gamma(c^0) \otimes 1_N & 0 & 0 & 0 \\ 0 & - & 0 & 0 \\ 0 & 0 & - & 0 \\ 0 & 0 & 0 & - \end{pmatrix} \otimes \mathbf{1}_3 \end{array}$$

$$+ \begin{array}{c} \begin{matrix} u_R & d_R & u_L & d_L \end{matrix} \\ \begin{pmatrix} \gamma(c^a) \otimes 1_N & 0 & 0 & 0 \\ 0 & - & 0 & 0 \\ 0 & 0 & - & 0 \\ 0 & 0 & 0 & - \end{pmatrix} \otimes \frac{\lambda_a}{2} \end{array} \begin{matrix} u_R \\ d_R \\ u_L \\ d_L \end{matrix}, \qquad (65a)$$

$$\theta_q = \begin{array}{c} \begin{matrix} u_R & d_R & u_L & d_L \end{matrix} \\ \begin{pmatrix} \gamma(g^0) \otimes 1_N & 0 & 0 & 0 \\ 0 & - & 0 & 0 \\ 0 & 0 & - & 0 \\ 0 & 0 & 0 & - \end{pmatrix} \otimes \mathbf{1}_3 \end{array}$$

$$+ \begin{array}{c} \begin{matrix} u_R & d_R & u_L & d_L \end{matrix} \\ \begin{pmatrix} \gamma(g^a) \otimes 1_N & 0 & 0 & 0 \\ 0 & - & 0 & 0 \\ 0 & 0 & - & 0 \\ 0 & 0 & 0 & - \end{pmatrix} \otimes \frac{\lambda_a}{2} \end{array} \begin{matrix} u_R \\ d_R \\ u_L \\ d_L \end{matrix}. \qquad (67a)$$

[10] *Combination of the electroweak and chromodynamics sectors. Modularity condition. Yang-Mills action. Fermionic action.* Putting together the results of the two last paragraphs is not satisfactory owing to two related unwanted features: the gauge group is $U(1)\times SU(2)\times U(1)\times U(1)\times SU(2)$, and the number of gauge fields is in excess by two. These two drawbacks can, however, be removed by imposing a "modularity condition" [1] leading to the following coalescence of the gauge fields $\mathbf{a}, \mathbf{a}'$ and $\mathbf{c}^0$:

$$\begin{cases} \mathbf{a}' = \mathbf{a} \\ \mathbf{c}^0 = -\frac{1}{3}\mathbf{a} \end{cases}, \tag{70}$$

with, correspondingly:

$$\begin{cases} \mathbf{f}' = \mathbf{f} \\ \mathbf{g}^0 = -\frac{1}{3}\mathbf{f} \end{cases}. \tag{71}$$

Effecting the identification (71) after addition of the electroweak and the chromodynamic curvatures then yields the following Yang-Mills action. The Yang-Mills action corresponding to the leptonic, resp. quark percentages $\alpha_1 = (1+x)/2$ and $\alpha_q = (1-x)/2$ equals

$$YM = \alpha_1 \tau_{D_1}(\theta_1^* \theta_1) + \alpha_q \tau_{D_q}(\theta_q^*, \theta_q) = \frac{1}{2}\pi^2 \int L_B \, dv, \tag{72}$$

with the Lagrangian density

$$\begin{aligned} L_B = &-\tfrac{1}{2}(1-x)N\mathbf{g}^a_{\mu\nu}\mathbf{g}^{\mu\nu}_a - \tfrac{1}{3}(10-x)N\mathbf{f}_{\mu\nu}\mathbf{f}^{\mu\nu} - \tfrac{1}{4}(2-x)N\mathbf{h}^s_{\mu\nu}\mathbf{h}^{\mu\nu}_s \\ &+2L(\mathbf{D}\phi_j)(\mathbf{D}\phi^j) + K(\phi_i\phi^i - 1)^2, \end{aligned} \tag{73}$$

where one has

$$\begin{cases} L = \mathrm{Tr}[\alpha_1\mu_e + 3\alpha_q(\mu_u + \mu_d)] \simeq \frac{3}{2}(1-x)m_t^2 \\ K = \frac{3}{2}\mathrm{Tr}[\alpha_1\mu_e^2 + 3\alpha_q(\mu_u^2 + \mu_d^2)] + 3\alpha_q\mathrm{Tr}[\mu_u\mu_d] \\ \quad -N^{-1}[2^{-1}(\alpha_e + 3\alpha_q)^{-1} + (\alpha_e + 6\alpha_q)] \\ \quad \simeq \frac{9}{4}(1-x)[1 - \frac{1}{2}\frac{3x^2-8x+5}{5x^2-17x+14}]m_t^4 \end{cases}, \tag{74}$$

and $\mu_u = M_u M_u^*, \mu_d = M_d M_d^*, \mu_e = M_e M_e^*, M_u M_d, M_e$ the respective mass matrices of the leptons, the upper quarks, and the lower quarks (we have indicated the approximate values of L and K in terms of the top quark mass m_t supposed dominant, for $x = 0$ and $N = 3$).

The fermionic action is

$$\frac{1}{2\pi^2}\int \mathcal{L}_F dv \tag{75}$$

with the ansatz

$$\mathcal{L}_F = \mathcal{L}_{F1} + \mathcal{L}_{Fq} = (\psi \mathcal{D}_{\nabla 1}\psi) + (\mathcal{Q}\mathcal{D}_{\nabla_q}\mathcal{Q}), \tag{76}$$

specified as follows:
– the leptonic field $\psi \in \mathbf{H}_1$, is

$$\psi = \psi = (\psi^R, \psi^L) \text{ with } \begin{cases} \psi^R = (\psi_f^R)_{f=1,\ldots,N} \in H_1^R \otimes \mathbf{C}^N \\ \psi^L = (\psi_f^{L1}, \psi_f^{L2})_{f=1,\ldots,N} \in H_1^L \otimes \mathbf{C}^2_{iso} \otimes \mathbf{C}^N \end{cases}; \tag{77}$$

– the quark field $\mathcal{Q} \in \mathbf{H}_q$:

$$\mathcal{Q} = (\mathcal{Q}^R, \mathcal{Q}^L) = (\mathcal{Q}_f^{Ru}, \mathcal{Q}_f^{Rd}, \mathcal{Q}_f^{L1}, \mathcal{Q}_f^{L2})_{f=1,\ldots,N}, \tag{78}$$

with

$$\begin{cases} \mathcal{Q}_f^{Ru} = \mathcal{Q}_f^{Rum} \otimes e_m \\ \mathcal{Q}_f^{Rd} = \mathcal{Q}_f^{Rdm} \otimes e_m \\ \mathcal{Q}_f^{L1} = \mathcal{Q}_f^{L1m} \otimes e_m \\ \mathcal{Q}_f^{L2} = \mathcal{Q}_f^{L2m} \otimes e_m \end{cases} \tag{79}$$

(summation on m from 1 to 3), with $\{e_m\}$ the canonical basis of $\mathbf{C}^3_{color}$. We recall that f is the fermion-family index, R and L stand respectively for right and left, and the spinor fields correspond to the following particle types:

$$\begin{cases} \text{Leptons: } \psi^R : e_R & \psi_f^{L1} : \nu_L \ \psi_f^{L2} : e_R \\ \text{Quarks: } \mathcal{Q}_f^{Ru} : u_R \ \mathcal{Q}_f^{Rd} : d_R & \mathcal{Q}_f^{L1} : u_L \ \mathcal{Q}_f^{L2} : d_L \end{cases}. \tag{80}$$

This leads to the fermionic Lagrangian $\mathcal{L}_F = \mathcal{L}_{F1} + \mathcal{L}_{Fq}$, with leptonic term:

$$\begin{aligned} \mathcal{L}_{F1} = & \textstyle\sum_{f=1,\ldots,N}\{\bar{\psi}_f^R i\gamma^\mu \mathbf{D}_\mu^{R1}\psi_f^R + \bar{\psi}_f^L i\gamma^\mu \mathbf{D}_\mu^{R1}\psi_f^L\} \\ & + \textstyle\sum_{f_1 f_2=1,\ldots,N}\{M^*_{ef_1 f_2}\bar{\psi}_{f_1}^R \bar{\phi}\gamma^5\psi_{f_2}^L + M_{ef_1 f_2}\bar{\psi}_{f_1}^L \phi\gamma^5\psi^R\} \end{aligned} \tag{81}$$

with the covariant derivatives

$$\begin{cases} \mathbf{D}_\mu^{R1} = \nabla_\mu - 2i\mathbf{a}_\mu \\ \mathbf{D}_\mu^{L1} = \nabla_\mu - i\mathbf{a}_\mu i\mathbf{b}_\mu = \nabla_\mu - i\mathbf{a}_\mu - i\mathbf{b}_\mu^i \frac{\tau_i}{2} \end{cases}, \tag{82}$$

and the quark term

$$\begin{aligned} \mathcal{L}_{Fq} = & \textstyle\sum_{f=1,\ldots,N}\{\bar{\mathcal{Q}}_{fm}^{Ru} i\gamma^\mu \mathbf{D}_\mu^{Ru}\mathcal{Q}_f^{Rum} + \bar{\mathcal{Q}}_{fm}^{Rd} i\gamma^\mu \mathbf{D}_\mu^{Rd}\mathcal{Q}_f^{Rdm} \\ & + \bar{\mathcal{Q}}_{fm}^{L} i\gamma^\mu \mathbf{D}_\mu^{Lq}\mathcal{Q}_f^{Lm}\} \\ & + \textstyle\sum_{f_1 f_2=1,\ldots,N}\{M^*_{df_1 f_2}\bar{\mathcal{Q}}_{f_1 m}^{Rd}\bar{\phi}\gamma^5\mathcal{Q}_{f_2}^{Lm} + M_{ef_1 f_2}\bar{\mathcal{Q}}_{f_1 m}^{L}\phi\gamma^5\mathcal{Q}_{f_2}^{Rdm}\} \\ & + \textstyle\sum_{f_1 f_2=1,\ldots,N}\{M^*_{uf_1 f_2}\bar{\mathcal{Q}}_{f_1 m}^{Ru}\bar{\tilde{\phi}}\gamma^5\mathcal{Q}_{f_2}^{Lm} + M_{uf_1 f_2}\bar{\mathcal{Q}}_{m f_1}^{L}\tilde{\phi}\gamma^5\mathcal{Q}_{f_2}^{Rum}\} \end{aligned} \tag{83}$$

with the covariant derivatives

$$\begin{cases} \mathbf{D}_\mu^{Ru} = \nabla_\mu + \frac{4}{3}i\mathbf{a}_\mu - i\mathbf{c}_\mu = \nabla_\mu + \frac{4}{3}i\mathbf{a}_\mu - i\mathbf{c}_\mu^a\frac{\lambda_a}{2} \\ \mathbf{D}_\mu^{Rd} = \nabla_\mu - \frac{2}{3}i\mathbf{a}_\mu - i\mathbf{c}_\mu = \nabla_\mu - \frac{2}{3}i\mathbf{a}_\mu - i\mathbf{c}_\mu^a\frac{\lambda_a}{2} \\ \mathbf{D}^{Lq} = \nabla_\mu - i\mathbf{b}_\mu + \frac{1}{3}i\mathbf{a}_\mu - i\mathbf{c}_\mu = \nabla_\mu - i\mathbf{b}_\mu^i\frac{\tau_i}{2} + \frac{1}{3}i\mathbf{a}_\mu - i\mathbf{c}_\mu^a\frac{\lambda_a}{2} \end{cases}. \tag{84}$$

References

[1] A. Connes, The metric aspects of non-commutative geometry, Preprint *Collège de France* (1991).

[2] D. Kastler, A detailed account of Alain Connes' version of the standard model in non-commutative geometry III.

[3] D. Kastler and Th. Schücker, Remarks on Alain Connes' approach to the standard model in non-commutative geometry, *Theoret. Math. Phys.* **92**, No.3, 523

[4] D. Kastler and D. Testard, Quantum forms of tensor products, to appear in *Commun.Math.Phys.*

Supersymmetric Extension of Quantum Scalar Field Theories

ASAO ARAI
Department of Mathematics, Hokkaido University, Sapporo 060, Japan

Dedicated to Professor Huzihiro Araki on the occasion of his 60th birthday

Abstract. A family of supersymmetric extensions of a class of quantum scalar field theories is constructed in an abstract form.

Key words: quantum field theory, supersymmetry, infinte dimensional analysis, De Rham operator, Dirac operator, Laplacian

1. Introduction

Supersymmetries are symmetries which relate bosons to fermions. Since Wess and Zumino [33,34] introduced a formalism of supersymmetric quantum field theory (SSQFT), numbers of researches have been made on supersymmetries and related aspects (see references in [18,32]). In these developments, there has been a growing belief that supersymmetry may play a fundamental role in unifying all the forces between elementary particles. As one of the evidences that support such a belief, one may mention the fact that (partial) cancelations of ultraviolet divergences occur in some models of SSQFT.

On the other hand, we have ordinary non-supersymmetric quantum theories, which also have played important roles in the developments of quantum physics. From the above mentioned point of view on supersymmetry in connection with the unification of the interactions between elementary particles, it would be natural to ask if such ordinary non-supersymmetric quantum theories have something to do with supersymmetry. This question leads us to investigate "hidden" supersymmetric structures that may exist in non-supersymmetric quantum theories. As for non-relativistic quantum mechanical models whose Hamiltonians are given by Schrödinger operators, it has been shown that they have supersymmetric (SUSY) extensions, which, roughly speaking, means that they can be regarded as part of supersymmetric quantum theories or they can be "embedded" in supersymmetric quantum theories ([4,6] and references therein). In the case of quantum field

H. Araki et al. (eds.), Quantum and Non-Commutative Analysis, 73–90.

theories, however, only a few examples are known to have SUSY extensions [4]. In the previous work [3], we outlined in a heuristic way how to construct a SUSY extension of a class of quantum scalar field theories. In this paper we give a mathematically rigorous formulation of the method suggested in [3] and show in an abstract form the existence of a family of SUSY extensions of a class of quantum scalar field models. In concrete realizations, the class contains models in two dimensions such as spatially cutoff $P(\phi)_2$ theories (e.g.,[19, 28-31]). The technical details of results in the present paper will be given in a separate paper [15].

In Section 2 we review (non-relativistic) supersymetric quantum theory in an abstract form and give a definition of SUSY extension of a quantum theory. In Section 3 we define a class of quantum scalar field models in an abstract form, where we use the Q-space formalism to describe Boson Fock space. Section 4 is devoted to analysis of further properties of the models. In Section 5 we present an infinite dimensional exterior differential calculus. We introduce operators of de Rham type and Laplacians associated with them, and summarize some of their properties. In the final section, developing further the mathematical framework given in Section 5, we construct a supersymmetric quantum field theory in an abstract form and show that the quantum scalar field models have SUSY extensions.

2. Supersymmetry and Supersymmetric Extension

In this section we review an abstract formulation of (non-relativistic) supersymmetric quantum theory [1,2,4,17,20,21,35] and introduce a notion of supersymmetric extension [4].

Definition 2.1. Let N be a positive integer. A supersymmetric quantum theory (SSQT) with N supersymmetry is a quadruple $\{\mathcal{X}, \{Q_n\}_{n=1}^N, H, N_F\}$ consisting of a Hilbert space $\mathcal{X}$, a set of self-adjoint operators $\{Q_n\}_{n=1}^N$ ("supercharges"), self-adjoint operators H ("supersymmetric Hamiltonian") and N_F ("Fermion number operator") in $\mathcal{X}$ with the following properties:

(S.1) $N_F^2 = I$ ($N_F \neq \pm I$), where I denotes identity.

(S.2)
$$H = Q_1^2 = \cdots = Q_N^2.$$

(S.3) For all $n = 1, \cdots, N$, N_F leaves the domain $D(Q_n)$ of Q_n invariant and
$$Q_n N_F + N_F Q_n = 0$$
on $D(Q_n)$.

(S.4) For $m \neq n$, Q_m and Q_n anticommute in the sense of quadratic form:
$$(Q_m\Psi, Q_n\Phi) + (Q_n\Psi, Q_m\Phi) = 0, \qquad \Psi, \Phi \in D(Q_m) \cap D(Q_n),$$
$$n \neq m, m, n = 1, \cdots, N,$$
where $(\cdot\,, \cdot)$ denotes the inner product of $\mathcal{X}$.

Remarks. (i) Property (S.1) means that N_F is a grading operator on $\mathcal{X}$, so that $\mathcal{X}$ admits the orthogonal decomposition

$$\mathcal{X} = \mathcal{X}_+ \oplus \mathcal{X}_-,$$

where $\mathcal{X}_\pm$ is the eigenspace of N_F with eigenvalue ± 1. Physically the Hilbert space $\mathcal{X}_+$ (resp. $\mathcal{X}_-$) is called the space of "bosonic" (resp. "fermionic") states.

(ii) It follows from (S.2) that H is nonnegative, i.e., $H \geq 0$. In the formulation of relativisitic SSQT's, (S.2) must be replaced by a more complicated anticommutation relations (cf. [18,32]).

(iii) Property (S.3) implies that, for each n, there exists a uniuqe densely defined closed linear operator $Q_{n,+} : \mathcal{X}_+ \to \mathcal{X}_-$ such that

$$Q_n = \begin{pmatrix} 0 & Q_{n,+}^* \\ Q_{n,+} & 0 \end{pmatrix},$$

so that we have by (S.2)

$$H = \begin{pmatrix} H_+ & 0 \\ 0 & H_- \end{pmatrix}$$

with $H_+ = Q_{n,+}^* Q_{n,+}$ and $H_- = Q_{n,+} Q_{n,+}^*$, which show that H is reduced by $\mathcal{X}_\pm$.

(iv) The supersymmetry is said to be *broken* if the supersymmetric Hamiltonian has no zero-energy states, i.e., $\operatorname{Ker} H = \{0\}$ [35,36]. A necessary condition for the supersymmetry breaking is given in terms of the "Witten index"

$$I_W := \dim \operatorname{Ker} H_+ - \dim \operatorname{Ker} H_-.$$

Indeed, it is obvious that, if the supersymmetry is broken, then $I_W = 0$. It is easy to see that

$$I_W = \operatorname{index}(Q_{n,+}) := \dim \operatorname{Ker} Q_{n,+} - \dim \operatorname{Ker} Q_{n,+}^*.$$

Thus a SSQT is related to index theory. Index theorems in connection with SSQT models have been discussed (e.g., [5,12,14,17,21,22] and references therein).

(v) The abstract formulation of SSQT is useful to study general mathematical aspects of SSQT's (see, e.g., [1,2,4,9,17,20-22,35]).

Definition 2.2. Let L be a self-adjoint operator in a Hilbert space $\mathcal{H}$. We say that L has a supersymmetric (SUSY) extension if there exist a SSQT $\{\mathcal{X}, \{Q_n\}_{n=1}^N, H, N_F\}$, a closed subspace $\mathcal{M}$ in $\mathcal{X}$ which reduces H, and a unitary operator $U : \mathcal{H} \to \mathcal{M}$ such that

$$ULU^{-1} = H|\mathcal{M},$$

where $H|\mathcal{M}$ denotes the reduced part of H in $\mathcal{M}$. In this case, we call the SSQT $\{\mathcal{X}, \{Q_n\}_{n=1}^N, H, N_F\}$ a SUSY extension of L.

Remark. For a self-adjoint operator to have a SUSY extension, it must be nonnegative, because a SUSY Hamiltonian is nonnegative (Property (S.2) in Definition 2.1) and so is any reduced part of it. A SUSY extension of a self-adjoint operator is not unique in general (cf. [4]). Typical examples for which one may consider SUSY extensions are Schrödinger operators in both finite and infinite dimensions. For the case of Schrödinger operators on $\mathbf{R}^n$, see [4] and references therein.

3. A Class of Quantum Scalar Field Models

We first review the Q-space formalism to describe Boson Fock space (e.g., [26, §X.7], [31, §1.3]). Let $\mathcal{H}$ be a real separable Hilbert space with inner product $(\cdot, \cdot)_{\mathcal{H}}$ and norm $\|\cdot\|_{\mathcal{H}}$. Let $\{\phi(f)|f \in \mathcal{H}\}$ be the Gaussian mean zero random process indexed by $\mathcal{H}$, i.e., $\{\phi(f)|f \in \mathcal{H}\}$ is a family of random variables on a probability measure space $(E, \mathbf{B}, \mu_0)$ such that the mapping $:f \to \phi(f)$ is linear,

$$\int_E e^{i\phi(f)} d\mu_0 = e^{-\|f\|_{\mathcal{H}}^2/2}, \quad f \in \mathcal{H},$$

and the Borel field $\mathbf{B}$ is generated by $\{\phi(f)|f \in \mathcal{H}\}$. The Hilbert space $L^2(E, d\mu_0)$ is unitarily equivalent in a "canonical" way to the Boson Fock space over $\mathcal{H}_c$ (e.g., [31, §1.3]), where $\mathcal{H}_c$ is the complexification of $\mathcal{H}$.

The Wick product $: \phi(f_1) \cdots \phi(f_n) :$ of the random variables $\phi(f_j), j = 1, \ldots, n$, is defined by the following recursion relations:

$$\begin{aligned} :\phi(f_1): &= \phi(f_1), \\ :\phi(f_1)\cdots\phi(f_n): &= \phi(f_1) :\phi(f_2)\cdots\phi(f_n): \\ &\quad - \sum_{j=2}^n (f_1, f_j)_{\mathcal{H}} :\phi(f_2)\cdots\check{\phi}(f_j)\cdots\phi(f_n):, \end{aligned}$$

where $\check{\phi}(f_j)$ indicates the omission of $\phi(f_j)$. Let $\Gamma_n(\mathcal{H})$ be the closed subspace generated by $: \phi(f_1) \cdots \phi(f_n) :$, $f_j \in \mathcal{H}, j = 1, \cdots, n$ $(\Gamma_0(\mathcal{H}) := \mathbf{C})$. Then we have the orthogonal decomposition

$$L^2(E, d\mu_0) = \bigoplus_{n=0}^{\infty} \Gamma_n(\mathcal{H}).$$

Let A be a strictly positive self-adjoint operator in $\mathcal{H}$. Then the *second quantization* $d\Gamma_b(A)$ of A is defined as the self-adjoint operator in $L^2(E, d\mu_0)$

which is reduced by each $\Gamma_n(\mathcal{H})$ with

$$d\Gamma_b(A)|\Gamma_0(\mathcal{H}) = 0,$$
$$d\Gamma_b(A) : \phi(f_1)\cdots\phi(f_n) : = \sum_{j=1}^{n} : \phi(f_1)\cdots\phi(Af_j)\cdots\phi(f_n) :,$$
$$f_j \in D(A), j = 1,\cdots,n,\ n \geq 1.$$

(e.g., [31,§1.4], [25, §VIII.10], [26, §X.7].) We set

$$H_0 := d\Gamma_b(A),$$

which describes in an abstrcat form the Hamiltonian of a free quantum scalar field with one-particle Hamiltonian A.

Let V be a real-valued measurable function on $(E, \mathbf{B})$ such that

(V.1) $V \in L^p(E, d\mu_0)$ for some $p > 2$;

(V.2) $e^{-tV} \in L^1(E, d\mu_0)$ for all $t \in \mathbf{R}$.

The quantum scalar field Hamiltonians we are concerned with are given in an abstract form by

$$H_V = H_0 + V.$$

The following result is known.

Theorem 3.1. (Segal [28,29], Simon-Hoegh-Krohn [30]). *The operator H_V is essentially self-adjoint on $D(H_0) \cap D(V)$ and bounded from below.*

By Theorem 3.1, the closure $\overline{H}_V$ of H_V is self-adjoint and

$$E_0 = \inf \sigma(\overline{H}_V)$$

is finite, where $\sigma(\cdot)$ denotes spectrum. Our purpose is to show that the nonnegative self-adjoint operator

$$\hat{H}_V := \overline{H}_V - E_0$$

has a SUSY extension. To do that, however, we need to know some further properties of $\overline{H}_V$.

An eigenvector of $\overline{H}_V$ with eigenvalue E_0 is said to be a ground state of $\overline{H}_V$. As for the existence and uniqueness of ground state of $\overline{H}_V$, we have the following.

Theorem 3.2. (Simon-Hoegh-Krohn [30]). *There exists a unique (up to constant multiples) ground state Ω of $\overline{H}_V$ with the following properties:*

(1) $\Omega > 0$ *a.e.*;

(2) $\Omega \in \bigcap_{0<q<\infty} L^q(E, d\mu_0)$;

(3) $\Omega \in D(H_0)$.

Another important property of $\overline{H}_V$ is the following:

Theorem 3.3. *For all $t > 0$, $\exp(-t\overline{H}_V)$ is positivity preserving, i.e., if $\Psi \geq 0$(a.e.), $\Psi \in L^2(E, d\mu_0)$, then $\exp(-t\overline{H}_V)\Psi \geq 0$ (a.e.).*

Proof. It is known that $\{e^{-tH_0}\}_{t\geq 0}$ is a hypercontractive semi-group [26, §X.9, Theorem X.61]. For each $n = 1, 2, \cdots$, we define V_n by $V_n = V$ if $V \geq -n$ and $V_n = -n$ if $V \leq -n$. Let $H_n = H_0 + V_n$. It is easy to see that $V_n \to V$ in $L^p(E, d\mu_0)$ as $n \to \infty$ and $\|e^{-tV_n}\|_{L^1(E,d\mu_0)}$ is uniformly bounded in n for each $t > 0$ fixed. Hence we can apply Theorem 2.21 in [30] to conclude that, for all $t > 0$,

$$e^{-tH_n} \to e^{-t\overline{H}_V} \tag{3.1}$$

in norm as maps of $L^2(E, d\mu_0)$ to $L^2(E, d\mu_0)$. We can apply the Trotter product formula [25, Theorem VIII.31] to e^{-tH_n} to obtain

$$e^{-tH_n} = \mathrm{s} - \lim_{N\to\infty} \left(e^{-tH_0/N} e^{-tV_n/N}\right)^N. \tag{3.2}$$

We recall that e^{-tH_0} is positivity preserving [31, §I.4, Theorem I.16]. Hence (3.2) implies that e^{-tH_n} is positivity preserving. Using this result and (3.1), we conclude that $\exp(-t\overline{H}_V)$ is positivity preserving. □

4. Perturbation of the Measure μ_0

Let $C_b^\infty(\mathbf{R}^n)$ be the space of infinitely differentiable functions on $\mathbf{R}^n$ with all the derivatives being bounded on $\mathbf{R}^n$. Given a linear operator T from $\mathcal{H}_c$ to another space, we define $D_{b,T}^\infty$ to be the subspace algebraically spanned by functions of the form

$$F(f_1, \cdots, f_n) := F(\phi(f_1), \cdots, \phi(f_n)), F \in C_b^\infty(\mathbf{R}^n), f_j \in D(T) \cap \mathcal{H}, n \geq 0.$$

If $D(T) \cap \mathcal{H}$ is dense in $\mathcal{H}$, then $D_{b,T}^\infty$ is dense in $L^2(E, d\mu_0)$. For the case where T is the idenity I on $\mathcal{H}_c$, we write

$$D_{b,I}^\infty = D_b^\infty.$$

For a Hilbert space $\mathcal{M}$, we denote by $L^2(E, d\mu_0; \mathcal{M})$ the Hilbert space of $\mathcal{M}$-valued square integrable functions on $(E, \mathbf{B}, \mu_0)$, which is identified with $L^2(E, d\mu_0) \otimes \mathcal{M}$. For two vector spaces $\mathcal{V}$ and $\mathcal{W}$, we denote their algebraic tensor product by $\mathcal{V} \hat{\otimes} \mathcal{W}$.

We define the gradient operator $\nabla : L^2(E, d\mu_0) \to L^2(E, d\mu_0; \mathcal{H}_c)$ with domain D_b^∞ by

$$\nabla F(f_1, \cdots, f_n) = \sum_{j=1}^n (\partial_j F)(f_1, \cdots, f_n) f_j$$

for $F(f_1,\cdots,f_n) \in D_b^\infty$ and extending it by linearity to all functions in D_b^∞, where $(\partial_j F)(x) = \partial F(x)/\partial x_j, j = 1,\cdots,n, x = (x_1,\cdots,x_n) \in \mathbf{R}^n, F \in C_b^\infty(\mathbf{R}^n)$.

Lemma 4.1. *The operator ∇ is well defined and closable. Moreover,*

$$D(\nabla^*) \supset D_b^\infty \hat{\otimes} \mathcal{H}_c \tag{4.1}$$

and

$$\nabla^*(\Psi \otimes f) = -(J_{\mathcal{H}} f, \nabla\Psi)_{\mathcal{H}_c} + \phi(f)\Psi, \quad \Psi \in D_b^\infty, f \in \mathcal{H}_c, \tag{4.2}$$

where $J_{\mathcal{H}}$ is the natural conjugation on $\mathcal{H}_c$.

Proof. The well-definedness of ∇ can be proven by using the following integration by parts formula with respect to (w.r.t.) μ_0: For all $\Psi, \Phi \in D_b^\infty, f \in \mathcal{H}_c$,

$$\int_E (f, \nabla\Psi)_{\mathcal{H}_c} \Phi d\mu_0 = -\int_E \Psi(f, \nabla\Phi)_{\mathcal{H}_c} d\mu_0 + \int_E \phi(J_{\mathcal{H}} f)\Psi\Phi d\mu_0. \tag{4.3}$$

This formula can be proven in the same way as in the proof of [19, Theorem 6.3.1]). Formula (4.3) also gives (4.1) and (4.2). Since $D_b^\infty \hat{\otimes} \mathcal{H}_c$ is dense in $L^2(E, d\mu_0; \mathcal{H}_c)$ and hence so is $D(\nabla^*)$, it follows that ∇ is closable. □

For each $f \in \mathcal{H}_c$, we introduce an operator $\tilde{\nabla}_f$ with domain D_b^∞ by

$$\tilde{\nabla}_f \Psi = (f, \nabla\Psi)_{\mathcal{H}_c}, \quad \Psi \in D_b^\infty.$$

In the same way as in the proof of Lemma 4.1, we can prove the following fact.

Lemma 4.2. *For all $f \in \mathcal{H}_c$, $\tilde{\nabla}_f$ is closable. Moreover,*

$$\tilde{\nabla}_f^* \Psi = \nabla^*(\Psi \otimes f), \quad \Psi \in D_b^\infty, f \in \mathcal{H}_c$$

We denote the closure of ∇ and of $\tilde{\nabla}_f$ by the same symbol, respectively.

Let Ω be as in Theorem 3.2. Without loss of generality, we can assume that Ω is normalized:

$$\int_E \Omega^2 d\mu_0 = 1.$$

Then we can define a probability measure μ on $(E, \mathbf{B})$ by

$$d\mu = \Omega^2 d\mu_0.$$

Lemma 4.3. (Integration by parts formula w.r.t.μ) *For all $f \in \mathcal{H}$, $\Omega \in D(\tilde{\nabla}_f)$ and*

$$\int_E \tilde{\nabla}_f \Psi d\mu = \int_E \lambda_f \Psi d\mu, \quad \Psi \in D_b^\infty, f \in \mathcal{H}, \tag{4.4}$$

where

$$\lambda_f = \phi(f) - \frac{2\tilde{\nabla}_f \Omega}{\Omega}.$$

Proof (Outline). Throughout the proof, $(\cdot\,,\,\cdot)$ and $||\cdot||$ denote the inner product and the norm of $L^2(E, d\mu_0)$, respectively. We first show that $D^\infty_{b,A}$ is a core of H_0. Then, by Theorem 3.2(3), there exists a sequence $\{\Omega_n\}_{n=1}^\infty$ in $D^\infty_{b,A}$ such that $\Omega_n \to \Omega$, $H_0\Omega_n \to H_0\Omega$ in $L^2(E, d\mu_0)$ as $n \to \infty$. By formula (4.3), we have

$$(\Omega_n, \tilde{\nabla}_f \Psi \Omega_n) = -2(\tilde{\nabla}_f \Omega_n, \Psi\Omega_n) + (\phi(f)\Omega_n, \Psi\Omega_n), \ \Psi \in D_b^\infty, f \in \mathcal{H}. \quad (4.5)$$

We can show that

$$||\tilde{\nabla}_f \Psi|| \le ||f||_{\mathcal{H}_c} ||\nabla\Psi||_{L^2(E,d\mu_0;\mathcal{H}_c)} \le C||f||_{\mathcal{H}_c} ||H_0^{1/2}\Psi||, \quad \Psi \in D^\infty_{b,A^{1/2}},$$

with a constant $C > 0$. Using this estimate, we can see that $\{\tilde{\nabla}_f \Omega_n\}_{n=1}^\infty$ is a Cauchy sequence in $L^2(E, d\mu_0)$. Hecne $\Omega \in D(\tilde{\nabla}_f)$ and $\tilde{\nabla}_f\Omega_n \to \tilde{\nabla}_f\Omega$ in $L^2(E, d\mu_0)$ as $n \to \infty$. Moreover we can show that

$$||\phi(f)\Psi|| \le C||f||_{\mathcal{H}_c} \left(||H_0^{1/2}\Psi|| + ||\Psi||\right), \quad f \in \mathcal{H}, \Psi \in D(H_0^{1/2}),$$

with a constant $C > 0$. Hence the second term on the RHS of (4.5) converges to $(\phi(f)\Omega, \Psi\Omega)$. In this way we obtain (4.4). For more details see [15]. □

In the same way as in the proof of Lemma 4.1, we can prove the following lemma.

Lemma 4.4. *The operator* $A^{1/2}\nabla : L^2(E, d\mu) \to L^2(E, d\mu; \mathcal{H}_c)$ *with domain* $D^\infty_{b,A^{1/2}}$ *is closable with* $D((A^{1/2}\nabla)^*) \supset D_b^\infty \hat{\otimes} D(A^{1/2})$ *and*

$$(A^{1/2}\nabla)^*(\Psi \otimes f) = -(J_{\mathcal{H}} A^{1/2} f, \nabla\Psi)_{\mathcal{H}_c} + \lambda_{A^{1/2}f}\Psi, \quad \Psi \in D_b^\infty, f \in D(A^{1/2}).$$

We define a nonnegative self-adjoint operator in $L^2(E, d\mu)$:

$$L_A = (\overline{A^{1/2}\nabla})^* \overline{A^{1/2}\nabla}.$$

Lemma 4.5. *For all* $F(f_1, \cdots, f_n) \in D^\infty_{b,A}$,

$$L_A F(f_1, \cdots, f_n) = - \sum_{j,k=1}^{n} (f_j, Af_k)_{\mathcal{H}} (\partial_j \partial_k F)(f_1, \cdots, f_n) + \sum_{j=1}^{n} \lambda_{Af_j} (\partial_j F)(f_1, \cdots, f_n).$$

Proof. Direct computations using Lemma 4.4. □

A self-adjoint contraction semi-group $\{T_t\}_{t\geq 0}$ on the Hilbert space $L^2(X, d\nu)$ with a probability measure space (X, ν) is said to be *Markovian* if $0 \leq T_t u \leq 1$ a.e. for all $t > 0$ whenever $0 \leq u \leq 1$ a.e. (e.g., [27]). A self-adjoint operator L in $L^2(X, d\nu)$ is said to be a *Markovian self-adjoint extension* of a symmetric operator M in $L^2(X, d\nu)$ if $M \subset L$ and $\{\exp(-tL)\}_{t\geq 0}$ is Markovian.

Lemma 4.6. *The semi-group $\{\exp(-tL_A)\}_{t\geq 0}$ on $L^2(E, d\mu)$ is Markovian.*

Proof. See [27] and references therein. □

The operator $U : L^2(E, d\mu_0) \to L^2(E, d\mu)$ defined by

$$U\Psi = \Omega^{-1}\Psi, \quad \Psi \in L^2(E, d\mu_0),$$

is unitary.

Lemma 4.7. *The semi-group $\{\exp(-tU\widehat{H}_V U^{-1})\}_{t\geq 0}$ on $L^2(E, d\mu)$ is Markovian.*

Proof. We set

$$T = U\widehat{H}_V U^{-1}.$$

Let $\Psi \in L^2(E, d\mu), 0 \leq \Psi \leq 1$ (a.e.). We have

$$e^{-tT}\Psi = Ue^{-t\widehat{H}_V}U^{-1}\Psi.$$

Since $\exp(-t\widehat{H}_V)$ is positivity preserving (Theorem 3.3) and $U > 0$ (a.e.), it follows that $\exp(-tT)$ is positivity preserving. In particular, $\exp(-tT)\Psi \geq 0$ and $\exp(-tT)(1 - \Psi) \geq 0$. Since $\exp(-t\widehat{H}_V)\Omega = \Omega$, it follows that

$$e^{-tT}1 = 1.$$

Hence

$$1 - e^{-tT}\Psi = e^{-tT}(1 - \Psi) \geq 0.$$

Thus $\exp(-tT)$ is Markovian. □

We are now ready to prove the main theorem in this section.

Theroem 4.8. *The following operator equality holds:*

$$U\widehat{H}_V U^{-1} = L_A. \tag{4.6}$$

Proof (Outline). Using Lemma 4.5, we first show that (4.6) holds on the dense subspace $D^\infty_{b,A}$. In [27] it is shown that $L_A | D^\infty_{b,A}$ has a unique Markovian self-adjoint extension. Combining these facts with Lemmas 4.6 and 4.7, we obtain (4.6) as an operator equality. □

By Theorem 4.8, the problem of showing the existence of a SUSY extension of $\widehat{H}_V$ is reduced to that of L_A.

5. Operators of De Rham type and Lalacians

Let $\mathcal{K}$ be a separable complex Hilbert space and $\bigwedge^p(\mathcal{K})$ $(p = 0, 1, 2, \cdots)$ be the p-fold antisymmetric tensor product of $\mathcal{K}$ ($\bigwedge^0(\mathcal{K}) = \mathbf{C}$). For $u_j \in \mathcal{K}, j = 1, \cdots, p$, we define their exterior product $u_1 \wedge \cdots \wedge u_p \in \bigwedge^p(\mathcal{K})$ by

$$\begin{aligned} u_1 \wedge \cdots \wedge u_p &= \frac{1}{p!} \sum_{\sigma \in \mathsf{S}_p} \varepsilon(\sigma) u_{\sigma(1)} \otimes \cdots \otimes u_{\sigma(p)} \\ &= A_p(u_1 \otimes \cdots \otimes u_p), \end{aligned}$$

where S_p is the symmetric group of order p, $\varepsilon(\sigma)$ is the sign of the permutation σ and A_p is the antisymmetrization operator on the p-fold tensor product of $\mathcal{K}$. We define

$$\bigwedge^p(E, \mathcal{K}) := L^2\left(E, d\mu; \bigwedge^p(\mathcal{K})\right) = L^2(E, d\mu) \otimes \bigwedge^p(\mathcal{K}).$$

Let S be a densely defined closed linear operator from $\mathcal{H}_c$ to $\mathcal{K}$ such that $J_{\mathcal{H}} D(S) \subset D(S)$ and $J_{\mathcal{H}} S^* S \subset S^* S J_{\mathcal{H}}$. Then $D(S) \cap \mathcal{H}$ is dense. Hence $D^\infty_{b,S}$ is dense in $L^2(E, d\mu)$. We introduce a dense subspace in $\bigwedge^p(E, \mathcal{K})$:

$$\mathsf{D}_{S,p} = D^\infty_{b,S} \hat{\otimes} A_p(\mathcal{K} \hat{\otimes} \cdots \hat{\otimes} \mathcal{K}).$$

We define an operator $d_{S,p} : \bigwedge^p(E, \mathcal{K}) \to \bigwedge^{p+1}(E, \mathcal{K})$ with domain $\mathsf{D}_{S,p}$ by

$$d_{S,p} F u_1 \wedge \cdots \wedge u_p = \sqrt{p+1} S \nabla F \wedge u_1 \wedge \cdots \wedge u_p, \quad F \in D^\infty_{b,S}, u_k \in \mathcal{K},$$

and extending it by linearity to all vectors in $\mathsf{D}_{S,p}$. Let

$$\mathsf{D}^*_{S,p} = D^\infty_b \hat{\otimes} A_p(D(S^*) \hat{\otimes} \cdots \hat{\otimes} D(S^*)).$$

Lemma 5.1. *For all $p \geq 0$, $d_{S,p}$ is well defined and closable with*

$$D(d^*_{S,p}) \supset \mathsf{D}^*_{S,p+1} \tag{5.1}$$

and

$$\begin{aligned} d^*_{S,p} G v_1 \wedge \cdots \wedge v_{p+1} = \frac{1}{\sqrt{p+1}} \sum_{j=1}^{p+1} & \left(-\tilde{\nabla}_{J_{\mathcal{H}} S^* v_j} G + \lambda_{J_{\mathcal{H}} S^* v_j} G\right) \\ & \times v_1 \wedge \cdots \check{v}_j \cdots \wedge v_{p+1} \end{aligned} \tag{5.2}$$

for $G v_1 \wedge \cdots \wedge v_{p+1} \in \mathsf{D}^*_{S,p+1}$. Moreover,

$$d_{S,p+1} d_{S,p} = 0, \quad p \geq 0. \tag{5.3}$$

Proof (Outline). Relation (5.1) and formula (5.2) can be proven by using Lemma 4.3. Since $\mathsf{D}^*_{S,p+1}$ is dense in $\bigwedge^{p+1}(E,\mathcal{K})$, it follows that $D(d^*_{S,p})$ is dense in $\bigwedge^{p+1}(E,\mathcal{K})$. Thus $d_{S,p}$ is closable. Equation (5.3) follows from direct comutations. □

We denote the closure of $d_{S,p}$ by the same symbol.

We may define the pth cohomology space $H_{S,p}$ for the complex $\{d_{S,p}, D(d_{S,p})\}_{p=0}^{\infty}$ by

$$H_{S,p} = \operatorname{Ker} d_{S,p}/\overline{R(d_{S,p-1})}, \quad p \geq 1,$$

where $\overline{R(d_{S,p-1})}$ denotes the closure of the range $R(d_{S,p-1})$ of $d_{S,p-1}$ in $\bigwedge^p(E,\mathcal{K})$. As in the case of analysis on finite dimensional manifolds, it is natural to consider the Laplacians associated with $\{d_{S,p}\}_{p=0}^{\infty}$, which may be defined by $d^*_{S,p}d_{S,p} + d_{S,p-1}d^*_{S,p-1}$. In the present case, however, we do not know whether $D(d^*_{S,p}d_{S,p}) \cap D(d_{S,p-1}d^*_{S,p-1})$ is dense in $\bigwedge^p(E,\mathcal{K})$. To avoid this difficulty, noting that $D(d_{S,p}) \cap D(d^*_{S,p-1})$ is dense in $\bigwedge^p(E,\mathcal{K})$, we employ the method of sesquilinear form [10]:

Lemma 5.2. *For each $p \geq 0$, there exists a unique nonnegative self-adjoint operator $\Delta_{S,p}$ in $\bigwedge^p(E,\mathcal{K})$ such that*

$$D(\Delta_{S,p}^{1/2}) = D(d_{S,p}) \cap D(d^*_{S,p-1})$$

and

$$(\Delta_{S,p}^{1/2}\Phi, \Delta_{S,p}^{1/2}\Psi) = (d_{S,p}\Phi, d_{S,p}\Psi) + (d^*_{S,p-1}\Phi, d^*_{S,p-1}\Psi), \Phi, \Psi \in D(\Delta_{S,p}^{1/2}),$$

where we set $d_{S,-1} = 0$.

Proof. See [10]. □

The pth cohomology space H^p_S can be identified in terms of the pth Laplacian:

Theorem 5.3. *For all $p \geq 1$, $H_{S,p}$ is isomorphic to* $\operatorname{Ker}\Delta_{S,p}$ *as a vector space. In particular,*

$$\dim H_{S,p} = \dim \operatorname{Ker} \Delta_{S,p}.$$

Proof. Apply the decomposition theorem given in [12, Appendix A]. □

For a class of S, we can identify L_A with $\Delta_{S,0}$:

Theorem 5.4. *Suppose that S satisfies*

$$S^*S = A.$$

Then

$$L_A = \Delta_{S,0}.$$

Proof (Outline). We need only to show that, on $D^\infty_{b,A^{1/2}}$, the sesquilinear form of L_A coincides with that of $\Delta_{S,0}$. Note that, in the present case, $D(S) = D(A^{1/2})$. □

6. Construction of a SSQT and SUSY Extension of the Models

We define a large Hilbert space:

$$\bigwedge(E,\mathcal{K}) = \bigoplus_{p=0}^{\infty} \bigwedge^p(E,\mathcal{K}),$$

which is identified as

$$\bigwedge(E,\mathcal{K}) = L^2(E, d\mu; \bigwedge(\mathcal{K})),$$

with

$$\bigwedge(\mathcal{K}) = \bigoplus_{p=0}^{\infty} \bigwedge^p(\mathcal{K}),$$

the Fermion Fock space over $\mathcal{K}$ (e.g., [25, §II.4]).

The Hilbert space $\bigwedge(E,\mathcal{K})$ admits the orthogonal decomposition

$$\bigwedge(E,\mathcal{K}) = \bigwedge_+(E,\mathcal{K}) \bigoplus \bigwedge_-(E,\mathcal{K}) \tag{6.1}$$

with

$$\bigwedge_+(E,\mathcal{K}) = \bigoplus_{p=0}^{\infty} \bigwedge^{2p}(E,\mathcal{K}), \quad \bigwedge_-(E,\mathcal{K}) = \bigoplus_{p=0}^{\infty} \bigwedge^{2p+1}(E,\mathcal{K}).$$

Let $P_\pm$ be the orthogonal projection from $\bigwedge(E,\mathcal{K})$ onto $\bigwedge_\pm(E,\mathcal{K})$ and define

$$\Gamma = P_+ - P_-.$$

Then it is easy to check that Γ is a grading operator on $\bigwedge(E,\mathcal{K})$ which corresponds to the decompsition (6.1).

The de Rham operators $\{d_{S,p}\}_{p=0}^\infty$ can be lifted in a natural way to the Hilbert space $\bigwedge(E,\mathcal{K})$. Namely we can define an operator d_S acting in $\bigwedge(E,\mathcal{K})$ by

$$(d_S\Psi)^{(0)} = 0,$$
$$(d_S\Psi)^{(p)} = d_{S,p-1}\Psi^{(p-1)}, \quad p \geq 1,$$

with

$$D(d_S) = \Big\{\Psi = \{\Psi^{(p)}\}_{p=0}^{\infty} \in \bigwedge(E,\mathcal{K}) \,\Big|\, \Psi^{(p)} \in D(d_{S,p}), p \geq 0, \sum_{p=0}^{\infty} ||d_{S,p}\Psi^{(p)}||^2 < \infty\Big\}.$$

Basic properties of d_S are summarized in the following lemma.

Lemma 6.1.
(1) *The operator d_S is densely defined and closed. Further, we have*

$$d_{\alpha S} = \alpha d_S, \quad \alpha \in \mathbf{C},$$
$$d_S^2 = 0.$$

(2) *The adjoint d_S^* is given by*

$$(d_S^*\Psi)^{(p)} = d_{S,p}^*\Psi^{(p+1)}, \quad p \geq 0,$$

with domain

$$D(d_S^*) = \Big\{\Psi = \{\Psi^{(p)}\}_{p=0}^{\infty} \in \bigwedge(E,\mathcal{K}) \,\Big|\, \Psi^{(p+1)} \in D(d_{S,p}^*), p \geq 0, \sum_{p=0}^{\infty} ||d_{S,p}^*\Psi^{(p+1)}||^2 < \infty\Big\}.$$

Proof. See [15]. □

The d_S-cohomology space may be defined by

$$H_S = \operatorname{Ker} d_S / \overline{R(d_S)}.$$

We define a *Lapacian*, acting in $\bigwedge(E,\mathcal{K})$, associated with $\{\Delta_{S,p}\}_{p=0}^{\infty}$ by

$$\Delta_S = \bigoplus_{p=0}^{\infty} \Delta_{S,p},$$

which is nonnegative and self-adjoint.

Theorem 6.2. $D(\Delta_S^{1/2}) = D(d_S) \cap D(d_S^*)$ *and*

$$(\Delta_S^{1/2}\Phi, \Delta_S^{1/2}\Psi) = (d_S\Phi, d_S\Psi) + (d_S^*\Phi, d_S^*\Psi), \quad \Phi, \Psi \in D(\Delta_S^{1/2}). \tag{6.2}$$

Proof (Outline). We first prove (6.2) on the dense domain

$$\Big\{\Psi = \{\Psi^{(p)}\}_{p=0}^{\infty} \,\Big|\, \Psi^{(p)} \in D(d_{S,p-1}^*) \cap D(d_{S,p}), p \geq 0, \Psi^{(p)} = 0 \text{ for all but finitely many } p\Big\},$$

which is a core of $\Delta_S^{1/2}$, and then use a limiting argument. □

Theorem 6.2 and the decomposition theorem in [12, Appendix A] gives the following result.

Theorem 6.3. *The d_S-cohomology space H_S is isomorphic to* $\operatorname{Ker}\Delta_S$. *In particular,*

$$\dim H_S = \dim \operatorname{Ker}\Delta_S.$$

We introduce

$$Q_S = d_S + d_S^*$$

with $D(Q_S) = D(d_S) \cap D(d_S^*)$.

Proposition 6.4.
(1) *The operator Q_S is closed and symmetric.*
(2) *For all $\Psi \in D(Q_S)$, $\Gamma\Psi \in D(Q_S)$ and*

$$Q_S\Gamma\Psi + \Gamma Q_S\Psi = 0.$$

(3) *For all $\alpha \in \mathbf{C}$,*

$$Q_{\alpha S} = \alpha d_S + \alpha^* d_S. \tag{6.3}$$

Proof. See [15]. □

Proposition 6.4 shows that Q_S is an abstract Dirac operator in the sense of [23]. Note that, putting $\alpha = i$ in (6.3), we have

$$Q_{iS} = i(d_S - d_S^*),$$

which shows that Q_{iS} is an operator of Kähler type [24]. Thus, if we regard S as a "parameter", we may call Q_S a *Dirac-Kähler operator.*

A relation between the Dirac-Kähler operator Q_S and the Laplacian Δ_S is given by the following theorem.

Theorem 6.5. *The operator equality*

$$\Delta_S = Q_S^* Q_S$$

holds.

Proof. This can be proven by using Lemma 6.1(1) and Theorem 6.2. □

Remark. Unfortunately we do not know whether Q_S is self-adjoint or not. In what follows, we show that Q_S has a self-adjoint extension.

By the decomposition (6.1), we can write each vector $\Psi \in \bigwedge(E, \mathcal{K})$ as

$$\Psi = \begin{pmatrix} \Psi_+ \\ \Psi_- \end{pmatrix}$$

with $\Psi_\pm \in \bigwedge_\pm(E,\mathcal{K})$. Then every linear operator in $\bigwedge(E,\mathcal{K})$ can be represented as a 2×2 matrix with entries being operators. Proposition 6.4 (1) and (2) imply that there exist a unique densely defined closed linear operator $Q_{S,+}$ (resp. $Q_{S,-}$) from $\bigwedge_+(E,\mathcal{K})$ (resp. $\bigwedge_-(E,\mathcal{K})$) to $\bigwedge_-(E,\mathcal{K})$ (resp. $\bigwedge_+(E,\mathcal{K})$) such that

$$Q_S = \begin{pmatrix} 0 & Q_{S,-} \\ Q_{S,+} & 0 \end{pmatrix}. \tag{6.4}$$

We have

$$Q_S^* = \begin{pmatrix} 0 & Q_{S,+}^* \\ Q_{S,-}^* & 0 \end{pmatrix}. \tag{6.5}$$

It follows from Theorem 6.5, (6.4) and (6.5) that

$$\Delta_S = \begin{pmatrix} Q_{S,+}^* Q_{S,+} & 0 \\ 0 & Q_{S,-}^* Q_{S,-} \end{pmatrix}. \tag{6.6}$$

We define

$$\widehat{Q}_S = \begin{pmatrix} 0 & Q_{S,+}^* \\ Q_{S,+} & 0 \end{pmatrix}. \tag{6.7}$$

Lemma 6.6. *The operator $\widehat{Q}_S$ is a self-adjoint extension of Q_S. Moreover, Γ leaves $D(\widehat{Q}_S)$ invariant and*

$$\widehat{Q}_S \Gamma \Psi + \Gamma \widehat{Q}_S \Psi = 0, \quad \Psi \in D(\widehat{Q}_S).$$

Proof. Since Q_S is symmetric, we have $Q_S \subset Q_S^*$, which implies that

$$Q_{S,-} \subset Q_{S,+}^*, \quad Q_{S,+} \subset Q_{S,-}^*.$$

Hence $Q_S \subset \widehat{Q}_S$. It is easy to see that $\widehat{Q}_S$ is self-adjoint. Thus $\widehat{Q}_S$ is a self-adjoint extension of Q_S. The second half of the lemma is easily proven. □

Remark. The idea of the above method to construct a self-adjoint extension of Q_S has already been used in [8,11] (cf. [23]). In the case where μ is a Gaussian measure, it can be shown that Q_S is self-adjoint [5,12]. As for perturbations of Q_S in a Gaussian measure case, see [16].

Let

$$\widehat{\Delta}_S := \widehat{Q}_S^2.$$

Lemma 6.7. *For all $p \geq 0$, $\bigwedge^{2p}(E,\mathcal{K})$ reduces $\widehat{\Delta}_S$ with*

$$\widehat{\Delta}_S \Big| \bigwedge^{2p}(E,\mathcal{K}) = \Delta_{S,2p}, \quad p \geq 0.$$

Proof. By (6.7) we have

$$\widehat{\Delta}_S = \begin{pmatrix} Q_{S,+}^* Q_{S,+} & 0 \\ 0 & Q_{S,+} Q_{S,+}^* \end{pmatrix},$$

which implies that $\widehat{\Delta}_S$ is reduced by $\bigwedge_+(E,\mathcal{K})$ with

$$\widehat{\Delta}_S \Big| \bigwedge\nolimits_+(E,\mathcal{K}) = Q_{S,+}^* Q_{S,+}.$$

Hence, by (6.6), we see that

$$\widehat{\Delta}_S \Big| \bigwedge\nolimits_+(E,\mathcal{K}) = \Delta_S \Big| \bigwedge\nolimits_+(E,\mathcal{K}) = \bigoplus_{p=0}^{\infty} \Delta_{S,2p},$$

from which the desired result follows. □

Theorem 6.8. *The quadruple $\{\bigwedge(E,\mathcal{K}), \widehat{Q}_S, \widehat{\Delta}_S, \Gamma\}$ is a SSQT. The supersymmetry is unbroken.*

Proof. The first half of the theorem follows from Lemma 6.6. It follows from the definition of $d_{S,p}$ that $d_{S,0}1 = 0$, which implies that $\Delta_{S,0}1 = 0$. Hence, by Lemma 6.7., the vector $\Psi_0 \in \bigwedge(E,\mathcal{K})$ such that $\Psi_0^{(0)} = 1, \Psi_0^{(p)} = 0, p \geq 1$, satisfies $\widehat{\Delta}_S \Psi_0 = 0$, so that $\text{Ker}\,\Delta_S \neq \{0\}$. Thus the supersymmetry is unbroken. □

Remark. The theory developed above can be extended to a more general class of measures μ and a general framework for an infinite dimensional analysis is obtained. See [15] for the details. Some related mathematical aspects are discussed in [5,7,10-14,16].

We now solve the problem of supersymmetric extension of the operator $\widehat{H}_V$.

Theorem 6.9. *Suppose that S satisfies*

$$S^* S = A. \tag{6.8}$$

Then the SSQT $\{\bigwedge(E,\mathcal{K}), \widehat{Q}_S, \widehat{\Delta}_S, \Gamma\}$ is a SUSY extension of $\widehat{H}_V$ such that $U\widehat{H}_V U^{-1} = \widehat{\Delta}_S | \bigwedge^0(E,\mathcal{K})$. In this SSQT, the supersymmetry is unbroken.

Proof. Apply Theorems 4.7, 5.4, Lemma 6.7 and Theorem 6.8. □

Remark. For a given $\{\mathcal{H}, A, \mathcal{K}\}$, the operator S satisfying (6.8) is not uniquely determined. In fact, for any partial isometry $W : \mathcal{H}_c \to \mathcal{K}$, $S := WA^{1/2}$ satisfies (6.8).

Acknowledgement

The author would like to thank Professor Shigeo Kusuoka for a helpful discussion about Theorem 4.8.

References

[1] Arai, A., Supersymmetry and singular perturbations, *J.Funct.Anal.*, **60**, 378-393 (1985).
[2] Arai, A., Some remarks on scattering theory in supersymmetric quantum systems, *J. Math. Phys.*, **28**, 472-476 (1987).
[3] Arai, A., Formal aspects of supersymmetric embedding of Hamiltonians in quantum scalar field theories, *Lett.Math.Phys.*, **15**, 275-279 (1988).
[4] Arai, A., Supersymmetric embedding of a model of a quantum harmonic oscillator interacting with infinitely many bosons, *J.Math.Phys.*, **30**, 512–520 (1989).
[5] Arai, A., Path integral representation of the index of Kähler-Dirac operators on an infinite dimensional manifold, *J.Funct.Anal.*, **82**, 330-369 (1989).
[6] Arai, A., A criterion for the boundedness from below with a class of symmetric operators and its applications, *J.Math.Anal.Appl.*, **145**, 539-554 (1990).
[7] Arai, A., A general class of infinite dimensional Dirac operators and related aspects : in *Functional Analysis and Related Topics* (Ed. S. Koshi) (World Scientific, Singapore, 1991.)
[8] Arai, A. and Ogurisu, O., Meromorphic $N = 2$ Wess-Zumino supersymmetric quantum mechanics, *J.Math.Phys.*, **32**, 2427-2434 (1991).
[9] Arai, A., Exactly solvable supersymmetric quantum mechanics, *J.Math.Anal.Appl.*, **158**, 63-79 (1991).
[10] Arai, A. and Mitoma, I., De Rham-Hodge-Kodaira decomposition in ∞-dimensions, *Math.Ann.*, **291**, 51-73 (1991).
[11] Arai, A., De Rham operators, Laplacians, and Dirac operators on topological vector spaces, *Hokkaido University Preprint Series in Math.* No.115, 1991.
[12] Arai, A., A general class of infinite dimensional Dirac operators and path integral representation of their index, *J.Funct.Anal.*, **105**, 342-408 (1992).
[13] Arai, A. and Mitoma, I., Comparison and nuclearity of spaces of differential forms on topological vector spaces, *J.Funct.Anal.* (in press).
[14] Arai, A., Dirac operators in Boson-Fermion Fock spaces and supersymmetric quantum field theory, *Hokkaido University Preprint Series* No.161, 1992 (to appear in the Proceedings of the XXVIII Karpacz Winter School of Theoretical Physics, Karpacz, Poland, February 17-29, 1992.)
[15] Arai, A., Analysis on an infinite dimensional exterior bundle and supersymmetric extension of quantum scalar fields, in preparation.
[16] Arai, A., On the self-adjointness of Dirac operators in Boson-Fermion Fock spaces, in preparation.
[17] Bollé, D., Gesztesy, F., Grosse, H., Schweiger, W., and Simon, B., Witten index, axial anomaly and Krein's spectral shift function in supersymmetric quantum mechanics, *J. Math. Phys.*, **28**, 1512-1525 (1987).
[18] Freund, P.G.O., Introduction to Supersymmetry (Cambridge University Press, Cambridge, 1986.)
[19] Glimm, J. and Jaffe, J., Quantum Physics (Second Edition, Springer, New York, 1987.)
[20] Grosse, H. and Pittner, L., Supersymmetric quantum mechanics defined as sesquilinear forms, *J.Phys.A:Math.Gen.*, **20**, 4265-4684(1987).
[21] Jaffe, A., Lesniewski, A., and Lewenstein, M., Ground state structure in supersymmetric quantum mechanics, *Ann.Phys. (N.Y.)*, **178**, 313-329 (1987).

[22] Jaffe, A. and Lesniewsjki,A., Supersymmetric quantum fields and infinite dimensional analysis : in *Nonperturbative Quantum Field Theory*, (Ed. by G.'t Hooft, A. Jaffe, G. Mack, P.K. Mitter and R. Stora, Plenum Press, New York, 1988.)

[23] Jorgensen, P.E.T., Spectral theory for self-adjoint operator extensions associated with Clifford algebras, Preprint, 1991.

[24] Kähler, E., Der innere Differentialkalkül, *Rend.Mat.*, **21**, 425-523(1962).

[25] Reed, M. and Simon, B., Methods of Modern Mathematical Physics I: Functional Analysis (Academic Press, New York, 1972.)

[26] Reed, M. and Simon, B., Methods of Modern Mathematical Physics II: Fourier Analysis, Self-adjointness (Academic Press, New York, 1975.)

[27] Röckner, M., and Zhang, T.S., Uniqueness of generalized Schrödinger operators and applications, *J.Funct.Anal.*, **105**, 187-231(1992).

[28] Segal, I., Notes towards the construction of nonlinear relativistic quantum fields. III. Properties of the C^*-dynamics for a certain class of interactions, *Bull.Amer.Math.Soc.*, **75**, 1390-1395 (1969).

[29] Segal, I., Construction of nonlinear local quantum processes I, *Ann. of Math.*, **92**, 462-481(1970).

[30] Simon, B. and Hoegh-Krohn, R., Hypercontractive semigroups and two dimensional self-coupled Bose fields, *J.Funct.Anal.*, **9**, 121-180(1972).

[31] Simon, B., The $P(\phi)_2$ Euclidean (Quantum) Field Theory (Princeton Univ. Press, Princeton, NJ, 1974.)

[32] Wess, J. and Bagger, J., Supersymmetry and Supergravity (Princeton University Press, Princeton, 1983.)

[33] Wess, J. and Zumino, B., Supergauge transformations in four dimensions, *Nucl.Phys.*, **B 70**, 39-50(1974).

[34] Wess, J. and Zumino, B., A Lagrangian model invariant under supergauge transformations, *Phys. Lett.*, **49B**, 52-54(1974).

[35] Witten, E., Supersymmetry and Morse theory, *J. Differential Geometry*, **17**, 661-692 (1982).

[36] Witten, E., Constraints on supersymmetry breaking, *Nuclear Phys.*, **B 202**, 253-316 (1982).

2. Statistical and Solid State Physics

Optimal Two–Uniform Convexity and Fermion Hypercontractivity

ERIC A. CARLEN
School of Mathematics, Georgia Institute of Technology, Atlanta Ga., 30332 U.S.A.

and

ELLIOTT H. LIEB
Departments of Mathematics and Physics, Princeton University, Princeton NJ, 08544 U.S.A.

Abstract. An optimal two-uniform convexity theorem (obtained together with K. Ball) is presented. This is shown to lead to optimal hypercontractivity bounds for the fermion oscillator semigroup, conjectured by L. Gross in 1975. These are the fermion analogs of the optimal hypercontractivity bounds for the boson oscillator semigroup obtained in 1973 by E. Nelson .

Key words: hypercontractivity, non-commutative integration, fermions

1. Introduction

This is a shortened version of a paper [5] which will appear in the issue of Communications in Mathematical Physics dedicated to Huzihiro Araki. We begin by motivating the use of non-commutative integration methods in the study of fermion systems.

Many questions in quantum mechanics concern spectral properties of a Hamiltonian H. While such questions are intrinsic to the Hilbert space setting, the tools used to answer them are often extrinsic to this setting, and require additional structure such as that which is present in the Schrödinger q-space representation in which the Hilbert space is L^2, the square integrable functions on configutation space.

For example, consider the Schrödinger operator $H = -\Delta + V$ acting on $L^2(\mathbb{R}^n, dx)$. Two standard tools for addressing the question – say – of existence and uniqueness of a ground state for such an operator are the Sobolev inequality and the Perron-Frobenius theorem. Applicability of these tools is predicated on the fact that since the Hilbert space in question is an L^2 space, it comes along with L^p norms and a natural notion of positivity.

H. Araki et al. (eds.), Quantum and Non-Commutative Analysis, 93–111.

In an abstract Hilbert space setting, neither of these would be available.

In field theory, depending on whether the basic configuration observables commute or anticommute, the Hilbert space on which observables act is conventionally realized as the boson or the fermion Fock space.

More specifically, suppose there are n modes to the underlying field. If these are boson modes, the state space for the system is the **boson Fock space** $\mathcal{B}$:

$$\mathcal{B} = \oplus_{j=0}^{\infty}(\otimes_{sym}^{j}(\mathbb{C}^n))$$

In the fermion case, the state space is the **fermion Fock space:**

$$\mathcal{F} = \oplus_{j=o}^{n}((\mathbb{C}^n)^{\wedge j}) \tag{1.1}$$

The basic free Hamiltonian on the Fock spaces – the analog of $-\Delta$ on $L^2(\mathbb{R}^n, \mathrm{d}x)$ – is the **number operator**

$$\widehat{H}_0 = \sum_{j=1}^{n} c_j^* c_j \tag{1.2}$$

where c_j and c_j^* are the usual anihilation and creation oprators (boson in the boson case and fermion in the fermion case of course). Since the Fock spaces are not L^2 spaces, it is not evident how one could apply analogs of the Sobolev inequality or the Perron-Frobenius theorem.

However, for n boson degrees of freedom, the state space may be realized *either* as the boson Fock space, or as $L^2(\mathbb{R}^n, (2\pi)^{-n/2}e^{-x^2/2}\mathrm{d}x)$. The natural isomorphism between these spaces was pointed out by Segal, and may be regarded as providing a Schrödinger q-space realization. The boson oscillator semigroup is the semigroup generated by the boson number operator, $\exp(-t\widehat{H}_0)$. Though we have just defined it in Fock space terms, it may also be considered as an operator semigroup on $L^2(\mathbb{R}^n, (2\pi)^{-n/2}e^{-x^2/2}\mathrm{d}x)$ where its action is given by the Mehler kernel. In this setting, it is **hypercontractive**; i.e., for any finite p greater than 2, there is t_p sufficiently large, for which the semigroup is a contraction from $L^2(\mathbb{R}^n, (2\pi)^{-n/2}e^{-x^2/2}\mathrm{d}x)$ to $L^p(\mathbb{R}^n, (2\pi)^{-n/2}e^{-x^2/2}\mathrm{d}x)$ for all $t \geq t_p$. Since t_p is independent of n, this result proved useful in treating perturbation theoretic problems for boson fields. In particular, it is equivalent to an analog of the Sobolev inequality appropriate for boson fields, and it permits the application of a Perron-Frobenius type theorem of Gross [10] to establish existence and uniqueness of ground states for certain perturbations of $\widehat{H}_0$.

For fermion fields, the configuration observables do not commute, and this q-space representation is not available. However, the non-commutative integration theory of Irving Segal [29] permits the construction of a suitable substitute, and in fact it was created with such a purpose in view. This approach to the study of fermion systems has been extensively developed by Gross [10] who, among other things, proved a version of the Perron-Frobenius theorem adapted to the setting and applied it to prove existence and uniqueness of ground states for certain fermion quantum field models.

In order to pass from the Fock space description for a system of n fermion degrees of freedom to a non-commutative analog of the Schrödinger q-space description (in which we have non-commutative analogs of L^p norms), we introduce configuration observables $Q_j = c_j + c_j^*$, and let $\mathcal{C}(\mathcal{Q})$ be the algebra with unit which they generate. This is an operator algebra on the 2^n–dimensional Hilbert space $\mathcal{F}$. As we explain in Section 2, it is a Clifford algebra naturally associated to $\mathbb{C}^n$ with its usual inner product.

What follows here shall be explained in more detail in Sections 2–4, but the key fact for our present considerations is that the 2^n distinct monomials in the Q_j are a basis for $\mathcal{C}(\mathcal{Q})$ as a *vector space*. Thus, any A in $\mathcal{C}(\mathcal{Q})$ can be uniquely written as

$$A = \alpha I + \sum_j \alpha_j Q_j + \sum_{j<k} \alpha_{j,k} Q_j Q_k + \cdots + \alpha_{1,\ldots,n} Q_1 \cdots Q_n \quad . \tag{1.3}$$

It is not hard to see that any vector v in $\mathcal{F}$ can be written as $v = A\Omega$ for some unique $A \in \mathcal{C}(\mathcal{Q})$ acting on the vacuum Ω. The operator $\exp(-t\widehat{H}_0)$ acts on such vectors v. Thus, $\exp(-t\widehat{H}_0)v = B\Omega$ for some unique $B \in \mathcal{C}(\mathcal{Q})$ (which depends on t and A, of course). The map $P_t : A \mapsto B$ is clearly linear, and a simple calculation (elucidated in Section 4) shows that

$$P_t A = \alpha I + e^{-t} \sum_j \alpha_j Q_j + e^{-2t} \sum_{j<k} \alpha_{j,k} Q_j Q_k + \cdots + e^{-nt} \alpha_{1,\ldots,n} Q_1 \cdots Q_n \quad . \tag{1.4}$$

Instead of considering Fock space, $\mathcal{F}$, and the operator $\exp(-t\widehat{H}_0)$, Segal's idea is to consider $\mathcal{C}(\mathcal{Q})$ and the linear operator P_t that acts on $\mathcal{C}(\mathcal{Q})$ as above. The operators P_t form a semigroup $exp(-tH_0)$; *this defines* H_0. Following Segal, we can construct L^p norms on $\mathcal{C}(\mathcal{Q})$, the most important case being $p = 2$, in terms of which one can do quantum mechanics without mentioning Fock space.

To construct the norms, define the linear functional $\tau : \mathcal{C}(\mathcal{Q}) \to \mathbb{C}$ by

$$\tau(A) = \langle \Omega, A\Omega \rangle_{\mathcal{F}} \quad . \tag{1.5}$$

It turns out, as we explain in Section 3 that $\tau(A)$ is, for $A \in \mathcal{C}(\mathcal{Q})$, nothing other than the usual normalized trace on $\mathcal{F}$ applied to A.

The basic ingredients of a non-commutative integration theory are an algebra of operators such as $\mathcal{C}(\mathcal{Q})$, and a linear functional on it such as τ. Norms analogous to the L^p norms of commutative integration theory can now be introduced by

$$\|A\|_p = (\tau[(A^*A)^{p/2}])^{1/p} \qquad \text{for} \qquad 1 \le p < \infty \quad .$$

$\|A\|_\infty$ denotes the operator norm of A. Let $\mathcal{C}^p(\mathcal{Q})$ denote $\mathcal{C}(\mathcal{Q})$ equipped with the corresponding norm.

Having expanded our horizons beyond Hilbert space, we can ask for bounds between $\|P_tA\|_q$ and $\|A\|_p$ for different q, p and t Our main result, Theorem 4, is the optimal **fermion hypercontractivity** inequality; i.e. for all $1 < p \le q < \infty$ and all A in $\mathcal{C}^p(\mathcal{Q})$,

$$\|P_tA\|_q \le \|A\|_p \qquad \text{when} \qquad e^{-2t} \le \frac{p-1}{q-1} \quad , \tag{1.6}$$

and the t saturating the inequality on the right is the smallest for which the inequality on the left always holds. We prove this for n degrees of freedom with n an arbitrary finite integer. Since the estimate is independent of n, a theorem of Gross [10] implies that it holds as well with infinitely many degrees of freedom.

The fermion result (1.6) was conjectured by Gross [11] who proved it [10] in the special case $p = 2$, $q = 4$. The cases $p = 2$, $q = 2m$ where m is an integer were proved by Lindsay and Meyer [19] following earlier work by Lindsay [18] on the case $p = 2$, $q = 2^m$. Since P_t is self adjoint, duality yields a corresponding family of results in which $q = 2$. Until now all other cases (except when $n = 1$ or 2) had remained open. The optimal relation between t, p and q found here for fermion hypercontractivity is the same as that found by Nelson [24] for boson hypercontractivity.

By now there are many proofs of Nelson's boson inequality. Neveu's elegant proof [26], like Nelson's original proof, is based on probabilistic methods. Proofs based on geometric methods have been given by Carlen and Loss [6] and by Lieb [17] who considers generalizations in which the Mehler kernel (the kernel for the boson oscillator semigroup) is replaced by an arbitrary Gaussian kernel. More references can be found in the bibliography to Gross's article [12]. However, none of the existing approaches to the boson problem has been found to solve the fermion problem.

As a corollary to the optimal fermion hypercontractivity inequality, we obtain the optimal **fermion logarithmic Sobolev inequality**:

$$\tau(|A|^2 \ln |A|^2) - (\|A\|_2^2 \ln \|A\|_2^2) \le 2\tau(A^* H_0 A) \tag{1.7}$$

where $|A| = (A^*A)^{1/2}$. The inequality (1.7) was conjectured by Gross, and proved by him in a weaker form with the constant on the right increased by a factor of $\ln 3$. In studying perturbations of H_0 by multiplication operators V, this inequality plays the same role as does the usual Sobolev inequality in studying perturbations of $-\frac{1}{2}\Delta$ by multiplication operators V on $L^2(\mathbb{R}^n, dx)$ [9]. (The multiplication operator associated to a self-adjoint element V of $\mathcal{C}(\mathcal{Q})$ is defined to be the average of left and right multiplication by V, and is denoted here by the same symbol V.) Again, in the standard Fock space setting, there is no natural way to formulate such an a-priori regularity inequality, or even to introduce the notion of a multiplication operator.

This paper is organized as follows: In Section 2 we study the structure of $\mathcal{C}(\mathcal{Q})$ for finite n. It is actually simpler, as well as technically advantageous, to consider it as a subalgebra of the algebra $\mathcal{C}(\mathcal{K})$ generated by the identity, the configuration observables $Q_1, \ldots, Q_n$ and their conjugate momenta $P_1 \ldots, P_n$. Since they both turn out to be certain Clifford algebras, their structure has been worked out long ago with the representation theory of the orthogonal groups. Thus, this section contains no new result but simply introduces notation and prepares the way for what follows. Of particular use are an explicit spin– chain representation of the operator algebra $\mathcal{C}(\mathcal{K})$, and the Jordan-Wigner transform identifying it with the algebra generated by n "hard core bosons".

Section 3 concerns properties of the spaces $\mathcal{C}^p(\mathcal{K})$ and their norms. The main result here is an optimal uniform convexity inequality for $\mathcal{C}^p(\mathcal{K})$, $1 < p \leq 2$, which is joint work with Keith Ball. We need only a special case of this inequality here, and a proof is provided in an appendix for the reader's convenience.

Section 4 introduces a convenient expression for P_t in terms of the conditional expectation $\pi_{\mathcal{Q}}$ of $\mathcal{C}(\mathcal{K})$ with respect to $\mathcal{C}(\mathcal{Q})$. The main result in this section is an inequality for conditional expectations which enables us to prove that

$$\sup\{||P_tA||_q \ : \ ||A||_p = 1\} \quad = \quad \sup\{||P_tA||_q \ : \ A \geq 0 \text{ and } ||A||_p = 1\}$$

Thus, to establish (1.6) in general, we need only consider positive A.

In Section 5 we establish the optimal fermion hypercontractivity bounds and the corresponding optimal fermion logarithmic Sobolev inequality. This is done in several steps. First, using results collected in Sections 2 and 3 we establish that (1.7) holds for $1 < p \leq 2$ and $q = 2$. At $t = 0$ this is an equality; differentiating it there yields the logarithmic Sobolev inequality (1.7). Gross showed that (1.6) would follow from (1.7), if it were true (as we show here), for all *self adjoint* A. His result rests on a deep inequality he established for positive operators. Since it is *not* in general true that $||P_tA||_q \leq ||P_t|A|||_q$, as is trivially true in the commutatative case, Gross's result does not allow us to conclude (1.6) for *general* (i.e. non-self adjoint) A from (1.7). The results of the Section 4 do, however, allow us to draw this conclusion.

Finally, in Section 6 we show that the same hypercontractivity relation (1.6) holds for a mixed system of bosons and fermions.

Non-commutative probability theory has grown into a substantial branch of analysis with a number of physical applications. The mathematical theory is reviewed and developed in [21] and [22], while other sorts of physical applications, besides those discussed here, are treated in [7] and [13] for example.

It is a pleasure to thank Leonard Gross for discussing his results and conjectures with us, and for encouraging us to take up the latter. We are indebted to Keith Ball for his collaboration on the subject of convexity inequalities that led to Theorem 1 [3], which is one of the key ingredients in the present work. Thanks are also due to G.-F. Dell'Antonio, A. Jaffe and A. Wightman for useful discussions.

2. Fermions and the Clifford Algebra

We begin by recalling for later use some well known facts about fermions.

The fundamental observables for a system of n fermion degrees of freedom are **configuration operators** $Q_1, Q_2, \ldots, Q_n$ together with their **conjugate momenta operators** $P_1, P_2, \ldots, P_n$ all acting as operators on a complex Hilbert space $\mathcal{H}$ and satisfying the **canonical anticommutation relations:**

$$Q_j Q_k + Q_k Q_j = 2\delta_{jk} \quad , \quad P_j P_k + P_k P_j = 2\delta_{jk} \quad , \quad P_j Q_k + Q_k P_j = 0 \quad . \tag{2.1}$$

We denote the complex algebra generated by the identity and the configuration observables by $\mathcal{C}(\mathcal{Q})$, and the complex algebra generated by the identity, the configuration observables and the momentum observables all together by $\mathcal{C}(\mathcal{K})$. $\mathcal{C}(\mathcal{Q})$ is the object of primary interest; but many aspects of its structure are most readily seen within the larger algebra $\mathcal{C}(\mathcal{K})$.

This algebra can be concretely represented as the algebra of observables for a spin–chain as follows. We define the matrices

$$I = \begin{bmatrix} 1 & 0 \\ 0 & 1 \end{bmatrix} \quad , \quad U = \begin{bmatrix} 1 & 0 \\ 0 & -1 \end{bmatrix} \quad , \quad Q = \begin{bmatrix} 0 & 1 \\ 1 & 0 \end{bmatrix} \quad , \quad P = \begin{bmatrix} 0 & i \\ -i & 0 \end{bmatrix} \quad .$$

Let $\mathcal{H}$ denote the n-fold tensor product of $\mathbb{C}^2$ with itself:

$$\mathcal{H} = \underbrace{\mathbb{C}^2 \otimes \ldots \otimes \mathbb{C}^2}_{n \text{ times}} \quad ,$$

and on $\mathcal{H}$ define the operators

$$Q_j = U \otimes \cdots U \otimes Q \otimes I \otimes \cdots \otimes I \quad , \quad P_j = U \otimes \cdots U \otimes P \otimes I \otimes \cdots \otimes I \tag{2.2}$$

where the Q and the P occur in the jth places.

The operators $Q_1, \ldots, Q_n, P_1 \ldots, P_n$ just defined are easily seen to satisfy the canonical anticommutaion relations. Of great use in studying the algebra $\mathcal{C}(\mathcal{K})$ that they generate is the fact that it is also the algebra generated by n "hard core boson" degrees of freedom. More explicitly, put

$$\widehat{Q}_j = I \otimes \cdots I \otimes Q \otimes I \otimes \cdots \otimes I \quad , \quad \widehat{P}_j = I \otimes \cdots I \otimes P \otimes I \otimes \cdots \otimes I \quad , \tag{2.3}$$

and call the algebra generated by the operators $\widehat{Q}_1,\ldots,\widehat{Q}_n,\widehat{P}_1\ldots,\widehat{P}_n$ the **hard core boson algebra**, and denote it by $\widehat{\mathcal{C}}(\mathcal{K})$. To see that the two algebras coincide, put

$$U_j = I\otimes\cdots I\otimes U\otimes I\otimes\cdots\otimes I \qquad \text{and} \qquad V_k = \prod_{j=1}^{k-1} U_j \ . \tag{2.4}$$

Then since $P_jQ_j = \widehat{P}_j\widehat{Q}_j = iU_j$, each V_k belongs to both the hard core boson algebra and $\mathcal{C}(\mathcal{K})$. Moreover

$$\widehat{Q}_k = V_kQ_k \quad \text{and} \quad \widehat{P}_k = V_kP_k \tag{2.5}$$

with the inverse relation given as well by

$$Q_k = V_k\widehat{Q}_k \quad \text{and} \quad P_k = V_k\widehat{P}_k \ . \tag{2.6}$$

Thus, Q_k,P_k are in $\widehat{\mathcal{C}}(\mathcal{K})$, and $\widehat{Q}_k$,$\widehat{P}_k$ are in $\mathcal{C}(\mathcal{K})$.

What we call the hard core boson algebra was initially introduced by Jordan and Klein [14] as a first attempt to implement the Pauli exclusion principle mathematically. The transformation of observables (2.6) was discovered by Jordan and Wigner [15] and used by them to write the algebraic relations characterizing the algebra in the familiar covariant form (2.1); today it is known as the Jordan–Wigner transform. It has been used many times since – for example, in a solution of the two dimensional Ising model by Schultz, Mattis and Lieb, whose paper [28] can be consulted for references to other applications. It is also the key to Brauer and Weyl's treatment [4] of the Clifford algebra on which the exposition in this section is largely based.

We close this section by specifying a useful orthonormal basis for $\mathcal{H}$. Let

$$e_1 = \begin{pmatrix}1\\0\end{pmatrix} \qquad \text{and} \qquad e_{-1} = \begin{pmatrix}0\\1\end{pmatrix}$$

be the standard basis of $\mathbb{C}^2$. For each $j = 1,\ldots,n$, let σ_j be either 1 or -1. Then the vectors

$$e_{\sigma_1,\ldots,\sigma_n} = e_{\sigma_1}\otimes\cdots\otimes e_{\sigma_n}$$

provide our orthonormal basis for $\mathcal{H}$. The vector $e_{-1,-1,\ldots,-1}$ plays a distinguished role; it is denoted by Ω and called the **ground state**:

$$\Omega = \underbrace{\begin{pmatrix}0\\1\end{pmatrix}\otimes\cdots\otimes\begin{pmatrix}0\\1\end{pmatrix}}_{n \text{ times}} \tag{2.7}$$

3. Analysis on the Clifford Algebra

Let $\mathcal{A}$ be a von Neumann algebra of operators on some finite dimensional Hilbert space. By a trace on $\mathcal{A}$ we shall mean a linear functional τ which is positive in the sense that $\tau(A^*A) > 0$ for all non zero A in $\mathcal{A}$, and cyclic in the sense that $\tau(AB) = \tau(BA)$ for all A and B in $\mathcal{A}$. Such a functional is evidently unitarily invariant in the sense that whenever A and U belong to $\mathcal{A}$, and U is unitary, then $\tau(U^*AU) = \tau(A)$. Since $\mathcal{C}(\mathcal{K})$ is a full matrix algebra [4], it contains all unitaries. Hence any trace on $\mathcal{C}(\mathcal{K})$ must assign the same value to all rank one projections, and thus must be a scalar multiple of the standard trace Tr on the matrix algebra. Henceforth, τ denotes this trace normalized by the condition that $\tau(I) = 1$ and Tr denotes the standard unnormalized trace.

In the non-commutative integration theories of Dixmier [8] and Segal [29], the trace functional τ is the non-commutative analog of the functional that assigns to an integrable function its integral. When the Hilbert space is infinite dimensional, some further regularity properties are required of τ in order to obtain a useful analog. Since all of our estimations will be carried out in the finite dimensional setting, we shall not go into this here, but simply refer the reader to these original papers as well as the accounts in [10] and [25].

Norms on $\mathcal{C}(\mathcal{K})$ which are the non commutative analogs of the L^p norms can now be introduced; namely for $1 \leq p < \infty$ we put

$$\|A\|_p = (\tau((A^*A)^{p/2}))^{1/p} \quad , \tag{3.1}$$

and denote the operator norm of A by $\|A\|_\infty$.

$\mathcal{C}^p(\mathcal{K})$ shall denote $\mathcal{C}(\mathcal{K})$ equipped with the norm $\|\cdot\|_p$; evidently $\mathcal{C}^2(\mathcal{K})$ is the Hilbert space of $2^n \times 2^n$ matrices equipped with the Hilbert-Schmidt norm.

Consider the monomials

$$E_{[\alpha_1\dots,\alpha_j;\beta_1,\dots,\beta_k]} = Q_{\alpha_1}\cdots Q_{\alpha_j}P_{\beta_1}\cdots P_{\beta_k} \tag{3.2}$$

where $\alpha_1 > \cdots > \alpha_j$ and $\beta_1 > \cdots > \beta_k$ and $j + k > 0$. Evidently

$$E^*_{[\alpha_1\dots,\alpha_j;\beta_1,\dots,\beta_k]}E_{[\alpha_1\dots,\alpha_j;\beta_1,\dots,\beta_k]} = I$$

and thus $\|E_{[\alpha_1\dots,\alpha_j;\beta_1,\dots,\beta_k]}\|_p = 1$ for all p. Moreover, an easy computation reveals that

$$\tau(E_{[\alpha_1\dots,\alpha_j;\beta_1,\dots,\beta_k]}) = 0 \quad , \tag{3.3}$$

from which it follows that

$$\tau(E^*_{[\alpha_1\dots,\alpha_j;\beta_1,\dots,\beta_k]}E_{[\gamma_1\dots,\gamma_m;\delta_1,\dots,\delta_n]}) = 0 \tag{3.4}$$

unless the two monomials coincide. Thus, together with the identity, the assemblage of such monomials forms an orthonormal basis for $\mathcal{C}^2(\mathcal{K})$. Finally observe that since $Qe_- = e_+$

$$\langle \Omega, E_{[\alpha_1 \ldots, \alpha_j]} \Omega \rangle = 0 \tag{3.5}$$

whenever $j \geq 1$, and, as indicated, $k = 0$. It now follows that, restricted to $\mathcal{C}(\mathcal{Q})$,

$$\tau(A) = \langle \Omega, A\Omega \rangle \tag{3.6}$$

for all A in $\mathcal{C}(\mathcal{K})$.

Formula (3.6) is very important for us. It permits us to calculate the "physically" relevant quantity $\langle \Omega, A\Omega \rangle$ in terms of the apparently mathematically simpler quantity $\tau(A)$.

Many familiar inequalities for L^p norms hold for the $\mathcal{C}^p$ norms as well [8]. This is true in particular of the Hölder inequality

$$\|AB\|_r \leq \|A\|_p \|B\|_q \qquad \frac{1}{r} = \frac{1}{p} + \frac{1}{q} \quad .$$

Certain optimal inequalities expressing the uniform convexity properties of the L^p norms also hold for the $\mathcal{C}^p$ norms, and this fact constitutes one cornerstone of our analysis.

The modulus of convexity δ_p of $\mathcal{C}^p(\mathcal{K})$ is defined by

$$\delta_p(\epsilon) = \inf\{1 - \frac{1}{2}\|A + B\|_p : \|A\|_p = \|B\|_p = 1 \ , \ \|A - B\|_p = \epsilon\} \tag{3.7}$$

for $0 < \epsilon < 2$. For $1 < p < \infty$, δ_p is always positive which means these norms are uniformly convex. Useful geometric information is contained in the rate at which $\delta_p(\epsilon)$ tends to zero with ϵ. It is known [32] that for $2 \leq p < \infty$, $\delta_p(\epsilon) \sim \epsilon^p$, but that for $1 < p \leq 2$, $\delta_p(\epsilon) \sim \epsilon^2$.

An optimal expression of this fact is given by the following theorem which was proved jointly with Keith Ball [3]:

Theorem 1: (Optimal 2-uniform convexity for matrices). *For all $m \times m$ matrices A and B and all p for $1 \leq p \leq 2$,*

$$\left(\frac{Tr|A+B|^p + Tr|A-B|^p}{2}\right)^{2/p} \geq (Tr|A|^p)^{2/p} + (p-1)(Tr|B|^p)^{2/p} \quad . \tag{3.8}$$

For $1 < p < 2$, there is equality only when $B = 0$.

This result, which we interpret here as a statement about $\mathcal{C}^p(\mathcal{K})$, will be proved in the special case that both $A + B$ and $A - B$ are positive; this is the only case in which we shall use it here, and the proof is considerably simpler in this case. The full result is proved in [3], in which other geometric inequalities for trace norms are proved as well.

Proof (when $A \pm B \geq 0$): Let Z and W be the $2m \times 2m$ matrices given by

$$Z = \begin{bmatrix} A & 0 \\ 0 & A \end{bmatrix} \quad , \quad W = \begin{bmatrix} B & 0 \\ 0 & -B \end{bmatrix} \quad .$$

Our goal is to establish that for all r with $0 \leq r \leq 1$,

$$\left(\frac{Tr(A + rB)^p + Tr(A - rB)^p}{2} \right)^{2/p} \geq (Tr(A)^p)^{2/p} + r^2(p-1)(Tr|B|^p)^{2/p} \quad ,$$

or what is the same,

$$Tr(Z + rW)^{2/p} \geq (Tr(Z)^p)^{2/p} + r^2(p-1)(Tr|W|^p)^{2/p} \quad . \tag{3.9}$$

First, note that the null space of $Z + rW$ is exactly the null space of Z for $0 \leq r < 1$. Thus by carrying out all of the following computations on the orthogonal complement of this fixed null space, we may freely assume that $Z + rW > 0$ for all $0 \leq r < 1$. Next, both sides of (3.9) agree at $r = 0$, and the first derivatives in r of both sides vanish there as well. We define $\psi(r)$ to be $Tr(Z + rW)^p$. Then the second derivative in r of the left side of (3.9) satisfies

$$\frac{d^2}{dr^2}(\psi(r))^{2/p} \geq \frac{2}{p}\psi(r)^{(2-p)/p}\frac{d^2}{dr^2}\psi(r) \quad .$$

The second derivative on the right side is just

$$2(p-1)(Tr|W|^p)^{2/p} \quad ,$$

and we are left with showing that

$$\frac{1}{p}\psi(r)^{(2-p)/p}\frac{d^2}{dr^2}\psi(r) \geq (p-1)(Tr|W|^p)^{2/p} \tag{3.10}$$

for all $0 < r < 1$. By redefining Z to be $Z + rW$, it suffices to establish (3.10) at $r = 0$.

Now $\frac{d}{dr}\psi(r) = p(Tr(Z + rW)^{(p-1)}W)$, since $A \pm B \geq 0$, $Z + rW \geq 0$ for small r, and we can use the integral representation

$$(Z + rW)^{(p-1)} = c_p \int_0^\infty t^{(p-1)} \left[\frac{1}{t} - \frac{1}{t + (Z + rW)} \right] dt$$

to conclude that

$$\frac{d^2}{dr^2}\psi(0) = pc_p \int_0^\infty t^{(p-1)} Tr\left[\frac{1}{t+Z} W \frac{1}{t+Z} W \right] dt \quad . \tag{3.11}$$

Consider the right side as a function, $f(Z)$, of Z for fixed W. It is easy to see that f is convex in Z. (Simply replace Z by $Z+sX$, with X self-adjoint, and then differentiate twice with respect to s; the positivity follows from the Schwarz inequality for traces.) Also,

$$f(UZU^*) = f(Z)$$

provided U is unitary and U commutes with W. In a basis in which W is diagonal, we form the set $\mathcal{U}$ consisting of the 2^{2m} distinct diagonal unitary matrices, each with $+1$ or -1 in each diagonal entry. Each of these clearly commutes with W. Then

$$f(Z) = 2^{-2m} \sum_{U \in \mathcal{U}} f(UZU^*) \geq f(2^{-2m} \sum_{U \in \mathcal{U}} UZU^*) = f(Z_{\text{diag}}),$$

where Z_{diag} is the matrix that is diagonal in the basis diagonalizing W, and whose diagonal entries are those of Z in this basis.

Replacing Z by Z_{diag} in (3.11), the integration can be carried out, and we obtain

$$\frac{\mathrm{d}^2}{\mathrm{d}r^2}\psi(0) \geq p(p-1)\left(\sum_{j=1}^{2m} z_j^{(p-2)} w_j^2\right)$$

where z_j and w_j, respectively, denote the jth diagonal entries of Z and W in a basis diagonalizing W.

Now consider $\psi(0) = Tr(Z^p)$ as a function of Z. It is clearly convex, and thus by the averaging method just employed, we obtain

$$\psi(0) \geq \left(\sum_{j=1}^{2m} z_j^p\right) \quad .$$

To establish (A.2), we are only left with showing that

$$\left(\sum_{j=1}^{2m} z_j^p\right)^{(2-p)/p} \left(\sum_{j=1}^{2m} z_j^{(p-2)} w_j^2\right) \geq \left(\sum_{j=1}^{2m} |w_j|^p\right)^{2/p} \quad , \tag{3.12}$$

but this follows immediately from Hölder's inequality.

To complete the proof, observe that equality in (3.9) for $r = 1$ and $1 < p < 2$ implies equality in (3.14) for almost every r in $[0,1]$. Here, recall that z_j in (3.14) really denotes the jth diagonal element of $Z + rW$; these are the numbers $z_j + rw_j$, where z_j denotes the jth diagonal element of Z.

Let us assume that $w_j \neq 0$ for some j. Then equality in Hölder's inequality (3.14) requires that the vector with positive components $z_j + rw_j$ be proportional to the vector with components $|w_j|$. Thus, for almost every r in $[0,1]$ we require

$$z_j + rw_j = c(r)|w_j|$$

for some number $c(r)$ that depends on r but not on j. The left side above is a linear function, and thus $c(r) = a + rb$ for some numbers a and b. But then clearly $b = w_j/|w_j|$, and all non-zero eigenvalues of W would necessarily have the same sign. This is impossible since $TrW = 0$.

Q.E.D.

The theorem implies that

$$\delta_p(\epsilon) \geq \frac{(p-1)}{2}\left(\frac{\epsilon}{2}\right)^2 \qquad \text{for} \qquad 1 < p \leq 2 \tag{3.13}$$

as one sees by considering $A = (C+D)/2$, $B = (C-D)/2$, $\|C\|_p = \|D\|_p = 1$ and $\|C - D\|_p = \epsilon$. It is easily seen that the constant $(p-1)/8$ cannot be improved.

Our main application of this result is made in Section 5. There we will also need to know that the norms on $\mathcal{C}^p(\mathcal{K})$ are continuously differentiable away from the origin for $1 < p < \infty$. This is known [11], but a simple proof can be based on inequalities of the form $\delta_p(\epsilon) \geq K_p \epsilon^{r(p)}$ such as we have found above for $1 < p < 2$. This proof, also yields the modulus of continuity of the derivative [5].

4. Conditional Expectations and the Fermi Oscillator Semigroup

We are concerned with the subalgebra $\mathcal{C}(\mathcal{Q})$ of $\mathcal{C}(\mathcal{K})$, and the conditional expectation [8][33] with respect to it plays a basic role in our investigation. For any A in $\mathcal{C}(\mathcal{K})$, the **conditional expectation** $\pi_{\mathcal{Q}}(A)$ of A with respect to $\mathcal{C}(\mathcal{Q})$ is defined to be the unique element of $\mathcal{C}(\mathcal{Q})$ such that $\tau(B^*\pi_{\mathcal{Q}}(A)) = \tau(B^*A)$ for all B in $\mathcal{C}(\mathcal{Q})$. Otherwise said, $\pi_{\mathcal{Q}}$ is the orthogonal projection from $\mathcal{C}^2(\mathcal{K})$ onto $\mathcal{C}^2(\mathcal{Q})$. It is well known that the conditional expectation is positivity preserving; a familiar argument shows that $\pi_{\mathcal{Q}}(A^*A) \geq \pi_{\mathcal{Q}}(A)^*\pi_{\mathcal{Q}}(A)$. We can use the conditional expectation to give a useful expression for the oscillator semigroup for fermion fields. Let $R_\theta : \mathcal{C}(\mathcal{K}) \to \mathcal{C}(\mathcal{K})$ be the linear map defined by the following action on P's and Q's, and then extended in the obvious way to products:

$$R_\theta(Q_j) = (\cos\theta)Q_j + (\sin\theta)P_j, \quad \text{and} \quad R_\theta(P_j) = (\cos\theta)P_j - (\sin\theta)Q_j \tag{4.1}$$

for each j. Since the "rotated" P's and Q's satisfy the same anticomutation relation (2.1), R_θ is an automorphism of $\mathcal{C}(\mathcal{K})$. Since the linear transformation in (4.1) is real, it is moreover a $*$-automorphism.

Next, for each $t \geq 0$, define $\theta(t) = \arccos(e^{-t})$ and define the operator P_t on $\mathcal{C}^2(\mathcal{K})$ by

$$P_t A = \pi_{\mathcal{Q}} \circ R_{\theta(t)} \circ \pi^*_{\mathcal{Q}} A \quad . \tag{4.2}$$

Note that π_Q^* is the **natural imbedding** of $\mathcal{C}(\mathcal{Q})$ into $\mathcal{C}(\mathcal{K})$, and regarded as such, it is a $*$-automorphism. Formula (4.2) is the analog of the familiar expression for the boson oscillator semigroup on $L^2(\mathcal{Q},(2\pi)^{-n/2}e^{-q^2/2}d^nq)$, i.e. the Mehler semigroup

$$P_t^{(boson)}A(\mathrm{q}) = \int_{\mathcal{Q}} A\left(e^{-t}\mathrm{q} + (1-e^{-2t})^{1/2}\mathrm{p}\right)(2\pi)^{-n/2}e^{-p^2/2}d^np \quad .$$

Note that since all of the operators on the right in (4.2) are positivity preserving, so is P_t. Also, since the first two operations on the right preserve the $\mathcal{C}^p$-norms, and since the conditional expectation is readily seen to be a contraction from $\mathcal{C}^p(\mathcal{K})$ to $\mathcal{C}^p(\mathcal{Q})$ for each p, it is readily seen that P_t possesses this property as well.

To obtain a more familiar expression for P_t, note that

$$R_{\theta(t)} \circ \pi_Q^*(E_{[\alpha_1,\dots,\alpha_k]}) = e^{-kt}E_{[\alpha_1,\dots,\alpha_k]} + (\text{terms annihilated by } \pi_Q) \quad .$$

Hence

$$P_t(E_{[\alpha_1,\dots,\alpha_k]}) = e^{-kt}E_{[\alpha_1,\dots,\alpha_k]}. \tag{4.3}$$

Evidently $\{P_t : t \geq 0\}$ is generated by H_0 with $H_0(E_{[\alpha_1,\dots,\alpha_k]}) = kE_{[\alpha_1,\dots,\alpha_k]}$. It is easy to see that under the unitary equivalence between $\mathcal{C}^2(\mathcal{Q})$ and fermion Fock space $\mathcal{F}$ described by Segal [30], H_0 is equivalent to the usual number operator, or in other words, oscillator Hamiltonian on $\mathcal{F}$.

Our primary goal is to prove optimal hypercontractivity bounds for P_t. That is, given $1 < p < q < \infty$ we want to show that for some finite t, P_t is a contraction from $\mathcal{C}^p(\mathcal{Q})$ to $\mathcal{C}^q(\mathcal{Q})$, and to find the smallest such t. Let

$$||P_t||_{p\to q} = \sup\{||P_tA||_q \ : \ ||A||_p = 1\} \quad . \tag{4.4}$$

As a first reduction, we shall show that the supremum on the right in (4.4) can be restricted to the positive operators A with $||A||_p = 1$. In the boson case this follows immediately from the fact that, in ordinary probability theory, the absolute value of a conditional expectation is no greater than the conditional expectation of the absolute value.

In general, matters concerning the absolute value in the non-commutative setting are more troublesome than in the commutative setting. An example is provided by the Araki-Yamagami inequality [2] which, specialized to our context, asserts that the map $A \mapsto |A|$ is Lipschitz continuous on $\mathcal{C}^2(\mathcal{K})$ with constant $\sqrt{2}$ instead of the constant 1 which we would have in the commutative setting.

Thus while the conditional expectation in an operator algebra has many properties analogous to those of the conditional expectation in ordinary probability theory [33], it is not in general true that $|\pi_Q(A)|$ will be a smaller operator than $\pi_Q(|A|)$. The following theorem expresses a useful property in this direction which does hold, and after proving it we shall show by example that stronger properties do not hold. The theorem and its proof are easily extended to a more general von Neumann algebra setting by the methods in [27].

Theorem 2: (A Schwarz inequality for conditional expectations). *For all A in $\mathcal{C}(\mathcal{K})$ and all p with $1 \le p \le \infty$,*

$$\|\pi_{\mathcal{Q}}(A)\|_p \le \|\pi_{\mathcal{Q}}(|A|)\|_p^{1/2}\|\pi_{\mathcal{Q}}(|A^*|)\|_p^{1/2} \quad . \tag{4.5}$$

Remark: See [5] for a discussion of the relationship of this result to other Schwarz type inequalities for matrix functions to be found in [16] and [20]. The fact that (4.5) really requires the A and the A^* can be seen from the example $A = \begin{bmatrix} 0 & 1 \\ 0 & 1 \end{bmatrix}$, (again, see [CaLi] for details).

Proof: Let $A = U|A|$ be the polar decomposition of A. Then $\|\pi_{\mathcal{Q}}(A)\|_p = \tau(CU|A|)$ for some C in $\mathcal{C}(\mathcal{Q})$ with $\|C\|_{p'} = 1$. Let $C = V|C|$ be the polar decomposition of C. Both V and $|C|$ belong to $\mathcal{C}(\mathcal{Q})$ as well. Thus

$$\|\pi_{\mathcal{Q}}(A)\|_p = \tau(CU|A|^{1/2}|A|^{1/2}) = \tau\left(|C|^{1/2}U|A|^{1/2}|A|^{1/2}V|C|^{1/2}\right)$$

$$\le \tau(|C|^{1/2}U|A|U^*|C|^{1/2})^{1/2}\tau(|C|^{1/2}V^*|A|V|C|^{1/2})^{1/2}$$

$$= \tau(|C|(U|A|U^*))^{1/2}\tau((V|C|V^*)|A|)^{1/2} \le \|\pi_{\mathcal{Q}}(U|A|U^*)\|_p^{1/2}\|\pi_{\mathcal{Q}}(|A|)\|_p^{1/2} \quad .$$

Finally, we note that $U|A|U^* = |A^*|$. (4.6)

Q.E.D.

Theorem 3: (P_t has positive maximizers). *The norm of P_t from $\mathcal{C}^p(\mathcal{Q})$ to $\mathcal{C}^q(\mathcal{Q})$ is achieved on the positive operators; i.e.*

$$\|P_t\|_{p\to q} = \sup\{\|P_tA\|_q \ : \ A \ge 0 \quad \text{and} \quad \|A\|_p = 1\,\} \quad . \tag{4.7}$$

Proof: Since R_θ and $\pi^*_{\mathcal{Q}}$ are $*$-automorphisms, $|R_{\theta(t)} \circ \pi^*_{\mathcal{Q}}A| = R_{\theta(t)} \circ \pi^*_{\mathcal{Q}}|A|$ and $|(R_{\theta(t)} \circ \pi^*_{\mathcal{Q}}A)^*| = R_{\theta(t)} \circ \pi^*_{\mathcal{Q}}|A^*|$. Thus $\|P_tA\|_q \le \|P_t|A|\|_q^{1/2}\|P_t|A^*|\|_q^{1/2}$, and of course $\||A^*|\|_p = \||A|\|_p = \|A\|_p$.

Q.E.D.

5. Hypercontractivity for Fermions

Our main result is the following theorem which is established in this section.

Theorem 4: (Optimal fermion hypercontractivity). *For all* $1 < p \leq q < \infty$, $\|P_t\|_{p \to q} = 1$ *exactly when*

$$e^{-2t} \leq \frac{p-1}{q-1} \quad .$$

The heart of the matter is the following lemma:

Lemma: *For all* $1 < p \leq 2$, $\|P_t\|_{p \to 2} = 1$ *exactly when* $e^{-2t} \leq (p-1)$.

Proof: Fix a positive element A of $\mathcal{C}(\mathcal{Q})$. Pick a basis $\{q_1, \ldots, q_n\}$ of $\mathcal{Q}$, and let $\mathcal{C}(\mathcal{Q}_{(n-1)})$ denote the Clifford algebra associated with the span of the first $n-1$ of these basis elements. It is evident from the form of our standard basis of $\mathcal{C}(\mathcal{Q})$ that A can be uniquely decomposed as $A = B + CQ_n$ where B and C belong to $\mathcal{C}(\mathcal{Q}_{(n-1)})$. Then using the Jordan-Wigner transform we can write $A = B + CV_n\widehat{Q}_n$. Now write

$$\mathcal{H} = \mathcal{H}_{(n-1)} \otimes \mathbb{C}^2 \tag{5.1}$$

so that B, C and V_nC can be considered as operators on the first factor $\mathcal{H}_{(n-1)}$. Let $u_\pm = (e_+ \pm e_-)/\sqrt{2}$, so that $Qu_\pm = \pm u_\pm$. Then if v is any vector in $\mathcal{H}_{(n-1)}$,

$$\langle (v \otimes u_\pm), A(v \otimes u_\pm) \rangle_{\mathcal{H}} = \langle v, (B \pm CV_n)v \rangle_{\mathcal{H}_{(n-1)}} \quad . \tag{5.2}$$

We see from this that since $A \geq 0$, so are both $B + CV_n \geq 0$ and $B - CV_n \geq 0$. By a similar argument, we see that the spectrum of A is the union of the spectra of $B + CV_n$ and $B - CV_n$.

Now let Tr_1 and Tr_2 denote the partial traces over the first and second factors in (5.1), so that with Tr still denoting the full trace, we have $Tr = Tr_1 Tr_2$. Now applying Theorem 1 in the special case which is proved in the appendix

$$\|A\|_p^2 = \left(\frac{1}{2^n} Tr|B + CV_n\widehat{Q}_n|^p\right)^{2/p}$$

$$= \left(\frac{1}{2^{(n-1)}}\right)^{2/p} \left(\frac{Tr_1|B + CV_n|^p + Tr_1|B - CV_n|^p}{2}\right)^{2/p}$$

$$\geq \left(\frac{1}{2^{(n-1)}}\right)^{2/p} \left((Tr_1|B|^p)^{2/p} + (p-1)(Tr_1|C|^p)^{2/p}\right)$$

since V_n is unitary.

Thus $\|A\|_p^2 \geq \|B\|_p^2 + (p-1)\|C\|_p^2$ where the norms on the right are all norms on $\mathcal{C}(\mathcal{Q}_{(n-1)})$.

Now we make the inductive assumption that the lemma has been established for $\mathcal{C}(\mathcal{Q}_{(n-1)})$. This is clearly the case when $n = 1$. We then have

$$\|A\|_p^2 \geq \|P_tB\|_2^2 + (p-1)\|P_tC\|_2^2$$

where $e^{-2t} = (p-1)$. But clearly from (4.3) $\|P_tCQ_n\|_2^2 = (p-1)\|P_tC\|_2^2$ where the norms are once again norms on $\mathcal{C}(\mathcal{Q})$. Moreover by (3.4) P_tB and P_tCQ_n are orthogonal. Thus

$$\|P_tB\|_2^2 + (p-1)\|P_tC\|_2^2 = \|P_t(B + CQ_n)\|_2^2 = \|P_tA\|_2^2$$

Q.E.D.

Theorem 5: (Optimal fermion logarithmic Sobolev inequality). *For all $A \in \mathcal{C}(\mathcal{Q})$,*

$$\tau(|A|^2 \ln |A|^2) - (\|A\|_2^2 \ln \|A\|_2^2) \leq 2\langle A, H_0 A\rangle \quad . \tag{5.3}$$

Proof: By the lemma, $\|P_tA\|_2^2 \leq \|A\|_{(1+e^{-2t})}^2$ and there is equality at $t = 0$. Both sides are continuously differentiable, and comparing derivatives at $t = 0$ we obtain the result. Indeed,

$$\frac{\mathrm{d}}{\mathrm{d}p}\|A\|_p = \frac{1}{p}\|A\|_p^{1-p}\Big(\tau(|A|^p \ln |A|) - \|A\|_p^p \ln \|A\|_p\Big) \tag{5.4}$$

and of course

$$\frac{\mathrm{d}}{\mathrm{d}t}\|P_tA\|_2^2 = \frac{\mathrm{d}}{\mathrm{d}t}\langle A, P_{2t}A\rangle = -2\langle A, H_0A\rangle \quad .$$

Q.E.D.

Gross refers to the quadratic form on the right side of (5.3) as the Clifford Dirichlet form since it shares many properties of Dirichlet forms in the ordinary commutative setting. An approach to the development of a theory of Dirichlet forms in the non-commutative setting can be found in [1].

Proof of Theorem 4: By a deep result of Gross, when $A \geq 0$ and $1 < p < \infty$,

$$\langle A^{p/2}, H_0A^{p/2}\rangle \leq \frac{(p/2)^2}{p-1}\langle A, H_0A^{p-1}\rangle \quad .$$

Replacing A in (5.3) by $A^{p/2}$ and using the inequality just quoted we obtain, following Gross's ideas [11],

$$\tau(A^p \ln A) - \|A\|_p^p \ln \|A\|_p \leq \frac{p/2}{p-1}\langle A^{p-1}, H_0 A\rangle \quad .$$

By combining this with (5.4) a differential inequality is obtained which implies that $\|P_t A\|_{q(t)}$ is a decreasing function of t when $q(t) = 1 + e^{2t}(p-1)$. This establishes the result for $A \geq 0$, and by Theorem 3 it is established in general. By (4.3), $P_t I = I$, and therefore $\|P_t\|_{p\to q}$ is always at least 1 for all p and q. That the inequality is best possible follows from a direct computation with one degree of freedom. To be precise, $\|P_t(I + Q_1)\|_q = \|I + e^{-t}Q_1\|_q$ is easily computed and compared with $\|I + Q_1\|_p$ [10]. The first quantity is greater than the second if $e^{-2t} > (p-1)/(q-1)$.

Q.E.D.

6. Hypercontractivity for Bosons and Fermions Together

As a result of the present work and of earlier work on bosons, we know that for t given by $e^{-2t} = (p-1)/(q-1)$, both the fermion and the boson oscillator semigroups are contractive from the appropriate p–spaces to the appropriate q–spaces, and that this value of t is optimal for each case *separately*.

It is natural to expect that the same condition governs hypercontractivity in a situation in which we have bosons and fermions *together*. This is indeed the case, as we now show using Minkowski's inequality in an argument based on Segal's method for showing that the optimal conditions for hypercontractivity with m boson degrees of freedom are the same as for one degree of freedom.

Let $\mu(dx) = (2\pi)^{-m/2} e^{-x^2/2} dx$ be the unit Gauss measure on $\mathbb{R}^m$. Then in our mixed setting, with m boson degrees of freedom and n fermion degrees of freedom, the relevant p–space is

$$\mathcal{B}^p = L^p(\mathbb{R}^m, \mu) \otimes \mathcal{C}^p(\mathcal{Q}_{(n)}),$$

which may be regarded as consisting of $\mathcal{C}^p(\mathcal{Q}_{(n)})$ valued measurable functions $x \mapsto A(x)$ such that

$$|||A|||_p^p = \int_{\mathbb{R}^m} \|A(x)\|_p^p \mu(dx)$$

is finite. This equation defines the norm on $\mathcal{B}^p$. For $p = 2$, $\mathcal{B}^p$ is naturally isomorphic to the tensor product of the symmetric tensor algebra over $\mathbb{C}^m$ and the antisymmetric tensor algebra over $\mathbb{C}^n$ as shown by Segal. On the latter space we have the mixed oscillator semigroup generated by the sum of the boson and fermion number operators

$$\exp\left\{-t\left[\sum_{j=1}^{m} a_j^* a_j + \sum_{j=1}^{n} c_j^* c_j\right]\right\}.$$

Considered as operators on $\mathcal{B}^p$, the operators $\mathcal{P}_t$ which constitute this semigroup are given by

$$\mathcal{P}_t A(x) = \int_{R^M} M_t(x,x') P_t[A(x')]\mu(\mathrm{d}x') \tag{6.1}$$

where $M_t(x,x')$ is the Mehler kernel; i.e., the positive integral kernel for the boson oscillator semigroup $P_t^{(boson)}$ discussed in Section 4. Of course P_t denotes the fermion oscillator semigroup studied throughout this paper.

Now successively applying Minkowski's inequality, our theorem on optimal fermion hypercontractivity, and Nelson's theorem on optimal boson hypercontractivity, we have for $e^{-2t} \leq (p-1)/(q-1)$:

$$\begin{aligned}
\|\mathcal{P}_t A\|_q^q &= \int_{\mathbb{R}^m} \left\| \int_{\mathbb{R}^m} M_t(x,x') P_t[A(x')]\mu(\mathrm{d}x') \right\|_q^q \mu(\mathrm{d}x) \\
&\leq \int_{\mathbb{R}^m} \left(\int_{\mathbb{R}^m} M_t(x,x') \|P_t A(x')\|_q \mu(\mathrm{d}x') \right)^q \mu(\mathrm{d}x) \\
&\leq \int_{\mathbb{R}^m} \left(\int_{\mathbb{R}^m} M_t(x,x') \|A(x')\|_p \mu(\mathrm{d}x') \right)^q \mu(\mathrm{d}x) \\
&\leq \left(\int_{\mathbb{R}^m} \|A(x)\|_p^p \mu(\mathrm{d}x) \right)^{q/p} = \|A\|_p^q \quad .
\end{aligned}$$

References

[1] Albeverio, S., Høegh-Krohn, Dirichlet forms and Markov semigroups on C^*-algebras, Commun. Math. Phys., **56** (1977) 173-187.

[2] Araki, H.: Yamagami, S.: An inequality for the Hilbert-Schmidt norm, Commun. Math. Phys., **81** (1981) 89-96.

[3] Ball, K., Carlen, E.A., Lieb, E.H.: preprint 1992.

[4] Brauer, R., Weyl, H.: Spinors in n dimensions, Am. Jour. Math., **57** (1935) 425-449.

[5] Carlen, E. A., Lieb, E. H.: Optimal Hypercontractivity for Fermi fields and related non-commutative integration inequalities, Commun. Math. Phys. (issue dedicated to Araki), in press (1992).

[6] Carlen, E.A., Loss, M.: Extremals of functionals with competing symmetries, Jour. Func. Analysis **88** (1991) 437-456

[7] Davies, E.B.: *Quantum Theory of Open Systems*, Academic Press, New York, 1976.

[8] Dixmier, J.: Formes linéaires sur un anneau d'opérateurs, Bull. Soc. Math. France **81** (1953) 222-245.

[9] Federbush, P.: A partially alternate derivation of a result of Nelson, Jour. Math. Phys., **10** (1969) 50-52.

[10] Gross, L.: Existence and uniqueness of physical ground states, Jour. Funct. Analysis, **10** (1972) 52-109.
[11] Gross, L.: Hypercontractivity and logarithmic Sobolev inequalities for the Clifford-Dirichlet form, Duke Math. J., **43** (1975) 383-396.
[12] Gross, L.: Logarithmic Sobolev inequalities for the heat kernel on a Lie group and a bibliography on logarithmic Sobolev inequalities and hypercontractivity, pp. 108-130 in *White Noise Analysis, Mathematics and Applications*, eds. Hida et al., World Scientific, Singapore 1990.
[13] Hudson, R, Parthasarathy, K.R.: Quantum Itô's formula and stochastic evolutions, Commun. Math. Phys., **93** (1984) 301-323.
[14] Jordan, P., Klein, O.: Züm Mehrkörperproblem der Quantentheorie, Zeits. für Phys., **45** (1927) 751-765.
[15] Jordan, P., Wigner, E.P.: Über das Paulische Äquivalenzverbot, Zeits. für Phys., **47** (1928) 631-651.
[16] Lieb, E.H.: Inequalities for some operator and matrix functions, Adv. Math. **20** (1976) 174-178
[17] Lieb, E.H.: Gaussian kernels have only Gaussian maximizers, Invent. Math. **102** (1990) 179-208
[18] Lindsay, M.: Gaussian hypercontractivity revisited, Jour. Funct. Analysis **92** (1990) 313-324.
[19] Lindsay, M., Meyer, P.A.: preprint, 1991.
[20] Merris, R.,Dias da Silva,J.A.: Generalized Schur functions, Jour. Lin. Algebra **35** (1975) 442-448.
[21] Meyer, P.A.: Eléments de probabilités quantiques, exposés I-V, pp.186-312 in Sem. de Prob. XX, Lecture notes in Math. **1204**, Springer, New York, 1985.
[22] Meyer, P.A.: Eléments de probabilités quantiques, exposés VI-VIII, pp. 27-80 in Sem. de Prob. XXI, Lecture notes in Math. **1247**, Springer, New York, 1986.
[23] Nelson, E.: A quartic interaction in two dimensions, in *Mathematical Theory of Elementary Particles*, R. Goodman ans I. Segal eds., MIT Press, Cambridge Mass., 1966.
[24] Nelson, E.: The free Markov field, Jour. Funct. Analysis, **12** (1973) 211-227.
[25] Nelson, E.: Notes on non-commutative integration, Jour. Funct. Analysis, **15** (1974) 103-116.
[26] Neveu, J.: Sur l'esperance conditionelle par rapport à un mouvement Brownien, Ann. Inst. H. Poincaré Sect. B. (N.S.) **12** (1976) 105-109
[27] Ruskai, M.B.: Inequalities for traces on Von Neumann algebras, Commun. Math. Phys., **26** (1972) 280-289.
[28] Schultz, T.D., Mattis, D.C., Lieb, E.H.: Two dimensional Ising model as a soluble problem of many fermions, Rev. Mod. Phys. **36** (1964) 856-871.
[29] Segal, I.E.: A non-commutative extension of abstract integration, Annals of Math., **57** (1953) 401-457.
[30] Segal, I.E.: Tensor algebras over Hilbert spaces II, Annals of Math., **63** (1956) 160-175.
[31] Segal, I.E.: Construction of non-linear local quantum processes: I, Annal of Math., **92** (1970) 462-481.
[32] Tomczak-Jaegermann, N.: The moduli of smoothness and convexity and Rademacher averages of trace classes $S_p(1 \leq p < \infty)$, Studia Mathematica **50** (1974) 163-182.
[33] Umegaki, H.: Conditional expectation in operator algebras I, Tohoku Math. J., **6** (1954) 177-181.

Renormalization Group and Nonlinear PDE's

J.BRICMONT *
Physique Théorique, UCL, 2, ch. du cyclotron,
B-1348 Louvain-la-Neuve, Belgium

and

A.KUPIAINEN**
Mathematics Department, University of Helsinki
Hallituskatu 15, 00100, Helsinki, Finland

Abstract. Renormalization Group was originally developed for the understanding of Critical Phenomena and continuum limit in Quantum Field Theory. We show how similair ideas turn out to be useful in the study of long time asymptotics of solutions of parabolic nonlinear or random Partial Differential Equations. We show how universality, i.e. independence of the asymptotics on initial data and the equation, emerges naturally. Concrete problems discussed include patterns and moving fronts in Ginzburg Landau equation and diffusion in random environment.

Key words: renormalization group, front, non linear PDE's, diffuison in random medium

1. Introduction

The purpose of this talk is to apply Renormalization Group ideas to the analysis of the long time behaviour of solutions of parabolic partial differential equations of the form:

$$\dot{u} = \Delta u + F(u, \partial u, \partial^2 u) \tag{1}$$

where $u = u(x,t)$ is real or complex, $x \in \mathbb{R}^d$, ∂ denotes the partial derivative(s) with respect to x and F is nonlinear and/or random.
Problems where (1) occurs include :

Diffusion in a random environment :

Consider (1) as a Fokker-Planck equation with $F = -\nabla \cdot (\mathbf{b}u)$ and $\mathbf{b}$ a random force whose covariance is :

$$\overline{b_\alpha(x) b_\beta(y)} = \epsilon^2 \delta_{\alpha\beta} \delta(x-y) \tag{2}$$

Here, $u(x,t)$ is the probability density to be at x, at time t, starting from 0. The problem is whether $u(x,t)$ behaves, for long times, like a Gaussian.

* Supported by EC Grant SC1-CT91-0695.
** Supported by NSF Grant DMS-8903041.

H. Araki et al. (eds.), Quantum and Non-Commutative Analysis, 113–118.

However, to make sense of the problem, one has to regularise $\delta(x-y)$ in (2) by formulating everything on a lattice. This gives rise to random walks with random transition probabilities [3,4,5].

Nonlinear diffusion :

Fairly general nonlinear (deterministic) F's are considered but particularly $F = \pm u^p$ (with $u(x,t) \geq 0$ so that the sign is fixed) or, for $x \in \mathbb{R}$, the Burgers'equation $F = u\partial u$ [6].

Blowup :

Without the Laplacian in (1) we have an ordinary differential equation, and, with $F = u^p$, $p > 1$, any positive initial data leads to a solution which diverges in a finite time. An extensive literature (see e.g. [11]) analyses what happens with the Laplacian, in particular the form of $u(x,t)$ near the blowup point.

Propagating fronts :

Consider the Ginzburg-Landau equation, in one dimension, i.e. u is complex in (1), $F = u - |u|^2 u$ and $x \in \mathbb{R}$. This equation has a two parameter family of stationary solutions :

$$u_{q\theta}(x) = \sqrt{1-q^2}e^{i(qx+\theta)} \tag{3}$$

and one inquires about the time development of initial data $u(x)$ which approach two such solutions at $\pm\infty$:

$$\lim_{x\to\infty} |u(x) - u_{q\pm\theta\pm}(x)| = 0 \tag{4}$$

See [9,10,7]. A particularly well known example is $q_- = 0$, $\theta_- = 0$, $q_+ = 1$ and $u(x)$ real, in which case $u(x,t)$ converges to a front moving at a speed depending only on the rate at which $u(x)$ tends to zero at $+\infty$ [1,2].

2. The Renormalization Group

It turns out that for long times and suitable initial data, the solution of (1) takes the form :

$$u(x,t) \simeq t^{-\alpha/2} f^*(xt^{-\frac{1}{2}}) \tag{5}$$

which is the analogue of *scaling* in the theory of critical phenomena. Moreover, α and f^* are independent of the initial data and of F in (1), in given classes. This is the phenomenon of *universality*. The presence of scaling and universality calls for a Renormalization Group approach. The latter was initiated in [12].

The basic idea of the Renormalization Group applied to equation (1) is to integrate the equation over a finite time, rescale and iterate. Since one is interested in the long time behaviour, this is clearly analogous to the Renormalization Group method in the theory of critical phenomena : integrate over short distance fluctuations, rescale and iterate.

More precisely, let us fix some space of initial data S (usually a ball in some Banach space), and, for convenience, set the initial time to be $t = 1$. Then, choose a number $L > 1$ and define

$$u_L(x,t) = L^{\alpha} u(Lx, L^2 t) \tag{6}$$

where u is the solution of (1) with $u(\cdot, 1) = f(\cdot) \in S$. The Renormalization Group map is simply

$$(R_L f)(x) \equiv u_L(x, 1) \tag{7}$$

The goal is to prove that R_L maps S into itself so that the transformation can be iterated. Then, one analyses the fixed points of R_L and their stability.

The iteration takes two forms. First, the equation may be scale invariant, for example:

$$\dot{u} = \partial^2 u - u^p \tag{8}$$

(with $x \in \mathbb{R}$). If we choose $\alpha = \frac{2}{p-1}$, u_L still satisfies (8) and R_L has the semigroup property :

$$R_{L^n} = (R_L)^n \tag{9}$$

Assume that R_L possesses a fixed point :

$$R_L f^* = f^* \tag{10}$$

and that $(R_L)^n f \to f^*$ as $n \to \infty$, for a class of initial data. Then, from (6,7,9), we get:

$$u(\sqrt{t}\cdot, t) \simeq t^{-\alpha/2} f^*(\cdot) \tag{11}$$

for $t = L^{2n} \to \infty$.

To illustrate the notion of fixed points, consider the trivial example :

$$\dot{u} = \Delta u \tag{12}$$

in $\mathbb{R}^d$ and let $\alpha = d$. Then, in Fourier transform,

$$\widehat{R_L f}(k) = e^{-k^2(1-\frac{1}{L^2})} \hat{f}(\frac{k}{L}) \tag{13}$$

which clearly has $\hat{f}^*(k) = e^{-k^2}$ as a fixed point. To investigate the stability of f^* at a formal level, write

$$\hat{f}(k) = \hat{f}(0) + ck + \cdots \tag{14}$$

Then, for $k = \mathcal{O}\ (1)$, we get :

$$\widehat{R_L f}(k) = \hat{f}(0)\hat{f}^*(k) + \mathcal{O}(L^{-1}) \tag{15}$$

corresponding to the well-known fact that, if $u(x,1) = f(x)$ is in $L^1(\mathbb{R})$,

$$u(x,t) \simeq \frac{\hat{f}(0)e^{-\frac{x^2}{4t}}}{\sqrt{4\pi t}} \tag{16}$$

as $t \to \infty$.

We have also to consider the non scale invariant situation : in general u_L will satisfy:

$$\dot{u}_L = \Delta u_L + F_L(u_L, \partial u_L, \partial^2 u_L) \tag{17}$$

where

$$F_L(a,b,c) = L^{2+\alpha} F(L^{-\alpha}a, L^{-1-\alpha}b, L^{-2-\alpha}c) \tag{18}$$

and the "semigroup" property becomes

$$R_{L^n,F} = R_{L,F_{L^{n-1}}} \circ \cdots \circ R_{L,F_L} \circ R_{L,F} \tag{19}$$

where we underlined the dependence of R on L and F. However, it is preferable to regard R_L as acting on a pair $\begin{pmatrix} f \\ F \end{pmatrix}$ and we can again ask for a fixed point $\begin{pmatrix} f^* \\ F^* \end{pmatrix}$ where (1) with $F = F^*$ is scale invariant (i.e. $F_L^* = F^*$) and

$$R_{L,F^*} f^* = f^* \tag{20}$$

To investigate the stability in the space of F's, consider:

$$\dot{u} = \partial^2 u + F^* + \lambda u^{p_0} (\partial u)^{p_1} (\partial^2 u)^{p_2} \tag{21}$$

and let F^* be invariant for a given α in (18). Then, from (18) we see that u_L satisfies (21) with $\lambda \to \lambda L^{-D}$ where $D = -(2+\alpha) + p_0\alpha + p_1(\alpha+1) + p_2(\alpha+2)$. Following the standard terminology we call a term with

$D > 0$ irrelevant

$D = 0$ marginal

$D < 0$ relevant

For example, in equation (8), (which is like (21) with $F^* = 0$) and for $\alpha = 1$, $p > 3$ is irrelevant, $p = 3$ marginal and $p < 3$ relevant.

In [6], we show that, for eq. (21) with $F^* = 0$, $\alpha = d$, $D > 0$ and λ small, $u(x,t)$ behaves as in (16), except that $\hat{f}(0)$ is replaced by some constant A, depending on f and F. Here A is analogous to a renormalised quantity in field theory. We also discuss the marginal cases, in particular $d = 1$, $F = -\lambda u^3$, where the solution

$$u(x\sqrt{t},t) \simeq (\frac{2\lambda t \log t}{\sqrt{3}})^{-\frac{1}{2}} e^{-\frac{x^2}{4}}$$

as $t \to \infty$. The "logarithmic correction" here is similar to what happens in asymptotically free theories.

One may also consider nontrivial fixed points: going back to (6, 7), one sees that $R_L f^* = f^*$ for eq. (8) with $\alpha = \frac{2}{p-1}$ amounts to having f^* solve the *ordinary* differential equation:

$$f'' + \frac{1}{2} x f' + \frac{f}{p-1} - f^p = 0 \tag{22}$$

A lot is known about the solutions of this equation (see the discussion and references in [6]). This establishes the existence of fixed points. Their stability is more difficult to analyse.

Concerning the problems mentioned in the beginning, the Renormalization Group method allows to prove:

- For the lattice version of (1,2), if $d > 2$ and if the disorder is weak (ϵ small in (2)) the probability $u(x,t)$ is Gaussian with an ϵ-dependent diffusion constant [5]
- For the Ginzburg-Landau equation, there are two types of results. One knows (see [9,10]) that the stationary solutions $u_{q\theta}$ are linearly stable if $q^2 < \frac{1}{3}$. One may inquire about the development of a solution with initial data (4) such that $q_\pm$ are both in the stable domain, or such that one is stable and the other unstable. In the first case one expects some kind of pattern to emerge while, in the second case, one expects the solution to be a moving front, since the stable phase invades the unstable one. We obtain results in that direction: For $q_\pm$, $\theta_\pm$ small, one finds the following kind of "diffusive repair" : for $x = \mathcal{O}(1)$,

  $$u(x,t) \simeq e^{i\sqrt{t}\varphi^*} u_{q^*\theta^*}(x)$$

  where the constants φ^*, q^*, θ^* depend only on $q_\pm$, $\theta_\pm$ i.e. the boundary conditions select a stationary solution around which the true solution oscillates [7]..

  For the front propagation, we consider only the extreme case $q_- = 0$, $q_+ = 1$ but with u complex, and we find that the solution tends to the real front analysed in [1,2] i.e. the phase of u diffuses away [8]

References

[1] Aronson, D.G., Weinberger, H.F., Adv. Math. **30**, 33 (1978).
[2] Bramson, M., Memoirs Amer Math. Soc. **44**, 1 (1983).
[3] Bricmont, J., Kupiainen, A., Phys. Rev. Lett. **66**, 1689 (1991).
[4] Bricmont, J., Kupiainen, A., Physica A **163**, 31 (1990).
[5] Bricmont, J., Kupiainen, A., Commun. Math. Phys. **142**, 345 (1991).
[6] Bricmont, J., Kupiainen, A., Lin, G., Renormalization Group and asymptotics of solutions of non-linear parabolic equations. To appear in Commun. Pure Appl. Math.
[7] Bricmont, J., Kupiainen, A., Renormalization Group and the Ginzburg-Landau equation, to appear in Commun. Math. Phys.
[8] Bricmont, J., Kupiainen, A., in preparation.
[9] Collet, P., Eckmann, J-P., Instabilities and fronts in extended systems, Princeton Univ. Press, 1990.
[10] Collet, P., Eckmann, J-P., Space-time behaviour in problems of hydrodynamic type: a case study, preprint (1992).
[11] Giga, Y., Kohn, R.V., Indiana Univ. Math. J., **36**, 1 (1987).
[12] Goldenfeld, N., Martin, O., Oono, Y., Lin, F., Phys. Rev. Lett. **64**, 1361 (1990).

Random Walk Rperesentations and Mayer Expansion

KEIICHI R. ITO
Department of Mathematics, College of Liberal Arts, Kyoto University, Kyoto 606, Japan * *and*
Department of Mathematics and Information, Konan College of Women, Konan 483, Japan

Abstract.
We represent the free energy of the N-vector models in terms of the correlation functions of random walks. This enables us to find an analytic region of the free energy with respect to the inverse temperature β. We apply this method to the block spin transformations.

Key words: random walk, Mayer expansion, block spin transformation

1. Introduction

This is a shortened version of a paper [7] which will appear in the issue of Communications in Mathematical Physics dedicated to H.Araki. Some applications to the renormalization group method are added.

It seems that we need a new method of cluster expansion to investigate non-Abelian statistical models by means of renormalization group methods. We start with the definition of the model, say ν dimensional Heisenberg model with $O(N)$ symmetry (N-vector mdel):

$$< . >\equiv \frac{1}{Z_\Lambda(J)}\int (.) \exp[-\frac{J}{2}\sum_{|x-y|=1}(\phi(x)-\phi(y))^2]\prod d\phi(x), \qquad (1)$$

where $\phi(x) \in S^{N-1}$ for all lattice points $x \in \Lambda(\subset Z^\nu)$, $J > 0$ is the coupling constant (=inverse temperature) and the integration $d\phi(x)$ is over the $N-1$ dimensional sphere S^{N-1}. Finally $Z_\Lambda(J)$ is the normalization constant chosen so that $< 1 >= 1$. There are several studies on the thermodynamic properties of this system. Among them is

Theorem (Brydge-Froehlich-Spencer-Sokal [3], see also Simon-Lieb [9]). *If $J < N/2\nu$, there exist strictly positive constants m and C such that*

$$< \phi(0)\phi(x) >\leq C\exp[-m|x|].$$

* e-mail:ito@kurims.kyoto-u.ac.jp

H. Araki et al. (eds.), Quantum and Non-Commutative Analysis, 119–135.

It is a long standing problem in modern physics if this holds for any (real) J provided $N > 2$ and $\nu=2$. We want to represent the free energy $\alpha(J) = \lim |\Lambda|^{-1} \log(Z_\Lambda)$ as a function of J and obtain the region in which α is analytic in J. Our main results are summarized in the following theorem:
Theorem(1) *The free energy $\alpha(J)$ is represented in terms of the turncated correlation (Ursell) functions of random walks.*
(2) *The free energy $\alpha(J)$ is analytic in J in the region*

$$\Omega(N,\nu) = \{J; |J| < J_c \equiv (1-\varepsilon)\frac{N}{2\nu}\}$$

where ν stands for the dimensions of the lattice space and $\varepsilon > 0$ tends to 0 as $\nu \to \infty$ uniformly in N.

For $\nu=2$, we have numerical bounds on J_c which are very close to $N/2\nu$. We also discuss an alternative method in order to improve the convergence radius. This is a sort of planar-diagram dominance which holds in the limit of $N \to \infty$.

In the final section of this paper, we show an application of the present method to the block spin transformations. We argue that several steps of the block spin transformations together with the present method enable us to extend the analytic region along the real axis.

2. Random Walk Representation of the Free Energy

After some standard calculations [3], we start with the following random walk representation of the partition function:

$$Z_\Lambda(J) = \int \exp[-\frac{J}{2}\sum(\phi(x)-\phi(y))^2]\prod d\phi(x)$$

$$= \sum_{n=0}^{\infty} \frac{1}{n!}(\frac{N}{2})^n \sum_{\omega_1,\dots,\omega_n} \frac{J_{\omega_1}\dots J_{\omega_n}}{2^{|\omega_1|}\dots 2^{|\omega_n|}} \exp[-V(\omega_1,\dots,\omega_n)] \tag{2}$$

(except for a trivial constant coefficient), where $\omega = \{\{b_0,\dots,b_L\};\ b_j = (x_j, x_{j+1}), x \in Z^\nu, |x_j - x_{j+1}| = 1,\quad x_{L+1} = x_0\}$ are closed oriented loops made by nearest neighbor bonds in Z^d, $|\omega_i|$ = length of the loop ω_i which equals $\sum_x v_i(x) \equiv |v_i|$, the sum of the visiting numbers. Here two loops ω and $\tilde{\omega}$ are identified if they differ only by starting points, and

$$J_\omega = \frac{1}{\tau_\omega}\prod_{b\in\omega} J_b, \tag{3a}$$

$$\exp[-V(\omega_1,...,\omega_n)] = \prod_{x\in\cup\omega_i} \frac{[N/2-1]!}{[v(x)+N/2-1]!}, \tag{3b}$$

where τ_ω is the winding number of loop ω defined by

$$\tau_\omega = \frac{|\omega|}{p} \in \{1,2,...,|\omega|/2\}, \tag{3c}$$

with p being the smallest integer $\in \{2,3,...,|\omega|\}$ such that $b_i = b_{[i+p]}$ for all i ($[i+p] = i+p \bmod |\omega|$), and finally

$$v(x) = \sum_i v_i(x) \tag{3d}$$

is the sum of the visiting numbers $v_i(x)$ of ω_i at $x \in Z^\nu$. The winding number τ_ω is is different from 1 only when ω has the period p less than $|\omega|$ and then $\tau_\omega = 1$ for most ω.

Now the standard Mayer expansion method implies that

$$\alpha(J) = \lim |\Lambda|^{-1} \log Z_\Lambda(J)$$

$$= \sum_{n=1}^{\infty} \frac{1}{n!}\left(\frac{N}{2}\right)^n \sum_{\omega_1,...,\omega_n} \frac{1}{|X|} \frac{J_{\omega_1}...J_{\omega_n}}{2^{|\omega_1|}...2^{|\omega_n|}} [\exp[-V(\omega_1,...,\omega_n)]]^T, \tag{4}$$

where $X = \cup \text{supp}\omega_i$, $0 \in X$ and $[...]^T$ mean the truncated correlation functions of random walks:

$$[\exp[-V(\omega_1,...,\omega_n)]]^T = \exp[-V(\omega_1,...,\omega_n)]$$

$$-\sum_{t_1\cup t_2=t} \exp[-V(t_1)-V(t_2)] + 2\sum_{t_1\cup t_2\cup t_3=t} \exp[-\sum V(t_i)] - ... \tag{5}$$

Here $t = \{1,..,n\}$ and $t = \cup_1^k t_i$ means the partitions of t into non-overlapping subsets t_i and for $t_i = \{a_1,..,a_j\}$

$$\exp[-V(t_i)] = \exp[-V(\omega_{a_1},...,\omega_{a_j})].$$

In order to represent the truncated correlation functions of the random walks, we introduce truncated potentials $V^T(\omega_1,...,\omega_n)$ defined by the following formulas:

$$V(\omega_1) = V^T(\omega_1),$$

$$V(\omega_1,\omega_2) = \sum_{t\subset\{1,2\}} V^T(t),$$

..................,

$$V(\omega_1,..,\omega_n) = \sum_{t\subset\{1,..,n\}} V^T(t). \tag{6a}$$

Therefore we conversely have

$$V^T(\omega_1,...,\omega_k) = \sum_{t\subset\{1,..,k\}} (-1)^{k-|t|}V(t), \tag{6b}$$

where t runs over all possible subsets of $\{1,2,..,k\}$. Given a subset t, we define the composite random walk denoted by $\omega_{[t]}$ or simply by $[t]$. This is the random walk that is equal to $\cup_{i\in t}\omega_i$ together with the visiting numbers. We similarly define many body potentials $V^T([t_1],..,[t_n])$ for the composite random walks:

$$V^T([t_1]) = V([t_1]),$$
$$V^T([t_1],[t_2]) = V([t_1],[t_2]) - V([t_1]) - V([t_2]),$$

and etc. Using eq.(6b), we easily find that

$$V^T([t_1],...,[t_n]) = \sum_{t'_1,..,t'_n} V^T(t'_1,...,t'_n), \tag{7}$$

where t'_i are subsets of t_i different from $\emptyset$ and the truncated functions on the right-hand-side are *the truncated functions with respect to all walks contained in* $\cup t'_i$.

The Ursell functions have complicated expressions when we have many-body interaction terms in the potentials. We first explain our notation following Refs.5 and 6.

For $t = \{1,2,...,n\}$, let $t_1 = \{1\}$, and let $\{2,3,...,n\} = \cup_{i=2}^k t_i$ be any partition of $\{2,...,n\}$. Given k, let η_k be a map $\{1,2,...,k-1\} \to \{1,2,..,k-1\}$ such that $\eta_k(i) \leq i$. Moreover let $X_1 = t_1 = \{1\}$, $X_i = X_{i-1}\cup t_i$, $i = 2,..,k$. Then for any subset $b \subset t$, define functions s_i so that

$$s_i(b) = \begin{cases} s_i & \text{if } X_i \cap b \neq \emptyset, \text{ and } X_i^c \cap b \neq \emptyset \\ 1 & \text{otherwise.} \end{cases} \tag{8}$$

Namely $s_i(b) = 1$ whenever $b \subset X_i$ or $b \subset X_i^c$ and $s_i(b) = s_i$ otherwise. Let

$$\begin{aligned} W(\{t_i\}_1^k;\{s_i\}_1^{k-1}) &\equiv \textstyle\sum_{b\subset t} s_1(b)...s_{k-1}(b)V^T(b) \\ &= \textstyle\sum_{1\leq i<j\leq k} s_i...s_{j-1}\mathcal{W}(i,j) + \sum V([t_i]), \end{aligned} \tag{9a}$$

where

$$\mathcal{W}(i,j) \equiv V^T([i],[j]) + V^T([i],[\cup_{i+1}^{j-1}t_k],[j]) \tag{9b}$$

and this form follows by the definition of the many body potential (6a) and we identified the subscript i with the set of loops t_i. (Note that $V([t]) =$

$V(t) = V(\omega_1, ..., \omega_n)$.) This is a convex linear combination of $V(t_i, t_{i+1}, ..., t_j)$, $1 \leq i \leq j \leq k$ and thus satisfies the bound (in fact we show $\mathcal{W} \geq 0$, see Lemma 2)

$$\exp[-W(\{t_i\}_1^k; \{s_i\}_1^{k-1})] \leq \exp[-\sum V(t_i)]$$

$$= \prod_i \prod_{x \in \cup_{j \in t_i} \omega_j} \frac{[\frac{N}{2} - 1]!}{[\frac{N}{2} + \sum_j v_j(x) - 1]!}, \tag{10}$$

where $\sum_{j \in t_i} v_j(x)$ is the sum of visiting numbers of $\{\omega_j; j \in t_i\}$ at x.

For a function η_k, we have the tree graph $T = T(\eta_k)$ over t regarding $(i+1, \eta_k(i))$ as the bond connecting two vertices $i+1$ and $\eta_k(i) \in \{1, 2, ..., i\}$. Set $\eta_k(i) = j$ for simplicity. We define the reduced two-body potential $\mathcal{V}(j, i+1)$, $(j < i)$ by

$$\mathcal{V}(j, i+1) \equiv \sum_{t'_j \subset t_j} \sum_{t'_{j+1} \subset t_{j+1}} \cdots \sum_{t'_i \subset t_i} V^T(t'_j, .., t'_i, t_{i+1}), \tag{11}$$

where $t'_{j-1}, .., t'_i$ may be empty but t'_j is not empty. If some of $t'_i, ..., t'_{j-1}$ are empty, we just neglect those in V^T. Thus we have

$$\mathcal{V}(j, i+1) = V^T([j], i+1) + V^T([j], [t], i+1), \tag{12}$$

where $[t]$ means the composite random walk made by single random walks in t_ℓ, $(\ell = j+1, .., i)$.

Then we have the following theorem whose proof is essentially in [2].

Theorem 1.

$$[\exp[-V(\omega_1, ..., \omega_n)]]_T = \sum_{k=2}^{n} \sum_{t_1, .., t_k} \sum_{\eta_k} \int_0^1 ds_1 ... \int_0^1 ds_{k-1}$$

$$\times f(\eta_k, s) \prod_i \mathcal{V}(\eta_k(i), i+1) \exp[-W(\{t_i\}; \{s_i\})], \tag{13}$$

where $t_1 = 1$ and $t_2 \cup ... \cup t_k$ changes over all partitions of $\{2, ..., n\}$ into $k-1$ subsets, and

$$f(\eta_k, s) = \prod_{i=1}^{k-1} s_{i+1} ... s_{\eta(i)}. \tag{14}$$

3. Truncated Correlation Functions

We here estimate $\mathcal{V}(i,j+1)$ and $\mathcal{W}(i,j+1)$ which are expressed in terms of V^T. To do so, we start with the estimate of more general $V^T(t) = V^T(\omega_1,...,\omega_n)$. Let $x \in Z^d$ be a point in $\cap \mathrm{supp}\, \omega_i$. Then if $n > 1$

$$V^T(t_k) = \sum f_N(x), \tag{15a}$$

$$f_N(x) = \sum_{t'} (-1)^{n-|t'|} \log[(\frac{N}{2} + v(t') - 1)!], \tag{15b}$$

$$v(t') = \sum_{i \in t'} v_i(x), \tag{15c}$$

where $t' \subset t$ including the case of $t' = \emptyset$. For example, we have

$$f_N(x) = \log \frac{[\frac{N}{2} + v_1(x) - 1]!}{[\frac{N}{2} - 1]!}$$

for $n = 1$ and

$$f_N(x) = \log \frac{[\frac{N}{2} + v_1(x) + v_2(x) - 1]![\frac{N}{2} - 1]!}{[\frac{N}{2} + v_1(x) - 1]![\frac{N}{2} + v_2(x) - 1]!}$$

for $n = 2$, respectively. Now we see

$$f_{N+2} - f_N = \sum_{t'} (-1)^{n-|t'|} \log(\frac{N}{2} + v(t'))$$

$$= \int_0^1 .. \int_0^1 ds_1...ds_n (\prod \frac{\partial}{\partial s_i}) \log(\frac{N}{2} + \sum v_i s_i)$$

$$= (-1)^{n-1}(n-1)! \int_0^1 ... \int_0^1 \prod ds_i \frac{\prod v_i}{[\frac{N}{2} + \sum v_i s_i]^n}.$$

Thus we have

$$f_N(x) = \sum_{k=0}^{\infty} (-1)^n (n-1)! \int_0^1 ds_1 ... \int_0^1 ds_n \frac{\prod v_i}{[\frac{N}{2} + \sum v_i s_i + k]^n}$$

$$= \sum_{k=0}^{\infty} (-1)^n (n-1)! \int_0^{2v_1/N} dy_1 ... \int_0^{2v_n/N} dy_n \frac{1}{[1 + \sum y_i + \frac{2k}{N}]^n} \tag{16a}$$

which converges absolutely for $n > 1$ and is equals to

$$(-1)^n (n-1)! \frac{N}{2(n-1)} (1 + O(\frac{n}{N})) \int_0^{2v_1/N} ... \int_0^{2v_n/N} \prod dy_i \frac{1}{[1 + \sum y_i]^{n-1}}, \tag{16b}$$

where $O(n/N) < 2(n-1)/N$ is the correction term caused by replacing the summation with the integral. See Remark 1 below. We let $t = \cup_2^i t_i$ and let $[t]$ stand for the composite random walk made by single random walks contained in t. Assume $x \in t_1 \cap t \cap t_{i+1}$, and assume that t_1 visits x v_1 times, t visits x v_2 times and t_{i+1} visits x v_3 times, and finaly each loop $\omega_1, ..., \omega_n$ in t_{i+1} visits x v_i' times. Then $v_3 = \sum v_i'$ and the following two lemmas and a corollary are immediate:

Lemma 2. *$\mathcal{V}(1, i+1) = \sum_x \mathcal{V}(1, i+1)(x)$ where*

$$\mathcal{V}(1,i+1)(x) = \sum_{k=0}^{\infty}(-1)^{n+1}(n+1)! \int_0^{2v_1/N} dy_1 \int_{2v_2/N}^{\infty} dy_2$$

$$\times \int_0^{2v_1'/N} ... \int_0^{2v_n'/N} \prod dy_i' \frac{1}{[1+y_1+y_2+\sum y_i' + \frac{2k}{N}]^{n+2}}. \tag{17}$$

Corollary 3. *Let $n_j \geq 1$ be the number of loops contained in the composite loop $\omega_{[t_j]}$ and let $v_j(x) \in N$ be the visiting number of the loop ω_j at x. Then*

$$|\mathcal{V}(i,j)(x)| \leq (1 + O(\frac{n_j}{N}))(\frac{2}{N})^{n_j}(n_j - 1)! v_{[t_i]}(x) \prod_{s=1}^{n_j} v_{j_s}(x). \tag{18}$$

Lemma 4. *$\mathcal{W}(1, i+1) = \sum_x \mathcal{W}(1, i+1)(x)$ where*

$$\mathcal{W}(1,i+1)(x) = \sum_{k=0}^{\infty} 2 \int_0^{2v_1/N} \int_{2v_2/N}^{\infty} \int_0^{2v_3/N} \frac{\prod dy_i}{[1+y_1+y_2+y_3+\frac{2k}{N}]^3}. \tag{19}$$

Remark 1. It is easy to see that the correction term $O(n/N)$ is bounded by $2(n-1)/N$. Then the additional factor $1 + O(n/N)$ is conpensated by the marginal term $1 + 2(n-1)/N$ in $(2/N)^n(1+2/N)^{-1}...(1+2(n-1)/N)^{-1}$ contained in $\exp[-V(\omega_{[t]})]$. This comes from the n-ple self-crossing points in $\omega_{[t]} = \cup_{i \in t}\omega_i$.

4. Convergence of the Mayer Expansion

Let $t = \{1, 2, ..., n\}$, $t_1 = \{1\}$ and let $t = t_1 \cup ... \cup t_k$ be partitions of t into sets of loops. Then the n-loop contribution K_n to α is given by

$$K_n = \frac{N^n}{n! 2^n} \sum_{\omega_1, .., \omega_n} \frac{1}{|X|} \prod \frac{J_{\omega_i}}{2^{|\omega_i|}} \sum_k \sum_{t_1, ..., t_k} (\prod_{i=1}^{k-1} \int_0^1 ds_i) \mathcal{U}(s) e^{-\mathcal{W}(s) - \sum V(t_i)} \tag{20}$$

where $0 \in X \equiv \cup \mathrm{supp}\omega_i$ and we set

$$\mathcal{U}(s) = \prod_{i=2}^{k} U(i), \mathcal{W}(s) = \sum_{i=1}^{k-1} s_i W(i+1)$$

together with the following notation ($i = 2, ..., k$):

$$\begin{aligned} U(i) &= \mathcal{V}(i-1,i) + s_{i-2}\mathcal{V}(i-2,i) + ... + s_1 s_2 ... s_{i-2}\mathcal{V}(1,i), \\ W(i) &= \mathcal{W}(i-1,i) + s_{i-2}\mathcal{W}(i-2,i) + ... + s_1 s_2 .. s_{i-2}\mathcal{W}(1,i). \end{aligned}$$

Here $s_0 \equiv 0$, sgn$(U(i))$=$(-1)^{n_i}$ and then

$$\mathrm{sgn}(K_n) = (-1)^{n-1}. \tag{21}$$

We define the following quantities (s_{k-1} is a dummy in eq.(23)):

$$|v_{[i]}| \equiv |v_{[t_i]}| = \sum_x v_{t_{[i]}}(x) = \sum_{j \in t_i} \sum_x v_j(x), \tag{22}$$

$$I_k \equiv \int_0^1 ... \int_0^1 \prod_1^{k-1} ds_i |v_{[1]}|(|v_{[2]}| + s_1|v_{[1]}|)...$$

$$\times(|v_{[k-1]}| + s_{k-2}|v_{[k-2]}| + ... + s_{k-2}s_{k-3}...s_2 s_1|v_{[1]}|). \tag{23}$$

Lemma 5 [1,2,4]. *Let* $M = \sum_1^n |v_i|$. *Then* $M \geq 2n$ *and*

$$I_k < (\frac{M}{k})^k e^k.$$

Proof. Set $\mathbf{v}(s, i-1) = |v_{[i-1]}| + s_{i-2}|v_{[i-2]}| + ... + s_{i-2}s_{i-3}..s_2 s_1|v_{[1]}|$. Multiplying $\exp[\alpha\mathbf{v}(s, k-1)s_{k-1}] \geq 1$ with $\alpha > 0$, we have

$$\int_0^1 ds_{k-1}\mathbf{v}(s, k-1)\exp[\alpha\mathbf{v}(s, k-1)s_{k-1}]$$

$$= \alpha^{-1}(\exp[\alpha|v_{[k-1]}| + \alpha\mathbf{v}(s, k-2)s_{k-2}] - 1)$$

$$< \alpha^{-1}\exp[\alpha|v_{[k-1]}|]\exp[\alpha\mathbf{v}(s, k-2)s_{k-2}],$$

which yields $I_k < (\alpha)^{-k}\exp[\sum_1^{k-1}\alpha|v_{[i]}|]$. Then set $\alpha = k/\sum|v_{[i]}|$.

Q.E.D.

Though this estimate may look crude, this inequality is almost saturated if all $|v_i|$ are equal (e^k is replaced by $k^k/k!$ [1].)

Suppose loops $\{\omega_1, ..., \omega_n\}$ and partitions $t_1 = \{1\}, t_2, ..., t_k$ of $\{1, 2, ..., n\}$ are given. Let η be tree graphs over $t_1, ...t_k$, $C_n = N^n/2^n n!$, and consider

$$C_n \sum_{\text{transl.}} \frac{1}{|X|} \prod \frac{J_{\omega_i}}{2^{|\omega_i|}} \sum_{\eta} \int .. \int \prod ds_i f(\eta, s) \prod \mathcal{V}(\eta(i), i+1) e^{-W(s) - \sum V(t_i)} \tag{24}$$

where $0 \in X \equiv \cup \text{supp}\omega_i$ and $\sum_{\text{transl.}}$ means the sum over all possible translations of $\omega_i, i > 1$. For the above quantity to be different from zero, all loops $\omega_{i'}$ contained in ω_{t_i} must cross each other at a same point, and all composite loops ω_{t_i} must cross each other. From eq.(18) in Corollary 3, we have

$$\sum_{\text{transl.}} |\mathcal{V}(i,j)| \leq (1 + O(\frac{n_j}{N}))(\frac{2}{N})^{n_j} (n_j - 1)! |v_{[i]}| \prod_{s=1}^{n_j} |v_{j_s}|$$

where we considered the translations of $\omega_{j_s}, j_s \in t_j$ only. Thus we finally have

$$|\text{eq.}(24)| \leq \frac{N}{2n!} \prod (n_i - 1)! \{\prod \frac{J_{\omega_i}}{2^{|\omega_i|}} |v_i| \exp[-V(\omega_i)]\} I_k$$

$$\leq \frac{N}{2n!} \prod \{(n_i - 1)! \frac{J_{\omega_i}}{2^{|\omega_i|}} |v_i| \exp[-V(\omega_i)]\} (\frac{\sum_{i=1}^{n} \sum_x v_i(x)}{k})^k e^k, \tag{25}$$

where we used Lemma 5 and we omitted for simplicity the correction term $O(n/N)$ canceled by the marginal term in $\exp[-V(\omega)]$, see Remark 1.

Suppose n_2, ..., n_k are given in such a way that $n_i \geq 1$ and $\sum_2^k n_i = n - 1$. The number of the partitions of {2,...,n} into these groups is

$$\frac{(n-1)!}{n_2! ... n_k!},$$

and the number of the solutions of the equation $\sum_2^k n_i = n - 1$ with $n_i \geq 1$ is $\binom{n-2}{k-2}$. Therefore $|K_n|$ is bounded by

$$\sum_{0 \in \omega_i} \frac{N}{n} \prod \frac{J_{\omega_i} |v_i| \exp[-V(\omega_i)]}{2^{|\omega_i|}} \{\sum_{k=2}^{n} H_k (\frac{\sum |v_i|}{k})^k e^k\} \tag{26a}$$

$$\leq \sum_{0 \in \omega_i} \frac{N}{n} \prod \frac{J_{\omega_i} |v_i| \exp[-V(\omega_i)]}{2^{|\omega_i|}} \{\sum_{k=2}^{n} \binom{n-2}{k-2} (\frac{\sum |v_i|}{k})^k e^k\}, \tag{26b}$$

where H_k is defined by

$$H_k \equiv \sum_{n_2 + ... + n_k = n-1} \frac{1}{n_2 ... n_k} \leq \binom{n-2}{k-2}. \tag{27}$$

Moreover setting $M = \sum |v_i|$ $(\geq 2n)$, we have

$$\sum_{k=2}^{n} \binom{n-2}{k-2} (\frac{M}{k})^k e^k = \{\sum_{k=2}^{n} \binom{n-2}{k-2} (\frac{n}{eM})^{n-k} (\frac{n}{k})^k\} \exp[n \log \frac{eM}{n}].$$

Here

$$\sum_{k=2}^{n} \binom{n-2}{k-2} (\frac{n}{eM})^{n-k} (\frac{n}{k})^k \leq (1+C_1)^n, \tag{28a}$$

$$\exp[n \log \frac{M}{n}] \leq \exp[C_2 \sum |v_i|], \tag{28b}$$

with $C_1 \leq 1/2, C_1 \to 0$ as $M/n \to \infty$, and $C_2 \leq 1/e, C_2 \to 0$ as $M/n \to \infty$, since $M = \sum |v_i| \geq 2n$. In order to improve C_i numerically, we estimate H_k more carefully [7]. This yields

$$\sum_{k=2}^{n} H_k (\frac{n}{eM})^{n-k} (\frac{n}{k})^k \leq C n^{\alpha} (1+C_1)^n,$$

where $C_1 < 0.2581$, and C and α are some positive constants.
Lemma 6.

$$|K_n| < CNn^{\alpha} \prod [\sum_{0 \in \omega_i} \frac{J_{\omega_i} |v_i| \exp[-V(\omega_i)]}{2^{|\omega_i|}} (1+C_1) \exp[1 + C_2 |v_i|]], \tag{29}$$

where $C = O(1)$, $C_1 < 0.2581$, $C_2 < 1/e$ uniformly in n and α is a positive constant. Moreover C_1 and C_2 converge to 0 as $\sum_i |v_i|/n \to \infty$.

Remark 2. Since $\exp[C_2|v_i|] = \exp[C_2|\omega_i|]$, the factor $\exp[C_2|v_i|]$ is absorbed by replacing J_b with $\exp[C_2]J_b = \exp[1/e]J_b = 1.4446 J_b$ (i.e. scaling) in the definition of J_{ω_i}.

5. Numerical Evaluation of the Convergent Radius

Let p_i and q_i be the numbers of unit walks $x \to x + e_i$ and $x - e_i \to x$ respectively, contained in ω. A walk ω of length $2m$ forms a loop if and only if $p_i = q_i = n_i$ and $\sum n_i = m$. Then the number of loops of length $2m$ is given by

$$\ell_m \equiv \sum_{n_1 + \ldots + n_\nu = m} \frac{(2m)!}{(n_1! \ldots n_\nu!)^2} = O(\nu^m). \tag{30}$$

For small m, they are:

$$\begin{aligned}\ell_1 &= 2\nu,\\ \ell_2 &= 12\nu^2 - 6\nu,\\ \ell_3 &= 120\nu^3 - 180\nu^2 + 80\nu,\\ \ell_4 &= 1680\nu^4 - 5040\nu^3 + 6740\nu^2 - 2326\nu,\end{aligned}$$

and it is easy to see

$$\ell_m \leq \frac{1}{\nu}(2\nu)^{2m} \text{ for } m \leq m(\nu)$$

for some constant $m(\nu)$ ($m(\nu) \to \infty$ as $\nu \to \infty$). By the Stirling formula, we have

$$\ell_m = O(1)2^{2m} \sum_{n_1+...+n_\nu=m} (\frac{m!}{n_1!...n_\nu!})^2$$

and then $O(1)2^{2m}\nu^m < \ell_m$. One can also show that $\ell_m = O(m^{-\alpha})(2\nu)^{2m}$. The number of random walks of length $2m$ is

$$\sum_{p_1+q_1+...+p_\nu+q_\nu=2m} \frac{(2m)!}{p_1!q_1!...p_\nu!q_\nu!} = (2\nu)^{2m}.$$

By collecting terms satisfying $p_i + q_i = 2n_i$, and noticing that

$$\sum_{p+q=2n} \frac{(n!)^2}{p!q!} = \frac{2^{2n}(n!)^2}{(2n)!} \geq 1 + \sqrt{n},$$

we have

$$\ell_m < \tau_m(2\nu)^{2m}, \quad \tau_m = \frac{1}{1+\sqrt{m}}.$$

Here we used the inequality $\prod(1+\sqrt{n_i}) \geq 1+\sqrt{m}$. The main contribution comes from a neighborhood of $n_1 = ... = n_\nu = m/\nu$. Thus if $\nu = 2$ we can improve the bound as follows (by an explicit calculation):

$$\tau_m \leq \frac{1}{1+m/2}. \tag{31}$$

Theorem 7.(1)*Let $\tilde{J}_b = \exp[C_2]J_b$. The Mayer expansion converges absolutely if*

$$\sum_{0\in\omega} \frac{\tilde{J}_\omega}{2^{|\omega|}} \exp[-V(\omega)]|v| < \frac{1}{e(1+C_1)} \tag{32}$$

where ω are random loops starting from the origin.
(2) *The Mayer expansion converges absolutely in the complex region*

$$\Omega(N,\nu) \equiv \{J; |J| < (1 - O(\frac{1}{\nu}))\frac{N}{2\nu}\} \tag{33}$$

where ν stands for the dimension of the lattice space.
(3) *Let $\nu = 2$. Then the Mayer expansion converges absolutely in the complex region*

$$\Omega(N,2) = \{J; |J| < 0.102N\} \tag{34}$$

Proof. (1) This is immediate from Lemma 6.
(2) By eq.(26b) and eq.(28a), it is enough to argue the convergence of the following series:

$$\sum_{n=1}^{\infty}\{\sum_{M=n}^{\infty}\{\sum_{m_1+m_2+...+m_n=M} h(m_1)...h(m_n)\}\exp[n\log\frac{2M}{n}]\} \tag{35}$$

where

$$h(m) \equiv \sum_{\omega:|\omega|=2m} \frac{|\omega|J_\omega}{2^{|\omega|}}\exp[-V(\omega)] < 2m\ell_m(\frac{J}{N})^{2m}.$$

Let $2m$ be the length of a random loop ω. At each point, the walk ω has 2ν directions to go, and one of them is backward. If ω visits $x + e_\mu$ from x and ω goes back, ω visits x twice at least. Then

$$h(m) \le 2m\sum_{p=0}^{\infty}\binom{2m}{p}(\frac{(2J-1)J}{N})^{2m-p}(\frac{J}{N})^p = 2mx^{2m}$$

where $x = (2\nu - 1)J/N + J/(N+2) < 2\nu J/N$. We use the fact that the number of loop diagrams of length $2m$ is much less than $(2\nu)^{2m}$ for large ν: $\ell_m = O(\nu^m)$. We take ν large keeping $x = 2\nu J/N < 1$ constant. Then the following inequalities hold:

$$\sum_{m=1}^{k} h(m) < O(\frac{1}{\nu}), \quad \sum_{m=k+1}^{k'} h(m) = O(1) \le \frac{2(k+1)x^{2(k+1)}}{(1-x^2)^2} \tag{36}$$

where k and k' are some positive integers and $k \to \infty$, $k' \to \infty$ as $\nu \to \infty$. Then $\sum h(m) <$ const yields the bound $x < 1 - \varepsilon$ for any const < 1 as $\nu \to \infty$, where $\varepsilon \to 0$ as $k \to \infty$. Since we are interested in the upper bound, we can assume

$$h(m) = \begin{cases} O(\nu^{-1}) & \text{for } m < k, \\ 2m(\frac{2\nu J}{N})^{2m} & \text{for } m \ge k. \end{cases}$$

Or we may assume $h(m) = (2\nu J/N)^{2m}$ for $m \geq k$ with a small change of J since m is sufficiently large. Thus we have

$$L(n, M) \equiv \sum_{m_1+...+m_n=M} h(m_1)...h(m_n)$$

$$= \sum_{\ell=0}^{n} \sum_{s=\ell}^{\ell k} \binom{n}{\ell} \{ \sum_{m_1+...+m_\ell=s; m_i \leq k} \prod_{1}^{\ell} h(m_i) \} \{ \sum_{m_{\ell+1}+...+m_n=M-s; m_i>k} \prod_{\ell+1}^{n} h(m_i) \}$$

$$\leq \sum_{\ell=0}^{n} \sum_{s} \binom{n}{\ell} \binom{M-s-(n-\ell)k-1}{n-\ell} (\frac{c}{\nu})^\ell (\frac{2\nu J}{N})^{2(M-s)}.$$

Applying the Stirling formula to the summand, we see that $L(n, M)$ takes its maximum at $M = M_0 = (1-O(\nu^{-1}))nk >> n$, and decreases exponentially in $M - M_0$ and the contribution from $M < M_0/2$ is exponentially small compared with the one from $M \geq M_0/2$. The factor $\exp[(n/2M)\log(2M/n)]$ tends to 1 as $M \to \infty$ and thus the assertion holds.

(3) For $\nu = 2$ we estimate eq.(32) explicitly up to $|\omega| = 10$ $(m = 5)$:

$$8\alpha^2 + 144\alpha^4 + 2400\alpha^6 + 39200\alpha^8 + 634860\alpha^{10}$$

$$+ \sum_{m=6}^{\infty} 2m\tau_m(4\alpha)^{2m} < \frac{1}{e(1+C_1)} \tag{37}$$

where $1/e(1 + C_1) > 0.2924$ and $\alpha = \tilde{J}/N = \exp[1/e]J/N$. We can numerically show that this is satisfied if $\alpha < 0.1483$, using the bound (31) for τ_m. Therefore $J/N < 0.102$.

Q.E.D.

6. Planar Diagram Dominance and Improvements of the Estimate

We here present a provisional argument which will be discussed elsewhere [7]. We remark that the theory is planar-diagram dominant in the limit of $N \to \infty$. In the present approach, this may be partly explicit by the factor N in the coefficient of each loop. But the planar diagram dominace is somewhat implicit: there exists a systematic cancelation in the sum of correlation functions of loops $\{\omega_1, ..., \omega_n\}$ such that $\cup\omega_i = \Omega$. Fix Ω. Then there are many loops such that $\cup_{i=1}^{n}\omega_i = \Omega$. As we have shown already, $\mathrm{sgn}(K_n(\Omega)) = (-1)^{n-1}$, where $K_n(\Omega)$ is the total contribution from n loops such that $\cup\omega_i = \Omega$.

We briefly sketch how this cancelation takes place: Assume $\cup_{i=1}^{n}\omega_i = \cup_{i=1}^{n+1}\tilde{\omega}_i = \Omega$ where $\omega_i = \tilde{\omega}_i, i = 1, 2, ..., n-1$. Assume $\Omega = \cup\tilde{\omega}_i$ is separated into two disconneceted parts Ω_1 and Ω_2 by resolving a loop (say) ω_n into two loops $\tilde{\omega}_n$ and $\tilde{\omega}_{n+1}$ at a vertex v of ω_n at which two lines cross each other. The vertex v of ω_n carries the factor $\alpha = \frac{1}{1+(2/N)}$. The factor $N/2$ is compensated by $\alpha - 1$ in the truncated function $K_{n+1}(\{\tilde{\omega}_1, ..., \tilde{\omega}_{n+1}\})$ since $(N/2)(\alpha - 1) = -\alpha$. One should note that the numbers of (oriented) loops in $\{\omega_i\}$ and $\{\tilde{\omega}_i\}$ are equal in this case. Then the contributions from these two groups of loops cancel each other.

If the multiplicity of the vertex v is more than 2, the cancelation is not complete but yields a small factor. In this way, we can show the following theorem [7]

Provisional Theorem.

$$\alpha = \sum_u \frac{N}{2} \frac{J_u}{|u|2^{|u|}} + \sum_\omega f_\omega \frac{N}{2} \frac{J_\omega}{|\omega|2^{|\omega|}},$$

where u are oriented non-self-crossing loops, and ω are self-crossing loops counted up to the equivalent classes of ω which have the same supports. Morever

$$|f_\omega| \leq c^{|\omega|} L(\omega),$$

where $L(\omega)$ are the numbers of oriented loops on supp ω *and c is a constant such that $c \in [1, 1+\epsilon]$, $\quad 0 < \epsilon << 1$.*

This thoeorem implies that α is essentially equal to the sum over single loops (not the sum over the loop correlation functions) except for their signatures. Thus the bound is expected to close to $N/2\nu$.

7. Application to the Block Spin Transformation

Consider two-dimensional Lattice models and let Λ be the square in Z^2 centered at the origin of size $L^K \times L^K$, where K is any large integer and L is a positive integer (≥ 2).

We set $\beta = J$ and $\sqrt{\beta}\phi(x) \equiv \phi(x) \in \sqrt{\beta}S^{N-1}$ so that eq.(1) is replaced by

$$< . > \equiv \frac{1}{Z} \int (.) \exp[-\frac{1}{2} \sum (\phi(x) - \phi(y))^2] \prod d\phi(x). \tag{38}$$

We define the block spin operator C by

$$(C\phi)(x) = \frac{1}{L^2} \sum_\zeta \phi(Lx + \zeta) \tag{39}$$

where $x \in \Lambda_1 \equiv L^{-1}\Lambda \cap Z^2$ and $\zeta \in \Lambda$ belong to the square of size $L \times L$ with center at the origin ($-L/2 \leq \zeta_\mu < L/2$). We fix $(C\phi)(x)$ to be $\phi^1(x)$, and integrate over the remaining degrees of freedom. We decompose the configuration space $\Xi_1 = \{\phi^1(x); x \in \Lambda_1\}$ of the block spins into several subsets.

$$\begin{aligned}
\mathcal{K}(D_1) &= \{\{\phi^1(x); x \in D_1\}; |\phi^1(x)|^2 \in (\beta_1 - cN, \beta_1 + cN), \\
&\quad |\phi^1(x) - \phi^1(x+e_\mu)|^2 < c_1 N\} \\
\mathcal{D}_L(D_2) &= \{\{\phi^1(x); x \in D_2\}; |\phi^1(x)|^2 > \beta_1 + cN\}, \\
\mathcal{D}_S(D_3) &= \Xi_1 \backslash \mathcal{K}(D_1) \cup \mathcal{D}_L(D_2).
\end{aligned}$$

Here c, c_1 are some positive constants and β_1 is the renormalized inverse temperature given by $\beta - \kappa N$, where κ is a positive constant determined later (compare our result with [8]). The region $\mathcal{D}_L$ is the large field region in which all spins take the same direction to make a large block spin value. The region $\mathcal{D}_S$ is the small field region in which there exist strong domain walls so that the block spins take smaller values, while $\mathcal{K}(D)$ is the smooth field region in which one observes block spins of reasonable strength with mild change of orientations. The configurations $\mathcal{D}_L$ and $\mathcal{D}_S$ are suppressed by the probabilistic and energetic reasons, respectively. The dominant configuration is $\mathcal{K}(D)$ in which one observes block spins with decreased magnitudes (by asymptotic freedom) and smooth field configurations. We expect that the interaction of the block spins in $\mathcal{K}$ are essentially equivalent to that of the original system.

We choose a rectangle of size 2×2, and assume for simplicity that all block spins take the same direction. Then we introduce following new variables $s(x) \in R$ and $u(x) \in R^{N-1}$:

$$\phi(x) =^t (\varphi(x) + s(x), u(x)),$$

$$\varphi(x) = \sum_{x' \in \Lambda_1} \mathcal{A}(x, x')\phi^1(x'),$$

where $\Lambda_1 = L^{-1}\Lambda \cap Z^2$ and the sums of $\{s(x)\}$ and $\{u(x)\}$ over the blocks are zero. $\mathcal{A}$ is the Kernel function intorduced in [5] and is almost local. The result approximately takes the following form:

$$\exp[-H^{(1)}] \simeq \exp[-\frac{1}{2}(\phi^1, G_1^{-1}\phi^1)]$$

$$\times \exp[-\sum_x \rho(x) + \sum_Y \alpha(Y) + \sum 2(N-2)\log(\beta - \phi^1(x)^2)], \qquad (40)$$

where G_1^{-1} is almost equivalent to the original lattice Laplacian, $\rho(x)$ and $\alpha(Y)$ are given by the convergent Mayer expansion if $r(x) \equiv (\beta - \phi^1(x)^2)^{1/2}$

$\simeq |u(x)|$ are small:

$$\exp[\sum_x \rho(x)+\sum_Y \alpha(Y)] = \int \exp[\sum\sum_i \beta r(x)r(x+e_i)\omega_x\omega_{x+e_i}]\prod d\omega_x, \quad (41)$$

where Y are sets of the bonds and block averages of $\omega_x \in S^{N-2}$ are zero. The factor $\rho(x)$ comes from the domain wall effect of $u(x)$ in each block and is then local in $\phi^1(x)$: approximately $\rho(x) = \text{const}(\beta - \phi^1(x)^2)$. Moreover $\text{const}(N-2)\log(\beta-\phi^1(x)^2))$ comes from the phase volume of the fluctuation fields $u(x)$, $|u(x)| \simeq (\beta-\phi^1(x)^2)^{1/2}$. If we neglect $\sum\alpha(Y)$, the effective action of ϕ^1 is

$$\sum_x\{-C_1\phi^1(x)^2 - C_2(N-2)\log(\beta-\phi^1(x)^2)\}$$

which takes the minimum at $\phi^1(x)^2 = \beta_1 \equiv \beta - \text{const}(N-2)$. Then the effective action is

$$-\frac{\text{const}}{N-2}\sum(\phi^1(x)^2-\beta_1)^2$$

in a neighborhhod of $(\phi^1)^2 = \beta_1$. In $\mathcal{K}(D_1)$, $\{r(x)\}$ are small and $\sum_Y \alpha(Y)$ are convergent and small. Especially non-local terms are exponentially small.

In this way, we continue several steps of the block spin transformations at least and we have interactions of short distances with effective β_k which are smaller, without much destroying the original form of the interaction. Obviously many difficult works remain in order to carry out the above scenario until the system reaches the high temperature (among them the convergent radius of the Mayer expansion is not large enough). These works are now in progress and will be discussed elsewhere.

References

[1] Battle,G.A. and Federbush,P.: A Phase Cell Cluster Expansion for Euclidean Field Theories, *Ann.Phys.*,**142**,95-139 (1982); A Note on Cluster Expansions, Tree Graph Identities, Extra 1/N! Factors !!!,*Lett. Math. Phys.*, **8**,55-58 (1984).; Battle, G.A. New Combinatoric Estimate for Cluster Expansion, *Commun.Math.Phys.* **94**,133-139 (1984).

[2] Brydge,D: A Short Cource on Cluster Expansion , in "Critical Phenomena, Random Systems, Gauge Theories " Les Housche, Session XLIII, ed. by K.Osterwalder and R.Stora, (1986) 129-183.

[3] Brydge,D., Froehlich,J. and Spencer,T., Random Walk Representations of Classical Spin Systems and Correlation Inequalities, *Commun.Math.Phys.*, **83**, 123-150 (1982); Brydge,D., Froehlich,J. and Sokal,A: Random Walk Representations of Classical Spin Systems and Correlation Inequalities, (II) ibid. **91**, 117-139 (1985).

[4] Brydge,D. and Federbush,P.: A New Form of the Mayer Expansions in Statistical Mechanics, *J.Math.Phys.*, **19**, 2064-2067 (1978).

[5] Gawedzki, K. and Kupiainen, A.: Renromalization Group Study of a Critical Latice Model, *Commun.Math.Phys.*, **82**,407-433 (1983).

[6] Ito,K.R., Renormalization Group Flow of Two-Dimenaional Hierarchical Heisenberg model of Dyson-Wilson Type,*Commun. Math.Phys.*, **139**, 45-70(1991).

[7] Ito,K.R.: Random Walk Representations and the Mayer Expansion in Non-Abelian Statistical Models, to be published in Commun.Math.Phys.; Ito,K.R.: Preprint in preparation.

[8] Polyakov, P.: Interactions of Golstone Particles in Two-Dimensions, *Phys. Letters.* **59B**, 79-81 (1975)

[9] Simon,B., Correlation Inequalities and the Decay of Correlations in Ferromagnets, *Commun.Math.Phys.*,**77**, 111-126 (1980); Lieb,E.:A Refinement of Simon's Correlation Inequality, ibid. **77**, 127-135 (1980).

The Free-energy Theorem

R.F.STREATER
King's College, London WC2R 2LS, England

Abstract. We explain what goes into the proof of the free-energy theorem, which says that in the non-equilibrium dynamics of a system in contact with a heat-bath at a given temperature T, the free energy $F = E - TS$ is a decreasing function of time and reaches its minimum at thermal equilibrium.

Key words: statistical dynamics, isothermal dynamics, entropy, free energy

1. Introduction

In order to understand the statement of the free-energy theorem, let alone prove it, we must give a concrete formulation. The classical statistical mechanics of particles is not a good place to start, because of the Gibbs paradox. This shows that entropy is not an extensive property in a classical theory of particles. There is a good mathematical reason for the paradox: the configuration space $\Omega(V)$ of classical particles in a volume V is not the product space of the configuration spaces associated with subvolumes V_1, $V - V_1$:

$$\Omega(V) = \Omega(V_1) \times \Omega(V - V_1) \tag{1}$$

does not hold.The physical reason for the paradox is the distinguishability of the particles.

It is not the use of classical probability as such that causes the paradox. We shall see that the statistical dynamics of classical discrete *fields* gives no problems and reproduces many of the main results of quantum statistical mechanics. But quantum mechanics underpins the theory, since it allows the energy-levels to be discrete.

Statistical dynamics is irreversible. It will not be fruitful to try to derive it rigorously from a Hamiltonian theory, be it classical or quantum. This is too hard. Instead, the theory uses a more general type of motion through the state-space Σ. A state p is a probability on Ω, the configuration space. It is useful to consider discrete time: the dynamics is defined by a map $\tau : \Sigma \rightarrow \Sigma$ giving the change in the state in one time-step. We shall distinguish between isolated and isothermal systems. For isolated systems τ increases the entropy S except at equilibrium (Boltzmann's H-theorem). Here, S is the Shannon entropy

H. Araki et al. (eds.), Quantum and Non-Commutative Analysis, 137–147.

$$S = -\sum_{\omega\in\Omega} p(\omega)\log p(\omega) \tag{2}$$

which is often finite for discrete systems. Since S is a function of p, and p is a function of time, we get a sound definition of a non-equilibrium entropy as a function of time; it is a Liapunov function for the dynamics.

In isothermal dynamics the system exchanges heat with a bath in such a way that after each time-step the system remains at the temperature T of the bath. In order to implement this exchange of heat, we introduce heat-particles which can flow in and out.These represent the kinetic energy of the particles or phonons, or even photons. Hence the notation γ for these particles. They are not really revolutionary: Einstein and then Debye used oscillators to represent the relative motion of the molecules. The heat-particles can be present even in states in which there are no other particles. Thus, the heat is not attached to any individual species of particle, but is shared. This is a good model of the chemistry in liquids; the heat-particles in states with no active chemicals represent the kinetic energy of the solvent. To get a realistic model of things like specific heats, the number of distinct heat particles should correspond to the number of degrees of freedom. That is, we should get the density of states right.We might also introduce *bond specific* heat-particles representing the energy of vibration of important molecular bonds. These can contribute a lot to the specific heat at a temperature comparable with the excitation-energy of the mode. Heat particles allow us to take account of kinetic energy without introducing momentum variables, and we can remain entirely within classical probability on configuration space.

The configuration space Ω is assumed to factorise:

$$\Omega = \Omega_c \times \Omega_\gamma \tag{3}$$

into a product of that of the chemicals Ω_c and that of the heat Ω_γ. To formulate the first law of thermodynamics we must specify the energy $\mathcal{E}$ of a configuration; $\mathcal{E}$ is a random variable which we assume is the sum of the chemical energy $\mathcal{E}_c$, a function on Ω_c, and the heat $\mathcal{E}_\gamma$, a function on Ω_γ. Let E denote the mean value of $\mathcal{E}$:

$$E = \sum_{\omega\in\Omega} p(\omega)\left(\mathcal{E}_c(\omega) + \mathcal{E}_\gamma(\omega)\right) \tag{4}$$

Clearly, E is a linear function of p and so is a function of time. The map τ will be defined in Section 2. Since it comes from the collision term of Boltzmann's equation, it conserves E so that in fact E is a constant in time. Thus we get the first law. In chemical kinetics we use the simple energy function

$$\mathcal{E}_c = \epsilon_A\mathcal{N}_A + \epsilon_B\mathcal{N}_B + \cdots \tag{5}$$

for the chemicals and

$$\mathcal{E}_\gamma = \omega_\gamma \hbar \mathcal{N}_\gamma \tag{6}$$

for each heat particle.

We shall see in Section 3 that the map τ determines the corresponding isothermal dynamics, which we call τ_β ; it is a map on the state space $\Sigma(\Omega_c)$ of the chemicals. The entropy S might not increase under $\tau_\beta \times id$, as there have been heat-flows in or out, to maintain the temperature. But we do show that

$$\Psi = S_c - \beta E \tag{7}$$

is an increasing function along the orbit, as a consequence of the H-theorem. If $\beta > 0$, this tells us that $F = E - TS$ decreases (the free- energy theorem) and so is a Liapunov function for the isothermal dynamics. Moreover, it is a strict Liapunov function, that is, it does not stand still except at a fixed point of the dynamics. Also, the configuration space splits into slices in which the orbits lie; each slice has a unique fixed point, which follows from the fact that F is a strictly convex function on the space Ω. One then shows that the iterated map $\{\tau_\beta^n\}_{n=0,1,...}$ converges to equilibrium for any initial state with finite E. This is the fundamental theorem of chemical kinetics [1].

In Section 4, we illustrate the theory for the model reaction

$$A + B \rightleftharpoons C + \gamma \tag{8}$$

and obtain the quantum Boltzmann equation with the Arrhenius factor, even though the theory uses only classical probability. It is found that states of many particles thermalise very fast. It is suggested that this could lead to an understanding of some of the divergences of quantum field theory.

2. The Stochastic Model

At each site x of a finite space Λ we introduce the sample space Ω_x. The total sample space is $\Omega = \times_{x\in\Lambda}\Omega_x$. A point ω in Ω is thus a *field*. That is, ω is a map $\omega : \Lambda \to \bigcup_x \Omega_x$ such that $\omega(x) \in \Omega_x$ for all x. Each space Ω_x is the product of the spaces $\Omega_A, \Omega_B, \ldots \Omega_\gamma$ of the particles (chemicals, ions, atoms, molecules, heat-particles) at x:

$$\Omega_x = \Omega_A \times \Omega_B \times \cdots \times \Omega_\gamma \tag{9}$$

Finally, each particle space $\Omega_A \ldots$ is the counting space

$$\Omega_A = \{0, 1, 2, \ldots\} \tag{10}$$

The points in Ω_A label the values of the numbers $\mathcal{N}_A \cdots$ of atom $A \cdots$ at the point x. $\mathcal{N}_A$ is thus a random variable on Ω_A.

A state on Ω_A at x is a probability $p_{x,A}$:

$$p_{x,A}(n) \geq 0 \; ; \qquad \sum_{n\geq 0} p_{x,A}(n) = 1 \tag{11}$$

These functions, for each chemical A, define a probability on Ω_x:

$$p(x) = p_{x,A} \otimes p_{x,B} \otimes \cdots \tag{12}$$

in which all the particles are independent. Then

$$p = \bigotimes_{x\in\Lambda} p(x) \tag{13}$$

defines a probability on Ω in which the numbers of chemicals are independent at different points. Conversely, given a probability p on Ω we can obtain the marginal probability $M_{x,A}\, p \in \Sigma\,(\Omega_{x,A})$. The means of $\mathcal{N}_{x,A}$ in the states p and $M_{x,A}\, p$ are the same.

The Stosszahlansatz of Boltzmann is the the map on $\Sigma(\Omega)$ given by

$$p \mapsto \bigotimes_{x,A} M_{x,A}\, p \tag{14}$$

It is known that this map increases entropy except at a fixed point. It is clearly non-linear.

The remnant of the collision term in Boltzmann's equation is given, in statistical dynamics, by a symmetric Markov matrix T acting on $\Sigma(\Omega)$. The (linear) map $p \mapsto Tp$ is then entropy non-decreasing. We choose T to link only states of the same energy and atomic number; this reflects their conservation in collisions. As a consequence, $\mathcal{E}$ and $\mathcal{N}_A$ both have the same mean in the state Tp as in the state p.

There is one more randomising map that is needed before the theory can describe chemistry or the interior of stars: we must incorporate the hypothesis of local thermodynamic equilibrium (LTE). We say that the state p_x is in LTE if there is a local beta $\beta(x) = 1/k_B T(x)$ such that for all heat particles γ

$$Prob\,\{\mathcal{N}_{x,\gamma} = n\} = \exp\left(-\beta(x) n\hbar\omega_\gamma\right)\left(1 - \exp\left(-\beta(x)\hbar\omega_\gamma\right)\right) \tag{15}$$

and for each ion A there exists a chemical potential $\mu_A(x)$ such that

$$Prob\,\{\mathcal{N}_{x,A} = n\} = \exp\{\beta(x)\left(\mu_A(x) - E_A\right) n\} Z_A^{-1} \tag{16}$$

where $Z_A(x)$ is the local partition function for the ion A:

$$Z_A(x) = \left(1 - \exp\{\beta(x)\left(\mu_A(x) - E_A\right)\}\right)^{-1} \tag{17}$$

The parameters $\beta(x)$, $\mu_A(x)\ldots$ are determined by the mean values of $\mathcal{N}_A, \ldots \mathcal{N}_\gamma$ and conversely. A system in LTE has well-defined local partial pressures $P_A(x) = \beta^{-1} \log Z_A(x)$.

It is known that the speed of thermalisation in a small enough local region is 10^{10} times faster than the rates of chemical reactions. Hence before each time-step of the dynamics we can assume that the state is in LTE. However in general the map T will upset the property of being in LTE. The only exception is if the whole system is in thermal and chemical equilibrium. We can implement the hypothesis of LTE, without introducing the thermalising forces explicitly, by using a further randomising map Q. Q replaces a probability $p_{x,\gamma}$ on $\Omega_{x,\gamma}$ by the equilibrium state given by Equation 15, and $p_{x,A}$ on $\Omega_{x,A}$ by the state given by Equation 16. The parameters $\beta(x)$ and $\mu_A(x)$ are fixed by the mean values of $\mathcal{N}_{x,\gamma}$ and $\mathcal{N}_{x,A}$ in the state p_x. These are denoted $n_\gamma(x)$ and $n_A(x)$ respectively. The map Q imitates the quasifree map of quantum statistical mechanics [2]. So, by definition, Q does not alter the mean-values of energy or number. One can prove that Q increases entropy unless p_x is a quasi-free state [3].

We are now in a position to define the dynamical map τ of the isolated system. Let $p \in \Sigma(\Omega)$. Then define the map τ by

$$p \mapsto Tp \mapsto \bigotimes_{x,A} M_{x,A}Tp \mapsto Q \bigotimes_{x,A} M_{x,A}Tp = \tau p \tag{18}$$

Each step is entropy non-decreasing, and the mean values of energy and atomic number are conserved. Hence τ has these properties too. If the initial state has finite energy, the motion occurs on a compact set with S as a Liapunov function. If T is ergodic in a suitable sense (easy to achieve) then general theorems [4], [5], [6], [7] assure us that $\tau^n p \to$ equilibrium as $n \to \infty$ whatever the initial state p. This is the *fundamental theorem of chemical kinetics* for isolated systems.

Some parts of the theory, especially for continuous time, have been developed by Alicki and Messer [8]; see also [9].

3. The Isothermal Dynamics

Suppose now that each point $x \in \Lambda$ is attached to a common heat-bath of beta β. Then all the heat-particles must have the distribution given by Equation 15, with the same beta for all x. We call this distribution s_β. This is the situation at time t. Most chemical reactions produce heat, or use it up. In statistical dynamics, this is modelled by choosing T with non-zero entries linking the relevant states which, if heat is produced, will have different numbers of heat-particles, or heat-particles of different energy. The creation of the γ in one time-step will upset the thermal equilibrium that was there at time t : after the dynamical map τ, the heat-particles are no longer in the state s_β. We now feed it heat, or remove heat, in just the right amounts at each point x, to readjust the population of heat-particles back to the state s_β. This gives the motion

$$p \mapsto M_c \tau p \otimes s_\beta = \tau_\beta p \tag{19}$$

This is called the isothermal dynamics at constant volume. We should not expect τ_β to increase the entropy: there have been heat flows. But τ itself increases the entropy, and τ and τ_β coincide on $\Sigma(\Omega_c)$. This gives us a Liapunov function for τ_β, the free energy. To see this, at time t we have the state $p = p_c \otimes s_\beta$, so the entropy is the sum $S(p) = S(p_c) + S(s_\beta)$. Let $\tau p = p'_c \otimes s'$ so that $\tau_\beta p = p'_c \otimes s_\beta$. Let us label the energy levels of Ω_γ by ω_n. Then

$$\begin{aligned} S(p'_c) &= S(p'_c) + S(s') - S(s') = S(\tau p) + \sum_n s'(n) \log s'(n) \\ &\geq S(p) + \sum_n s'(n) \left(\log s'(n) - \log s_\beta(n)\right) + \sum_n s'(n) \log s_\beta(n) \\ &\quad \text{and so by Klein's inequality,} \\ &\geq S(p) + \sum_n s'(n) \log s_\beta(n) \\ &= S(p_c) - \sum_n s_\beta(n) \log s_\beta(n) + \sum_n s'(n) \log s_\beta(n) \\ &= S(p_c) - \sum_n \left(s_\beta(n) - s'(n)\right) \left(-\hbar\beta\omega_n - \log Z_\gamma(\beta)\right) \end{aligned} \tag{20}$$

Since $\sum_n s_\beta(n) = \sum_n s'(n) = 1$, the $\log Z_\gamma$ cancels. Also, by energy conservation

$$E_c + \sum_n \hbar\omega_n s_\beta(n) = E'_c + \sum_n \hbar\omega_n s'_\beta(n) \tag{21}$$

so

$$\hbar\beta \sum \left(s_\beta(n) - s'_\beta(n)\right) \omega_n = \beta(E'_c - E_c). \tag{22}$$

Hence we have

$$S(p'_c) \geq S(p_c) + \beta(E'_c - E_c) \tag{23}$$

which on rearranging gives the chemical free-energy theorem

$$\Psi' = S(p'_c) - \beta E'_c \geq S(p_c) - \beta E_c = \Psi. \tag{24}$$

The Liapunov function Ψ then enables us to prove [1] that the isothermal chemical kinetic equations converge to equilibrium. This is the *fundamental theorem of isothermal chemical kinetics*. In thermodynamics, the free energy includes the kinetic energy; in our model the kinetic energy is not changed by the isothermal dynamics, and so could be added to both sides of Equation 24 to get the usual statement of the free-energy theorem.

4. The Quantum Boltzmann Equation

The quantum Boltzmann equation was written down by Uehling and Uhlenbeck in 1933 [10]. For the reaction

$$A \rightleftharpoons B + \gamma \tag{25}$$

between Bosons A, B, andγ, this equation is

$$\frac{dn_A}{dt} = \kappa\left(-n_A\left(1+n_B\right)\left(1+n_\gamma\right) + n_B n_\gamma\left(1+n_A\right)\right) \tag{26}$$

The forward term $-\kappa n_A(1+n_B)(1+n_\gamma)$ comes from second-order perturbation theory combined with the Fermi Golden Rule, applied to the reaction $A \rightarrow B + \gamma$. Similarly, one can get the other term from the back reaction $B + \gamma \rightarrow A$. The interaction Hamiltonian needed is simply

$$H_I = e(a_A^* a_B a_\gamma + a_B^* a_\gamma^* a_A) \tag{27}$$

and Equation 26 corresponds to a matrix element coming from all diagrams with one vertex and any number of spectator particles, the latter being disconnected lines. Quantum field theory automatically gives the factors $(1+n)$, which are the sums of the "spontaneous" and "stimulated" parts predicted with uncanny insight in Einstein's seminal paper [11]. Some books [12] confuse mean values of $\mathcal{N}$ with probabilities, or random variables $\mathcal{N}$ with mean values [13]. For example in [12] the stimulated emission is attributed to the process $A+\gamma \rightarrow B+2\gamma$ while the inverse of this is wrongly omitted. In fact as is well known in chemistry the principle of detailed balance requires that at equilibrium each process be separately in balance.

The fixed points of Equation 26 obviously obey

$$\frac{n_A}{1+n_A} = \frac{n_B}{1+n_B}\frac{n_\gamma}{1+n_\gamma} \tag{28}$$

The mean numbers, n_A etc. are related in thermodynamic (but not necessarily chemical) equilibrium to the chemical activities λ_A etc.by

$$\begin{aligned} n_A &= \lambda_A/(1-\lambda_A) \\ n_B &= \lambda_B/(1-\lambda_B) \\ n_\gamma &= \lambda_\gamma/(1-\lambda_\gamma) \end{aligned}$$

where

$$\begin{aligned} \lambda_A &= \exp\left(-\beta\left(\epsilon_A - \mu_A\right)\right) \\ \lambda_B &= \exp\left(-\beta\left(\epsilon_B - \mu_B\right)\right) \\ \lambda_\gamma &= \exp\left(-\beta\omega\hbar\right) \end{aligned}$$

These relations can be got from Equations 15 and 16 by omitting the x-dependence and computing the means of $\mathcal{N}_A$, $\mathcal{N}_B$ and $\mathcal{N}_\gamma$. Then Equation28 says simply $\lambda_A = \lambda_B\lambda_\gamma$, well known as the condition for equilibrium in chemistry; in view of $\epsilon_A = \epsilon_B + \omega\hbar$ we get $\mu_A = \mu_B$ at equilibrium. The fixed point is therefore the Bose-Einstein-Planck grand canonical state at some beta and with the chemicals at a common potential. The dynamics can be regarded as motion through the space of parameters $\mu_A(t)$, $\mu_B(t)$, and $\beta(t)$ or equivalently through the space of parameters $n_A(t)$, $n_B(t)$, and $n_\gamma(t)$. The factor λ_γ is known as the *Arrhenius factor.*

It would be wrong to conclude that one can work backwards from the knowledge of the Planck distribution, via detailed balance, to Equation 26. In "activity-led" chemical reactions [14] the dynamics of $A \rightleftharpoons B+C$ is given by the equation

$$dn_A/dt = \kappa(-\lambda_A + \lambda_B\lambda_C) \tag{29}$$

which obviously has the same fixed points as Equation 26. We get Equation 29 from a stochastic model including a dynamical assumption expressing the "cage effect" of chemicals in liquids. It is a collective effect not easily seen in perturbation theory.

We shall construct a stochastic model which without further approximations leads to Equation 26. The main idea is still that of Einstein [11]. An interesting problem turns up: whereas the continuous-time equations can be defined and have solutions for all initial states of finite energy, the discrete-time equations with a chosen value of the time-step only make sense for states of sufficiently low energy. We shall return to this, but first let us construct the model. All the ideas show up in Equation 25 so we shall do this case in detail. We take Λ to contain one point, and have two chemicals and one heat-particle. The sample space is therefore

$$\Omega = \Omega_A \times \Omega_B \times \Omega_\gamma \tag{30}$$

and the typical point is $\omega = (i,j,k)$ where $i,j,k = 0,1,2,\ldots$. The state at time t will be in LTE; thus $p = p_A \otimes p_B \otimes p_\gamma$ where

$$p_\gamma(n) = Z_\gamma^{-1}\lambda_\gamma^n \tag{31}$$

and so on for p_A and p_B.

Any random variable on Ω is a function $f(\omega) = f(i,j,k)$; it can be written as a sum of the projections $P_{i,j,k}$

$$\begin{aligned} P_{i',j',k'}(\omega) &= 1 \text{ if } i = i', j = j', k = k'; \\ &= 0 \text{ otherwise.} \end{aligned} \tag{32}$$

In Equation 32, ω is the point (i,j,k). Then we can write

$$f(\omega) = \sum_{\omega'} f(\omega') P_{\omega'}(\omega) \tag{33}$$

In particular, $\mathcal{N}_A$ is given by

$$\mathcal{N}_A = \sum_{\omega'} i' P_{\omega'} \tag{34}$$

A state is also a function on Ω. Thus the state p in LTE has the form

$$p = Z_A^{-1} Z_B^{-1} Z_\gamma^{-1} \sum_{i',j',k'=0}^{\infty} \lambda_A^{i'} \lambda_B^{j'} \lambda_\gamma^{k'} P_{i',j',k'} \tag{35}$$

Because of the map Q, in one time-step we move from the LTE (n_A, n_B, n_γ) to another, say (n'_A, n'_B, n'_γ). Since Q and the marginal maps conserve $\mathcal{N}_A$, $\mathcal{N}_B$ and $\mathcal{N}_\gamma$, the change in mean-values is caused by T:

$$\begin{aligned}(\mathcal{N}_A, p') &= \sum_{\omega} \mathcal{N}_A(\omega) Q\left(M_A Tp \otimes M_B Tp \otimes M_\gamma Tp\right)(\omega) &(36)\\ &= (\mathcal{N}_A, Tp) &(37)\end{aligned}$$

where (,) is the ℓ^2 scalar product on Ω. The action of T on p (or any random variable) is given by its action on all the $P_{i',j',k'}$, which we now choose, following Einstein's intuition. T can be split into a forward part T_+ associated with $A \to B + \gamma$, and a backward part T_-, associated with $B + \gamma \to A$. We then put $T = (T_+ + T_-)/2$. If a state has k photons then the rate for $B + \gamma \to A$ must be proportional to k. Similarly, if there are j B-particles, it must be proportional to j. So the other process, $A \to B + \gamma$, which goes into a state with $(j+1)$ B-particles and $(k+1)$ gammas, must be proportional to $(j+1)(k+1)$, by microscopic reversibility. It must also be proportional to i, the number of A-particles. Since T must be a symmetric operator on $\ell^2(\Omega)$, we get the following for $T_\pm$:

$$\begin{aligned}T_+ P_{i,j,k} &= \left(1 - i(j+1)(k+1)\kappa\right) P_{i,j,k}\\ &\quad + i(j+1)(k+1)\kappa P_{i-1,j+1,k+1} &(38)\\ T_- P_{i,j,k} &= \left(1 - (i+1)jk\kappa\right) P_{i,j,k} + (i+1)jk\kappa P_{i+1,j-1,k-1} &(39)\end{aligned}$$

In order for the discrete map T to be stochastic, for a given i, j, k we must have

$$\left(i(j+1)(k+1) + (i+1)jk\right)\kappa \leq 2 \tag{40}$$

Thus the time-step must be taken smaller, the greater the number of particles in the state. The forward rate of change for n_A is then given by inserting T in Equation 37:

$$\begin{aligned}(\mathcal{N}_A, T_+ p) &= \sum_{\omega,\omega'} i' p(i,j,k) \left(P_{i',j',k'}\, , \{P_{i,j,k}(1 - \kappa i(j+1)(k+1))\right.\\ &\quad \left. + P_{i-1,j+1,k+1} \kappa i (j+1)(k+1)\}\right) &(41)\end{aligned}$$

which reduces to

$$n'_A = n_A - \kappa n_A(n_B + 1)(n_\gamma + 1) \tag{42}$$

giving the forward part of Equation 26, except for a factor of 2 in the definition of κ. Similarly, T_- gives the backward part. We are thus able to exhibit a purely classical random field which obeys the quantum Boltzmann equation. The particles, as well as the heat-particles, are Bosons; our condition for equilibrium Equation 28 should not be confused with the condition used by Einstein [11]. Our condition involves the *average* number, whereas Einstein used the *probability* that the upper level of a two-level atom is occupied. It is the probability that obeys the Boltzmann law at equilibrium. Probability and mean number are proportional for Fermions and other two-level systems, but not for Bosons.

5. Discussion

The models constructed in the previous section have a curious property; the discrete-time version with a chosen value of κ does not make sense on all initial states of finite energy. The higher the energy, the smaller κ must be. The map T with κ given makes sense on the states obeying Equation 40, but iterates of T will eventually lead out of these states: the domain of definition of T is not invariant under T. Indeed, even if the initial state has finitely many particles, one application of the map Q will take us to a state with an infinite tail. If we apply T outside it domain, we are led to chaos through the period-doubling route [6], [7]. The first bifurcation occurs well past the allowed limit, which we called the *bistochastic limit.* About half-way along the way to this limit, the system virtually thermalises in just one step $p \to \tau p$. The equation in continuous time is therefore unstable as a map on the space of states; no iterative scheme with fixed time-step will work. The states of very many particles thermalise very fast. This is connected with the infra-red problem in quantum electrodynamics: the very large number of soft photons cannot be described by Hamiltonian dynamics, because the tail of many particles is thermalising. This suggests that the divergences might be less severe if we assumed a Hamiltonian dynamics only for states containing fewer particles than some large number M; the dynamical map applied to states with more than M particles should not be treated as a Hamiltonian system, but as a rapidly thermalising one. A simple way to implement this idea is to replace the tail of the state p by a sub-probability with the same ℓ_1-norm as the tail of p, and having the same mean energy and number as the tail of p, such that the new tail is the tail of a grand canonical state. The dynamics is then Hamiltonian for few particles, and a small heat bath for large numbers, with a controllable leakage between the two descriptions caused by the interaction. One could then try letting M go to ∞. The infra-red problem of QCD could be treated in the same way. In constructive quantum field theory, the ultraviolet cut-off does not control the number of particles, only the energy of the individual bare quanta. The quantum Boltzmann equation suggests that here too we should thermalise the states of too many particles.

The quantum Boltzmann equation does make sense as a dynamics on the space of macroscopic variables $n_A, n_B \ldots$ since, for any given state of finite energy, these variables remain bounded.

For continuous time, one can get parallel results, such as the free-energy theorem, convergence to equilibrium, and so on, as has been shown by Koseki [15].

References

[1] R.F.Streater, *The F-Theorem for Stochastic Models* , Annals of Physics, **218**, 255-278, (1992).

[2] R.F.Streater, *A Boltzmann Map for Quantum Oscillators* J.Stat. Phys.**48**, 753-767 (1987).

[3] R.F.Streater, *Entropy and the Central Limit Theorem in Quantum Mechanics*, J.Phys.A, Math.Gen. **20**, 4321-4330 (1987).

[4] R.F.Streater, *Convergence of the Iterated Boltzmann map*, Publ.RIMS.Kyoto**20**,913-927 (1984).

[5] R.F.Streater, *Convergence of the Quantum Boltzmann Map*, Commun. Math. Physics, **98**, 177-185 (1985).

[6] L.Rondoni and R.F.Streater, *Chemical Reactions as Dynamical Systems on the Interval*, J.Stat.Phys.**66**, 1557-1574 (1992).

[7] L.Rondoni, *A Stochastic Treatment of Reaction and Diffusion*, Thesis, Ph.D., Virginia Polytechnic Institute, (1991).

[8] R.Alicki and J.Messer,J. Stat. Phys. **32**, 299-312 (1983).

[9] R.Alicki and K.Lendl, *Quantum Dynamical Semigroups and Applications*, Lecture Notes in Physics,**286**, Springer, 1987.

[10] E.A.Uehling and G.E.,Uhlenbeck, *Transport Phenomena in Einstein-Bose and Fermi-Dirac Gas*, Phys. Rev.bf 43, 552-561 (1933).

[11] A. Einstein *Zur Quantentheorie der Strahlung* , Physik Z. **18**, 121 (1917). Translated into English in: B.L.van der Waerden, *Sources of Quantum Mechanics* (Dover,N.Y. 1969), pp63-78.

[12] D.Clayton, *Principles of Stellar Evolution and Nucleosynthesis* (McGraw Hill, N.Y., 1968) p28,p32.

[13] G.Baym *Lectures in Quantum Mechanics* , (Benjamin, N.Y., 1969) p276.

[14] S.Koseki and R.F.Streater, *The Statistical Dynamics of Activity-led Reactions*, submitted to J.Phys.Condensed Matter.

[15] S.Koseki, *Stochastic Models of Non-equilibrium Statistical Mechanics for Open Systems*, Thesis, P.h.D., King's College,to appear.

Dynamical Entropy of Quasi–local Algebras

YONG MOON PARK
Department of Mathematics, Yonsei University Seoul 120–749, Korea

Dedicated to Professor Huzihiro Araki on his sixtieth birthday

Abstract. We give a brief survey on the recent developments of the dynamical entropy of automorphisms on quasi-local algebras in quantum statistical mechanics. We first describe a Kolmogorov-Sinai type continuity and then give a brief description of the computation of the dynamical entropy of space translations of CAR and CCR algebras with respect to an invariant quasi-free state.

Key words: Dynamical entropy, Quasi-local algebras, CAR algebras, CCR algebras, Mean entropy, Entropy defects

1. Introduction

In this article we review recent developments of the dynamical entropy of automorphisms on quasi–local algebras in quantum statistical mechanics. We first describe a Kolmogorov–Sinai type theorem for quasi–local algebras [18,19] and then give a brief description of the computation of the dynamical entropy of space translations of CAR and CCR algebras in ν–dimensional continuous space with respect to an invariant quasi–free state[19].

The concept of the Kolmogorov–Sinai (KS) entropy [9,20] of measure preserving transformations became a key notion in ergodic theory [5] and it played important roles in classical statistical mechanics [19]. In [7] Connes, Narnhofer and Thirring extended the notion of KS entropy to the case of automorphisms of C^*–algebras with respect to an invariant state, and also established the continuity of the KS type for AF–algebras. As in the classical ergodic theory [5], Connes, Narnhofer and Thirring (CNT) entropy should be useful to a classification of quantum chaotic evolutions [12]. In order to obtain a deep understanding of CNT entropy and find a classification of automorphisms of C^*–algebras, it should be important to develop the methods which enable to compute the entropy for quantum systems. There have been several results in this field. The dynamical entropy of space translations for the Gibbs state of quantum lattice systems [7,11] and quasi–free states of Fermi lattice gases [7], and the dynamical entropy of the quasi–free evolu-

[1] Supported in part by Basic Science Research Program, Ministry of Education. 1992

H. Araki et al. (eds.), Quantum and Non-Commutative Analysis, 149–165.

tions of CAR algebras [22] have been computed. Recently the chaotic behavior of automorphisms on the rotation algebra [10] and non–commutative 2–shift [13] were investigated. To find possible applications of the CNT theory to unbounded lattices systems and particle systems in quantum statistical mechanics, we have extended the KS type theorem to quasi–local algebras [16,17] and then computed the dynamical entropy of space translations of CAR and CCR algebras in $\nu-$ dimensional continuous spaces with respect to an invariant quasi–free state [17]. We mention that a result similar to ours for CAR algebras was obtained by Narnhofer and Thirring [14] by using a method which differs to ours.

In this survey we give a brief description of the results and the main methods used in [16,17]. For the details we refer to [16,17]. In Section 2, we review the definition and general results on the CNT entropy [7]. In Section 3, we give a KS type theorem for quasi–local algebras and basic ideas of the proof. In Section 4, we sketch a computation of the dynamical entropy of space translations of the CAR and CCR algebras with respect to a quasi–free state.

2. Dynamical Entropy of C^*–algebras

We review the definition and some general results on the dynamical entropy of C^*–algebras in [7]. Throughout this section we consider a C^*–dynamical system $(\mathcal{A}, \theta, \phi)$ where $\mathcal{A}$ is a unital C^*–algebra, θ an automorphism on $\mathcal{A}$, ϕ a state which is invariant with respect to θ.

Let $\mathcal{A}$ be a finite dimensional C^*–algebra and ϕ and ψ be states on $\mathcal{A}$. The density operators corresponding to ϕ and ψ are denoted by ρ_ϕ and ρ_ψ respectively. The *relative entropy* for the state ϕ and ψ is defined by

$$S(\phi|\psi) = Tr(\rho_\psi(\log \rho_\psi - \log \rho_\phi)) \tag{2.1}$$

where Tr denotes the trace.

A *completely positive map* γ between two unital C^*–algebras $\mathcal{A}$ and $\mathcal{B}$ is defined by a positive unity preserving map between $M_n(\mathcal{A})$ and $M_n(\mathcal{B})$ the $n \times n$ matrices with elements from $\mathcal{A}$ (respectively $\mathcal{B}$), $(\gamma(a))_{ij} = \gamma(a_{ij})$, is positive for any $n \in N$. For them positivity is strengthened to $\gamma(a^*)\gamma(a) \leq \gamma(a^*a)$. With respect to composition they form a semi–group which contains $*$–homomorphism. If $\mathcal{A}$ or $\mathcal{B}$ are abelian, any positive map is completely positive [8, 23]. If $\mathcal{B} \subset \mathcal{A}$, a positive unital map with $\gamma(b_1 a b_2) = b_1\gamma(a)b_2, b_i \in \mathcal{B}, a \in \mathcal{A}$, is called unital conditional expectation. It is automatically completely positive [23].

Let ϕ be a state on a (finite dimensional) C^*–algebra $\mathcal{A}$. An *abelian model* for $(\mathcal{A}, \phi)$ is a completely positive unital map P into a finite abelian algebra $\mathcal{B}$ with a state μ such that $\phi = \mu \circ P$. If $\{e_1, e_2, \cdots, e_r\}$ is a basis (minimal projections) of $\mathcal{B}$, then there are states $\widehat{\phi}_i, i = 1, 2, \cdots, r$ of $\mathcal{A}$ such that $P(x) = \sum \widehat{\phi}_i(x)e_i$ for any $x \in \mathcal{A}$ [4]. Since $\phi = \mu \circ P$, we have

$$\begin{aligned} \phi &= \sum \mu(e_i)\widehat{\phi}_i \\ &= \sum \mu_i \widehat{\phi}_i \end{aligned} \tag{2.2}$$

Conversely, to any decomposition of the state ϕ one can associate an abelian model. For an abelian model with the corresponding decomposition of the state ϕ as in (2.2), define

$$\mathcal{E}(P) = \sum_{i=1}^{r} \mu_i S(\phi|\widehat{\phi}_i). \tag{2.3}$$

The *entropy defect* of an abelian model $P : \mathcal{A} \to \mathcal{B}$ with the corresponding state decomposition (2.2) is defined to be

$$S_\mu(P) = S(\mu) - \mathcal{E}_\mu(P) \tag{2.4}$$

where $S(\mu)$ is the entropy of $(\mathcal{B}, \mu)$; $S(\mu) = -\sum \mu_i \log \mu_i$.

Consider a unital C^*–algebra $\mathcal{A}$ with a state ϕ. Let $\mathcal{N}_1, \cdots, \mathcal{N}_n$ be finite dimensional C^*–algebras and $\gamma_i : \mathcal{N}_i \to \mathcal{A}$ a completely positive unital map, $j = 1, \cdots, n$. Let $(\mathcal{B}, \mu)$ be an abelian model for $(\mathcal{A}, \phi)$ with a completely positive unital map $P : \mathcal{A} \to \mathcal{B}$. Let $\mathcal{B}_j, j = 1, \cdots, n$, be subalgebras of $\mathcal{B}$ and $E_j : \mathcal{B} \to \mathcal{B}_j$ the μ–invariant conditional expectation ($\mu = \mu \circ E_j$). Consider the following schem:

$$\begin{array}{ccc} \mathcal{N}_1 & & \mathcal{B}_1 \\ \vdots & & \vdots \\ \mathcal{N}_j & \xrightarrow{\gamma_i} (\mathcal{A}, \phi) \xrightarrow{P} (\mathcal{B}, \mu) \xrightarrow{E_j} & \mathcal{B}_j \\ \vdots & & \vdots \\ \mathcal{N}_n & & \mathcal{B}_n \end{array} \tag{2.5}$$

The quadruple $(\mathcal{B}, E_j, P, \mu)$ is called an *abelian model* for $(\mathcal{A}, \phi, \gamma_1, \cdots, \gamma_n)$. The model (2.5) provides abelian models

$$\rho_j \equiv E_j \circ P \circ \gamma_j : \mathcal{N}_j \to \mathcal{B}_j$$

for $(\mathcal{N}_j, \phi \circ \gamma_j), j = 1, \cdots, n$. For given completely positive unital maps $\gamma_j : \mathcal{N}_j \to \mathcal{A}$ from finite dimensional C^*–algebras $\mathcal{N}_j, j = 1, \cdots, n$ to C^*–algebra $\mathcal{A}$ with a state ϕ, put

$$H_\phi(\gamma_1, \cdots, \gamma_n) = \sup\{S(\mu|_{\vee_{j=1}^n \mathcal{B}_j}) - \sum_{j=1}^{n} S_\mu(\rho_j)\} \tag{2.6}$$

where the sup is taken over all abelian model for $(\mathcal{A}, \phi, \gamma_1, \cdots, \gamma_n)$.

Indeed using the various properties of the relative entropy one can show [7].

Proposition 2.1 [7, Proposition III.6]. (a) *Let $\theta_j' : \mathcal{N}_j' \to \mathcal{N}_j$, be completely positive unital maps, then the* monotonicity

$$H_\phi(\gamma_1 \circ \theta_1, \cdots, \gamma_n \circ \theta_n) \leq H_\phi(\gamma_1, \cdots, \gamma_n)$$

holds.

(b) *$H_\phi(\gamma_1, \cdots, \gamma_n)$ depends only upon the set $\{\gamma_1, \cdots, \gamma_n\} = X$, that is, $H_\phi(\gamma, \gamma) = H_\phi(\gamma)$.*

(c) *With the notation of* (b) *the* subadditivity

$$\max\{H_\phi(X), H_\phi(Y)\} \leq H_\phi(X \cup Y) \leq H_\phi(X) + H_\phi(Y)$$

holds.

Let $(\mathcal{A}, \theta, \phi)$ be a C^*–dynamical system. By the subadditivity, the following limit

$$h_{\phi,\theta}(\gamma) = \lim_{n\to\infty} \frac{1}{n} H_\phi(\gamma, \theta \circ \gamma, \cdots, \theta^{n-1} \circ \gamma) \tag{2.7}$$

exists for any completely positive unital map $\gamma : \mathcal{N} \to \mathcal{A}$. the *dynamical entropy* of θ with respect to the state ϕ is
defined by

$$h_\phi(\theta) = \sup_\gamma h_{\phi,\theta}(\gamma), \tag{2.8}$$

where the sup is taken over all completely positive unital map $\gamma : \mathcal{N} \to \mathcal{A}$.

We collect entropy results in [7]. Let $\gamma, \gamma' : \mathcal{N} \to \mathcal{A}$ be completely positive unital maps and let

$$\|\gamma - \gamma'\| = \sup_{\|x\|=1} \phi(((\gamma - \gamma')(x))^*(\gamma - \gamma')(x)).$$

The following result is a consequence of Proposition IV.3 and Theorem VI.3 of [7].

Proposition 2.2. *Let $\gamma_n : \mathcal{N} \to \mathcal{A}$ be a sequence of completely positive unital maps which converges to $\gamma : \mathcal{N} \to \mathcal{A}$ in either the uniform norm or else $\| \ \|_\phi$–norm. Then the following result*

$$h_{\phi,\theta}(\gamma) = \lim_{n\to\infty} h_{\phi,\theta}(\gamma_n)$$

holds.

The simplest example to which Proposition 2.2 applies is that of an AF–algebra, i.e., a C^*–algebra which is the norm closure of an increasing union $\cup_{n=1}^\infty \mathcal{A}_n$ of finite dimensional unital subalgebras $\mathcal{A}_n$.

Theorem 2.3 **[7, Corollary V.4].** *Let $\mathcal{A} = (\bigcup \mathcal{A}_n)^-$ be an AF–algebra. Then for any state ϕ on $\mathcal{A}$ and automorphism $\theta : \mathcal{A} \to \mathcal{A}$ such that $\phi = \phi \circ \theta$, one has*

$$h_\phi = \lim_{n \to \infty} h_{\phi,\theta}(\mathcal{A}_n).$$

Here the subalgebra $\mathcal{A}_n$ is standing for the inclusion map
$\mathcal{A}_n \to \mathcal{A}$.

3. Dynamical Entropy of Quasi–local Algebras

In order to apply the CNT theory to more general situations such as unbounded spin systems and particle systems in quantum statistical mechanics [3,15,16], it is necessary to extend Theorem 2.3 to quasi–local algebras which are not necessarily AF–algebras. The extension has been carried out in [17]. In this section we describe a slightly improved version of the extension briefly.

We work within the framework of quantum statistical mechanical systems where we start from strictly local algebras $\mathcal{A}_\Lambda$ indexed by bounded regions Λ in $\mathbf{R}^\nu$ (or $\mathbf{Z}^\nu$). The norm closure of $\cup \mathcal{A}_\Lambda$ is the C^*–algebra $\mathcal{A}$ of quasi–local operators. We remark that if $\sigma : \mathcal{A} \to \mathcal{A}$ be an automorphism which satisfies $\sigma^2 = id$, i.e., $\sigma(\sigma(x)) = x$ for all $x \in \mathcal{A}$, then each element $x \in \mathcal{A}$ has a unique decomposition into odd and even parts with respect to σ. This decomposition is defined by

$$x = x^+ + x^-, \quad x^\pm = \frac{1}{2}(x \pm \sigma(x)).$$

It follows that $\sigma(x^\pm) = \pm x^\pm$.

Definition 3.1. *A* Quasi-local algebra *is a C^*–algebra and a net $\{\mathcal{A}_\Lambda : \Lambda \subset \mathbf{R}^\nu(or\mathbf{Z}^\nu)\}$ such that the following properties are valid :*

(a) $\Lambda \subset \Lambda' \Rightarrow \mathcal{A}_\Lambda \subset \mathcal{A}_{\Lambda'}$;

(b) $\mathcal{A} = (\bigcup_\Lambda \mathcal{A}_\Lambda)^-$, *where the bar denotes the norm closure;*

(c) *The algebras $\mathcal{A}_\Lambda$ have a common identity* $\mathbf{1}$;

(d) *There exists an automorphism σ such that $\sigma^2 = id, \sigma(\mathcal{A}_\Lambda) = \mathcal{A}_\Lambda$, and $[\mathcal{A}^e_\Lambda, \mathcal{A}^e_{\Lambda'}] = \{0\}, [\mathcal{A}^e_\Lambda, \mathcal{A}^o_{\Lambda'}] = \{0\}.\{\mathcal{A}^o_\Lambda, \mathcal{A}^o_{\Lambda'}\} = \{0\}$ whenever $\Lambda \cap \Lambda' = \phi$, where $\mathcal{A}^e_\Lambda \subseteq \mathcal{A}_\Lambda$ and $\mathcal{A}^o_\Lambda \subseteq \mathcal{A}_\Lambda$ are even and odd elements with respect to σ.*

We have used the notation $[x, y] = xy - yx$ and $\{x, y\} = xy + yx$. In quantum mechanics $\sigma = id$ corresponds to Bose statistics, but for Fermi statistics $\sigma \neq id$. **Assumption 3.2.** *Let $(\mathcal{A}, \{\mathcal{A}_\Lambda\})$ be a quasi-local algebra indexed by bounded region $\Lambda \subset \mathbf{R}^\nu(or\mathbf{Z}^\nu)$ and let ϕ be a state on $\mathcal{A}$. We assume that the following properties are valid :*

(a) *For each $\Lambda, \mathcal{A}_\Lambda$ is a von Neumann algebra in a separable Hilbert space $\mathcal{H}_\Lambda$.*

(b) *Product property : Let* $\Lambda_1 \subset \Lambda$. *Then there exists* $\Lambda_2 \subset \Lambda$(*e.g.*, $\Lambda_2 = \Lambda \setminus \Lambda_1$) *such that* $\Lambda_1 \cap \Lambda_2 = \phi$ *and* $\mathcal{A}_\Lambda$ *is generated by* $\{x_1 x_2 : x_1 \in \mathcal{A}_{\Lambda_1},\ x_2 \in \mathcal{A}_{\Lambda_2}\}$.

(c) *Locally normal property :* ϕ *is locally normal and for each* Λ, *the corresponding density matrix* ρ_Λ *belongs to* $\mathcal{A}_\Lambda$.

We remark that in the most quantum statistical mechanical systems the equilibrium states obtained via thermodynamic limit of local Gibbs states are turned out to be locally normal [3,15,16].

For given Λ, denotes the spectral decomposition of ρ_Λ by

$$\rho_\Lambda = \sum_{i=1}^{\infty} \lambda_i p_i, \tag{3.1}$$

where λ_i, $i = 1, 2, \cdots$, are the eigenvalues of ρ_Λ listed in decreasing order (counting multiplicities). Put

$$P_n = \sum_{i=1}^{n} p_i, \qquad P_n^\perp = 1 - P_n$$

and

$$\mathcal{A}_{\Lambda,n} = P_n \mathcal{A}_\Lambda P_n \oplus \mathbf{C} \cdot P_n^\perp. \tag{3.2}$$

Then by Assumption 3.2, $\mathcal{A}_{\Lambda,n}$ is a finite dimensional subalgebra of $\mathcal{A}_\Lambda$. We adapt the convention that $\Lambda \uparrow \mathbf{R}^\nu$ indicates that Λ increases to $\mathbf{R}^\nu$ so that Λ eventually contains any bounded region of $\mathbf{R}^\nu$. The following result is an improved version of Theorem 3.2 of [17] and Theorem 2.4 of [18] :

Theorem 3.3. *Let* $(\mathcal{A}, \{\mathcal{A}_\Lambda\})$ *be a quasi-local algebra indexed by bounded regions* $\Lambda \subset \mathbf{R}^\nu$ *(or* $\mathbf{Z}^\nu$*),* θ *an automorphism on* $\mathcal{A}$ *and* ϕ *a state on* $\mathcal{A}$ *invariant under* θ. *Under Assumption 3.2, one has*

$$h_\phi(\theta) = \lim_{\Lambda \uparrow \mathbf{R}^\nu (\mathbf{Z}^\nu)} \lim_{n \to \infty} h_{\phi,\theta}(\mathcal{A}_{\Lambda,n}).$$

Here the $\mathcal{A}_{\Lambda,n}$ *is standing for the inclusion map* $\mathcal{A}_{\Lambda,n} \to \mathcal{A}$.

The following result is also useful in the computation of the dynamical entropy.

Proposition 3.4. *For every completely positive unital map* $\gamma_\Lambda : \mathcal{N} \to \mathcal{A}_\Lambda$, *there exists a completely positive unital map* $\sigma_{\Lambda,n} : \mathcal{A}_\Lambda \to \mathcal{A}_{\Lambda,n}$ *such that*

$$H_\phi(\tau_\Lambda \circ \gamma_\Lambda) = \lim_{n \to \infty} H_\phi(\tau_\Lambda \circ \tau_{\Lambda,n} \circ \sigma_{\Lambda,n} \circ \gamma_\Lambda)$$

where $\tau_{\Lambda,n} : \mathcal{A}_{\Lambda,n} \to \mathcal{A}_\Lambda$ *and* $\tau_\Lambda : \mathcal{A}_\Lambda \to \mathcal{A}$ *are the inclusion map.*

The proof of Theorem 3.3 (and Proposition 3.4) turned out to be lengthy. Here we describe the main ideas of the proof. For details we refer to [17,18].

Sketch of the proof of Theorem 3.3 (and Proposition 4.6). First we assert that for each $\Lambda \subset \mathbf{R}^\nu$ there exists a completely positive unital map $\sigma_\Lambda : \mathcal{A} \to \mathcal{A}_\Lambda$ such that for any $x \in \mathcal{A}$

$$\lim_{\Lambda \uparrow \mathbf{R}^\nu} \|\tau_\Lambda \circ \sigma_\Lambda(x) - x\| = 0, \tag{3.3}$$

where $\tau_\Lambda : \mathcal{A}_\Lambda \to \mathcal{A}$ is the inclusion map. Then it follows from (3.3) and Proposition 2.2 that for any finite dimensional C^*–algebra $\mathcal{N}$ and any completely positive unital map $\gamma : \mathcal{N} \to \mathcal{A}$ one has

$$h_{\phi,\theta}(\gamma) = \lim_{\Lambda \uparrow \mathbf{R}^\nu} h_{\phi,\theta}(\tau_\Lambda \circ \sigma_\Lambda \circ \gamma)$$

and so

$$h_\phi(\theta) = \lim_{\Lambda \uparrow \mathbf{R}^\nu} \sup_{\gamma_\Lambda : \mathcal{N} \to \mathcal{A}_\Lambda} h_{\phi,\theta}(\tau_\Lambda \circ \gamma_\Lambda). \tag{3.4}$$

Next we also assert that for a given completely positive map $\gamma_\Lambda : \mathcal{N} \to \mathcal{A}_\Lambda$ there exist completely positive unital maps $\sigma_{\Lambda,n} : \mathcal{A}_\Lambda \to \mathcal{A}_{\Lambda,n}$ such that for any $\epsilon > 0$ there exists $n_0 \in \mathbf{N}$ such that for $n \geq n_0$

$$|H_\phi((\theta^i \circ \tau_\Lambda \circ \gamma_\Lambda)_{i=0,\cdots,k-1}) - H_\phi((\theta^i \circ \tau_\Lambda \circ \tau_{\Lambda,n} \circ \sigma_{\Lambda,n} \circ \gamma_\Lambda)_{i=0,\cdots,k-1})| \leq k\epsilon, \tag{3.5}$$

where θ is an automorphism on $\mathcal{A}$ preserving the state ϕ, and $\tau_{\Lambda,n} : \mathcal{A}_{\Lambda,n} \to \mathcal{A}_\Lambda$ is the inclusion map. Using (3.5) and Proposition 2.1 (a) one has

$$\begin{aligned} h_{\phi,\theta}(\tau_\Lambda \circ \gamma_\Lambda) &= \lim_{n \to \infty} h_{\phi,\theta}(\tau_\Lambda \circ \tau_{\Lambda,n} \circ \sigma_{\Lambda,n} \circ \gamma_\Lambda) \\ &\leq \lim_{n \to \infty} h_{\phi,\theta}(\tau_\Lambda \circ \tau_{\Lambda,n}). \end{aligned}$$

Thus Theorem 3.3 follows from (3.4) and the above result. Proposition 3.4 is a consequence of (3.5).

We give a brief discussion on the proof of our assertions (3.3) and (3.5). Let us consider the assertion (3.3). We first define σ_Λ on the norm dense subalgebra $\bigcup_{\Lambda'} \mathcal{A}_{\Lambda'}$ as follows: For each Λ' with $\Lambda \subset \Lambda'$, there exists $\Lambda'' \subset \Lambda'$ such that $\Lambda \bigcap \Lambda'' = \phi$ and $\mathcal{A}_{\Lambda'}$ is generated by $\{yz \,|\, y \in \mathcal{A}_\Lambda,\, z \in \mathcal{A}_{\Lambda''}\}$. Denote by $\mathcal{S}_{\Lambda'}$ the family of finite linear combinations of elements in $\{yz \,|\, y \in \mathcal{A}_\Lambda,\, z \in \mathcal{A}_{\Lambda''}\}$. Let ω be any (fixed) locally normal state on $\mathcal{A}$ satisfying $\omega(x) = 0$ for any odd element $x \in \mathcal{A}^o$. For each $y \in \mathcal{A}_\Lambda$ and $z \in \mathcal{A}_{\Lambda''}$, define

$$\sigma_\Lambda(yz) = \omega(z)y.$$

Since $\omega(z) = 0$ for any odd element $z \in \mathcal{A}_{\Lambda''}$, it follows that $\sigma_\Lambda(yz) = \sigma_\Lambda(zy)$, and so σ_Λ is well defined. By linearity, σ_Λ extends to $\mathcal{S}_{\Lambda'}$. An inspection shows that $\sigma_\Lambda(b_1 a b_2) = b_1\sigma_\Lambda(a)b_2$, $b_i \in \mathcal{A}_\Lambda, a \in \mathcal{S}_{\Lambda'}$. Thus σ_Λ is a unital conditional expectation from $\mathcal{S}_{\Lambda'}$ to $\mathcal{A}_\Lambda$. As a corollary of Hahn–Banach theorem it turns out that σ_Λ is a contraction. By using the Kaplansky density theorem and the continuity it can be shown that σ_Λ extends to $\mathcal{A}_{\Lambda'}$ and then to $\mathcal{A}$ uniquely. The convergence (3.3) follows from the quasi–locality of $\mathcal{A}$. For the details, see [17, 18].

We next consider the convergence (3.5). As a candidate, define the maps $\sigma_{\Lambda,n} : \mathcal{A}_\Lambda \to \mathcal{A}_{\Lambda,n}$ by

$$\sigma_{\Lambda,n}(y) = P_n y P_n + \{\phi(P_n^\perp y P_n^\perp)/\phi(P_n^\perp)\}P_n^\perp$$

where P_n is the projection defined in the above of (3.2). The proof to show that the map $\sigma_{\Lambda,n}$ defined above indeed satisfies the result (3.5) was the heart of [17]. Since the proof is lengthy and too technical, we omit it here and refer to [17].

4. Dynamical Entropy of Space Translations of CAR and CCR with Respect to Quasi–free States

In this section we give a brief discussion on a computation of the dynamical entropy of space translations of CAR and CCR algebra with respect to an invariant quasi–free state. For details we refer to [18]. We first recall the notion of quasi–free states on CAR and CCR algebras [2,3,24]. Since CAR and CCR algebras are uniquely determined by the appropriate form of commutation relations [3], we shall deal with the CAR and CCR algebras in the Fock spaces.

Let $\mathcal{F}_+$ (resp. $\mathcal{F}_-$) be the symmetric (resp. antisymmetric) Fock space over $L^2(\mathbf{R}^\nu)$. For each bounded region $\Lambda \subset \mathbf{R}^\nu$, let $\mathcal{F}_{\Lambda,\pm}$ be the corresponding Fock space over $L^2(\Lambda)$. Denote by $a_\pm(f)$ and $a_\pm^*(g)$, $f, g \in L^2(\mathbf{R}^\nu)$, the annihilation and creation operators on $\mathcal{F}_\pm$ respectively. These operators satisfy the canonical commutation relations (CCRs)

$$[a_+(f),\, a_+(g)] = 0,\ [a_+^*(f), a_+^*(g)] = 0 \tag{4.1}$$

$$[a_+(f),\, a_+^*(g)] = (f, g)\mathbf{1}$$

and the canonical anticommutation relations (CARs)

$$\{a_-(f), a_-(g)\} = 0,\ \{a_-^*(f), a_-^*(g)\} = 0$$

$$\{a_-(f), a_-^*(g)\} = (f, g)\mathbf{1}, \tag{4.2}$$

where we used the notation $[A, B] = AB - BA$ and $\{A, B\} = AB + BA$. Notice that $\|a_-(f)\| = \|a_-^*(f)\| = \|f\|$. For the details, see [3]. From now on, we suppress $\pm$ in the notation if there is no confusion involved.

The CAR algebra is defined as follows: For each bounded region $\Lambda \subset \mathbf{R}^\nu$, define $\mathcal{A}_\Lambda$ as the von Neumann subalgebra generated by $\{a(f), a^*(g) \mid f, g \in L^2(\Lambda)\}$, where $a(f)$ and $a^*(f)$ satisfy the CARs in (4.2). Let $\mathcal{A}$ be the norm closure of $\bigcup_\Lambda \mathcal{A}_\Lambda$. It then follows that $(\mathcal{A}, \{\mathcal{A}_\Lambda\})$ is a quasi–local algebra.

In order to describe the CCR algebra, let $\Phi(f)$ be the closure of $(a(f) + a^*(f))/2$, where $a(f)$ and $a^*(f)$ satisfy the CCRs in (4.1). Then for each $f \in L^2(\mathbf{R}^\nu)$, $\Phi(f)$ is a self–adjoint operator on the symmetric Fock space $\mathcal{F}$. Let $W(f)$ denote the unitary operator $\exp\{i\Phi(f)\}, f \in L^2(\mathbf{R}^\nu)$. These operators satisfy the Weyl form of the CCRs:

$$W(f)W(g) = \exp\{-iIm(f,g)/2\}W(f+g). \tag{4.3}$$

For each bounded region $\Lambda \subset \mathbf{R}^\nu$, define $\mathcal{A}_\Lambda$ as the von Neumann subalgebra generated by $\{W(f) \mid f \in L^2(\Lambda)\}$, and let $\mathcal{A}$ be the norm closure of $\bigcup_\Lambda \mathcal{A}_\Lambda$. Then $(\mathcal{A}, \{\mathcal{A}_\Lambda\})$ is a quasi–local algebra.

If $0 \le A \le \mathbf{1}$ is an operator on $L^2(\mathbf{R}^\nu)$, then the quasi–free state on the CAR algebra is defined by

$$\omega_A(a^*(f_m)\cdots a^*(f_1)a(g_1)\cdots a(g_n)) = \delta_{nm}\det((g_i, Af_j)). \tag{4.4}$$

If A is a bounded positive operator on $L^2(\mathbf{R}^\nu)$, the quasi–free state on the CCR algebra is defined by

$$\omega_A(W(f)) = \exp\{-\frac{1}{4}(f,(1+2A)f)\}. \tag{4.5}$$

An simple calculation shows that

$$\omega_A(a_\pm^*(f)a_\pm(g)) = (g, Af), \tag{3.6}$$

for any $f, g \in L^2(\mathbf{R}^\nu)$, and any higher order truncated functionals are equal to zero [3].

For given bounded positive operator $A\,(0 \le A < \mathbf{1}$ for the CAR algebra) on $L^2(\mathbf{R}^\nu)$, let B be the operator defined by

$$B = A/(\mathbf{1} + \sigma A), \qquad (\sigma = \pm 1). \tag{4.7}$$

In the rest of this section we shall use the convention that $\sigma = +1$ (resp. $\sigma = -1$) stands for the CCR (resp. CAR) algebra. Equivalently, for given bounded positive operator $B\,(0 \le B < 1$ for the CCR algebra) on $L^2(\mathbf{R}^\nu)$, let A be the operator defined by

$$A = B/(1 - \sigma B), \qquad (\sigma = \pm 1) \tag{4.8}$$

and let ω_A be the quasi–free state corresponding to A. It will be much convenient to give conditions on B, instead of A. We assume that there exists L^1–function $K_B : \mathbf{R}^\nu \to \mathbf{R}$ such that

$$(Bf)(x) = \int K_B(x-y)f(y)d^\nu y \tag{4.9}$$

and let

$$\widehat{K_B}(k) = \int K_B(x)e^{-ik\cdot x}d^\nu x. \tag{4.10}$$

Similarly, let K_A be the kernel function of A :

$$(Af)(x) = \int K_A(x-y)f(y)d^\nu y \tag{4.11}$$

and $\widehat{K_A}$ the Fourier transform of K_A :

$$\widehat{K_A}(k) = \int K_A(x)e^{-ik\cdot x}d^\nu x. \tag{4.12}$$

We impose some conditions on B (and on $B^{1/2}$) :

Assumption 4.1. *We assume that the following conditions are satisfied :*
(a) *There exist constants* $0 \le M_1 < 1$ *and* $0 \le M_2$ *such that* $0 \le \widehat{K_B} \le M_1$ *for the CCR algebra and* $0 \le \widehat{K_B} \le M_2$ *for the CAR algebra.*
(b) *There exist constants* $0 \le M_3$ *and* $0 < \alpha$ *such that the bound*

$$|\widehat{K_B}(k)| \le M_3 \prod_{j=1}^{\nu}(1+|k_j|)^{-(1+\alpha)}$$

holds.
(c) *Let* $K_{B^{1/2}}(x)$ *be the kernel function of the operator* $B^{1/2}$, *i.e., the inverse Fourier transform of* $(\widehat{K_B})^{1/2}$. *There exist constants* $0 \le M_4$ *and* $0 < \alpha$ *such that the bound*

$$|K_{B^{1/2}}(x)| \le M_4 \prod_{j=1}^{\nu}(1+|x_j|)^{-(1+\alpha)}$$

holds.

Remark 4.2. (a) Assumption 4.1 (a) implies that $0 \le B \le M_1 < \mathbf{1}$ for the CCR algebra and $0 \le B \le M_2$ for the CAR algebra.
(b) Since

$$\widehat{K_A} = \widehat{K_B}/(1-\sigma\widehat{K_B}), \qquad (\sigma = \pm 1).$$

Assumption 4.1 (a) and (b) imply that $\widehat{K_A}$ is integrable.
(c) The states for ideal Fermi and Bose gases at inverse temperature $\beta = 1/T$ with activity $z = e^\mu$ are the quasi–free states corresponding to

$$\widehat{K_B}(k) = \exp(-\beta k^2 + \mu) \qquad (\mu < 0 \text{ for the CCRs}).$$

Obviousely all conditions in Assumption 4.1 are satisfied in these cases.

We consider the unit step space translations ($\mathbf{Z}^\nu$–action) on the CAR and CCR algebras. We will use the following notation:

$$\mathbf{n} = (n_1, n_2, \cdots, n_\nu) \in \mathbf{Z}^\nu \tag{4.13}$$

$$B(\mathbf{k}) = B(k_1, \cdots, k_\nu) = \{\mathbf{n} \in \mathbf{Z}^\nu \mid 0 \leq n_j < k_j, j = 1, \cdots, \nu\}.$$

For each $\mathbf{n} \in \mathbf{Z}^\nu$, let $U(\mathbf{n})$ be the unitary operator on $L^2(\mathbf{R}^\nu)$ defined by

$$(U(\mathbf{n})f)(x) = f(x - \mathbf{n}).$$

Then $U(\mathbf{n})$ induces an automorphism $\theta^{\mathbf{n}}$ on the CAR and CCR algebras by

$$\theta^{\mathbf{n}}(a^*(f)) = a^*(U(\mathbf{n})f). \tag{4.14}$$

Obviously the quasi–free states ω_A are invariant under $\theta^{\mathbf{n}}$, $\mathbf{n} \in \mathbf{Z}^\nu$. For a completely positive unital map $\gamma : \mathcal{N} \to \mathcal{A}$ from a finite dimensional C^*–algebra $\mathcal{N}$ to $\mathcal{A}$, and for the $\mathbf{Z}^\nu$–action $\theta : \mathcal{A} \to \mathcal{A}$, denote that

$$h_{\omega_A,\theta}(\gamma) = \lim_{\substack{k_j \to \infty \\ j = 1, \cdots, \nu}} \frac{1}{k_1 \cdots k_\nu} H((\theta^{\mathbf{n}} \circ \gamma) : \mathbf{n} \in B(\mathbf{k})). \tag{4.15}$$

The above limit exists by the subadditivity of the entropy functional $H_{\omega_A}(\gamma_1, \cdots, \gamma_n)$. The *dynamical entropy* of the space translations with respect to the quasi–free states ω_A on the CAR and CCR algebras is defined by

$$h_{\omega_A}(\theta) = \sup_\gamma h_{\omega_A,\theta}(\gamma). \tag{4.16}$$

It may be worth to compare the above definition to that in (2.7) and (2.8). For any bounded region $\Lambda \subset \mathbf{R}^\nu$, let χ_Λ denote the projection operator from $L^2(\mathbf{R}^\nu)$ onto $L^2(\Lambda)$, and let

$$A_\Lambda = \chi_\Lambda A \chi_\Lambda \tag{4.17}$$

and let $|\Lambda|$ be the volume of Λ. From (4.11), (4.12), Assumption 4.1 (b) and Remark 4.2 (b), it follows that

$$Tr_{L^2(\Lambda)}(A_\Lambda) = (2\pi)^{-\nu}|\Lambda| \int \widehat{K}_A(k) d^\nu k < \infty$$

and so the state ω_A is locally normal [3]. Let ρ_Λ be the density operator on $\mathcal{F}_\Lambda$ corresponding to $\omega_\Lambda = \omega_A|_{\mathcal{A}_\Lambda}$. The *local entropy* for the state ω_Λ is defined by

$$S(\omega_\Lambda) \equiv -Tr_{\mathcal{F}_\Lambda}(\rho_\Lambda \log \rho_\Lambda) \tag{4.18}$$

and the *mean entropy* is defined by

$$s(\omega_A) \equiv \lim_{\Lambda \uparrow \mathbf{R}^\nu} \frac{1}{|\Lambda|} S(\omega_\Lambda). \tag{4.19}$$

The above limit exists by the subadditivity of $S(\omega_\Lambda)$ [3]. We now state our main results:

Theorem 4.3. *Let $(\mathcal{A}, \{\mathcal{A}_\Lambda\})$ denote the CAR and CCR quasi-local algebras, θ the space translation automorphism on $\mathcal{A}$ and ω_A the quasi-free state on $\mathcal{A}$. Under the assumptions in Assumption 4.1, the dynamical entropies $h_{\omega_A}(\theta)$ for the CAR and CCR algebras are equal to*

$$h_{\omega_A}(\theta) = \frac{1}{(2\pi)^\nu} \int \{\eta(\widehat{K_A}(k) + \sigma\eta(1 - \sigma\widehat{K_A}(k))\} d^\nu k \qquad (\sigma = \pm 1)$$

where $\eta(x) = -x \log x$.

Remark 4.4. We remark that the right hand side of the expression in the above theorem is the mean entropy of the system under consideration [18].

The proof of Theorem 4.3 turned out to be complicate [18] and so we give only a bare description of basic ideas and refer to [18] for the details. For a notational simplification we only consider the case for $\nu = 1$. The extension to the case $\nu > 1$ will be obvious. We also suppress A from ω_A.

We first show that the dynamical entropy $h_\omega(\theta)$ is bounded above by the mean entropy $s(\omega)$. Let $\Lambda = (-L/2, L/2)$ and denoted by

$$\Lambda(k) = \{x + m : x \in \Lambda, 0 \leq n < k\} \tag{4.20}$$

$$B(k) = \{n \in Z : 0 \leq n < k\}.$$

Denote by $\mathcal{A}(\Lambda, B(k))$ the C^*-algebra generated by $\{\theta^m(\mathcal{A}_{\Lambda,n}) : m \in B(k)\}$, where $\mathcal{A}_{\Lambda,n}$ has defined in (3.2). It can be shown that $\mathcal{A}(\Lambda, B(k))$ is a finite dimensional [18]. From Proposition 2.1 (a) and (b), and Proposition 3.4 (in that order) it follows that

$$\begin{aligned}
& H_\omega((\theta^m(\mathcal{A}_{\Lambda,n})) : m = 0, \cdots, k-1) \\
& \quad \leq \; H_\omega(\mathcal{A}(\Lambda, B(k)), \cdots, \mathcal{A}(\Lambda, B(k))) \\
& \quad = \; H_\omega(\mathcal{A}(\Lambda, B(k))) \\
& \quad \leq \; \lim_{l \to \infty} H_\omega(\mathcal{A}_{\Lambda(k),l}) \\
& \quad \leq \; \lim_{l \to \infty} S(\omega|_{\mathcal{A}_{\Lambda(k),l}}) \\
& \quad = \; S(\omega_{\Lambda(k)}).
\end{aligned}$$

Here we have used the fact that $H_\phi(\mathcal{N}) \leq S(\phi)$[7]. We use Theorem 3.3 to conclude that

$$\begin{aligned}
h_\omega(\theta) &= \lim_{\Lambda \uparrow \mathbf{R}} \lim_{n \to \infty} (\lim_{k \to \infty} \frac{1}{k} H_\omega((\theta^m(\mathcal{A}_{\Lambda,n})) : m = 1, \cdots, k-1)) \\
&\leq s(\omega).
\end{aligned}$$

This gives the upper bound. It remains to show that $h_\omega(\theta)$ is bounded below by $s(\omega)$. Before discussing the proof, we need some preparations. We consider the entropy functional H defined in Section 2 in more details. For multi–index $I = (i_1, \cdots, i_n)$, let $\{\omega_I\}_I$ be a decomposition of the state ω:

$$\omega_I = \omega_{i_1,\cdots,i_n}, \qquad \sum_I \omega_I = \omega.$$

Denote that $\eta(x) = -x \log x$. Let

$$\omega^{(k)}_{i_k} \equiv \sum_{\substack{i_1,\cdots,i_n \\ i_k\,:\,\text{fixed}}} \omega_{i_1,\cdots,i_n}, \qquad \widehat{\omega}^{(k)}_{i_k} = \omega^{(k)}_{i_k} / \omega^{(k)}_{i_k}(1).$$

Let $\mathcal{A}_i, i = 1, \cdots, n$, be finite dimensional subalgebras of $\mathcal{A}$. Then the entropy function can be written as

$$H_\omega(\mathcal{A}_1, \cdots, \mathcal{A}_n) = \sup_{\sum_I \omega_I = \omega} [\sum_I \eta(\omega_I(I)) - \sum_{k=1}^n \sum_{i_k} \eta(\omega^{(k)}_{i_k}(1)) \tag{4.21}$$

$$+ \sum_{k=1}^n S(\omega|_{\mathcal{A}_k}) - \sum_{k=1}^n \sum_{i_k} \eta(\omega^{(k)}_{i_k}(1)) S(\widehat{\omega}^{(k)}_{i_k}|_{\mathcal{A}_k})].$$

For the details, see [7].

In order to get a lower bound of the entropy functional, one has to choose a decomposition of the state, $\omega = \sum \omega_I$, as much as possibly close to the optimal decomposition. In general a decomposition can be written

$$\omega_I(a) = \omega(x'_I a) = \omega(\sigma_{i/2}(x_I)a)$$

with $x'_I \in \pi_\omega(\mathcal{A})', x_I \in \pi_\omega(\mathcal{A})'', \sigma_t$ the modular automorphism of ω [1,3]. In our case, we shall choose x_I to be the projection operator P_I consisting of the spectral projection of the local density operator ρ_I.

Recall that χ_Λ is the projection operator defined in (4.17). Let B^Λ be the operator on $L^2(\Lambda)$ defined by

$$B^\Lambda = A_\Lambda/(1 + \sigma A_\Lambda), \qquad (\sigma = \pm 1) \tag{4.22}$$

where $A_\Lambda = \chi_\Lambda A \chi_\Lambda$. Let σ_t and σ^Λ_t be the modular automorphisms of ω and $\omega|_{\mathcal{A}_\Lambda}$ respectively. Let h and h_Λ be self–adjoint operators defined by

$$\sigma_t(a^*(f)) = a^*(e^{-ith} f), \quad f \in L^2(\mathbf{R}^\nu) \tag{4.23}$$

$$\sigma^\Lambda_t(a^*(g)) = a^*(e^{-ith_\Lambda} g), \quad g \in L^2(\Lambda)$$

By using the KMS condition, it can be check that [18]

$$B = \exp(-h), \quad B^\Lambda = \exp(-h_\Lambda). \tag{4.24}$$

Let $\{\gamma_k\}_{k=1}^{\infty}$ be the eigenvalues of h_Λ listed in increasing order and counting multiplicities, and let $\{f_k\}_{k=1}^{\infty}$ be corresponding normalized eigenvectors. Denote by

$$N(f_k) = a^*(f_k)a(f_k). \tag{4.25}$$

We describe the family of projection operator $\{P_I\}_I$, which will give a decomposition of ω, separately for the CAR algebra and for CCR algebra.

The case of the CAR algebra : For given $k \in \mathbf{N}$, put

$$P_k(0) = 1 - N(f_k), \qquad P_k(1) = N(f_k). \tag{4.26}$$

By the CARs, $P_k(0)$ and $P_k(1)$ are projections with $P_k(0) + P_k(1) = 1$. For a given (fixed) $N \in \mathbf{N}$, let

$$I = (i_1, \cdots, i_N) \in \{0,1\}^N \tag{4.27}$$

be a multi–index, and let

$$P_I = P_1(i_1)P_2(i_2)\cdots P_N(i_N). \tag{4.28}$$

Then $\{P_I : I \in \{0,1\}^N\}$ is a family of mutually commuting orthogonal projections satisfying $\sum_I P_I = 1$. Denote by $\mathcal{A}_\Lambda^{(N)}$ the finite dimensional abelian C^*–algebra generated by $\{P_I : I \in \{0,1\}^N\}$. Clearly one has that $\mathcal{A}_\Lambda^{(N)} \subset \mathcal{A}_\Lambda$.

The case of the CCR algebra : For each $k \in \mathbf{N}, N(f_k)$ has pure point spectrum $\mathbf{Z}_+ = \{0\} \cup \mathbf{N}$. Let $\{Q_k(n) : n \in \mathbf{Z}_+\}$ be the family of the spectral projections of $N(f_k)$. For a given (fixed) $N \in \mathbf{N}$, denote by

$$P_k(i) = Q_k(i), \qquad 0 \le i < N, \quad k \in \mathbf{N} \tag{4.29}$$

$$P_k(N) = 1 - \sum_{i=0}^{N-1} Q_k(i).$$

For a given (fixed) $N \in \mathbf{N}$, let

$$I = (i_1, \cdots, i_N) \in \{0, 1, \cdots, N\}^N \tag{4.30}$$

be multi–index, and let P_I be defined as in (4.28). Then $\{P_I : I \in \{0,1,\cdots, N\}^N\}$ is a family of mutually commuting projections satisfying $\sum_I P_I = 1$. Denote by $\mathcal{A}_\Lambda^{(N)}$ the C^*–algebra generated by $\{P_I\}_I$.

We come back to the functional in (4.21). We choose Λ as $\Lambda = (0, L)$, where L is a positive integer. Let $B(k)$ be defined as in (4.20) and let

$$B(Lk) = \{n \in \mathbf{Z} : 0 \le n < Lk\}.$$

From Proposition 2.1 (c) it follows that

$$H_\omega((\theta^n(\mathcal{A}_\Lambda^{(N)})) : n \in B(Lk)) \geq H_\omega((\theta^{Ln}(\mathcal{A}_\Lambda^{(N)})) : n \in B(k))$$

From (4.15), (4.16) and the above inequality, we conclude that the bound

$$h_\omega(\theta) \geq \lim_{L\to\infty} \lim_{k\to\infty} \frac{1}{Lk} H_\omega((\theta^{Ln}(\mathcal{A}_\Lambda^{(N)})) : n \in B(k)). \tag{4.31}$$

holds. We choose the decomposition of the state $\omega, \sum_J \omega_J = \omega$, as follows : Let $J = (I_n : n \in B(k))$ be a multi-index, where for each $n \in B(k), I_n \in \{0,1\}^N$ for the CARs and $I_n \in \{0, 1, \cdots, N\}^N$ for the CCRs. Let

$$\omega_J(a) = \omega(\sigma_{i/2}(x_J)a) \tag{4.32}$$

where for $J = (I_n : n \in B(k))$

$$x_J = \prod_{n\in B(k)} (\theta^{Ln}(P_{I_n})).$$

Notice that $\sum_J x_J = 1$, and so (4.32) is a decomposition of the state ω. Using the decomposition in (4.32), we obtain from (4.21) that

$$\begin{aligned} &H_\omega((\theta^{Ln}(\mathcal{A}_\Lambda^{(N)})) : n \in B(k)) \\ &\geq \sum_{J=(I_n:n\in B(k))} \eta(\omega(\prod_{n\in B(k)} (\theta^{Ln}(P_{I_n})))) - k\sum_I \omega(P_I)S(\widehat{\omega}_I|_{\mathcal{A}_\Lambda^{(N)}}) \\ &\equiv E(\Lambda, N, k) - kD(\Lambda, N) \end{aligned} \tag{4.33}$$

where $\widehat{\omega}_I(a) = \omega(\sigma_{i/2}(P_I)a)/\omega(P_I)$. Here we have used the fact that

$$S(\omega|_{\mathcal{A}_\Lambda^{(N)}}) = \sum \eta(\omega(P_I)).$$

The quantity $E(\Lambda, N, k)$ is the entropy of the state ω on the finite dimensional abelian C^*–algebra generated by $\{\theta^{Ln}(\mathcal{A}_\Lambda^{(N)}) : n \in B(k)\}$ and $D(\Lambda, N)$ is the entropy defect of the decomposition (4.32) of the state ω.

Proposition 4.5. (a) *For given $\Lambda = (0, L)$ and $\epsilon > 0$ there is $N_0 \in \mathbf{N}$ such that for $N \geq N_0$ the bound*

$$S(\omega_{\Lambda(Lk)}) \leq E(\Lambda, N, k) + k\epsilon$$

holds, where $\Lambda(Lk)$ is defined as in (4.20).

(b) *There exists a constant* $c > 0$ *independent of* $\Lambda = (0, L)^\nu$ *and* N *such that the bound*

$$L^{-\nu}|D(\Lambda, N)| \le cL^{-1/8}$$

holds.

The above result (Prop. 3.5 (a)) can be generalized to the case for $\nu > 1$. Now Theorem 4.3 follows from (4.31), (4.33) and the above proposition. The proof of the proposition is too lengthy and so we close this article with a few remarks. The proof of the part (a) of Proposition 3.5 is relatively easier than that of the part (b). See section 4 of [18]. The proof of the part (b) turned out to be very complicate and relied on several apriori estimates. Let us give a reason why Proposition 3.5 (b) should hold. Notice that

$$\begin{aligned} D(\Lambda, N) &= \sum_I \omega(P_I) S(\widehat{\omega}_I|_{\mathcal{A}_\Lambda^{(N)}}) \\ &= \sum_I \sum_{I'} \omega(P_I) \eta\Big(\frac{\omega(\sigma_{i/2}(P_I) P_{I'})}{\omega(P_I)}\Big) \end{aligned}$$

Let σ_t^Λ be the modular automorphism of $\omega|_{\mathcal{A}_\Lambda}$ on $\mathcal{A}_\Lambda$ and let

$$\widehat{\omega}_I^\Lambda(a) = \omega(\sigma_{i/2}^\Lambda(P_I)a)/\omega(P_I)$$

Since $\sigma_{i/2}^\Lambda(P_I) = P_I$, we have that $S(\widehat{\omega}_I|_{\mathcal{A}_\Lambda^{(N)}}) = 0$. One may expect that the difference, $(\widehat{\omega}_I - \widehat{\omega}_I^\Lambda)(a), a \in \mathcal{A}_\Lambda^{(N)}$, should be bounded by the term which depends only on the surface $\partial\Lambda$ of Λ. For the detailed proof, we refer the reader to [18].

References

[1] Accardi, L. and Cecchini, C. ; Conditional expectations in von Neuman algebras and a theorem of Takesaki, *J. funct. Anal.* **45** (1982) pp 245–273.

[2] Araki, H and Shiraishi, M. ; On quasi–free states of the canonical commuyation relations I.*Publ. R.I.M.S., Kyoto Univ.* **7** (1971) pp 105–120.

[3] Bratteli, O. and Robinson, D.W. ; Operator algebras and quantum statistical mechanics vol. I and II (Springer–Verlag 1979).

[4] Choi, M. D. and Effros, E. G. ; *Ann. Math* **104** (1976) 585.

[5] Confeld, I. P., Fomin, S. V. and Sinai, Ya. G. ; Ergodic theory, (Springer–Verlag 1980).

[6] Connes, A. and Störmer, E. ; *Acta. Maty.* **134** (1975) 289.

[7] Connes, A., Narnhofer, H. and Thirring, W. ; Dynamical entropy of C^*–algebras. *Commun. Math. Phys.* **112** (1987) 691.

[8] Kadison, R. V. and Ringrose, J. R. ; Fundmentals of the theory of operator algebras Vol I and II (Academic press 1986).

[9] Kolmogorov, A. V. ; *Dokal. Akad. Nauk.* **119** (1958) 851.

[10] Narnhofer, H. ; Quantized Arnold cat maps can be entropic K systems, *UWThPh*

[11] Narnhofer, H. and Thirring, W. ; Dynamical entropy and third law of thermodynamics, *Lett. Math. Phys.* **15** (1988) pp 261 - 273.
[12] Narnhofer, H. and Thirring, W. ; Quantum K–systems, *Commun. Math. Phys.* **125** (1989) pp 565–577.
[13] Narnhofer, H. and Thirring, W. ; Chaotic properties of the noncommutative 2–shift, *UwThPh* (1991).
[14] Narnhofer, H. and Thirring, W. ; Dynamical Entropy of Quantum Systems and Their Abelian Counterpart, *UWThPh* (1991).
[15] Park, Y.M. ; Quantum statistical mechanics for superstable interactions; *J. Stat. Phys.* **40** (1984) 259.
[16] Park, Y.M. ; Quantum statistical mechanics of unbounded continuous spin system, *Journal of the Korean Mathematical Society* **22**.1 (1985) pp 43–74.
[17] Park, Y.M. and Shin, H.H. ; Dynamical entropy of quasi–local algebras in quantum statistical mechanics, *Commun. Math. Phys.* **144** (1992) pp 149–161.
[18] Park, Y.M. and Shin, H.H. ; Dynamical entropy of Space Translations of CAR and CCR Algebras with respect to Quasi–free States, preprint (1992).
[19] Ruelle, D. ; Thermodynamic Formalism, (Addison–Wesley, Reading, Mà. 1978).
[20] Sinai, Ya. G. ; *Dokal. Akad. Nauk.*, **124** (1959) 768.
[21] Störmer, E.; Entropy of Some Automorphisms of II_1-factor off Free Group in Infinite Number of Generators, Preprint (1992).
[22] Störmer, E. and Voiculescu, D. ; Entropy of Bogoliubov Automorphisms of Canonical Anticommutation Relations, *Commun. Math. Phys.* **133** (1990) pp 521–542.
[23] Takesaki, M. ; Theory of Operator Algebras. I. (Springer–Verlag, Berlin, Heidelberg, New York 1979).
[24] Verbeure, A, ; Normal and locally normal quasi–free states of Fermi systems, (Cargése Lecture in Physics, Vol. 4. 1970).

The Weak Coupling Limit for a Fermi Gas in a Random Potential

L.J.LANDAU
Mathematics Department, King's College London, Strand, London WC2R 2LS

Abstract. Under certain conditions on the initial state S of the Fermi gas and on the random potential V, the weak coupling limit (Van Hove limit) yields increase of entropy, propagation of chaos, convergence of the state for sufficiently small values of the rescaled time parameter τ to a gauge-invariant and quasi-free asymptotic state, and the semigroup describing the evolution of the two-point function. Results are obtained not only for the average over the random potential but also with probability one.

Key words: Van Hove limit, random potential, irreversibility

1. General Framework

This is a report on joint work with T.G.Ho and A.J.Wilkins [1]. It is a contribution to the general program of deriving irreversible behaviour for macroscopic quantum systems with underlying reversible dynamics, an important part of the foundations of non-equilibrium statistical mechanics. Just as the precise mathematical characterization of phase transitions requires the idealization to infinite systems, so the precise treatment of irreversibility requires limiting procedures. The infinite volume limit eliminates Poincarè recurrences and the infinitely rescaled time of the Van Hove limit leads to irreversibility due to the flow of particles to and from infinity.[1]

The system under consideration is the infinite volume non-relativistic Fermi gas of non-interacting particles in the presence of random static impurities represented as a random one-particle potential V: Let $v(x)$ be a random field (a random variable-valued function of x). Then the random potential V is the operator on $\mathcal{H} = L^2(\mathbb{R}^3, d^3x)$ given as multiplication by v.

The framework for the study is algebraic quantum theory, the C^* algebra $\mathfrak{A}$ of quasilocal observables being the CAR algebra over $\mathcal{H}$. For a given random field v, the (reversible) time-evolution is given by the one-parameter group of automorphisms α_t^λ defined by

[1] Other models might lead to different, perhaps more realistic, mechanisms for irreversible behavior.

H. Araki et al. (eds.), Quantum and Non-Commutative Analysis, 167–178.

$$\alpha_t^\lambda a(f) = a(U_t^\lambda f)$$

where $U_t^\lambda = \exp[itH^\lambda]$, $H^\lambda = -\frac{1}{2}\Delta + \lambda V$.

For a given random field v and initial state S of the Fermi gas, the multi-time correlation functions are

$$S\,(A_1(t_1)\cdots A_N(t_N)) \tag{1}$$

where $A_1, \ldots, A_N \in \mathfrak{A}$ and $A(t) = \alpha_t^\lambda A$. The average over v of this correlation function is

$$\langle S\,(A_1(t_1)\cdots A_N(t_N))\rangle \tag{2}$$

More generally, the average of products of such correlation functions or of truncated correlation functions may be considered:

$$\left\langle S^T(A_1(t_1),\cdots,A_N(t_N))\cdots S^T(A_1'(t_1'),\cdots,A_{N'}'(t_{N'}'))\right\rangle \tag{3}$$

A study is made of the weak coupling limit

$$\begin{aligned} \lambda &\longrightarrow 0 \\ t &\longrightarrow \infty \\ \lambda^2 t &= \tau \end{aligned}$$

of (1), (2), and (3).

2. Main Results

2.1. Existence of the Dynamics

Theorem 1. *Let the random field v be translation invariant and satisfy*

$$\langle \exp[-\delta v(0)]\rangle < \infty \quad \textit{and} \quad \left\langle |v(0)|^{2+\epsilon}\right\rangle < \infty$$

for some $\epsilon, \delta > 0$. Then $-\frac{1}{2}\Delta + \lambda V$ is essentially self-adjoint on C_0^∞ with probability one.

Proof.

$$\begin{aligned} &\left\langle \int d^3x \exp[-x^2]\{|v(x)|^{2+\epsilon} + \exp[-\delta v(x)]\}\right\rangle \\ &= \int d^3x \exp[-x^2]\left\langle |v(x)|^{2+\epsilon} + \exp[-\delta v(x)]\right\rangle \\ &= \int d^3x \exp[-x^2]\left\langle |v(0)|^{2+\epsilon} + \exp[-\delta v(0)]\right\rangle < \infty \end{aligned}$$

Hence $\int d^3x \exp[-x^2]\{|v(x)|^{2+\epsilon} + \exp[-\delta v(x)]\} < \infty$ with probability one. The proof is now completed using the following theorem. Q.E.D.

Theorem (Klein-Landau[2]). *Let $v : \mathbb{R}^3 \to \mathbb{R}$ be measurable and let V be the self-adjoint operator on $\mathcal{H}$ given as multiplication by v. If for some $\epsilon, \delta, c > 0$*

$$\int d^3x \exp[-cx^2] \left(|v(x)|^{2+\epsilon} + \exp[-\delta v(x)]\right) < \infty$$

then $-\Delta + V$ is essentially self-adjoint on $C_0^\infty(\mathbb{R}^3)$.

Hence the existence with probability one of a reversible continuous unitary one-particle dynamics U_t^λ is demonstrated, rather than only the existence of the averaged dynamics, as in [3], [4], and [5]. Thus the automorphism group α_t^λ of the algebra $\mathfrak{A}$ is well-defined with probability one.

2.2. Convergence of the Perturbation Expansion

In order to control the perturbation expansion for the (in general unbounded) random potential V, it is necessary to impose stronger conditions on the random field v than the conditions in Theorem 1. In addition to a stronger condition on $v(0)$, a clustering condition is required on the truncated functions. Let γ_n denote the truncated function of order $n+1$:

$$\langle v(x_1) \cdots v(x_{n+1}) \rangle^T = \gamma_n(x_1 - x_{n+1}, \ldots, x_n - x_{n+1})$$

Theorem 2. *Let the translation-invariant random field v satisfy $\langle \exp[tv(0)] \rangle < \infty \quad \forall t \in \mathbb{R}$ and the cluster property for all n:*

$$||\gamma_n||_1 < \infty \quad \textit{and} \quad ||\hat{\gamma}_n|| < C^n$$

for some $C < \infty$. Then the expectation over the random potential of products of time-evolved functions (3), where A_j is linear in creation,destruction operators, has an absolutely convergent perturbation expansion for all values of t.

Note that the perturbation expansion is derived only for the average over the random potential.

The proof of Theorem 2 is based on the following steps.

A. Introduction of the δ cut-off

Definition 1.

$$v_\delta(x) = \delta^{-3} \int d^3y \exp[-\delta y^2/2] h(\delta^{-1}(x-y)) v(y)$$

where $0 \leq h \in C_0^\infty$ with support in the ball of radius one in $\mathbb{R}^3$ and $||h|| = 1$.

With probability one, the regularized random field $v_\delta \in L^p(\mathbb{R}^3, d^3x)$ for all p, $1 \le p \le \infty$. One consequence of this is that with probability one the corresponding random potential V_δ is a bounded operator. Letting

$$\begin{aligned} H^{\lambda,\delta} &= -\frac{1}{2}\Delta + \lambda V_\delta \\ U_t^{\lambda,\delta} &= \exp[itH^{\lambda,\delta}] \end{aligned}$$

the boundedness of V_δ leads to a norm convergent perturbation expansion of $U^0_{-t}U_t^{\lambda,\delta}$ for all $t \in \mathbb{R}$.

B. Removal of the δ cut-off

Control of the perturbation expansion for the random potential without cut-off is obtained by means of the following lemmas.

Lemma 1. *With probability one,*

$$U_t^{\lambda,\delta} \longrightarrow U_t^\lambda \quad \text{strongly on } \mathcal{H}$$

as $\delta \to 0$.

The lemma is a consequence of the fact that, with probability one, $H^{\lambda,\delta} f$ converges in $\mathcal{H}$ to $H^\lambda f$ for $f \in C_0^\infty$ (which is a domain of essential self-adjointness for H^λ).
The expectation of an individual term of the perturbation expansion is controlled in the $\delta \to 0$ limit using the cluster properties of the random potential. Expanding a correlation function of the random potential into products of truncated functions, estimates are obtained based on

Lemma 2. *Let* $\phi(k_1, \ldots, k_n)$ be continuous and bounded. Then

$$\lim_{\delta \to 0} \int dk_1 \cdots dk_n \, \langle \hat{v}_\delta(k_1) \cdots \hat{v}_\delta(k_n) \rangle^T \, \phi(k_1, \ldots, k_n) =$$

$$\int dk_1 \cdots dk_n \hat{\gamma}(k_1, \ldots, k_{n-1})(2\pi)^{\nu/2} \delta(k_1 + \cdots + k_n) \phi(k_1, \ldots, k_n)$$

The limit process obeys the uniform bound

$$\left| \int dk_1 \cdots dk_n \, \langle \hat{v}_\delta(k_1) \cdots \hat{v}_\delta(k_n) \rangle^T \, \phi(k_1, \ldots, k_n) \right| \le (2\pi)^{\nu/2} \|\hat{\gamma}\|_1 \|\phi\|_\infty$$

Proof. The integral

$$\int dk_1 \cdots dk_n \, \langle \hat{v}_\delta(k_1) \cdots \hat{v}_\delta(k_n) \rangle^T \, \phi(k_1, \ldots, k_n)$$

may be re-expressed in the form

$$(2\pi)^{\nu/2}\int du_1\cdots du_{n-1}dwdk_1\cdots dk_{n-1}$$

$$\times\exp[-(-u_1^2+\cdots+u_{n-1}^2+w^2)/2]\hat{\gamma}(k_1,\ldots,k_{n-1})$$

$$\times\hat{h}(\delta[k_1-\delta^{1/2}u_1])\cdots\hat{h}(\delta[k_{n-1}-\delta^{1/2}u_{n-1})\hat{h}(\delta\{\delta^{1/2}w-[k_1+\cdots+k_{n-1}]\})$$

$$\times\phi(k_1-\delta^{1/2}u_1,\ldots,k_{n-1}-\delta^{1/2}u_{n-1},\delta^{1/2}w-[k_1+\cdots+k_{n-1}])$$

Since $|\hat{h}(p)|\leq(2\pi)^{-\nu/2}$ and $\int due^{-u^2/2}=(2\pi)^{\nu/2}$, the integral is bounded by $(2\pi)^{\nu/2}\|\hat{\gamma}\|_1\|\phi\|_\infty$. The limit $\delta\to 0$ is obtained using a dominated convergence argument.

Q.E.D.

2.3. The Weak Coupling Limit

In order to control the perturbation expansion in the weak coupling limit, it is necessary to restrict to *Gaussian* random fields, due to the rapidly increasing number of higher order truncated functions as the order of the perturbative expansion increases. *Mildly* non-Gaussian random fields (absolutely continuous with respect to a Gaussian field) are treated in §3. Conditions must also be imposed on the initial state of the Fermi gas, which will here be taken translation invariant and rapidly clustering. *Local* perturbations of such states will be treated in §3. We shall thus impose the following requirements.

a. The random field $v(x)$ shall be translation invariant and Gaussian with mean zero and covariance $\langle v(x)v(y)\rangle=\gamma(x-y)$, where $\|\gamma\|_1<\infty$, $\|\hat{\gamma}\|_1<\infty$.

b. The initial state S of the Fermi gas shall be uncorrelated with the random potential and translation invariant, with two-point function

$$S\left(a(f)^*a(g)\right)=(f,\Gamma g)=\int d^3p\overline{f(p)}\Gamma(p)g(p)$$

c. The state S is uniformly L^1-clustering:
For each $N=1,2\ldots$ there is a constant C_N such that for all $\hat{f}_1,\ldots,\hat{f}_N\in C_0^\infty$,

$$\int dx_2\cdots dx_N\left|S^T\left(a^\#(f_1),a^\#(f_{2,x_2}),\ldots,a^\#(f_{N,x_N})\right)\right|$$

$$\leq C_N \|f_1\|_1 \cdots \|f_N\|_1$$

where $f_{j,x}(y) = f_j(y-x)$ and the $a^\#(f_j)$ are either creation or destruction operators.

Remark. The imposed clustering condition follows if the truncated function is a bounded complex measure Γ_N:

$$S^T\left(a^\#(f_1),\ldots,a^\#(f_N)\right)$$

$$= \int dx_1 d\Gamma_N(x_2,\ldots,x_N) f_1(x_1) f_2(x_2+x_1)\cdots f_N(x_N+x_1)$$

in which case

$$C_N = \int_{x_2,\ldots,x_N} d|\Gamma_N|(x_2,\ldots,x_N).$$

Theorem 3. *There exists a $\tau_0 > 0$ and a translation invariant, gauge invariant, quasi-free state S_τ defined for $\tau < \tau_0/2$ such that:*

- *Let $A \in \mathfrak{A}$ be of order N (a product of N creation and destruction operators). Then*

  $$S(\alpha^\lambda_{\lambda^{-2}\tau} A) \xrightarrow{\lambda\to 0} S_\tau(A) \quad \textit{for } 0 < \tau < \tau_0/N$$

  as L^p functions of the random field v for all p, $1 \leq p < \infty$.

- *If S is quasi-free the convergence holds for $0 < \tau < \tau_0/2$, and in this case extends to all $A \in \mathfrak{A}$. Hence*

  $$weak^* \lim \alpha^\lambda_{\lambda^{-2}\tau}{}^* S = S_\tau$$

- *A consequence of the L^p convergence is the existence of a sequence $\lambda_n \to 0$ such that the convergence holds with probability one.*

- *There exists a positivity preserving contraction semigroup T_τ on $L^p(\mathbb{R}^3)$, $1 \leq p \leq \infty$, such that if $\Gamma \in C'$ (continuous functions going to zero at infinity) then the two-point function Γ_τ of S_τ is given by*

  $$\Gamma_\tau = T_\tau \Gamma$$

Notice that the asymptotic state S_τ is *sure*. (It is not a random variable.) Hence, with probability one, the weak coupling limit does not depend on the particular sample field v, but depends only on the *class* of random fields specified by the two-point function γ.

Theorem 3 is proved in the following way. Let A be of order N and expand $S(\alpha^\lambda_{\lambda^{-2}\tau} A)$ into truncated functions:

$$S(\alpha^{\lambda}_{\lambda^{-2}\tau}A) = \sum S^T(\cdots)\cdots S^T(\cdots) \tag{4}$$

We must show:

1. Each higher order truncated function $S^T(\cdots)$ tends to zero in the weak coupling limit.
2. Each non-gauge invariant two-point function $S^T(\cdots)$ tends to zero in the weak coupling limit.
3. Each gauge invariant two-point function $S^T(\cdots)$ converges to the asymptotic two-point function $S_\tau(\cdots)$.

We show the above convergence in the L^2 sense. That is, in points (1.) and (2.) above, $\phi = S^T(\cdots)$ is a function of the random field v. We show $\langle|\phi|^2\rangle \to 0$ in the weak coupling limit. Now ϕ is uniformly bounded and it therefore follows that $\phi \to 0$ as an L^p function of the random field, for all p, $1 \leq p < \infty$. In point (3.) above, we show $\|\phi - S_\tau(\cdot)\|_2 \to 0$ by showing $\langle|\phi|^2\rangle - |\langle\phi\rangle|^2 \to 0$ and $\langle\phi\rangle \to S_\tau(\cdot)$. Again the L^p-convergence follows.

Consequently, in equation (4) each term on the right-hand-side will tend to zero in all L^p, $1 \leq p < \infty$, unless it contains only gauge-invariant two-point functions, and such terms will converge in all L^p to the corresponding product of two-point functions $S_\tau(\cdot)$ which is just $S_\tau(A)$.

The restriction to $\tau < \tau_0/N$ arises from the fact that the convergence in point (1.) of a higher truncated function of order M can be controlled only for $\tau < \tau_0/M$, and if A is of order N then in expansion (4) the highest order truncated function which occurs is of order N. However, if the initial state is quasi-free then, independently of the order of A, only two-point functions appear in (4) and consequently the convergence may be controlled for $\tau < \tau_0/2$. As elements of $\mathfrak{A}$ of finite order are dense in $\mathfrak{A}$ we may extend the weak coupling limit to all $A \in \mathfrak{A}$ and conclude the weak* convergence of $\alpha^{\lambda}_{\lambda^{-2}\tau}{}^*S$ to S_τ.

The technique used to prove points (1.), (2.), and (3.) is discussed in §4.

3. Additional Results

3.1. The Asymptotic System is Bernoulli

Definition 2. *Let $A_1, \ldots, A_N \in \mathfrak{A}$. If $\Delta = \{j_1, \ldots, j_k\}$ and $j_1 < \cdots < j_k$, then*

$$A_\Delta = A_{j_1} \cdots A_{j_k}$$

If Δ is empty, $A_\Delta = 1$.

Theorem 4.

- *Let $A_1, \ldots, A_N$ be linear in creation, destruction operators. For* $\tau_1 + \cdots + \tau_N < \tau_0$,

$$S\left([\alpha^{\lambda}_{\lambda^{-2}\tau_1}A_1]\cdots[\alpha^{\lambda}_{\lambda^{-2}\tau_N}A_N]\right)$$

converges in $L^p(\Omega, d\mu)$, $1 \le p < \infty$, as $\lambda \to 0$ to

$$\sigma S(A_{\Delta_0})S_{\tau^{(1)}}(A_{\Delta_1})\cdots S_{\tau^{(l)}}(A_{\Delta_l})$$

where Δ_j is the set of all k with $\tau_k = \tau^{(j)}$, and $\tau^{(0)} = 0$. σ is the sign of the permutation bringing $\{\Delta_0, \Delta_1, \ldots, \Delta_l\}$ to natural order.

- *If the initial state S is quasi-free, this result holds for $\tau_k < \tau_0/2$, $k = 1, \ldots, N$.*

Theorem 4 is proved in a similar way to theorem 3. Note that the gauge invariant two-point function

$$S\left(\left[\alpha^{\lambda}_{\lambda^{-2}\tau_1}a(f_1)\right]^*\left[\alpha^{\lambda}_{\lambda^{-2}\tau_2}a(f_2)\right]\right) \longrightarrow 0$$

as $\lambda \to 0$, if $\tau_1 \neq \tau_2$.

Theorem 4 expresses the independence of observables at different values of τ in the weak coupling limit. It is a consequence of the fact that a non-zero difference in τ-values leads, in the limit $\lambda \to 0$, to an infinitely large time difference, causing the observables to become independent.

3.2. Propagation of Chaos

Initially the state S of the Fermi gas is uncorrelated with the random potential: $\langle S(A)F(v)\rangle = S(A)\langle F(v)\rangle$. The time-evolved state will no longer satisfy this independence, but asymptotically the independence is restored.

Theorem 5. *Let $F(v)$ be an L^1-function of the random field. Let $A \in \mathfrak{A}$ be of order N. Then*

$$\left\langle S(\alpha^{\lambda}_{\lambda^{-2}\tau}A)F(v)\right\rangle \longrightarrow S_\tau(A)\langle F(v)\rangle \quad as \ \lambda \to 0$$

if $\tau < \tau_0/N$ ($\tau_0/2$ is S is quasi-free).

Proof. Let $F(v) \in L^p$ for some $p > 1$. By Theorem 3, $S(\alpha^{\lambda}_{\lambda^{-2}\tau}A)$ converges to $S_\tau(A)$ in $L^{p'}$, where $1/p + 1/p' = 1$, and hence Theorem 5 holds in this case. By the boundedness of $S(\alpha^{\lambda}_{\lambda^{-2}\tau}A)$ and the denseness of L^p in L^1 the proof of convergence extends to all $F(v) \in L^1$.

Q.E.D.

3.3. Mild Non-Translation Invariance

A. Random fields absolutely continuous with respect to v

Let the random field v' be absolutely continuous with respect to the translation invariant clustering Gaussian random field v. Then expectations with respect to v' are given by

$$\langle G\rangle' = \langle GF\rangle / \langle F\rangle$$

for some $F(v) \in L^1$. From Theorem 5 we conclude that the weak coupling limit with the random field v' is the same as the weak coupling limit with v.

B. Local perturbations of the initial state S

Let $B \in \mathfrak{A}$ and let S' be defined by

$$S'(A) = S(B^*AB)/S(B^*B)$$

Then S' is said to be a local perturbation of the state S. More generally, a local perturbation of S is a mixture of such states. It follows from Theorem 5 that the weak coupling limit for the initial state S' is the same as for S. Indeed, let A be of order N, B of order M, B' of order M'. By Theorem 5, if N is odd,

$$S(B^*(\alpha^\lambda_{\lambda^{-2}\tau}A)B') \longrightarrow \pm S(B^*B')S_\tau(A) = 0$$

since S_τ is even; and if N is even,

$$S(B^*(\alpha^\lambda_{\lambda^{-2}\tau}A)B') \longrightarrow S(B^*B')S_\tau(A)$$

3.4. Increase of Entropy Density

The entropy density of the translation invariant gauge invariant quasi-free state S_τ with two-point function Γ_τ is

$$s_\tau = -(2\pi)^{-3}\int d^3p[\Gamma_\tau(p)\ln\Gamma_\tau(p) + (1-\Gamma_\tau(p))\ln(1-\Gamma_\tau(p))]$$

If $\Gamma \in C_0$ (continuous with compact support) then $\Gamma_\tau \in C_0$ and s_τ is finite.

$$\Gamma_\tau = T_\tau\Gamma = e^{-\tau\mathcal{L}}\Gamma$$

and

$$\frac{d}{d\tau}\Gamma_\tau = -\mathcal{L}\Gamma_\tau$$

where

$$(\mathcal{L}\Gamma_\tau)(p) = 4\pi \int d^3k\delta(k^2 - p^2)\hat{\gamma}(p-k)[\Gamma_\tau(p) - \Gamma_\tau(k)]$$

Using $\hat{\gamma}(k) = \hat{\gamma}(-k) \geq 0$ and $(\ln x - \ln y)(x - y) \geq 0$ we conclude that

$$\frac{ds_\tau}{d\tau} \geq 0$$

Thus the entropy density increases with increasing values of the rescaled time parameter τ (for τ positive). The reversibility of the underlying dynamics leads in this case to symmetry in $\tau \to -\tau$: evolving the initial state to negative values of τ leads to the same asymptotic state $S_{|\tau|}$. Therefore, the entropy density is also increasing as $\tau \to -\infty$. Hence the entropy density decreases as τ increases to zero and then increases as τ increases to infinity. The state at time zero takes on the appearance of a fluctuation.

4. Method of Analysis

An individual term of the perturbation expansion may be expressed as a momentum space integral involving an imaginary quadratic exponential, due to the quadratic nature of the free Hamiltonian. An important role is therefore played by the standard estimate:

Proposition (Spreading of the Wave-Packet). *If $\phi, \hat{\phi} \in L^1(\mathbb{R}^n)$ and Q is an invertible real symmetric $n \times n$ matrix, then for all $k \in \mathbb{R}^n$*

$$\left| \int d^n p \, \phi(p) e^{\frac{i}{2}(p,Qp) + i(k,p)} \right| \leq \frac{\|\hat{\phi}\|_1}{\sqrt{|\det Q|}}$$

Of course we also have

$$\left| \int d^n p \, \phi(p) e^{\frac{i}{2}(p,Qp) + i(k,p)} \right| \leq \|\phi\|_1$$

Definition 3. *Let $\mathbb{M}_n(\mathbb{C})$ denote the set of complex $n \times n$ matrices.*

- *Let $m \leq n$, $Q \in \mathbb{M}_n(\mathbb{C})$, $P \in \mathbb{M}_m(\mathbb{C})$. Write $P \leq Q$ if P is a submatrix of Q; that is,*

$$P_{jk} = Q_{\alpha(j)\alpha(k)} \quad j,k = 1,2,\ldots,m$$

where $1 \leq \alpha(1) < \cdots < \alpha(m) \leq n$.

- *For $Q \in \mathbb{M}_n(\mathbb{C})$,*

$$\max Q \stackrel{\text{def}}{=} \max\left\{1, \max_{P \leq Q} |\det P|\right\}$$

Now consider $\phi(p)$ in the above proposition of the form $\phi(p) = \gamma(p_1)\cdots\gamma(p_n)$, where $\gamma, \hat{\gamma} \in L^1(\mathbb{R})$. For each subset $p_{\alpha(1)}, \ldots, p_{\alpha(m)}$ of the variables $p_1, \ldots, p_n$ we associate the corresponding submatrix $P \leq Q$. Applying the proposition to the integral over $p_{\alpha(1)}, \ldots, p_{\alpha(m)}$ we obtain

$$\begin{aligned}\left|\int d^n p\, \gamma(p_1)\cdots\gamma(p_n) e^{\frac{i}{2}(p,Qp)+i(k,p)}\right| &\leq |\det P|^{-1/2}\|\gamma\|_1^{n-m}\|\hat{\gamma}\|_1^m \\ &\leq |\det P|^{-\frac{1}{2}}\|\gamma\|_0^n\end{aligned}$$

for every submatrix $P \leq Q$, where $\|\gamma\|_0 = \max\{\|\gamma\|_1, \|\hat{\gamma}\|_1\}$. This leads then to the additional standard estimate:

Proposition (Spreading of the Wave-Packet II).

$$\left|\int d^n p\, \gamma(p_1)\cdots\gamma(p_n) e^{\frac{i}{2}(p,Qp)+i(k,p)}\right| \leq [\max Q]^{-1/2}\|\gamma\|_0^n$$

An nth order term in perturbation theory is estimated by a time-ordered multiple time integral

$$\int_0^t dt_1 \int_0^{t_1} dt_2 \cdots \int_0^{t_{n-1}} dt_n\, [\max Q(t_1, \ldots, t_n)]^{-3/2}$$

An inductive approach to estimating such integrals may be developed on the basis of a useful matrix identity. Represent $Q \in \mathbb{M}_n(\mathbb{C})$ in the form

$$Q = \begin{pmatrix} a & \overline{v} \\ u & \widehat{Q} \end{pmatrix}$$

where $\widehat{Q} \in \mathbb{M}_{n-1}(\mathbb{C})$, $u, v \in \mathbb{C}^{n-1}$, and $a \in \mathbb{C}$. Then if $\widehat{Q}$ is invertible,

$$\det Q = \left[a - (v, \widehat{Q}^{-1}u)\right] \det \widehat{Q}$$

On the basis of this identity the following proposition may be proved.

Proposition 1. *For $s \in \mathbb{R}$, let $H(s) = \max(1, |s|)$. Then*

$$\max Q \geq H(a + f(Q)) \max \widehat{Q}$$

where $f(Q)$ does not depend on a.

Control of the weak coupling limit requires a generalization of Proposition 1 which is obtained by noting that $\max Q \geq |a|$. Hence

Proposition 2. *For any ϵ, $0 \leq \epsilon \leq 1$*

$$\max Q \geq |a|^{\epsilon}[H(a + f(Q))]^{1-\epsilon}[\max \widehat{Q}]^{1-\epsilon}$$

For details of the application of these inequalities to the control of the weak coupling limit see [1].

References

[1] Ho,T.G., Landau,L.J., Wilkins,A.J., On the Weak Coupling Limit for a Fermi Gas in a Random Potential, to appear in *Reviews in Mathematical Physics*

[2] Klein,A., Landau,L.J., Construction of a Unique Self-Adjoint Generator for a Symmetric Local Semigroup, *J. Funct. Anal.*, **44**, 121 (1981)

[3] Martin,P., Emch,G.G., A Rigorous Model Sustaining van Hove's Phenomenon, *Helv. Phys. Acta*, **48**, 59 (1975)

[4] Spohn,H., Derivation of the Transport Equation for Electrons Moving Through Random Impurities, *J. Stat. Phys.* **17**, 385 (1977)

[5] Dell'Antonio,G.F., Large Time, Small Coupling Behaviour of a Quantum Particle in a Random Potential, *Ann. Inst. Henri Poincaré -Sec. A*, **39**, 339 (1983)

The Gap Labelling Theorem: the case of Automatic Sequences

JEAN BELLISSARD
Université Paul Sabatier, Laboratoire de Physique Quantique
118, Route de Narbonne, 31062 Toulouse Cedex
FRANCE

Abstract. Gaps in the spectrum of a Schrodinger operator are labelled by elements of an abelian countable discrete group: the K-group of the C*-algebra generated by the Hamiltonian and its space-translated. The calculation of these labels is performed explicitely in 1D lattice case for potentials generated by automatic sequences.

Key words: gap labelling, automatic sequences.

Let $\mathcal{H}$ be a separable Hilbert space. Let G be a locally compact group acting on $\mathcal{H}$ by a projective unitary representation U. A selfadjoint operator H acting on $\mathcal{H}$ is called homogeneous with respect to the G action, if the family of its translated is precompact in the strong resolvent topology. The closure Ω of this set is called the "hull" of the pair (H, G), and G acts on it by homeomorphisms. Ω is a compact metrizable space. We will denote by H_ω the self adjoint operator corresponding to $\omega \in \Omega$. It satisfies the covariance relation:

$$U(g)H_\omega U(g)^{-1} = H_{g\omega} \qquad \omega \in \Omega, \qquad g \in G,$$

and the map $\omega \in \Omega \to H_\omega$ is continuous for the strong resolvent topology. It follows that the spectrum of H_ω is contained in the spectrum of H for all $\omega \in \Omega$.

Let $\mathcal{A}$ be the C^*-algebra generated in $\mathcal{H}$ by bounded functions of the H_ω's. Given any pair Γ_1, Γ_2 of gaps in the spectrum of H, the eigenprojection $P(\Gamma_1, \Gamma_2)$ corresponding to the part of the spectrum between these two gaps, defines an element of the countable abelian group $K_0(\mathcal{A})$. This is the "Abstract Gap Labelling Theorem". The labelling is invariant under a norm-resolvent perturbation of H.

Let us consider now the special case for which G is either $\mathcal{R}^D$ or $\mathcal{Z}^D$, and $\mathcal{H}$ is $L^2(G)$. Let us assume in addition that H is bounded from below and that the resolvent $(z - H)^{-1}$ admits a distribution kernel $G_z(x, y)$ which decreases fast enough as $| x - y | \to \infty$. Then one can show that $\mathcal{A}$ is contained in the crossed-product C^*-algebra $\mathcal{C}(\Omega) \times G$. In addition, given an invariant

H. Araki et al. (eds.), Quantum and Non-Commutative Analysis, 179–181.

probability measure $\mathbf{P}$ on Ω the trace per unit volume exists $\mathbf{P}$-almost surely and defines a trace τ on $\mathcal{A}$.

The "integrated density of states" (IDS) of H is defined by:

$$\mathcal{N}(E) = lim_{|\Lambda|\to\infty} \frac{\#\{eigenvalues\ of\ H_\omega\ |_\Lambda \leq E\}}{|\Lambda|},$$

where $|\Lambda|$ denotes the volume of Λ, and $H_\omega\ |_\Lambda$ denotes the restriction of the operator H_ω to the (open) set Λ with some boundary condition. In most examples of interest, it can be shown that the limit exists and is independent of ω, $\mathbf{P}$-almost surely. Moreover, the "Shubin formula" holds namely:

$$\mathcal{N}(E) = \tau(\chi(H \leq E)),$$

where $\chi(H \leq E)$ is the eigenprojection of H in the W^*-algebra $\mathcal{L}^\infty(\mathcal{A}, \tau)$ on the part of the spectrum contained in the interval $(-\infty, E]$.

If E belongs to a gap of H, the eigenprojection $\chi(H \leq E)$ belongs to $\mathcal{A}$ and therefore the IDS takes on values in the image by τ of the group $K_0(\mathcal{A})$. This image will be called the set of "gap labels".

A one dimensional chain is described by the discrete Schrödinger operator H_V acting on $l^2(\mathbf{Z})$ by

$$H_V\psi(n) = \psi(n+1) + \psi(n-1) + V(n)\psi(n), \qquad \in \mathbf{Z}, \qquad \psi \in l^2(\mathbf{Z}),$$

where the "potential" $V = \{V(n)\}_{n\in\mathbf{Z}}$ is a bounded sequence of real numbers. We will assume that V takes on finitely many values in the set $\aleph = \{V_1, V_2, \ldots, V_N\}$, called the alphabet of V. For such a potential, the gap label theorems read:

Theorem 1: *For a one dimensional chain with potential V taking values in a finite alphabet, the set of gap labels of H_V is the $\mathbf{Z}$-module generated by the probabilities of occurence of finite words contained in the sequence $\{V(n)\}_{n\in\mathbf{Z}}$.*

A substitution σ is a map from the finite alphabet $\aleph$ into the set of finite words $\aleph^*$ made with letters of $\aleph$. It can be extended as a map from the set of words into itself by concatenation. We assume that there is a letter denoted by 0 such that $\sigma(0)$ begins by zero, and that the length of the iterated $\sigma^n(0)$ goes to ∞ as $n \to \infty$. In the limit, we get an infinite word called a "substitution sequence". This sequence V will be extended on $\mathbf{Z}$ by symmetry around $n = 0$.

We assume σ to be "primitive", namely given any pair a, b of letters, there is an integer n such that b occurs in $\sigma^n(a)$. The matrix $M(\sigma)$ is indexed by letters and its element $M(\sigma)_{a,b}$ is the number of occurence of b into the word $\sigma(a)$. By the Perron-Frobenius theorem, it has a maximal eigenvalue θ, a positive algebraic number, corresponding to a unique eigenvector $v = \{v_a; a \in \aleph\}$ with positive coordinates and normalized by:

$$\sum_{a \in \aleph} v_a = 1.$$

Given a substitution sequence V, let $\aleph_2$ be the set of words of length 2 occuring in V. It can be considered as a new alphabet on which σ induces a new substitution σ_2 in a canonical way. Correspondingly we get a matrix M_2 and a Perron-Frobenius vector v_2 with the same eigenvalue θ. In this situation, the gap labelling theorem reads:

Theorem 2: *For a potential V given by a substitution sequence σ, the set of gap labels of H_V is the $\mathbf{Z}(\theta^{-1})$-module generated by the coordinates of the two Perron-Frobenius vectors v, v_2 corresponding to the eigenvalue θ of the matrix of that substitution.*

References

[1] J.BELLISSARD, A.BOVIER, J.M.GHEZ, *Rev. Math. Phys.*, 4, (1992),1-37.
[2] J.BELLISSARD, in *From Number Theory to Physics*, Chap.12, M.Waldschmidt, P.Moussa, J.M.Luck, C.Itzykson Eds., Springer Verlag (Berlin, Heidelberg, New-York, London, Paris, Tokyo, Hong-Kong), *to appear in* (1992).

Operator Algebra Approach to Soluble Models of Quantum Spin Lattice Systems

HUZIHIRO ARAKI
Research Institute for Mathematical Sciences
Kyoto University, Kyoto 606-01, Japan

Abstract.
Operator algebra methods for study of exactly soluble one-dimensional quantum spin lattice models are briefly reviewed.

Key words: Operator algebras, exactly soluble models, quantum statistical mechanics, spin lattice systems, XY model, XXZ model, Jordan Wigner transformation.

1. Introduction

Some of main features of operator algebra methods, in comparison with the conventional approach of studying a finite lattice problem first and then taking the infinite volume limit later, are as follows:

(1) Infinitely extended systems are directly studied. This leads to the following two features.

(a) Ground states defined there (with the condition that localized excitation always increase the energy) include all accumulation points of finite volume ground states but might not be restricted to them. All ground states in this sense are determined for the one-dimensional XY-model, yielding an interesting phase diagram in terms of parameters [3], and their analytic properties are obtained [4], [7].

(b) Infinite time asymptotic behavior of equilibrium states is easily analyzed without the problem of simultaneously taking the infinite volume limits [1], [2].

(2) The problem of defining a non-converging infinite product of z-spin operators at half of the lattice sites (encountered in the Jordan-Wigner transformation from the algebra of spins to the algebra of canonical anticommutation relations (CAR)) is solved by the notion of the crossed product, simply by enlarging the algebra of spins with the desired infinite product as a new element [2]. In this scheme, operator algebra methods are utilized to relate the ground states of the spin algebra and those of the CAR algebra with an intermediary of the ground states of the intersection of the two algebras via restriction-extension relations [3].

H. Araki et al. (eds.), Quantum and Non-Commutative Analysis, 183–196.

(3) Master symmetries can be fully analyzed by operator algebra methods for the XY-model. [5]

The analysis has been completely carried out for the XY model

$$H_{XY} = -J\sum\{(1+\gamma)\sigma_x^{(j)}\sigma_x^{(j+1)} + (1-\gamma)\sigma_y^{(j)}\sigma_y^{(j+1)} + 2\lambda\sigma_z^{(j)}\}, \qquad (1)$$

while the analysis is started for the XXZ model for the simplest case of $J > 0, 1 < \Delta$ [6]:

$$H_{XXZ} = -J\sum\{\sigma_x^{(j)}\sigma_x^{(j+1)} + \sigma_y^{(j)}\sigma_y^{(j+1)} + \Delta\sigma_z^{(j)}\sigma_z^{(j+1)}\}. \qquad (2)$$

2. C^*-dynamical System

We consider the C^*-algebra $\mathfrak{A}$ generated by the algebras of 2×2 matrices (Pauli spins)

$$\mathfrak{A}_j = \{c_0 1 + c_x\sigma_x^{(j)} + c_y\sigma_y^{(j)} + c_z\sigma_z^{(j)}\},$$
$$\sigma_x^{(j)} \approx \begin{pmatrix} 0 & 1 \\ 0 & 0 \end{pmatrix}, \quad \sigma_0^{(j)} \approx \begin{pmatrix} 0 & -i \\ i & 0 \end{pmatrix}, \quad \sigma_2^{(j)} \approx \begin{pmatrix} 1 & 0 \\ 0 & -1 \end{pmatrix}$$

associated with each site $j = 0, \pm 1, \pm 2, \ldots$ of infinite one-dimensional lattice as their C^* tensor product

$$\mathfrak{A} = \bigotimes_{j=-\infty}^{\infty} \mathfrak{A}_j,$$

in which $\mathfrak{A}_j$ is identified with the subalgebra

$$\Big(\bigotimes_{k<j} 1_k\Big) \otimes \mathfrak{A}_j \otimes \Big(\bigotimes_{l>j} 1_l\Big) \subset \mathfrak{A}$$

and hence $\sigma_\alpha^{(j)}$ $(\alpha = x, y, z)$ as elements of $\mathfrak{A}$.

A (pointwise norm continuous) one-parameter group of automorphisms α_t of $\mathfrak{A}$ $(\alpha_s\alpha_t = \alpha_{s+t},\ t \to \alpha_t(A) \in \mathfrak{A}$ continuous) is specified in terms of Hamiltonian such as (1) and (2) as follows:

$$\alpha_t(A) = \lim_{N\to\infty} e^{itH_N} A\, e^{-itH_N}, \quad A \in \mathfrak{A} \qquad (3)$$
$$H_N = \sum_{|j|<N} H^{(j)}.$$

Here each $H^{(j)}$ belongs to the subalgebra

$$\mathfrak{A}^0 = \text{ the algebra generated by } \cup_j \mathfrak{A}_j \subset \mathfrak{A} \qquad (4)$$

consisting of strictly local elements of $\mathfrak{A}$, such as

$$H_{XY}^{(0)} = -J\{(1+\gamma)\sigma_x^{(0)}\sigma_x^{(1)} + (1-\gamma)\sigma_y^{(0)}\sigma_y^{(1)} + \lambda\sigma_z^{(0)} + \lambda\sigma_z^{(1)}\}, \quad (5)$$

$$H_{XXZ}^{(0)} = -J\{\sigma_x^{(0)}\sigma_x^{(1)} + \sigma_y^{(0)}\sigma_y^{(1)} + \Delta\sigma_z^{(0)}\sigma_z^{(1)}\}. \quad (6)$$

$H^{(j)}$ for different j are related by

$$H^{(j+1)} = \tau H^{(j)}$$

where the τ is the automorphism of $\mathfrak{A}$ shifting the site by one:

$$\tau\sigma_\alpha^{(j)} = \sigma_\alpha^{(j+1)} \quad (\alpha = x, y, z). \quad (7)$$

The generator of the dynamics α_t is defined by

$$\delta A = (d/dt)\alpha_t(A)|_{t=0} \quad (8)$$

whenever the limit exists in $\mathfrak{A}$, the set of such A being the domain D_δ of δ, and is given for $A \in \mathfrak{A}^{(0)}$ by

$$\delta A = i\sum_j [H^{(j)}, A] \qquad \text{(automatically a finite sum)}. \quad (9)$$

The basis for such definitions is the following

Theorem 2.1. *(1) The limit in* (3) *exists for all* $A \in \mathfrak{A}$.

(2) The generator δ *defined by* (8) *is the closure of the derivation on* $\mathfrak{A}^{(0)}$ *defined by* (9).

3. States

A state φ of a C^*-algebra $\mathfrak{A}$ is a complex-valued linear functional on $\mathfrak{A}$ which is positive ($\varphi(A^*A) \geq 0$ for any $A \in \mathfrak{A}$), and normalized.

For the given C^* dynamical system $(\mathfrak{A}, \alpha_t)$, we are interested in the following two types of states.

Definition 3.1 A state φ of $\mathfrak{A}$ is an (α, β) KMS-state if

$$\varphi(AB) = \varphi(\alpha_{i\beta}(B)A) \quad (10)$$

for any A and B in $\mathfrak{A}$ such that $\alpha_t(B)$ has an entire analytic extension $\alpha_z(B)$ as an $\mathfrak{A}$-valued function of z.

There is a dense set of B in $\mathfrak{A}$ satisfying this entire analyticity condition (as a consequence of the continuity of α_t). Any such φ is automatically stationary:

$$\varphi(\alpha_t(A)) = \varphi(A), \qquad A \in \mathfrak{A}. \quad (11)$$

All such states are called equilibrium states.

For any state φ of a C^*-algebra $\mathfrak{A}$, there is a Hilbert space, $\mathcal{H}_\varphi$, a representation π_φ of $\mathfrak{A}$ on $\mathcal{H}_\varphi$ and a vector Ω_φ (the triple being unique up to unitary equivalence), such that

$$\varphi(A) = (\Omega_\varphi, \pi_\varphi(A)\Omega), \qquad A \in \mathfrak{A}$$

and the set of $\pi_\varphi(A)\Omega$, $A \in \Omega$ is dense in $\mathcal{H}_\varphi$.

If φ is stationary (i.e. if φ satisfies (11)), then there exists a (unique) selfadjoint operator H_φ on $\mathcal{H}_\varphi$ satisfying

$$H_\varphi \Omega_\varphi = 0, \tag{12}$$

$$e^{itH_\varphi} \pi_\varphi(A) e^{-itH_\varphi} = \pi_\varphi(\alpha_t(A)), \qquad A \in \mathfrak{A}. \tag{13}$$

Definition 3.2. A state φ is called a ground state of $(\mathfrak{A}, \alpha_t)$ if φ is stationary and $H_\varphi \geq 0$.

The operator H_φ is interpreted as representing energy and the condition $H_\varphi \geq 0$ is interpreted as showing φ as the state of the lowest energy (see (12)) among those states obtained from φ by application of elements of $\mathfrak{A}$ (i.e. under a local excitation).

In a simple example of the one-dimensional Ising model

$$H = -\sum_j \sigma_z^{(1)} \sigma_z^{(j+1)} \tag{14}$$

there are 4 different types of (pure) groud states. If we denote by $\varphi_\pm^{(j)}$ the eigenstate of $\sigma_z^{(j)}$ with the eigenvalue ± 1, respectively, then the product states

$$\varphi_+ = \bigotimes_k \varphi_+^{(k)}, \qquad \varphi_- = \bigotimes_k \varphi_-^{(k)} \tag{15}$$

as well as the so-called soliton states

$$\varphi_{+-j} = \Big(\bigotimes_{k \leq j} \varphi_+^{(k)}\Big) \otimes \Big(\bigotimes_{l > j} \varphi_-^{(l)}\Big), \tag{16}$$

$$\varphi_{-+j} = \Big(\bigotimes_{k \leq j} \varphi_-^{(k)}\Big) \otimes \Big(\bigotimes_{l > j} \varphi_+^{(l)}\Big), \tag{17}$$

are typical ground states, where j is arbitrary in (16) and (17).

Soliton states (16) and (17) are ground states according to our definition because any local excitation will raise its energy. However, physicists would consider these states to have a higher energy (by a finite amount) than the ground states (15) and might hesitate calling them ground states, the energy excess over (15) being caused by the neighbouring opposite spin state $\varphi_\pm^{(j)}$ and $\varphi_\mp^{(j+1)}$ for which $-\sigma_z^{(j)}\sigma_z^{(j+1)}$ gives $+1$ instead of -1 for (15).

4. Pure Phase

A convex combination of states φ_k

$$\varphi = \sum \lambda_k \varphi_k, \quad \lambda_k \geq 0, \quad \sum \lambda_k = 1$$

is another state, called mixture of $\{\varphi_k\}$ and the set S of all states is a covex set. An extremal point φ of S, i.e. a state which is not a convex combination of other states, is called a pure state and is characterized by the irreducibility of the corresponding representation π_φ.

The set $\mathrm{Eq}_\beta S$ of (α, β) KMS states for each fixed β is also a convex set. Its extremal point φ, i.e. an (α, β)-KMS state which is not a mixture of other (α, β) KMS states is called a pure phase and is characterized by $\pi_\varphi(\mathfrak{A})''$ being a factor.

The set $\mathrm{Gr}\, S$ of all ground states is also a convex set and its extremal point is a pure (ground) state.

In all these cases, the convex hull of extremal points is dense in the set and hence it is sufficient to determine all pure phases and all pure ground states for most purposes.

In the case of one-dimensional spin lattice system with dynamics given as above, it is known that there exists a unique pure phase for each β.

We shall see below that this is not necessarily the case for ground states and the dependence on the parameters gives rise to an interesting phase diagram for ground states.

5. Return to Equilibrium

For an (α, β) KMS state φ_β and $x \in \mathfrak{A}$, the new state

$$\varphi(A) = \varphi_\beta(x^* A x)/\varphi_\beta(x^* x), \qquad A \in \mathfrak{A} \tag{18}$$

is interpreted as an (approximately) local excitation of the state φ_β by the action of x. Return to equilibrium is defined by the validity of the following:

$$\lim_{t \to \infty} \varphi(\alpha_t(A)) = \varphi_\beta(A). \tag{19}$$

The validity of return to equilibrium for all $x \in \mathfrak{A}$ is equivalent to the following weak asymptotic abelian property of α_t:

$$w - \lim_{t \to \infty} \pi_\varphi([B, \alpha_t(A)]) = 0, \qquad A, B \in \mathfrak{A} \tag{20}$$

where the limit is in the weak operator topology.

For the XY-model, these properties hold if $(\lambda, \gamma) \neq (0, \pm 1)$. The excluded case is equivalent to the classical one-dimensional Ising model mentioned earlier (σ_z should be replaced by σ_x (if $\gamma = 1$) or by σ_y (if $\gamma = -1$)) and there are many α_t-invariant elements in $\mathfrak{A}$, which automatically violate (5.2) and (5.3).

The (norm) asymptotic abelian property of α_t

$$\lim_{t\to\infty} \|[B, \alpha_t(A)]\| = 0$$

does not hold for a general elements A and B even for $(\lambda, \gamma) \neq (0, \pm 1)$.

6. Ground States of the XY-model

For $|\lambda| \geq 1$, as well as for $\gamma = 0$, there is a unique ground state, namely the number of pure ground states is one.

For $|\lambda| < 1$, $\gamma \neq 0$ and $(\lambda, \gamma) \neq (0, \pm 1)$, there are exactly 2 pure ground states.

For $(\lambda, \gamma) = (0, \pm 1)$, which is the case of Ising model, there are 2 pure ground states corresponding to (15), which is analytic continuation of the above two pure ground states for surrounding values of the parameters λ, γ. In addition, there are 2 additional irreducible representations, each containing infinite number of pure ground state vectors corresponding to (16) or (17) for all values of j and their superpositions. (Again σ_z earlier has now to be replaced by σ_x or σ_y.) Therefore for these isolated values $(\gamma, \lambda) = (0, \pm 1)$ of parameters, there are an infinite number of pure ground states, giving rise to four inequivalent ground state representations. Those corresponding to (16) and (17) are called soliton sectors.

The phase diagram describing the above situation briefly is in Fig.1.

For each fixed local observables $A \in \mathfrak{A}^\circ$, its expectation value $\varphi(A)$ by a pure ground state is an analytic function of the parameters γ, λ, J except at the boundary lines of the phase diagram $|\lambda| = 1$ and $|\gamma| = 0$, $|\lambda| \leq 1$, where they are not holomorphic for a general A.

The splitting of the pure ground states can be related to the following broken symmetry. Let Θ be the unique automorphism of $\mathfrak{A}$ satisfying

$$\Theta(\sigma_x^{(j)}) = -\sigma_x^{(j)}, \quad \Theta(\sigma_y^{(j)}) = -\sigma_y^{(j)}, \quad \Theta(\sigma_z^{(j)}) = \sigma_z^{(j)}, \tag{21}$$

for all j. Then α_t commutes with Θ and hence the unique ground state φ satisfies

$$\varphi(\Theta(A)) = \varphi(A), \qquad A \in \mathfrak{A}. \tag{22}$$

However the Θ-invariance (22) does not hold for two pure ground states for the cases $|\lambda| < 1$, and $\gamma \neq 0$.

Actually, there are still some invariance remaining for the latter cases. Let j_x and j_y be the unique conjugate linear automorphisms of $\mathfrak{A}$ satisfying

$$j_x(\sigma_x^{(j)}) = -\sigma_x^{(j)}, \quad j_x(\sigma_y^{(j)}) = \sigma_y^{(j)}, \quad j_x(\sigma_z^{(j)}) = \sigma_z^{(j)}, \tag{23}$$

$$j_y(\sigma_x^{(j)}) = \sigma_x^{(j)}, \quad j_y(\sigma_y^{(j)}) = -\sigma_y^{(j)}, \quad j_y(\sigma_z^{(j)}) = \sigma_z^{(j)}, \tag{24}$$

for all j. Then the pure ground states for $|\lambda| < 1$, $\gamma > 0$ are j_y invariant in the sense

$$\varphi(j_y(A)) = \overline{\varphi(A)}, \qquad A \in \mathfrak{A} \tag{25}$$

while the pure ground states for $|\lambda| < 1$, $\gamma < 0$ are j_x invariant in the sense

$$\varphi(j_x(A)) = \overline{\varphi(A)}, \qquad A \in \mathfrak{A}, \tag{26}$$

where the bar denotes the complex conjugate. Since $\Theta = j_x j_y$, the j_x invariance (26) is broken in the first case and the j_y invariance (25) is broken in the second case.

Excitation spectrum in these ground states are explicitly known for all the above ground states.

7. Ground States of the XXZ Model

Our study in this case is just beginning. We consider the easiest case of $J > 0$ and $\Delta > 1$. Since the value of J is then irrelevant, we set $J = 1/2$ and

$$\Delta = \frac{1}{2}\left(q + \frac{1}{q}\right), \qquad q > 1. \tag{27}$$

We have obtained a complete excitation spectrum of ground states (15) for this system, as follows.

The shift automorphism τ defined by (7) commutes with α_t and the ground states $\varphi_\pm$ given by (15) are invariant under τ and α_t. Thus there are commuting family of unitary operators $U_\varphi(t) = e^{itH_\varphi}$ (H_φ satisfying (12) and (13)) and T_φ satisfying

$$T_\varphi \Omega = \Omega, \qquad T_\varphi \pi_\varphi(A) T_\varphi^* = \pi_\varphi(\tau A). \tag{28}$$

The spectral decomposition

$$\mathcal{H}_\varphi = \int_0^{2\pi} \mathcal{H}_\varphi(K) d\mu(K), \quad T_\varphi = \int_0^{2\pi} e^{iK} d\mu(K) \tag{29}$$

reduces $U_\varphi(t)$ and H_φ:

$$H_\varphi = \int_0^{2\pi} H_\varphi(K) d\mu(K). \tag{30}$$

We now describe the spectrum of $H_\varphi(K)$ for each fixed (momentum) K (mod. 2π).

It has a point spectrum for the following value $E_N(K)$ for each $N = 1, 2, \ldots$, representing a single spin wave ($N = 1$) and the bound state of N spin waves.

$$E_1(K) = 2\Delta + 2\cos K, \tag{31}$$

$$E_N(K) = 2\Delta + \frac{-2(q^{N-1} - q^{-(N-1)}) + 2(-1)^{N-1}(q - q^{-1})\cos K}{q^N - q^{-N}}. \tag{32}$$

In addition, there are continuous spectrum at the value E, interpreted as the scattering state of d excitations, each excitation being the bound state of m_j spin waves with total momentum k_j $(j = 1, \cdots, d)$ such that

$$K = k_1 + \cdots + k_d \qquad (\text{mod. } 2\pi), \tag{33}$$

$$E = E_{m_1}(k_1) + \cdots + E_{m_d}(k_d). \tag{34}$$

These scattering states are explicitly obtained.

8. Jordan Wigner Transformations

The classical method solving these models on a finite one dimensional lattice is the use of Jordan Wigner transformation:

$$c_j = \cdots \sigma_z^{(j-2)} \sigma_z^{(j-1)} \frac{1}{2}(\sigma_x^{(j)} - i\sigma_y^{(j)}), \tag{35}$$

$$c_j^* = \cdots \sigma_z^{(j-2)} \sigma_z^{(j-1)} \frac{1}{2}(\sigma_x^{(j)} + i\sigma_y^{(j)}), \tag{36}$$

which can be solved backward as

$$\sigma_z^{(j)} = 2c_j^* c_j - 1, \tag{37}$$

$$\sigma_x^{(j)} = \cdots \sigma_z^{(j-2)} \sigma_z^{(j-1)} (c_j + c_j^*), \tag{38}$$

$$\sigma_y^{(j)} = \cdots \sigma_z^{(j-2)} \sigma_z^{(j-1)} i(c_j - c_j^*). \tag{39}$$

The operators c_j and c_j^* so introduced satisfy the CAR (canonical anti-commutation relations):

$$[c_j, c_k]_+ = [c_j^*, c_k^*]_+ = 0, \tag{40}$$

$$[c_j, c_k^*]_+ = \delta_{jk} \ (= 1 \text{ if } j = k, = 0 \text{ if } j \neq k), \tag{41}$$

where $[x, y]_+ = xy + yx$. For a finite system, the spin algebra $\mathfrak{A}$ generated by $\sigma_\alpha^{(j)}$ $(j \in \mathbf{Z}, \alpha = x, y, z)$ can be identified with the CAR algebra $\mathfrak{A}^{CAR}$ generated by c_j and c_j^* $(j \in \mathbf{Z})$ through the above relations (35)–(39).

The merit of introducing $\mathfrak{A}^{CAR}$ is that the Hamiltonian becomes a quadratic expression of c_j and c_j^*, $j \in \mathbf{Z}$ (up to a constant term) for the XY-model:

$$H_{XY} = 2J \sum_j \{(c_j^* c_{j+1} + c_{j+1}^* c_j) \ + \ \gamma(c_j^* c_{j+1}^* + c_{j+1} c_j) \\ - \ \lambda(2c_j^* c_j - 1)\}, \tag{42}$$

and becomes a particle number preserving, discrete Schrödinger Hamiltonian with a contact interaction for the XXZ model:

$$H_{XXZ} = 2J \sum \{(a_j^* a_{j+1} + a_{j+1}^* a_j - 2\Delta a_j^* a_j a_{j+1}^* a_{j+1} + 2\Delta a_j^* a_j\}. \tag{43}$$

Their spectral analysis can be carried out explicitly even for an infinite system. (In the case of XY-model, the dynamics $\alpha_t(c_j)$ and $\alpha_t(c_j^*)$ can be explicitly solved. In the case of XXZ-model, scattering and bound states can be explicitly solved because of a factorizable S matrix.)

A problem for an infinite system arises because the product

$$T_j \equiv \cdots \sigma_z^{(j-2)} \sigma_z^{(j-1)} \tag{44}$$

arizing in (35), (36), (38) and (39) does not converge in $\mathfrak{A}$ for an infinite system. This difficulty can be solved by an operator algebra method.

We introduce a new element T from outside $\mathfrak{A}$, representing the infinite product

$$\cdots \sigma_z^{(-1)} \sigma_z^{(0)} \tag{45}$$

and demands the following properties which would be satisfied by (45) if it would converge:

$$T^2 = 1, \qquad T^* = T, \tag{46}$$

$$TAT = \Theta_-(A), \tag{47}$$

where Θ_- is the unique automorphism of $\mathfrak{A}$ satisfying

$$\Theta_-(\sigma_\alpha^{(j)}) = \sigma_\alpha^{(j)} \qquad \text{if } j > 0, \tag{48}$$

$$\Theta_-(\sigma_\alpha^{(j)}) = \Theta(\sigma_\alpha^{(j)}) \qquad \text{if } j \leq 0, \tag{49}$$

where $\alpha = x, y, z$ and Θ is defined by (21).

The new algebra generated by $\mathfrak{A}$ and T is

$$\widehat{\mathfrak{A}} = \mathfrak{A} + T\mathfrak{A}$$

which is the C^*-algebra called the crossed product of $\mathfrak{A}$ by Θ_--action of the two elements group $\mathbf{Z}_2$. In terms of T, all other T_j of (44) can be written as

$$T_j = T\sigma_z^{(1)} \cdots \sigma_z^{(j-1)}, \qquad j > 1, \tag{50}$$

$$T_1 = T, \qquad j = 1, \tag{51}$$

$$T_j = T\sigma_z^{(0)} \cdots \sigma_z^{(j)}, \qquad j < 1, \tag{52}$$

and hence c_j and c_j^* can be defined in $\widehat{\mathfrak{A}}$, giving rise to a subalgebra $\mathfrak{A}^{CAR}$ in $\widehat{\mathfrak{A}}$, which differs from the subalgebra $\mathfrak{A}$ of $\widehat{\mathfrak{A}}$. (Note that the $\sigma_z^{(j)}$ mutually commute and $(\sigma_z^{(j)})^2 = 1$.)

We can extend the automorphism Θ of $\mathfrak{A}$ given by (21) to $\widehat{\mathfrak{A}}$ uniquely by the requirement

$$\Theta(T) = T. \tag{53}$$

Any $A \in \widehat{\mathfrak{A}}$ can be decomposed as a sum of Θ-even and Θ-odd elements due to $\Theta^2 = id$:

$$A = A_+ + A_-, \qquad A_\pm = \frac{1}{2}(A \pm \Theta A), \tag{54}$$
$$\Theta(A_\pm) = \pm A_\pm.$$

Hence Θ invariant algebras also split into even and odd parts:

$$\mathfrak{A} = \mathfrak{A}_+ + \mathfrak{A}_-, \tag{55}$$
$$\widehat{\mathfrak{A}} = \mathfrak{A}_+ + \mathfrak{A}_- + T\mathfrak{A}_- + T\mathfrak{A}_+, \tag{56}$$
$$\mathfrak{A}^{CAR} = \mathfrak{A}_+ + T\mathfrak{A}_-, \quad (\mathfrak{A}_+^{CAR} = \mathfrak{A}_+,\ \mathfrak{A}_-^{CAR} = T\mathfrak{A}_-). \tag{57}$$

Relations of these algebras are indicated in Fig.2.

9. Operator Algebra Methods for Ground States of XY-model

(1) Analysis of ground states for $(\mathfrak{A}^{CAR}, \alpha_t)$

The finite volume Hamiltonian H_N of (3) for the XY-model belongs to $\mathfrak{A}_+^{CAR} = \mathfrak{A}_+$, α_t can be extended to $\widehat{\mathfrak{A}}$ and it leaves $\mathfrak{A}^{CAR} \subset \widehat{\mathfrak{A}}$ invariant. The C^* dynamical system $(\mathfrak{A}^{CAR}, \alpha_t)$ is the so-called quasifree dynamics which can be solved explicitly as indicated in the preceding section. Its ground state can be obtained explicitly as a quasifree state and is unique in the present case of the XY-model for all values of parameters, without showing any hint of the phase diagram in Fig.1, as far as the number of ground states are concerned. The only hint of the phase diagram at this stage is the existence of excitations with vanishing energy at the parameter values on the phase boundary.

(2) The analysis of ground states for $(\mathfrak{A}_+^{CAR}, \alpha_t)$.

The ground states of $(\mathfrak{A}_+^{CAR}, \alpha_t)$ can be found by an operator algebraic abstract arguments from the analysis of $(\mathfrak{A}^{CAR}, \alpha_t)$.

(i) The restriction of the unique ground state of $(\mathfrak{A}^{CAR}, \alpha_t)$ to $\mathfrak{A}_+^{CAR}$ is the ground state of $(\mathfrak{A}_+^{CAR}, \alpha_t)$ due to an easy argument.

(ii) On the other hand, the unique Θ invariant extention of a ground state of $(\mathfrak{A}_+^{CAR}, \alpha_t)$ to $\mathfrak{A}^{CAR}$ is not necessarily a ground state of $(\mathfrak{A}^{CAR}, \alpha_t)$, but the associated representation of this extention, has to be a ground state representation, that is, it contains a cyclic vector giving rise to a ground state of $(\mathfrak{A}^{CAR}, \alpha_t)$. A criterion for the occurrence of the latter and exact nature of such ground states of $(\mathfrak{A}_+^{CAR}, \alpha_t)$ can be given in terms of the spectral data for the ground state of $(\mathfrak{A}^{CAR}, \alpha_t)$. The analysis in this step utilizes the specific structure of $\mathfrak{A}^{CAR}$.

If $(\lambda, \gamma) \neq (0, \pm 1)$, the latter possibility does not occur and the ground state of $(\mathfrak{A}_+^{CAR}, \alpha_t)$ is still unique for all parameter values except $(0, \pm 1)$.

(iii) If $(\lambda, \gamma) = (0, \pm 1)$, then the unique ground state representation for $(\mathfrak{A}^{CAR}, \alpha_t)$ splits into 2 irreducible ground state representations of $(\mathfrak{A}_+^{CAR}, \alpha_t)$.

One of them contains the restriction of the unique ground state of $(\mathfrak{A}^{CAR}, \alpha_t)$ as the only ground state of $(\mathfrak{A}_+^{CAR}, \alpha_t)$ in that representation.

The other irreducible representation of $(\mathfrak{A}_+^{CAR}, \alpha_t)$ contains infinitely many ground state vectors forming an infinite dimensional subspace. In the next stage, this will split into two groups of soliton ground states (16) and (17) and there superpositions.

(3) Ground states of $(\mathfrak{A}_+, \alpha_t)$.

Since $\mathfrak{A}_+^{CAR} = \mathfrak{A}_+$, this is already analyzed in (2).

(4) Ground states of $(\mathfrak{A}, \alpha_t)$.

The restriction of a ground state of $(\mathfrak{A}, \alpha_t)$ to $\mathfrak{A}_+$ has to be a ground state of $(\mathfrak{A}_+, \alpha_t)$ and hence must be one of the states obtained in (2) above although different ground states of $(\mathfrak{A}, \alpha_t)$ may have the same restriction to $\mathfrak{A}_+$

Starting from a ground state φ of $(\mathfrak{A}_+, \alpha_t)$, any possible pure ground state of $(\mathfrak{A}, \alpha_t)$ can be obtained by the pure state decomposition of the unique Θ-invariant extension $\bar{\varphi}$ of the ground state φ of $(\mathfrak{A}_+, \alpha_t)$ to $\mathfrak{A}$. There is the dichotomy between $\bar{\varphi}$ being already a pure ground state of $(\mathfrak{A}, \alpha_t)$ and $\bar{\varphi}$ being split into an average of two pure ground states of $(\mathfrak{A}, \alpha_t)$ with mutually inequivalent, associated irreducible representations which are mapped to each other by Θ and hence are individually breaking the Θ-invariance. This dichotomy yeilds the one phase and two phase regions in Fig.1 depending on the values of the parameter.

The two alternatives can be decided for each parameter value by computing an index of a certain operator on the l_2 space over the lattice (the test function space of the CAR algebra). However, a certain automorphism method show that the choice is the same in the region of analyticity of the state with respect to the parameters, namely in the inside of open sets separated by the H shaped phase boundary in Fig.1. Thus the choice can be simply determined (except for the phase boundary) by checking for easily computable cases with some specific value of the parameter.

10. Master Symmetries of the XY-model

The master symmetries of the one-dimensional XY model proposed by Barouch and Fuchssteiner [8] by a computer computation without a mathematical proof (and in fact with some mathematical difficulty for a finite system), are explicitly obtained for an infinite system and some linear relations among the constant of motion obtained from them are found. They can be described on the abstract level as follows:

Let $H_0 = H$ be the Hamiltonian for the XY-model given by (1) or by (3) and (5) and interpreted as derivation of the form (9) rather than non-converging infinite sum in (1). (The following operators are all considered as derivations and they can be shown to have their (pure imaginary) exponentials as automorphisms of $\mathfrak{A}^{CAR}$ and of $\mathfrak{A}$, forming an infinite dimensional group.) There are another $S_0 = S$ explicitly given and commuting with H. In terms of c_j and c_j^*, S is also a simple quadratic expression.

$$S = \sum S(j) \qquad S(j) = \sigma_x^{(j)}\sigma_y^{(j+1)} - \sigma_y^{(j)}\sigma_x^{(j+1)} = 2i(c_{j+1}^* c_j - c_j^* c_{j+1}). \tag{58}$$

The master symmetry is another such expression P:

$$P = -\sum P(j), \qquad P(j) = \left(j + \frac{1}{2}\right) H^{(j)}. \tag{59}$$

It has the following remarkable properties. Let

$$H_{n+1} = i[P, H_n], \tag{60}$$
$$S_{n+1} = i[P, S_n], \tag{61}$$

recursively for $n = 0, 1, 2, \ldots$. Then $H_0, H_1, \ldots$ and $S_0, S_1, \ldots$ all commute with each other.

$$[H_j, H_k] = [S_j, H_k] = [S_j, S_k] = 0. \tag{62}$$

Since they commute with $H = H_0$, They are symmetries of the dynamical system $(\mathfrak{A}^{CAR}, \alpha_t)$, and $(\mathfrak{A}, \alpha_t)$ also. Since P gives rise to an infinite number of mutually commuting symmetries, it is called a master symmetry, although P itself does not commute with H.

A further surprising fact is the existence of another master symmetry P_1 which is again a quadratic expression in terms of c_j and c_j^* but is a somewhat complicated expression in terms of σ's. (We refer to [5] for an explicit expression.) P and P_1 together produce the following series of master symmetries recursively:

$$P_{j+1} = i[P, P_j], \qquad j = 1, 2, \ldots. \tag{63}$$

Let us define

$$H_{k+1,j} = i[P_j, H_{k,j}], \tag{64}$$
$$S_{k+1,j} = i[P_j, S_{k,j}], \tag{65}$$

recursively for $k = 0, 1, 2, \ldots$ for each j, starting with

$$H_{0,j} = H, \qquad S_{0,j} = S. \tag{66}$$

Each P_j ($j = 1, 2, \ldots$) is a master symmetry in the sense that $H_{k,j}$ and $S_{k,j}$ all commute with each other among the same index j, as well as among different index j, in addition:

$$[H_{k,j}, H_{l,m}] = [S_{l,m}, H_{k,j}] = [S_{k,j}, S_{l,m}] = 0, \tag{67}$$

for all $k = 0, 1, 2, \ldots$ and $j = 0, 1, 2, \ldots$, where we set

$$H_{k,0} = H_k, \qquad S_{l,0} = S_l. \tag{68}$$

They commute with $H = H_{0,j}$ and hence are a commuting family.

These are claims of Barouch and Fuchssteiner, which we not only verify mathematically but also clarify how such a miracle occurs. In fact we can clarify the strucure of an infinite dimensional Lie group of automorphisms of $\mathfrak{A}^{CAR}$, of which the above derivations $P, P_j, H_{k,j}$ and $S_{k,j}$ are genrators.

We also found that $H_{k,j}$ and $S_{l,m}$ for $k, l, j, m = 1, 2, \ldots$ are all finite linear combinations of H_p and S_q, $p, q = 0, 1, 2, \ldots$.

The latter set is linearly independent and generate a maximal abelian subgroup of the group $\mathrm{Aut}\,\mathfrak{A}^{CAR}$ of all automorphisms of $\mathfrak{A}^{CAR}$.

References

[1] Araki, H. and Barouch, E., On the dynamics and ergodic properties of the XY-model, *J. Statist. Phys.* **31** (1983), 327–345.

[2] Araki, H. On the XY-model on two-sided infinite chain, *Publ. RIMS Kyoto Univ.*, **20** (1984), 277–296.

[3] Araki, H. and Matsui, T., Ground states of the XY-model, *Commun. Math. Phys.* **101** (1985), 213–245.

[4] Araki, H. and Matsui, T., Analyticity of ground states of the XY-model, *Lett. Math. Phys.* **11** (1986), 87–94.

[5] Araki, H., Master symmetries of the XY model, *Commun. Math. Phys.* **132** (1990), 155–176.

[6] Araki, H., in preparation.

[7] Matsui, T., Explicit formulas for correlation functions of ground states of the 1 dimensional XY model, *Ann. Inst. Henri Poincaré,* **45** (1986), 49–59.

[8] Barouch, E. and Fushssteiner, B., Master symmetries and similarity equations, *Stud. Appl. Math.* **73** (1985), 221–237.

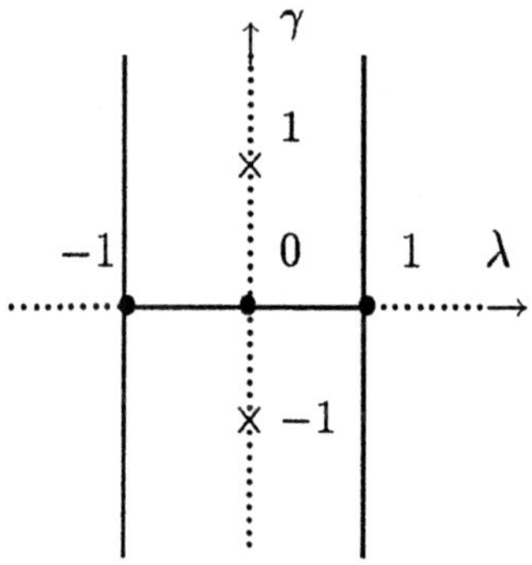

Fig. 1. Phase diagram for ground states of XY-model

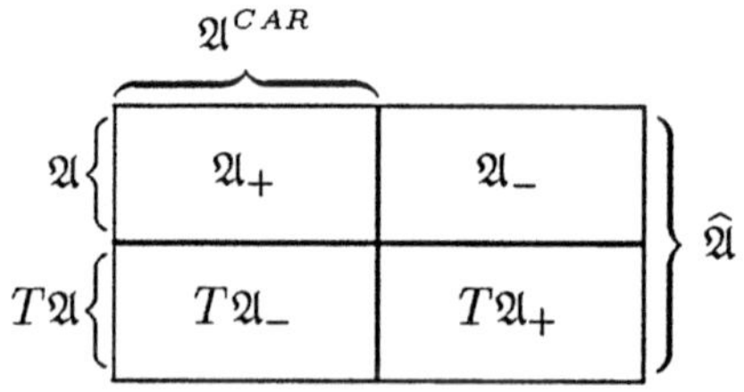

Fig. 2. Relations between $\widehat{\mathfrak{A}}$, $\mathfrak{A}$ and $\mathfrak{A}^{CAR}$

On Conservation Laws of the XY Model

TAKU MATSUI
Universityof Leuven, Belgium and Tokyo Metropolitan University, Tokyo Japan

(Received:October 3,1992)

Abstract. We consider the master symmetry of the 1 dimensional XY model found by E.Barouch and B.Fuchssteiner.We show that the conservation laws obtained by them generates a maximal family of mutually commuting conserved currents.

Key words: XYmodel,master symmetry,Bogoliubov automorphism

1. Introduction

We consider the master symmetry of the 1 dimensional XY model found by E.Barouch and B.Fuchssteiner in [5].The XY model is a simple exactly soluble model of quantum spin sysem.In the finite volume,it is equivalent to a free Fermion model via Jordan Wigner transformation.It is known that it exhibits a quite differnt physical behaviour from the free Fermion in the theromodynamic limit.(See [2] and references therein.)

An important feature of exactly soluble models is the existence of a infinite number of conservation laws.We are interested in the mathematically precise notion of quatum completely integrable system.In the case of infinite degree of freedom,this is rather ambiguous mathematically even though so many models are claimed to be completely integrable.The question whether obtained commuting currents for a fixed model is a complete set of conserved quantities has never been considered seriously.This is what we are going to discuss here for the XY model.We will use the language of operator algebras. (See [6].)

Let H be the infinite volume Hamiltonian of the 1 diemnsional XY model.

$$H = -\sum_{k=-\infty}^{\infty} \{(1+\gamma)\sigma_x^{(k)}\sigma_x^{(k+1)} + (1-\gamma)\sigma_y^{(k)}\sigma_y^{(k+1)} + 2\lambda\sigma_z^{(k)}\} \qquad (1)$$

where $\sigma_x^{(k)}, \sigma_y^{(k)}, \sigma_z^{(k)}$ are Pauli spin operators on the site k.(the spin is 1/2.) and γ, λ are real parameters.

Let A_{loc} be the set of local observables generated by Pauli spin operators(polynomials of Pauli spin operators) and Let A be the C^* algebra completion of A_{loc}. The time evolution α_t of an observable Q in A_{loc} is

H. Araki et al. (eds.), Quantum and Non-Commutative Analysis, 197–204.

determined by the following equation.

$$\frac{d}{dt}\alpha_t(Q) = i[H, \alpha_t(Q)] \tag{2}$$

We next give the conservation laws of [5].

$$\begin{aligned} S &= \sum_{k=-\infty}^{\infty} S(k), S(k) = \sigma_x^{(k)}\sigma_y^{(k+1)} - \sigma_y^{(k)}\sigma_x^{(k+1)} \\ P &= \sum_{k=-\infty}^{\infty} P(k), P(k) = (j+\frac{1}{2})H(k) - \lambda\sigma_z^{(k)} \\ H(k) &= (1+\gamma)\sigma_x^{(k)}\sigma_x^{(k+1)} + (1-\gamma)\sigma_y^{(k)}\sigma_y^{(k+1)} + 2\lambda\sigma_z^{(k)} \end{aligned} \tag{3}$$

Then we set $S_0 = S, H_0 = H$ and S_n and H_n are defined recursively by the equation,

$$H_n = i[P, H_{n-1}], S_n = i[P, S_{n-1}] \tag{4}$$

S_n and H_n generate 1 parameter groups of automorphisms of A.We introduce derivations of A_{loc},

$$\delta_n(Q) = i[H_n, Q], \epsilon_m(Q) = i[S_n, Q] \tag{5}$$

In [2] and [5],it was shown that these derivations commutes mutually.

$$\begin{aligned} \delta_n \circ \delta_m(Q) &= \delta_m \circ \delta_n(Q), \delta_n \circ \epsilon_m(Q) = \epsilon_m \circ \delta_n(Q) \\ \epsilon_n \circ \epsilon_m(Q) &= \epsilon_m \circ \epsilon_n(Q) \end{aligned} \tag{6}$$

Let $\alpha_t^{(n)}$ (resp.$\beta_t^{(n)}$) be the 1 parameter group of automorphism generated by δ_n (resp. ϵ_n). We will consider the following questions

1. Is the group of automorphisms generated by $\alpha_t^{(n)}$ and $\beta_s^{(m)}$ maximally abelian in the set of all automorphisms of A ?
2. Is the local conserved current commuting with any S_n and H_m the linear combination of S_n and H_m ?

These are questions raised in [4].(See the last section of [4].) We give a partial answer in the end of the next section.

2. Master Symmetry and Fermion Algebra

2.1. Jordan Wigner Transformation

As is well known, the XY model is transformed to a free Fermion model by Jordan Wigner transformation.However,this transformation is non local and ill-defined in the infinite volume limit.Following [2],we present here its treatment using the crossed product of C^*algebra. Let Θ and Θ_- be automorphisms determined by

$$\Theta\left(\sigma_x^{(k)}\right)=-\sigma_x^{(k)},\Theta\left(\sigma_y^{(k)}\right)=-\sigma_y^{(k)},\Theta\left(\sigma_z^{(k)}\right)=\sigma_z^{(k)} \tag{7}$$

$$\begin{aligned}\Theta_-\left(\sigma_x^{(k)}\right)=-\sigma_x^{(k)},\Theta_-\left(\sigma_y^{(k)}\right) &= -\sigma_y^{(k)},\Theta_-\left(\sigma_z^{(k)}\right)=\sigma_z^{(k)}\\ &\text{for } k\geq 1,\\ \Theta_-\left(\sigma_x^{(k)}\right)=\sigma_x^{(k)},\Theta_-\left(\sigma_y^{(k)}\right) &= \sigma_y^{(k)},\Theta_-\left(\sigma_z^{(k)}\right)=\sigma_z^{(k)}\\ &\text{kor } k\leq 0\ .\end{aligned} \tag{8}$$

Let B be the crossed product of A by the Z_2 action of Θ_-. B is generated by A and a selfadjoint unitary T satisfying

$$TQT=\Theta_-(Q)\ ,\ \ Q\in A \tag{9}$$

We also set

$$A_\pm=\{Q\in A,\Theta(Q)=\pm Q\} \tag{10}$$

$$\begin{aligned}C_j^*=T\prod_{k=1}^{j-1}\sigma_z^{(k)}\frac{1}{2}\left(\sigma_x^{(j)}+i\sigma_y^{(j)}\right),C_j &= T\prod_{k=1}^{j-1}\sigma_z^{(k)}\frac{1}{2}\left(\sigma_x^{(j)}-i\sigma_y^{(j)}\right)\\ &\text{for } j\geq 1\\ C_j^*=T\prod_{k=0}^{j}\sigma_z^{(k)}\frac{1}{2}\left(\sigma_x^{(j)}+i\sigma_y^{(j)}\right),C_j &= T\prod_{k=0}^{j}\sigma_z^{(k)}\frac{1}{2}\left(\sigma_x^{(j)}-i\sigma_y^{(j)}\right)\\ &\text{for } j\leq 0.\end{aligned} \tag{11}$$

Then we have the following canonical anitcommutation relation(CAR).

$$\{C_j^*,C_k^*\}=\{C_j,C_k\}=0,\{C_j,C_k^*\}=\delta_{j,k}1 \tag{12}$$

Let A^F be the CAR algebra generated by C_j and C_k^*. Θ and Θ_- are extendible to B if we set $\Theta(T)=T,\Theta_-(T)=T$.We also denote

$$A_+^F=A_+,\quad A_-^F=A_-T \tag{13}$$

We also use the selfdual formulation of the CAR algebra(=looked as an infinite dimensional Clifford algebra with involution). Let $\mathcal{H}$ be the Hilbert space of pairs of squrare summable sequences.

$$\mathcal{H}=l_2(Z)\oplus l_2(Z) \tag{14}$$

An element h of $\mathcal{H}$ consists of two vectors f and g of $l_2(Z)$ denoted by $h = \binom{f}{g}$ Next we set $Jh = \binom{\overline{g}}{\overline{f}}$ where $\overline{g}$ is the componentwise complex conjugation of g, and

$$B(h) = \sum_j f_j C_j^* + \sum_j g_j C_j \tag{15}$$

We have

$$B(h)^* = B(Jh), \{B(h_1)^*, B(h_2)\} = (h_1, h_2)1 \tag{16}$$

Let u be a unitary of $\mathcal{H}$ which commutes with J. The Bogoliubov automorphism is an automorphism of the CAR algebra determined by

$$\alpha_u(B(h)) = B(uh) \tag{17}$$

For example , let θ_- be the unitary determined by

$$(\theta_- f)_j = \begin{cases} f_j, & \text{for } j > o \\ -f_j, & \text{for } j < -1 \end{cases} \tag{18}$$

Then

$$\alpha_{\theta_-} = \Theta_-$$

LEMMA 1. *Let u be a unitary of $\mathcal{H}$ which commutes with J. Suppose that $\theta_- u \theta_- u^{-1} - 1$ is of trace class and*

$$det(\theta_- u \theta_- u^{-1}) = 1 \tag{19}$$

Then α_u is extendible to B .

This follows from Theorem4.4 of [7] and results of [1].

2.2. Ground States of the XY model

Our Hamiltonian gives rise to a quasifree motion on A^F ,

$$[H, B(h)] = B(2Kh) \tag{20}$$

where K is a bounded operator on $\mathcal{H}$. (The precise form of K can be found in [2].) Therefore

$$\alpha_t^0(B(h)) = B(e^{i2Kt}h) \tag{21}$$

A ground state is by definition a state φ of A satisfying

$$\varphi(Q^*[H, Q]) \geq 0 \text{ for any } Q \text{ in } A_{loc} \tag{22}$$

In [2], we obtained a complete list of ground states of the XY model. We collect some of the results which we will use in our proof.

1. The Θ invariant ground state is unique if $(\lambda, \gamma) \neq (0, \pm 1)$
2. The ground state for the quasifree motion (2.13) of the CAR algebra is unique. We denote it by ψ.
3. The states ψ and $\psi \circ \Theta_-$ are equivalent if $|\lambda| > 1$ or $\gamma \neq 0$ and $|\lambda| < 1$. Here the equivalence of states means that of their GNS representations.

The GNS representation for the Θ invariant ground state φ of A is realized on the Fermion Fock space associated with ψ of A^F as follows.

LEMMA 2. *Let $(\mathcal{F}, \Omega, \pi_F)$ be the Fock representation for ψ of A^F where $\mathcal{F}$ is the Hilbert space and Ω is the vacuum vector . Let V be the unitary of $\mathcal{F}$ implementing Θ .*

$$V\pi_F(Q)\Omega = V\pi_F\Theta(Q)\Omega \tag{23}$$

Let $P_\pm$ be the projection defined by

$$P_\pm = 1/2(1 \pm V) \tag{24}$$

Set

$$\begin{aligned} \pi_s(Q) &= \pi_F(Q)P_+ + \pi_F(\Theta_-(Q))P_- \ Q \text{ in } A_+ \\ \pi_s(Q) &= \pi_F(\Theta_-(Q))P_+ + \pi_F(Q)P_- \text{ for } Q \text{ in } A_- \end{aligned} \tag{25}$$

Then π_s and the vector Ω gives rise to the GNS representation of the Θ invariant ground state φ of A.

2.3. Master Symmetry

In [4],it is proved that the master symmetry (1.3) yields quasifree derivations.

$$\delta_j(B(f)) = B(h^{(j)}f), \epsilon_k(B(f)) = B(s^{(k)}f) \tag{26}$$

The explicit form of $h^{(j)}, s^{(k)}$ is in [4]. After Fourier transform,we can see that $h^{(j)}, s^{(k)}$ generate a maximally abelain von Neuman algebra in the Hilbert space $\mathcal{H}$. Recalling the result of Kishimoto in [8] ,we may expect that the master symmetry gives rise to the largest commuting family of conserved currents and that $\alpha_t^{(n)}$ and $\beta_s^{(m)}$ generate a maximally abelain group of automorphisms of A. We present here a partial answer to this question. Let G_0 be the closed abelian group of automorphisms of A generated by $\alpha_t^{(n)}$ and $\beta_s^{(m)}$ where the topology is determined by pointwise convergence.

PROPOSITION 3. *Assume either (i) $|\lambda| > 1$ or (ii) $\gamma \neq 0$, $|\lambda| < 1$ and $(\gamma, |\lambda|) \neq (\pm 1, 0)$. Let W be an automorphism of A which commutes with any element of G_0 and $\Theta_- W \Theta_- W^{-1}$ is an inner automorphism of A. Then*

W belongs to G_0. Futhermore,there exists a Bogoliubov automorphism α_w such that

$$\alpha_w(Q) = W(Q) \text{ where } Q \text{ is in } A_+. \tag{27}$$

The above proposition can be applied to any 1 parameter group generated by the translationally invariant local conserved current. More precisely,let α_u be the translation on the lattice,

$$\alpha_u\left(\sigma_\beta^{(j)}\right) = \sigma_\beta^{(j+1)}, \alpha_u(B(h)) = B(uh)$$
$$\text{where u is the shift on } l_2(Z), (uf)_j = f_{j-1}. \tag{28}$$

Let c be an selfadjoint element in A_{loc} and set

$$C = \sum_k \alpha_{u^k}(c) \quad \delta_C(Q) = [C, Q] \tag{29}$$

where Q is in A_{loc}.

COROLLARY 4. *Assume either (i) $|\lambda| > 1$ or (ii) $\gamma \neq 0$ and $|\lambda| < 1$, $(\gamma, \lambda) \neq (\pm 1, 0)$. Suppose that C commutes with S_j and H_k for any k,j as derivation.C gives rise to a quasifree derivation on A^F which is a (possibly infinite)linear combination of derivations δ_j, ϵ_k.(the sum is convergent on the local elements of A^F .)*

Roughly speaking this corollary implies that C is a linear combination of S_j and H_k. As these are unbounded operators , we stated the result in terms of derivation to avoid the question of domain.

3. Proof of Proposition 3

In this section,we sketch our proof of Proposition3.We assume the condition of Proposition 3 from now on. Let E be the projection defined by

$$E = \frac{1}{2}(1 + \frac{K}{|K|}) \tag{30}$$

where K is the operator of (20). Recall that our Fermion Fock space $\mathcal{F}$ is the exterior algebra for $E\mathcal{H}$. We consider the number operater of this space now.Set

$$v(s) = e^{is}E + e^{-is}(1 - E) \tag{31}$$

LEMMA 5. *Under the condition of Proposition 3,the Bogoliubov automorphism $\alpha_{v(s)}$ associated with $v(s)$ is extendible to B .*

This lemma follows from the fact that $\theta_- E\theta_- - E$is of trace class and results of [7]. (See [3].) Next we work in the representation of A realized in Lemma2. Let α_v be the Bogoliubov automorphism associated with a unitary v. We denote its implementor on $\mathcal{F}$ by $\Gamma(v)$.

$$\Gamma(v)\pi_F(Q)\Gamma(v)_{-1} = \pi_F(\alpha_v(Q)) \tag{32}$$

We fix the phase of $\Gamma(v)$ as

$$\Gamma(\theta_-)^2 = 1. \tag{33}$$

If v commutes with E,we set

$$\Gamma(u)\Omega = \Omega. \tag{34}$$

Let W be an automorphism of A which is in the commutant of G_0. As W commutes with Θ we have

$$\varphi \circ W = \varphi \tag{35}$$

so there exist a unitary T(W) such that

$$T(W)\pi_s(Q)T(W)_{-1} = \pi_s(W(Q)) \tag{36}$$

for Q in A.

LEMMA 6. *Let*

$$T^F(W) = T(W)P_+ + \Gamma(\theta_-)T(W)\Gamma(\theta_-)_{-1}P_- \tag{37}$$

Then for Q in A_+ ,we have

$$T(W)\pi_F(Q)T(W)_{-1} = \pi_F(W(Q)) \tag{38}$$

If W and W' commute,$T^F(W)$ and $T^F(W')$ commute.

LEMMA 7. *For the automorphism of Proposition3 there exists a unitary w on $\mathcal{H}$ such that*

$$JwJ = w \quad , \quad wEw_{-1} = E \tag{39}$$

$$T^F(W)\pi(B(Eh))\Omega = \pi(B(wEh))\Omega \tag{40}$$

This follows from the fact that W commutes with $\alpha_{v(s)}$ and the previous lemma. To complete the proof of our proposition,we use our assumption that there exists a unitary Z in A such that

$$\Gamma(\theta_-)T(W)\Gamma(\theta_-)T(W)^{-1} = \pi_s(Z). \tag{41}$$

The right-hand side of this identity is an even element so Z is in A_+ . We can show that the adjoint action of $T^F(W)$ induce the automorphism $\Xi(\cdot)$ of A^F.

$$T^F(W)\pi_F(Q)T^F(W)^{-1} = \pi_F(\Xi(Q)) \tag{42}$$

for any Q in A^F. For example,

$$T^F(W)\pi_F(B(h))T^F(W)^{-1}$$
$$= \pi_F(\Theta_-(Z))\pi_F(\Theta_-((W(B(\theta_-h)T)T))$$

Thus we can reproduce the proof of Theorem 1 of [8] to conclude that $T^F(W) = \Gamma(w)$.This implies the claim of Proposition 3.

Acknowledgement. The final part of this work was done while the author was visiting Institute for Theoretical Physics,University of Leuven as a research fellow of *Canon Foundation in Europe* .The author thanks to *Canon Foundation in Europe* for the financial support and to Institute for Theoretical Physics,University of Leuven for hospitality.

References

[1] Araki H.,On quasifree states of CAR and Bogoliubov automorphisms, *Publ.RIMS Kyoto Univ.***6**,385-442(1970).

[2] Araki H.,Matsui T.,Ground states of the XYModel, *Commun.Math.Phys.***101**,213-245(1985).

[3] Araki H.,Matsui T.,Analyticity of the ground states of the XY model, *Lett.Math.Phys.***11**,87-94(1986).

[4] Araki H.,Master symmetry of the XY model,*Commun.Math.Phys.* bf 132,155-176(1990).

[5] Barouch E. Fuchssteiner B., Master symmetries and similarities equations, *Stud.Appl.Math.***13**,221-275(1985)

[6] Bratteli O. Robinson D., Operator algebras and quantum statistical mechanics, (Springer,Berlin,Heidelberg,New York.1979).

[7] Evans D.E. Lewis J.,On a C^*algebraic approach to phase transition in the two-dimensional Ising model II, *Commun.Math.Phy.***102**,532-535(1986).

[8] Kishimoto A.,On invariant states and the commutant of a group of quasifree automorphisms of the CAR algebra, *Rep.Math.Phys.***15**,21-26(1979).

3. Quantum Groups

Reflection Equation Algebras and Quantum Groups

P.P. KULISH
Euler International Mathematical Institute and Yukawa Institute for Theoretical Physics

Abstract. Quadratic algebras related to the reflection equations are introduced. They are quantum group comodule algebras. The quantum group $F_q(GL(2))$ is taken as the example. The properties of the algebras (center, representations, realizations, real forms, fusion procedure etc) as well as the generalizations are discussed.

Key words: quantum group comodule algebra, non-commutative homogeneous space

1. Introduction

Progress in understanding the algebraic roots of the quantum integrability achieved in the last few decades has already resulted in the introduction of several new algebraic objects, such as the Yang-Baxter equation (YBE), quantum groups and quantum algebras, exchange and quadratic algebras. It seems that the list should be enriched with another item: the reflection equation (RE), which has arisen recently in several independent contexts.

The RE reads

$$R(u-v)\overset{1}{K}(u)R(u+v)\overset{2}{K}(v) = \overset{2}{K}(v)R(u+v)\overset{1}{K}(u)R(u-v) \qquad (1)$$

where $K(u)$ is a square $N \times N$ matrix, $\overset{1}{K} \equiv K \otimes \mathrm{id}_N$, $\overset{2}{K} \equiv \mathrm{id}_N \otimes K$ (the notation is usual in the quantum inverse scattering method), and $R(u)$ is a solution to the YBE. One can also consider (1) as defining relations for the associative algebra $\mathcal{A}$ generated by the elements of the matrix $K(u)$ [1].

Being introduced in [2] as the equation describing factorized scattering on a half-line, the RE and the related algebra $\mathcal{A}$ had found soon quite different applications in quantum current algebras [3] and integrable models with non-periodic boundary conditions [1, 4].

Since the constant (*i.e.* not including the spectral parameter u) solutions to the YBE are extensively used in the quantum group theory [5], it is quite natural to study a version of the RE without spectral parameter. Though such an equation and the related algebras (for different R matrices) has already appeared in several papers [7–14], they were not distinguished as separate objects of study until recently [15, 23].

In the present report we collect the basic facts which are necessary for any systematic study of the RE. We restrict ourselves to the case of the

H. Araki et al. (eds.), Quantum and Non-Commutative Analysis, 207–220.

simplest quantum group $F_q(GL(2))$ and choose the following form of the RE

$$R\overset{1}{K}R^{t_1}\overset{2}{K}=\overset{2}{K}R^{t_1}\overset{1}{K}R \tag{2}$$

The report is organized as follows. After introducing the basic definitions and notation the general properties of the quadratic algebra $\mathcal{A}$ defined by (2) are described in section 2. They are discussed and partially proved in section 3. In section 4 some equivalent variants of RE are pointed out as well as the relations between $\mathcal{A}$ and other algebras. In conclusion, we discuss some generalizations of the RE and its applications.

2. Definitions

The quantum group $F\equiv F_q(GL(2))$ can be defined as the associative algebra generated by four elements a, b, c, d and the relations

$$\begin{array}{lll} ab=qba, & bd=qdb, & [a,d]=\omega bc, \\ ac=qca, & cd=qdc, & [b,c]=0, \end{array} \tag{3}$$

q being a complex parameter and $\omega=q-q^{-1}$. Introducing the matrix T

$$T=\begin{pmatrix} a & b \\ c & d \end{pmatrix}$$

and using the notation $\overset{1}{T}\equiv T\otimes \mathrm{id}$, $\overset{2}{T}\equiv \mathrm{id}\otimes T$ one can rewrite the relations (3) in the compact form [5]

$$R\overset{1}{T}\overset{2}{T}=\overset{2}{T}\overset{1}{T}R \tag{4}$$

where the R matrix is given by

$$R=\begin{pmatrix} q & & & \\ & 1 & & \\ & \omega & 1 & \\ & & & q \end{pmatrix}, \qquad R^{t_1}=\begin{pmatrix} q & & & \omega \\ & 1 & & \\ & & 1 & \\ & & & q \end{pmatrix} \tag{5}$$

and t_1 means transposition with respect to the first space in $\mathbf{C}^2\otimes\mathbf{C}^2$.

Remark. If we replace the R matrix in (4) by $\mathcal{P}R^{-1}\mathcal{P}$, where $\mathcal{P}$ is the permutation operator in $\mathbf{C}^2\otimes\mathbf{C}^2$, the commutation relations (3) remain the same. The consequences of this fact are discussed in section 5.2.

The Hopf algebra structure is defined on F after introducing comultiplication map $\Delta: F\to F\otimes F$, counit map $\varepsilon: F\to\mathbf{C}$ and coinverse map $s: F\to F$ by the formulae

$$\Delta(T)=T_1T_2, \qquad \varepsilon(T)=\mathrm{id}, \qquad s(T)=T^{-1}$$

where $(T_1T_2)_{ij} \equiv \sum_{k=1}^{2} T_{ik} \otimes T_{kj}$.

Now define the associative algebra $\mathcal{A}$ by four generators $\alpha, \beta, \gamma, \delta$, which can be thought of as the elements of the matrix K

$$K = \begin{pmatrix} \alpha & \beta \\ \gamma & \delta \end{pmatrix},$$

and by the quadratic relations (2) or, more explicitely,

$$[\alpha, \beta] = \omega\alpha\gamma, \quad \alpha\gamma = q^2\gamma\alpha, \quad [\alpha, \delta] = \omega(q\beta + \gamma)\gamma,$$

$$[\beta, \gamma] = 0, \quad [\beta, \delta] = \omega\gamma\delta, \quad \gamma\delta = q^2\delta\gamma. \tag{6}$$

The algebra $\mathcal{A}$ has the following properties.

1. $\mathcal{A}$ is a Poincaré-Birkhoff-Witt (PBW) algebra which means that the linear space spanned by the monomials of order p in the generators $\alpha\beta\gamma\delta$ has the same dimension as in the commutative case that is $(p+3)!/p!3!$

2. $\mathcal{A}$ is an F-comodule-algebra that is there exists a map φ (coaction of F) $\varphi : \mathcal{A} \to F \otimes \mathcal{A}$ which is consistent with the comultiplication Δ in F

$$(\Delta \otimes \mathrm{id}) \circ \varphi = (\mathrm{id} \otimes \varphi) \circ \varphi \tag{7}$$

and the counit ε

$$(\varepsilon \otimes \mathrm{id}) \circ \varphi = \mathrm{id} \tag{8}$$

and, besides, is an algebra homomorphism.

By virtue of the duality between $F = F_q(GL(2))$ and the quantum algebra $\mathcal{U}_q(sl(2))$ the dual map $\varphi^* : \mathcal{U}_q(sl(2)) \otimes \mathcal{A} \to \mathcal{A}$ defines the structure of $\mathcal{U}_q(sl(2))$-module algebra on $\mathcal{A}$.

3. The center of $\mathcal{A}$ is generated by two elements (for generic q, not root of unity)

$$c_1 = \beta - q\gamma, \qquad c_2 = \alpha\delta - q^2\beta\gamma \tag{9}$$

4. There are three real forms of $\mathcal{A}$ consistent with three known real forms of F: $F_q(U(2))$, $F_q(U(1,1))$ and $F_q(GL(2,\mathbf{R}))$.

5. The two-sided ideal generated in $\mathcal{A}$ by the relation $c_1 = 0$ is invariant under the coaction φ and the corresponding quotient algebra is isomorphic to the quantum homogeneous space.

3. Discussion and Proofs

1. In order to prove the PBW-property of $\mathcal{A}$ it is necessary to verify that any monomial in $\alpha\beta\gamma\delta$ can be expanded uniquely into the sum of alphabetically ordered monomials. The possibility of alphabetical ordering is easy to establish using the commutation relations (6). As for the linear independence of

alphabetically ordered monomials, it is sufficient to verify it only for cubic monomials [16].

2. The (left) coaction of F on $\mathcal{A}$ is defined on the generators by the formula

$$\varphi(K) = TKT^t \tag{10}$$

where t stands for the matrix transposition, and is extended to the whole algebra $\mathcal{A}$ as an algebra homomorphism. For example,

$$\varphi(\beta) = ac\alpha + bc\gamma + ad\beta + bd\delta$$

Verification of the F-comodule algebra axioms (7), (8) is a matter of direct calculation based on the commutation relation (4) and the equivalent relations

$$R\overset{1}{T}{}^t\overset{2}{T}{}^t = \overset{2}{T}{}^t\overset{1}{T}{}^t R, \qquad \overset{1}{T}{}^t R^{t_1}\overset{2}{T} = \overset{2}{T} R^{t_1}\overset{1}{T}{}^t, \qquad \overset{1}{T} R^{t_1}\overset{2}{T}{}^t = \overset{2}{T}{}^t R^{t_1}\overset{1}{T}$$

obtained from (4) using transpositions and the symmetries

$$\mathcal{P}R^t\mathcal{P} = R, \qquad [\mathcal{P}, R^{t_1}] = 0 \tag{11}$$

of the R matrix (5). Here superscript t means total transposition and $\mathcal{P}$ is the permutation operator in $\mathbf{C}^2 \otimes \mathbf{C}^2$.

The duality between the quantum group $F = F_q(GL(2))$ and the quantum algebra $\mathcal{U}_q(sl(2))$ [5, 17] implies that $\mathcal{A}$ is also a $\mathcal{U}_q(sl(2))$-module algebra.

The algebra $\mathcal{U}_q(sl(2))$ is generated by three elements H, X_+, X_- and the relations

$$[H, X_\pm] = \pm X_\pm, \qquad [X_+, X_-] = \frac{q^{2H} - q^{-2H}}{q - q^{-1}}$$

or in the matrix form

$$R^\pm \overset{1}{L}{}^\pm \overset{2}{L}{}^\varepsilon = \overset{2}{L}{}^\varepsilon \overset{1}{L}{}^\pm R^\pm \qquad \forall \varepsilon \in \{+,-\}$$

where

$$L^+ = \begin{pmatrix} q^H & \omega X_- \\ 0 & q^{-H} \end{pmatrix}, \qquad L^- = \begin{pmatrix} q^{-H} & 0 \\ -\omega X_+ & q^H \end{pmatrix} \tag{12}$$

and

$$R^+ = q^{-1/2}\mathcal{P}R\mathcal{P}, \qquad R^- = q^{1/2}R^{-1}$$

The pairing $\langle\, , \rangle$ between F and $\mathcal{U}_q(sl(2))$ is described by the relations [5, 6]

$$\left\langle \overset{1}{L}{}^\pm, \overset{2}{T} \right\rangle = R^\pm_{12}, \qquad \left\langle \overset{1}{L}{}^\pm, \overset{2}{T}\overset{3}{T} \right\rangle = R^\pm_{12}R^\pm_{13}, \ldots$$

where, as usual, the subscripts mark the spaces where the corresponding R matrices act nontrivially.

The above formulae allow to calculate the corresponding action φ^* of $\mathcal{U}_q(sl(2))$ on algebra $\mathcal{A}$:

$$\varphi^*(\overset{1}{L}{}^{\pm})\overset{2}{K} \to R_{12}^{\pm}\overset{2}{K}(R_{12}^{\pm})^{t_2}$$

or, more explicitely, (cf. [18])

$$\varphi^*(X_-)\begin{pmatrix} \alpha & \beta \\ \gamma & \delta \end{pmatrix} = \begin{pmatrix} 0 & \alpha \\ q^{-1}\alpha & \beta + q^{-1}\gamma \end{pmatrix}$$

$$\varphi^*(X_+)\begin{pmatrix} \alpha & \beta \\ \gamma & \delta \end{pmatrix} = \begin{pmatrix} q\beta + \gamma & q\delta \\ \delta & 0 \end{pmatrix}$$

$$\varphi^*(q^H)\begin{pmatrix} \alpha & \beta \\ \gamma & \delta \end{pmatrix} = \begin{pmatrix} q\alpha & \beta \\ \gamma & q^{-1}\delta \end{pmatrix}$$

3. Before we proceed to the discussion of the center of $\mathcal{A}$ let us recall few facts about the quantum group F. The quantum determinant $\det_q T$ generating the center of F can be constructed via the fusion procedure technique [5, 19]

$$P_-\overset{1}{T}\overset{2}{T} = P_-\overset{1}{T}\overset{2}{T}P_- = \det{}_q T P_- \tag{13}$$

The rank one projector P_- and the complementary rank three projector P_+ in $\mathbf{C}^2 \otimes \mathbf{C}^2$ are defined by the spectral decomposition of the modified R matrix [20]

$$\hat{R} \equiv \mathcal{P}R = qP_+ - q^{-1}P_- \tag{14}$$

The commutativity of $\det_q T$ with T and its group-like property $\Delta(\det_q T) = \det_q T \otimes \det_q T$ follow from the relation (4), e.g.

$$\Delta(\det{}_q T) = P_-\overset{1}{T}{}_1\overset{1}{T}{}_2\overset{2}{T}{}_1\overset{2}{T}{}_2 = P_-\overset{1}{T}{}_1\overset{2}{T}{}_1P_-\overset{1}{T}{}_2\overset{2}{T}{}_2 = \det{}_q T \otimes \det{}_q T$$

The quantum group $F_q(GL(2))$ is well known to possess the invariant quadratic form ε_q

$$\varepsilon_q = \begin{pmatrix} 0 & 1 \\ -q & 0 \end{pmatrix}, \qquad T\varepsilon_q T^t = T^t\varepsilon_q T = \varepsilon_q \det{}_q T \tag{15}$$

The last relation, rewritten as $\sum_{mn}(\varepsilon_q^t)_{mn}T_{ni}(T^t)_{jm} = (\varepsilon_q^t)_{ji}\det_q T$, allows to introduce the trace operation for K matrices which is invariant under the quantum group coaction (10) up to $\det_q T$ factor

$$\mathrm{tr}_q K \equiv \mathrm{tr}\varepsilon_q^t K = \mathrm{tr}_q TKT^t/\det{}_q T, \qquad T \in F_q(GL(2)) \tag{16}$$

Now we can prove that c_1 and c_2 defined by (9) lie in the center of $\mathcal{A}$. Note that, by definition, $c_1 = \mathrm{tr}_q K$. Take tr_q of the relation (2) with respect to the first space in $\mathbf{C}^2 \otimes \mathbf{C}^2$. Then, using the fact that the R matrix, as a solution to the YBE, is a representation of the algebra F (4) having the quantum determinant $\det_q R = q$, and the identities (15), (16) one obtains from the left-hand-side of (2) the equality

$$\mathrm{tr}_q^{(1)}(R\overset{1}{K}R^{t_1})\overset{2}{K} = qc_1(K)\overset{2}{K}$$

and from the r.h.s. of (2) respectively

$$\overset{2}{K}\,\mathrm{tr}_q^{(1)}(R^{t_1}\overset{1}{K}R) = \overset{2}{K}\,c_1(K)q$$

which establishes the commutativity of $c_1(K)$ with the generators K of $\mathcal{A}$.

In order to obtain the expression for c_2 analogous to (13) multiply the RE (2) from the left by by the permutation operator $\mathcal{P}$

$$\hat{R}\overset{1}{K}R^{t_1}\overset{2}{K} = \overset{1}{K}R^{t_1}\overset{2}{K}\hat{R}$$

and then by the rank one projector P_-. Taking into account (14) one gets

$$P_-\overset{1}{K}R^{t_1}\overset{2}{K} = P_-\overset{1}{K}R^{t_1}\overset{2}{K}P_- = c_2(K)P_-$$

The proof of the commutativity of $c_2(K)$ with the generators K of $\mathcal{A}$ uses the same fusion technique as the above proof for $\det_q T$. The fact that the quantum determinants of R and R^{t_1} are numbers

$$P_-^{(12)}R_{32}R_{31} = qP_-^{(12)}, \qquad P_-^{(12)}R_{31}^{t_3}R_{32}^{t_3} = qP_-^{(12)}$$

again plays the crucial role.

Rewriting $q^2c_2(K)\overset{3}{K}$ as

$$q^2P_-^{(12)}\overset{1}{K}R^{t_1}\overset{2}{K}\overset{3}{K} = P_-^{(12)}\overset{1}{K}R_{12}^{t_1}\overset{2}{K}R_{31}^{t_3}R_{32}^{t_3}\overset{3}{K}R_{32}R_{31}$$

using then the YBE for R and R^{t_1} and RE for several times one arrives finally to

$$q^2\overset{3}{K}P_-^{(12)}\overset{1}{K}R^{t_1}\overset{2}{K} = q^2\overset{3}{K}c_2(K)$$

which proves the assertion.

Note that the central elements are transformed homogeneously under the quantum group coaction (10)

$$\varphi : c_1(K) \to \det_q T c_1(K), \qquad c_2(K) \to (\det_q T)^2 c_2(K). \qquad (17)$$

The transformation law for $c_1(K)$ follows from (16). The corresponding formula for $c_2(K)$ follows from (15), (16) and the identity

$$\mathrm{tr}_q K\varepsilon_q K = (1+q^2)c_2(K) - q(c_1(K))^2$$

Furthermore, one has matrix identity

$$K\varepsilon_q K = c_2(K)\varepsilon_q - qc_1(K)K.$$

The formula (17) and the following inversion formula for K matrix

$$K^{-1} = \frac{1}{c_2}\begin{pmatrix} \delta & -\beta+\omega\gamma \\ -q^2\gamma & \alpha \end{pmatrix}$$

allow to think of c_2 as the quantum deteminant of K.

We have no proof that for generic q the center of $\mathcal{A}$ is generated by c_1, c_2 though it is highly plausible conjecture taking into account the analogous results for the quantum group F.

4. Let us discuss now the real forms of $\mathcal{A}$. It is well known that for $F_q(GL(2))$ there are three real forms [5]. For $F_q(U(2))$ the parameter q is real $\bar{q} = q$ and the $*$-antiinvolution is given by

$$\{a^*, b^*, c^*, d^*\} = \{d, -qc, -q^{-1}b, a\}$$

For $F_q(U(1,1))$ again $\bar{q} = q$ and

$$\{a^*, b^*, c^*, d^*\} = \{d, qc, q^{-1}b, a\}$$

For $F_q(GL(2,\mathbf{R}))$ the parameter q is unitary $\bar{q} = 1/q$ and

$$\{a^*, b^*, c^*, d^*\} = \{a, b, c, d\}$$

The corresponding real forms of $\mathcal{A}$ are:

$$F_q(U(2)): \qquad \{\alpha^*, \beta^*, \gamma^*, \delta^*\} = \{q\delta, -\beta, -\gamma, q^{-1}\alpha\}$$

$$F_q(U(1,1)): \qquad \{\alpha^*, \beta^*, \gamma^*, \delta^*\} = \{q\delta, \beta, \gamma, q^{-1}\alpha\}$$

$$F_q(GL(2,\mathbf{R}): \qquad \{\alpha^*, \beta^*, \gamma^*, \delta^*\} = \{\alpha, \gamma + qc_1, q\gamma, \delta\}$$

It is easy to verify that these real forms of $\mathcal{A}$ are consistent with those of F that is $\varphi(K_{ij}^*) = \varphi(K_{ij})^*$.

5. The invariance of the ideal $c_1 = 0$ under the coaction φ follows immediately from (17). The resulting quotient algebra is generated by three generators α, γ, δ ($\beta = q\gamma$) and the relations

$$\alpha\gamma = q^2\gamma\alpha, \qquad \gamma\delta = q^2\delta\gamma, \qquad [\alpha,\delta] = q(q^2 - q^{-2})\gamma^2$$

The algebra is isomorphic to the quantum homogeneous space for the quantum group $F_q(SL(2))$. It is also isomorphic to the subalgebra of F generated by the elements of the matrix TT^t [5] (see [21, 22] for the $F_q(GL(n))$ case).

The representation theory for the algebra $\mathcal{A}$ is a topic deserving special investigation. We restrict ourselves with few examples.

There are two one-dimensional representations of $\mathcal{A}$:

$$K^{(0)} = \varepsilon_q = \begin{pmatrix} 0 & 1 \\ -q & 0 \end{pmatrix}, \qquad K^{(1)} = \begin{pmatrix} \lambda & \mu \\ 0 & \nu \end{pmatrix} \tag{18}$$

One-dimentional representations of comodule algebras related to the reflection equations with R matrices corresponding to other quantum (super)groups such as $F_q(GL(m|n))$ were studied in [23].

4. Variants of RE and related algebras

1. Consider the algebra $\mathcal{X}$ defined by the generators x, y, u, v and by the relations

$$xy = qyx, \qquad xu = qux, \qquad xv = vx + \omega uy,$$
$$uv = qvu, \qquad yu = uy, \qquad yv = qvy,$$

Introducing the column X and row Y

$$X = \begin{pmatrix} x \\ y \end{pmatrix}, \qquad Y = \begin{pmatrix} u & v \end{pmatrix}$$

one can represent the above commutation relations as the exchange algebras [8, 10, 13]

$$R\overset{1}{X}\overset{2}{X} = q\overset{2}{X}\overset{1}{X}, \qquad \overset{2}{Y}\overset{1}{Y}R = q\overset{1}{Y}\overset{2}{Y} \tag{19}$$

$$\overset{1}{Y}R^{t_1}\overset{2}{X} = \overset{2}{X}\overset{1}{Y}, \qquad \overset{2}{Y}R^{t_1}\overset{1}{X} = \overset{1}{X}\overset{2}{Y} \tag{20}$$

The algebra $\mathcal{X}$ coincides with (3) and has a central element $\zeta = Y\varepsilon_q X = uy - qvx$. The subalgebras of $\mathcal{X}$ spanned by X or Y are isomorphic to the "quantum plane" [16] with coactions

$$\varphi : \; X \to TX, \;\; Y \to YT^t.$$

By virtue of the relations (19), (20) the matrix

$$K = XY = \begin{pmatrix} xu & xv \\ yu & yv \end{pmatrix}$$

satisfies RE (2) the above formula giving thus a homomorphism $\mathcal{A} \to \mathcal{X}$. Note that for the obtained realization of $\mathcal{A}$ one has $c_2(K) = 0$ and $K\varepsilon_q K = \zeta K = -qc_1(K)K$.

2. In the papers [7, 8] another version of RE was obtained

$$R\overset{1}{M}R^{-1}\overset{2}{M} = \overset{2}{M}\widetilde{R}^{-1}\overset{1}{M}\widetilde{R}, \qquad \widetilde{R} \equiv \mathcal{P}R\mathcal{P}, \qquad M = \begin{pmatrix} \xi & \eta \\ \theta & \tau \end{pmatrix} \tag{21}$$

The corresponding algebra $\mathcal{B}$ is an F-comodule algebra with the coaction

$$\psi : \mathcal{B} \to F \otimes \mathcal{B}, \qquad \psi(M) = TMT^{-1}$$

In the case of $F = F_q(GL(2))$ the F-comodule algebra $\mathcal{B}$ is isomorphic to $\mathcal{A}$ because of the relation (15). The isomorphism is given by the formula

$$K = \begin{pmatrix} \alpha & \beta \\ \gamma & \delta \end{pmatrix} = \begin{pmatrix} -q\eta & \xi \\ -q\tau & \theta \end{pmatrix} = M\varepsilon_q$$

which implies the relations between the central elements

$$c_1(K) = \beta - q\gamma = \xi + q^2\tau \equiv z_1(M)$$

$$c_2(K) = \alpha\delta - q^2\beta\gamma = q^3(\xi\tau - q^{-2}\eta\theta) \equiv q^3 z_2(M)$$

Under this isomorphism the q-trace (16) maps into $\mathrm{tr}DM$, $D = \mathrm{diag}(q^{-1}, q)$ introduced in [5], see also [12].

The algebra $\mathcal{B}$ is connected closely to the quantum algebra $\mathcal{U}_q(sl(2))$ [6,7]. Namely, the matrix M can be realized in terms of $L^{\pm}$ (12)

$$M = s(L^+)L^- \tag{22}$$

where s is the antipode in $\mathcal{U}_q(sl(2))$: $s(H) = -H$, $s(X_\pm) = -q^{\mp 1}X_\pm$.

The formula (22) describes the algebra homomorphism $\chi : \mathcal{B} \to \mathcal{U}_q(sl(2))$

$$\chi(M) = \begin{pmatrix} q^{-2H} + q\omega^2 X_- X_+ & -\omega q X_- q^H \\ -\omega q^H X_+ & q^{2H} \end{pmatrix}$$

for which $z_1(M)$ is proportional to the well known Casimir operator for $\mathcal{U}_q(sl(2))$ and $z_2(M) = 1$ [12].

It is easy to see that the inversion of the homomorphism χ needs invertibility of the element τ (which is not the case for the one-dimensional representation $M = K^{(1)}\varepsilon_q$, see (18)). If the element τ is supposed to be invertible one can introduce in $\mathcal{B}$ a coproduct induced from $\mathcal{U}_q(sl(2))$ [8].

3. We conclude this section with describing the classical limit of $\mathcal{A}$. Let $q = e^h$. As $h \to 0$ or $q \to 1$ the algebra $\mathcal{A}$ becomes commutative, its commutator giving rise to the Poisson bracket $[\,,\,] = -h\{\,,\,\}$. The commutation relations (2) are transformed into the Poisson brackets relations

$$\{\overset{1}{K}, \overset{2}{K}\} = [r, \overset{1}{K}\overset{2}{K}] + \overset{1}{K}r^{t_1}\overset{2}{K} - \overset{2}{K}r^{t_1}\overset{1}{K} \tag{23}$$

where r is the classical r matrix $R = 1 + hr + O(h^2)$

$$r = \begin{pmatrix} 1 & & & \\ & & & \\ & 2 & & \\ & & & 1 \end{pmatrix}, \qquad r^{t_1} = \begin{pmatrix} 1 & & 2 & \\ & & & \\ & & & \\ & & & 1 \end{pmatrix},$$

Fixing values of the central functions $c_1 = \beta - \gamma$ and $c_2 = \det K = \alpha\delta - \beta\gamma$ one obtains the foliation of the 4-dimensional space spanned by $\alpha\beta\gamma\delta$ into 2-dimensional manifolds on which the Poisson bracket (23) is nondegenerate.

5. Generalizations

1. Let us discuss now the fusion procedure for RE. It does not differ much from the case of the YBE [5, 19]. The peculiarities of the RE case are well seen in the simplest spin-1 case. The R matrix in (4) $R \in \mathbf{C}^2 \otimes \mathbf{C}^2$ intertwines two spin-1/2 corepresentations of $F_q(GL(2))$. The R matrix in $\mathbf{C}^2 \otimes \mathbf{C}^3$ is

$$R_{1(2)} = P_+ R_{12'} R_{12} = P_+ R_{12'} R_{12} P_+$$

where P_+ is the rank 3 projector (14) from $\mathbf{C}^2 \otimes \mathbf{C}^2$ onto $\mathbf{C}^3$ and the corresponding spaces are labelled respectively by the indices 2, $2'$ and (2). We shall use also

$$R^{t_1}_{1(2)} = P_+ R^{t_1}_{12} R^{t_1}_{12'} P_+$$

The 3×3 matrix J

$$J = P_+ \overset{2}{K} R^{t_1}_{22'} \overset{2'}{K} P_+$$

satisfy the RE in $\mathbf{C}^2 \otimes \mathbf{C}^3$

$$R_{1(2)} \overset{1}{K} R^{t_1}_{1(2)} \overset{(2)}{J} = \overset{(2)}{J} R^{t_1}_{1(2)} \overset{1}{K} R_{1(2)}$$

and also the RE in $\mathbf{C}^3 \otimes \mathbf{C}^3$

$$R_{(1)(2)} \overset{(1)}{J} R^{t_1}_{(1)(2)} \overset{(2)}{J} = \overset{(2)}{J} R^{t_1}_{(1)(2)} \overset{(1)}{J} R_{(1)(2)}$$

The entries of the matrix J generate a subalgebra $\mathcal{A}'$ of $\mathcal{A}$ which is also an F-coideal that is $\varphi : \mathcal{A}' \to F \otimes \mathcal{A}'$. The analogous fusion procedure for the spectal parameter dependent RE (1) was developed in [24].

There is no doubt that, being properly generalized, the above described fusion procedure should be able to produce solutions to RE in matrices of any dimension corresponding to higher finite-dimensional irreducible corepresentations of $F_q(GL(2))$. At least for the one-dimensional representations of $\mathcal{A}$ (18) the expression for the "universal K matrix" can be written explicitely [14, 23, 25].

2. Throughout this paper we considered the algebra $\mathcal{A}$ defined by the relations (2) corresponding to the R matrix (5) of the quantum group $F_q(GL(2))$. However, the proof of the comodule-algebra property does not use anything besides the commutation relations (4) and the symmetries (11) of the R matrix. Therefore, to any R matrix satisfying (11) and the related quantum group F_R defined by relations (4) there corresponds some F_R-comodule-algebra $\mathcal{A}_R$ defined by relations (2).

The conditions (11), however, are not essential and can be disposed of. Let us extend the algebra F_R introducing another matrix of generators S in addition to T and imposing the commutation relations

$$\begin{aligned} R^{(1)}\overset{1}{T}\overset{2}{T} &= \overset{2}{T}\overset{1}{T}R^{(1)}, &\quad \overset{1}{S}R^{(2)}\overset{2}{T} &= \overset{2}{T}R^{(2)}\overset{1}{S}, \\ \overset{1}{T}R^{(3)}\overset{2}{S} &= \overset{2}{S}R^{(3)}\overset{1}{T}, &\quad R^{(4)}\overset{1}{S}\overset{2}{S} &= \overset{2}{S}\overset{1}{S}R^{(4)} \end{aligned} \tag{24}$$

parametrized by four R matrices. It is easy to verify that if K obeys the relation

$$R^{(1)}\overset{1}{K}R^{(2)}\overset{2}{K} = \overset{2}{K}R^{(3)}\overset{1}{K}R^{(4)} \tag{25}$$

then the matrix TKS obeys the same relation provided matrix elements of K commute with those of T and S.

Suppose now that $S = \sigma(T)$ where σ is an antiautomorphism of matrix algebra that is $\sigma(xy) = \sigma(y)\sigma(x)$ for any number matrices x, y. Then all the R matrices in (25) can be expressed in terms of the matrix R defining the quantum group (4). The resulting algebra is an F_R-comodule-algebra with the coaction $K \to TK\sigma(T)$.

Note that there exists a remarkable ambiguity in the choice of the R matrices. Namely, one can make the substitution $R \to \mathcal{P}R^{-1}\mathcal{P}$ independently in any of the matrices $R^{(i)}$ which does not affect the commutation relations (24), *cf.* the remark immediately after formula (5). Consequentely, there exist at least $2^4 = 16$ *a priori* nonequivalent algebras sharing the same comodule structure. The interrelations of these algebras will be described in a separate paper.

In the case of $\sigma(T) = T^t$ the R matrices are

$$\begin{aligned} R^{(1)} &= R &\quad &\text{or } \widetilde{R}^{-1} \\ R^{(2)} &= R^{t_1} &\quad &\text{or } (\widetilde{R}^{-1})^{t_1} \\ R^{(3)} &= (\widetilde{R})^{t_2} &\quad &\text{or } (R^{-1})^{t_2} \\ R^{(4)} &= \widetilde{R}^{t} &\quad &\text{or } (R^{-1})^{t} \end{aligned}$$

where, as in (21), the notation $\widetilde{R} \equiv \mathcal{P}R\mathcal{P}$ is used. Choosing the first option for every $R^{(i)}$ one obtains from (25) the equation [23]

$$R\overset{1}{K}R^{t_1}\overset{2}{K} = \overset{2}{K}\widetilde{R}^{t_2}\overset{1}{K}\widetilde{R}^{t} \tag{26}$$

coinciding with the RE (2) if R has the symmetries (11).

Another example is provided by $\sigma(T) = T^{-1}$. In this case

$$\begin{array}{ll} R^{(1)} = R & \text{or } \widetilde{R}^{-1} \\ R^{(2)} = R^{-1} & \text{or } \widetilde{R} \\ R^{(3)} = \widetilde{R}^{-1} & \text{or } R \\ R^{(4)} = \widetilde{R} & \text{or } R^{-1} \end{array}$$

The choice of the first option for every $R^{(i)}$ leads to the algebra $\mathcal{B}$ described in the section 4.2.

Another interesting question refers to the structure of the relations between the generators, which give rise to the comodule algebra. The matrix form of coactions on the generators (X, Y, K, M) (10), (20), (21) easily garanties the compatability of the coaction map with the coproduct and the counit of the Hopf algebra. Is it possible to change (to deform) relations (2), (19), (21) or even (4) (considering the letter one as the left F_q-comodule algebra relations) preserving the comodule algebra property? One can find the examples of the deformation of this kind: for (19), it is the covariant system of the q-oscillators (cf. [13]); for (21), one can add to the r.h.s. the 4×4 unit matrix (an analog of the central extension); for (4), one can add to the r.h.s. the rank one 4×4 matrix

$$RW_1W_2 = W_2W_1R + p\, h^t \otimes f,$$

where $h = (0, 1, -1/q, 0)$, $f = (0, q, 1, 0)$ and p is additional non-central element. The coaction is given by $W \to TW$; the term in the r.h.s. is invariant due to (15). The relations between the generators are [26]

$$\begin{array}{lll} uv = qvu, & vx = qxv, & [u, x] = \omega wv + q\omega p, \\ uw = qwu, & wx = qxw, & [v, w] = \omega p, \end{array}$$

$$W = \begin{pmatrix} u & v \\ w & x \end{pmatrix}$$

Analogous extensions exist for the quantum groups SO and Sp [6].

3. For quantum super-groups many relations and equations are almost the same as in the non-graded case provided that minus signs are put into appropriate places. In particular, despite of the fact that supertranspositons st_1 and st_2 are non-trivial one gets for the R-matrix: $st_1 \circ st_2 = t$ for all the non-zero entries of R are even. The super-analogue of the RE is

$$R\overset{1}{K}R^{st_1}\overset{2}{K} = \overset{2}{K}R^{st_1}\overset{1}{K}R$$

where signs are included into the definition of the Z_2-graded tensor product.

4. We conclude our discussion of RE with a remark concerning the very form of the equation. Note that formula (10) for the coaction of F is equivalent to

$$\varphi : K^{i_1 i_2} \to T^{i_1}_{j_1} T^{i_2}_{j_2} K^{j_1 j_2} \tag{27}$$

where the indices are shown explicitely. Here we use the commutativity of T and K and assume summation over repeated indices.

The formula (27) suggests that K can be thought of as a bivector (contravariant tensor of the second order). In a compact form one can write (27) as

$$\varphi : K_{12} \to \overset{1}{T}\overset{2}{T} \cdot K_{12} \tag{28}$$

(the notation is self-evident).

Analogously, the equation (26) can be rewritten as

$$R^{i_1 i_2}_{j_1 j_2} K^{j_1 j_1'} R^{i_1' j_2}_{j_1' k_2} K^{k_2 i_2'} = K^{i_2 j_2'} R^{k_2' i_1}_{j_2' j_1} K^{j_1 j_1'} R^{i_2' i_1'}_{k_2' j_1'}$$

or, using the same notation as in (28)

$$R_{12} R_{1'2} \cdot K_{11'} K_{22'} = \widetilde{R}_{1'2'} \widetilde{R}_{12'} \cdot K_{22'} K_{11'} \tag{29}$$

Moving two R matrices from the right hand side of (29) to the left hand side one obtains

$$\widetilde{R}^{-1}_{12'} \widetilde{R}^{-1}_{1'2'} R_{12} R_{1'2} \cdot K_{11'} K_{22'} = K_{22'} K_{11'} \tag{30}$$

Combining the two indices in $K^{i_1 i_2}$ into one composite index and the four R matrices in (30) into one composite R matrix we observe that the equation (30) describes a sort of exchange algebra (19).

The case $\sigma(T) = T^{-1}$ leads in the same way to a mixed tensor having one upper and one lower index. One can consider also consistent set of q-tensors of different ranks $K^{i_1 i_2 \ldots i_r}$.

This observation opens a road to a far-going generalization of the algebra $\mathcal{A}$. Acting in the same spirit one can introduce algebras corresponding to the quantum tensors with any number of upper and lower indices. The papers [16, 27, 28] where the quantum multilinear algebras are studied might be of relevance in performing this program.

6. Acknowledgments

We would like to thank E.K.Sklyanin for the collaboration on the material presented here. We are grateful to organizers of Oji and Non-Commutative Analysis International Seminars for the warm hospitality in Kyoto and our colleagues for useful discussions.

References

[1] Sklyanin,E., Boundary conditions for integrable quantum systems, *J. Phys. A: Math. Gen.*, **21**, 2375-2389 (1988).

[2] Cherednik,I., Factorized particles oon a half-line and root systems, *Theor. Math. Phys.*, **61**, 55-61 (1984).

[3] Reshetikhin,N. and Semenov-Tian-Shansky,M., Quantum current algebras, *Lett. Math. Phys.*, **19**, 133-141 (1989).

[4] Kulish,P. and Sklyanin,E., The general U(sl(2)) invariant XXZ integrable quantum spin chain, *J.Phys.A: Math.Gen.*, **24** L435-L439 (1991).

[5] Faddeev,L.,Reshetikhin,N. and Takhtajan,L., Quantization of Lie groups and Lie algebras, *Algebra i Analiz*, **1**, 178-191 (1989).

[6] Takhtajan,L., Introduction to quantum groups, *Lecture Notes Phys.*, **370**, 3-21 (1990).

[7] Moore,G. and Reshetikhin,N., Quantum group symmetry of CFT, *Nucl. Phys.*, **B328**, 557-571 (1990).

[8] Alekseev,A.,Faddeev,L. and Semenov-Tian-Shansky,M., Hidden quantum groups inside Kac-Moofy algebras, *Lecture Notes Math.*, **1510** 148- 158 (1992).

[9] Alekseev,A. and Faddeev,L., Toy model of CFT and quantum top, *Commun. Math. Phys.*, **141** 413 (1991).

[10] Babelon,O., Liouville theory on the lattice, *Commun. Math. Phys.*, **139**, 619-631 (1991).

[11] Freidel,L. and Maillet,J., Quadratic algebras and integrable systems, *Phys. Lett.*, **B262**, 278-281; **B263**, 403-407 (1991).

[12] Zumino,B., Introduction to the differential geometry of quantum groups. Preprint UCB-PTH-62/91 (1991).

[13] Kulish,P., Finite-dimentional Zamolodchikov-Faddeev algebra and q-oscillators *Phys. Lett.*, **A161**, 50-53 (1991).

[14] Kulish,P., Quantum groups and quantum algebras as symmetries of dynamical systems. Preprint YITP/K-959 (1991).

[15] Kulish,P. and Sklyanin,E., Algebraic structures related to reflection equations, Preprint YITP/K-980; *J.Phys.A: Math. Gen.*, **25**, (1992).

[16] Manin,Yu., Topics in noncommutative geometry, (Princeton Univ. Press, 1991)

[17] Majid,S., Hopf algebras and Yang-Baxter equations, *I. J. M. P.*, **A5**, 1-90 (1990).

[18] Majid,S., Examples of braided groups, *J. Math. Phys.*, **32**, 3246-3253 (1991).

[19] Kulish,P.,Reshetikhin,N. and Sklyanin,E., Yang-Baxter equation and representation theory, *Lett. Math. Phys.*, **5**, 393-399 (1981).

[20] Jimbo,M., A q-analogue of U(gl(N+1)), Hecke algebra and Yang- Baxter equation, *Lett. Math. Phys.*, **11**,247-252(1986).

[21] Jing,N. and Yamada,H., Polynômes zonaux pour le groupe linéaire général quantique. Preprint IAS (1990).

[22] Noumi,M., Macdonald's symmetric polynomials as zonal spherical functions on some quantum homogeneous spaces. Preprint UT-Komaba (1992).

[23] Kulish,P.,Sasaki,R. and Schwiebert,C., Constant solutions of reflection equations. Preprint YITP/U-92-07 (1992).

[24] Mezincescu,L. and Nepomechie,R., Fusion procedure for open chains. Preprint CERN-TH. 6152/91 (1991).

[25] Cremmer,E. and Gervais J-L., The quantum strip, *Commun. Math. Phys.*, **144**, 279-291 (1991).

[26] Noumi,M. and Mimachi,K., Big q-Jacobi polynomials, q-Hahn polynomials and a family of quantum 3-spheres, *Lett. Math. Phys.*, **19**, 299-305 (1990).

[27] Hashimoto,M. and Hayashi,T., Quantum multilinear algebra. Preprint Nagoya Univ. (1991).

[28] Gurevich,D. and Rubtsov,V., Yang-Baxter equation and deformation of algebras, *Lecture Notes Math.*, **1510**, 47-55 (1992).

Quantum Groups, Star Products and Cyclic Cohomology

MOSHÉ FLATO and DANIEL STERNHEIMER
Physique Mathématique, Université de Bourgogne, B.P. 138, 21004 DIJON Cedex, FRANCE

Abstract. After some historical remarks, we start with a rapid overview of the star-product theory (deformation of algebras of functions on phase space) and its applications to deformation-quantization. We then concentrate on Poisson-Lie groups and their "quantization", give a star-product realization of quantum groups and discuss uniqueness and the rigidity as bialgebra of a universal model for the quantum $SL(2)$ groups. In the last part we develop the notion of closed star-product (for which a trace can be defined on the algebra), show that it is classified by cyclic cohomology, permits to define a character and that there always exists one ; finally we show that the pseudodifferential calculus on a compact Riemannian manifold defines a closed star product whose character contains the Todd genus of the manifold.

Lectures presented in part by M. FLATO at the Oji International Seminar on Quantum Analysis (Kyoto, June 25-29, 1992) and in part by D. STERNHEIMER at the RIMS Symposium on Non-Commutative Analysis (Kyoto, June 29-July 2, 1992).
Dedicated to Huzihiro ARAKI with affection and appreciation.

Key words: Quantum groups, deformation-quantization, pseudodifferential operators, cyclic cohomology, index theorem.

1. BACKGROUND

Traditionally quantization is understood as a passage from "classical" observables, functions on a phase-space (with an evolution governed by differential equations, e.g. Hamilton equations) to their quantum counterparts, operators acting on states of a system, vectors in a Hilbert space (with an evolution described by a Schrödinger equation or the equivalent Heisenberg picture). Quantum analysis is thus traditionally a non-commutative analysis of operator algebras. We shall see here that quantum analysis can also be seen as non-commutative analysis for algebras of classical observables, endowed with a deformed (non-commutative) product, and how many frontier developments of mathematics motivated by mathematical physics - such as quantum groups and some parts of non-commutative geometry - relate to this framework.

At the beginning of quantum mechanics, Hermann Weyl [1] introduced

H. Araki et al. (eds.), Quantum and Non-Commutative Analysis, 221–237.

his famous mapping from classical functions to operators by expressing the functions as Fourier transforms and replacing the exponential kernel by the corresponding (unitary) group element in the representation of the canonical commutation relations (possibly multiplied by a weight). A few years later Eugene Wigner [2] inverted that map, finding a symbol for an operator - but could not interpret this symbol as a probability density, since it had no reason to be positive. Much later Moyal [3] and Groenewold [4] found the expression of the symbol of a bracket or a product of operators as sine or exponential of the Poisson bracket of the symbols - but again couldn't go much further in their physical interpretation. Many authors worked around those ideas, in various fields of physics. In particular semi-classical approximations were introduced as a first (and often easier to compute) step towards quantization. Clearly the idea of deformation was in the back of many minds, but remained subconscious.

In the early 70's we [5] had studied deformations of the Poisson bracket Lie algebra of functions on phase space by 1-differentiable differential operators, and in 1975 J. Vey [6] extended this to differentiable deformations, rediscovering in the flat case the bracket of Moyal (of which he was obviously unaware). Then - Eureka - everything fell in place, and in 1977 we could [7] not only express quantum mechanics as a deformation of classical mechanics, but also develop it in an autonomous manner (without operators) using mathematical analysis and deformed (non-commutative) products (called star-products) on the algebras of classical observables. This has lead to numerous developments (including a parallel approach to group representation theory), some of which we shall present here.

Mathematicians followed a somewhat parallel course. In 1957, Calderon and Zygmund [8] published their seminal paper on singular integral operators. The idea is, using Fourier transform, to generalize partial differential operators (with variable coefficients) from polynomials in the differentiations to more general functions (of polynomial growth). It is a Weyl transform of their symbols, with a weight to express that coefficients are placed on the left of the differentiations, what physicists call standard ordering (in contradistinction with the symmetric ordering appearing in Moyal). They are now called pseudo-differential operators, and their popularity since the 60's went far beyond mathematical analysis when the index theorems showed that, on manifolds, their index (dimension of kernel minus codimension of image) is expressed with topological invariants. As one can expect, the symbol of a product is a star-product of symbols, and even a "closed" one [9], that allows to define a trace by integration on the manifold.

In the 80's another mathematical object has known tremendous development. In the quantization of 2-dimensional integrable models, the school of Leningrad (starting with [10]) discovered a strange modification of the $\mathcal{SL}(2)$ Lie algebra. Together with Drinfeld they systematized this approach in the framework of Hopf algebras, leading to the important paper on "quantum groups" by Drinfeld [11] at the Berkeley IMU congress. Essentially the Hopf algebra of functions on a Lie group endowed with a compatible Poisson structure is deformed, a star-product replacing the commutative product of functions. The approach of Jimbo [12] is dual to it (deforming the coproduct on the Lie algebra level). Both can be realized using the independent formulation of Woronowicz [13] where, inspired by A. Connes' non-commutative geometry [14], he "twisted" the basic representation of compact groups by taking their coefficients in C^*-algebras with some relations so as to get e.g. the commutation relations found by the Leningrad school. By now this field is developing in many directions, in spite of the somewhat misleading name: there are "quantized" enveloping algebras, but no groups ; and the deformation parameter in the corresponding star-product has no reason to be related to the Planck constant $\hbar$; in fact [15] it seems to require an underlying deformation-quantization with $\hbar$ as a (different) parameter. One has also discovered quantum groups in various domains of physics, and even some of their special functions a century ago !

In this paper we shall first survey star-products and their use in quantization and quantum groups ("quantum analysis"). In the last chapter we shall deal with closed star-products and the cyclic cohomology of non-commutative geometry, in particular in connection with the pseudodifferential calculus ("non-commutative analysis"). We shall concentrate on the most recent developments in these areas : the realization of quantized universal enveloping (QUE) algebras with 2 parameters (q in the algebra and $\hbar$ in the background) and its uniqueness [15], the existence of a universal and rigid Fréchet Hopf algebra containing densely the original and the QUE algebras [16], and [9] the notion of closed star product and its relation with cyclic cohomology and the topological invariants of the Atiyah-Singer index formulas in pseudodifferential calculus.

As can be expected in such a paper, which is a "localized" review but not an exhaustive one on the various subjects (a whole book would be needed for that), the list of references contains only what is needed for our purpose and can serve as a starting point for finding further references, but is by no means meant to be exhaustive.

2. STAR-PRODUCTS AND QUANTIZATION

2.1. Star products : definition and basic properties

a. A Poisson manifold is a (connected, paracompact differentiable) manifold W endowed with a contravariant skew-symmetric 2-tensor Λ satisfying $[\Lambda, \Lambda] = 0$ (in the sense of the Schouten brackets). When it is everywhere non degenerate, W is a symplectic manifold (of even dimension

$2l$) and the corresponding 2-form ω is closed. For $u, v \in C^\infty(W) = N$ one defines the Poisson bracket :

$$P(u,v) = i(\Lambda)\,(du \wedge dv) \tag{1}$$

b. Star-product. When N is an associative (resp. Lie) algebra, one defines the Hochschild and Chevalley cohomologies, denoted (resp.) $\tilde{H}^p(N)$ and $H^p(N)$. For $N = C^\infty(W)$, it is consistent [7] to restrict oneself to differentiable cohomologies. Here we shall call "star-product" an associative deformation of the algebra (for the pointwise product) $N = C^\infty(W)$ with a complex parameter ν, of the form :

$$u * v = \sum_{r=0}^{\infty} \nu^r\, C_r(u,v)\ ;\ u, v \in N \tag{2}$$

$$C_0(u,v) = uv, \qquad C_1(u,v) - C_1(v,u) = 2P(u,v)$$

The cochains C_r are here only supposed to be bidifferential operators.

Nevertheless, since the corresponding bracket has the form :

$$\frac{1}{2\nu}\,(u * v - v * u) \equiv [u,v]_\nu = P(u,v) + \sum_{r=1}^{\infty} \nu^r\, C'_r(u,v) \tag{3}$$

we get in this way a deformation of the Poisson bracket Lie algebra N, for which the relevant Chevalley cohomologies are finite-dimensional, which permits to carry over most of the theory developed in [7] (where the cochains C_r were assumed to have same parity as r). All this extends in a natural manner to the algebra $\mathcal{A} = N[[\nu]]$ of formal power series in ν with coefficients in N, so that we can require associativity for the star-product and Jacobi identity for the bracket. At first order this means that the cochain is a cocycle (in particular, P is a non trivial Hochschild cocycle). We can also speak of equivalence between two star-products (or brackets) : we say that $* \simeq *'$ il there exists a formal series T (of differential operators T_s [7]) such that :

$$T(u *' v) = Tu * Tv,\ T = I + \sum_{s=1}^{\infty} \nu^s\, T_s \tag{4}$$

and the same for brackets. As in general for deformations [7,17], the obstructions to extend the existence (resp. the equivalence) of deformations from each level t to level $t+1$ lie in the 3-cohomology (resp. the 2-cohomology).

c. Example. Take $W = \mathbb{R}^{2l}$ and write the rth power of the bidifferential operator P :

$$r!\, C_r(u,v) = P^r(u,v) = \Lambda^{i_1 j_1} \ldots \Lambda^{i_r j_r} (\partial_{i_1 \ldots i_r} u)(\partial_{j_1 \ldots j_r} v) \tag{5}$$

so that we have the corresponding "Moyal" products and brackets :

$$u * v = \exp(\nu P)(u, v) \tag{6}$$

$$M(u, v) = \nu^{-1} \sinh(\nu P)(u, v) \tag{7}$$

More generally, we shall call **Vey product** (or bracket) a star-product for which the C_r 's are odd or even together with r, vanish on constants and have locally $(1/r!)$ P^r for principal part. From [7] and [19] (see subsection *d*) there always exists a Vey product on a symplectic manifold. The skew-symmetry of Λ gives (by integration by parts), for $r \geq 1$, whenever defined:

$$\int_W P^r(u, v)\, \omega^l = 0 \tag{8}$$

so that

$$\int_W u * v\, \omega^l = \int_W v * u\, \omega^l \tag{9}$$

d. Definition. A star-product satisfying (9) is called **strongly closed** [9].

Theorem [18]. *On any symplectic manifold W there exists a strongly closed star-product.*

The idea of the proof is to "glue" together Moyal products on canonical charts in a way that preserves the integrals of functions. This is achieved by showing that there exists a consistent way to define globally on W a formal completion of the Heisenberg enveloping algebra, called Weyl manifold (a locally trivial algebra bundle), and to extend the functions in $\mathcal{A}$ to sections of this bundle ("Weyl functions") in an integration-preserving way.

This result is very recent (September 1992). Previously the same authors had proved it without the closedness condition, and their result was simplified [19] (even for regular Poisson manifolds) by M. De Wilde and P. Lecomte, who had the original existence proof on any symplectic manifold, by showing with a covering of W by canonical charts that the local Moyal products can be transformed locally by equivalence so as to coincide on the intersections of charts ; but the existence proof of the equivalences could not show closedness. Former existence proofs for star-products or brackets required $b_3(W) = 0$ (vanishing third Betti number) to control the intersections of more than 2 charts.

e. Uniqueness and stability. One has $dim H^2(N) = 1 + b_2(W)$. Therefore, if the second Betti number of the symplectic manifold W vanishes, the Moyal-Vey bracket is unique up to (mathematical) equivalence. The physical interpretation of equivalence can however be subtle.

In particular, on $\mathbb{R}^{2l}$, all "orderings" [20] are equivalent (the equivalence is given by the Fourier transform in $\mathbb{R}^{2l}$ of the weight appearing in the Weyl mapping) but obviously the operators corresponding to the same classical function are in general different with different orderings. One can even prove [21] that if two equivalent star-products are isospectral (see below) for a large enough class of observables, they are identical.

Concerning stability, one can show that a 2-parameter Vey product

$$u *_{\nu\rho} v = \sum \nu^r \rho^s C_{rs}(u,v), \quad C_{rs}(u,v) = (-1)^r C_{rs}(v,u) \qquad (10)$$

is equivalent to a one-parameter $*_\nu$ (the ρ parameter deforms only Λ) [22] . However a $*$-similitude transformation on a Vey product can give a non-equivalent product (starting with a Jacobi bracket instead of a Poisson bracket) with 2 parameters, and permit e.g. a simple expression of KMS conditions in statistical mechanics [23].

f. Invariance and covariance. By definition, P is invariant under all symplectomorphisms of W. From their expression the P^r for $r \geq 2$, in the Moyal case on $\mathbb{R}^{2l}$, are invariant only under affine symplectomorphisms, i.e. vector fields generated by polynomials of degree ≤ 2 ; for other orderings one has an even smaller algebra (only the Heisenberg algebra in the generic case, possibly supplemented by a second degree polynomial). In general one has only a finite-dimensional Lie algebra $\mathcal{O}$ of "preferred" observables :

$$\mathcal{O} = \{a \in N; [a,u]_\nu = P(a,u)\, \forall u \in N\}. \qquad (11)$$

The star-product will be said **invariant** under $\mathcal{O}$.

A more general notion is that of **covariance** : a $*$-product is said **covariant** if (11) holds only when both a and u are restricted to the covariance Lie algebra. If we denote by G the corresponding Lie group, there exists then a representation τ of G into $Aut(\mathcal{A};*)$ such that

$$\tau_g u = (Id_N + \sum_{r=1}^{\infty} \nu^r \tau_g^r)(gu) \qquad (12)$$

where $g \in G$, $u \in N$, the τ_g^r are differential operators on W, and gu is given by the geometrical action of G on $W : gu(x) = u(g^{-1}x)$. **Invariance** means:

$$gu * gv = g(u * v). \qquad (13)$$

This has lead to an alternative approach to group representation theory, using star-products on coadjoint orbits instead of operators, with many results already obtained. Due to lack of space we shall not develop this approach here (see [25] for a partial summary).

2.2. The deformation-quantization approach

a. Definition. Let W be a phase space (symplectic or Poisson manifold with Poisson bracket P), and $N = C^\infty(W)$ the algebra of observables. In classical mechanics the evolution of an observable $u \in N$ in a system governed by a Hamiltonian H is described by the Hamilton equations, that give $\frac{du}{dt} = P(H, u)$. We shall call **quantization** of that system a **star-product** on N, invariant under an algebra $\mathcal{O} \subset N$ of "preferred observables", with parameter $\nu = \frac{1}{2}i\hbar$.

b. Example. $W = \mathbb{R}^{2l}$ with $\mathcal{O} = \mathcal{SP}(\mathbb{R}^{2l}).\mathcal{H}_l$ (where $\mathcal{H}_l$ is the 2l+1 dimensional Heisenberg algebra) and Moyal product (6). If $\mathcal{F}$ denotes the inverse Fourier transform and (P_α, Q_α, I) are the generators of the Heisenberg Lie algebra $\mathcal{H}_l$ in the Von Neumann representation on $L^2(\mathbb{R}^l)$, satisfying $[P_\alpha, Q_\beta] = i\hbar\ \delta_{\alpha\beta} I \quad (\alpha, \beta = 1, ..., l)$ then the Weyl transform $\Omega_w(u)$ of the observable $u \in N$ with weight $w(= 1$ in the Moyal case) writes :

$$\Omega_w(u) = \int_{\mathbb{R}^{2l}} (\mathcal{F}u)(\xi,\eta) w(\xi,\eta)\ \exp((\xi.P + \eta.Q)/i\hbar)\ d^l\xi\ d^l\eta. \tag{14}$$

Ω_w maps functions into operators in $L^2(\mathbb{R}^l)$, star-products into products, star-commutators into commutators. The so-called "standard ordering" (all $q's$ on the left) is given [20] by $w(\xi,\eta) = \exp(\frac{1}{2}i\xi\eta)$ (and corresponds to the pseudodifferential calculus) while the "normal ordering" (first "$p + iq$", then "$p - iq$") is given by $w(\xi,\eta) = \exp(-\frac{1}{4}(\xi^2 + \eta^2))$. Ω_w is one-to-one on large classes of functions and operators, in particular (for $w = 1$) between $L^2(\mathbb{R}^{2l})$ and Hilbert-Schmidt operators on $L^2(\mathbb{R}^l)$; the inverse is then given by a trace :

$$u = (2\pi\hbar)^{-l}\ \mathrm{Tr}(\Omega_1(u)e^{-(\xi.P+\eta.Q)/i\hbar}). \tag{15}$$

c. Spectrality. In exact parallel with usual quantum theory, the evolution of observables in a system described by a Hamiltonian $H(\in N)$ will be here given by the star-exponential (where $(H*)^n$ means the nth $*$-power of H) :

$$Exp(Ht) = \sum_{n=0}^{\infty} \frac{1}{n!} (\frac{t}{i\hbar})^n\ (H*)^n \tag{16}$$

It is in general a distribution (and the series converges in that sense), which is locally a function. We can take (in the distribution sense) its Fourier-Stieltjes transform $d\mu$ (a distribution on W and in λ) :

$$Exp(Ht) = \int e^{\lambda t/i\hbar}\ d\mu(\lambda) \tag{17}$$

Definition. The **spectrum** of $(H/\hbar)$ is the **support** (in λ) of $d\mu(\lambda)$.

(That is exactly the definition of L. Schwartz for the spectrum of a distribution). In particular, in the case of discrete spectrum, one has :

$$Exp(Ht) = \sum_{\lambda \in S} \Pi_\lambda \, e^{\lambda t/i\hbar}, \; \Pi_\lambda \in \mathcal{A} \text{ (or } N). \tag{18}$$

S is the spectrum of $(H/\hbar)$. One has :

$$H * \Pi_\lambda = \Pi_\lambda * H = \hbar\lambda\Pi_\lambda \tag{19}$$

$$\Pi_\lambda * \Pi_\mu = \delta_{\lambda\mu}\, \Pi_\lambda, \; \sum \Pi_\lambda = 1. \tag{20}$$

which mimic the similar relations from spectral theory ; since (19) does not define uniquely Π_λ and since (20) is somewhat hard to check, it is better to use the evolution star-exponential , which contains everything "built in".

d. Applications. [7] The typical example is the harmonic oscillator (in fact, many other examples can, after some functional manipulations, be brought to it or to other generators of the invariance algebra $\mathcal{SP}(\mathbb{R}^{2l})$). Take $W = \mathbb{R}^{2l}$, $X = \alpha p^2 + \beta\, pq + \gamma\, q^2$ $(\alpha, \beta, \gamma \in \mathbb{R})$ and set $d = \alpha\gamma - \beta^2$ ($d > 0$ gives compact generators, $d < 0$ gives hyperbolic - non compact semi-simple - elements, and $d = 0$ nilpotent generators). Then the star-exponential is given by (with $\delta = |d|^{1/2}$), for the Moyal product :

$$Exp(Xt) = \begin{cases} \cos(\delta t)^{-l} \exp((X/i\hbar\delta)\tan(\delta t)) & \text{for } d > 0 \\ \exp(Xt/i\hbar) & \text{for } d = 0 \\ \cosh(\delta t)^{-l} \exp((X/i\hbar\delta)\tanh(\delta t)) & \text{for } d < 0 \end{cases} \tag{21}$$

with Fourier decomposition :

$$Exp(Xt) = \begin{cases} \sum_{n=0}^{\infty} \Pi_n^{(l)}\, e^{-(n+\frac{l}{2})t} & \text{for } d > 0 \\ \int_{-\infty}^{\infty} e^{\lambda t/i\hbar}\, \Pi(\lambda, X) & \text{for } d < 0 \end{cases}$$

We get the spectrum $(n + \frac{l}{2})\hbar$ as expected ; the "eigenprojectors" $\Pi_n^{(l)}$ are given by some special functions (cf. [7]). It is the term $\frac{l}{2}$ which gives divergences in the infinite-dimensional case, with the Moyal product ; it disappears with normal ordering (and some other "nearby" orderings), hence its use in field theory. In fact, in the **infinite-dimensional case,** one has already troubles with the definition of P^2 (a Poisson bracket P can be defined), but **normal ordering** (and some others) can be defined and permit [26] to get free fields and some models of interacting fields.

Other examples [7] include the angular momentum and the hydrogen atom ; for the latter a covariant star-product on T^*S^3 has to be introduced. It is worth mentioning [27] that **path integrals** are nothing but Fourier transform over momentum of the star-exponential $Exp(Ht)$, and that for compact groups a global coherent state formalism exists, that allows to represent the star-exponential in terms of unitary characters.

3. STAR-PRODUCTS AND QUANTUM GROUPS

3.1. Poisson-Lie groups and their "quantization"

As is well known, if G is a group, the algebra N of functions on G (with the pointwise product) is a Hopf algebra, with comultiplication $\Delta = N \to N \otimes N$ defined by the multiplication $\mu : G \times G \to G$ of the group by $\Delta f(g_1, g_2) = f(g_1 g_2)$ for $f \in N$, $g_1, g_2 \in G$ (and antipode given by the inverse on G, counit by the value at the identity). The group is called a Poisson-Lie group (cf. e.g. [11]) if the group manifold is a Poisson manifold for which μ is a Poisson map, or equivalently if

$$\Delta P(f_1, f_2) = P(\Delta f_1, \Delta f_2) \tag{22}$$

where P denotes the Poisson bracket on G or $G \times G$ defined by the Poisson structure.

By duality the enveloping algebra $\mathcal{U}(\mathcal{G})$ (the elements of which can be considered as distributions with support at the identity of G) has also a Hopf algebra structure. It is equivalent to say that G is a Poisson-Lie group and that its Lie algebra $\mathcal{G}$ is a Lie bialgebra, i.e. that there exists a Lie bracket $\varphi^* : \mathcal{G}^* \wedge \mathcal{G}^* \to \mathcal{G}^*$ (on the dual $\mathcal{G}^*$ of $\mathcal{G}$) whose dual $\varphi : \mathcal{G} \to \mathcal{G} \times \mathcal{G}$ is a 1-cocycle.

It is by now well understood that a "**quantum group**" [11] is given by a star-product on N, the algebra of functions on a Poisson-Lie group G, i.e. is a **(non-commutative) deformation of the Hopf algebra** N. By **duality** one gets [12] the so-called "**quantized universal enveloping (QUE) algebras**" $\mathcal{U}_q(\mathcal{G})$, where the coproduct is deformed. On a such an algebra N there exists an invariant star-product, defined by invariant bidifferential operators $F_r(x, y)$, i.e. by a formal series $F = I + \sum_{r=1}^{\infty} \nu^r \, F_r$; the associativity condition for the star-product gives a solution $S \in End(\mathcal{U}(\mathcal{G}) \otimes \mathcal{U}(\mathcal{G}) \, ; \, \nu)$ of the quantum Yang-Baxter equation by $S(x, y) = F^{-1}(y, x) \, F(x, y)$, and conversely [28]. Moreover, from such an invariant star-product one can [29] build an equivalent one ($*'$) such that (with the above defined coproduct Δ on N) :

$$\Delta(u *' v) = \Delta u *' \Delta v. \tag{23}$$

3.2. Star-realization of quantum groups and the uniqueness of this structure

We shall here show that, in fact, "quantum groups" are neither "groups" (which is commonly understood) nor "quantum", in the sense that the QUE algebras can be realized as star-product algebras with a star-product depending on a parameter ($\hbar$) associated with quantization on phase-space (as defined in section 2) different from the so-called quantum parameter (t)

of the QUE algebra, and that this realization fixes essentially uniquely the relations of the QUE algebra.

a. Let us start with the $\mathcal{U}_t(\mathcal{SL}(2))$ algebra. If, on $\mathbb{R}^2$, we define the functions [15, 30]

$$H = pq \quad , \quad X_+ = q/\cosh(\frac{1}{2}\hbar t)$$

$$X_- = 2(1-\cos(\eta pq))/\eta^2 \; q\hbar t \operatorname{cotanh}(\frac{1}{2}t\hbar) \quad , \quad \hbar\eta = 2\tanh(\frac{1}{2}\hbar t) \tag{24}$$

we get, with the Moyal bracket M (with parameter $\nu = \frac{1}{2}i\hbar$) :

$$M(H, X_\pm) = \pm\, X_\pm \quad , \quad M(X_+, X_-) = -\frac{1}{t}(\sin *(tH)) \tag{25}$$

The first 2 relations follow from the fact that H is a preferred observable, and the last from (21) which implies a similar relation for the sine function. The limit $\hbar \to 0$ in (25) gives a "classical" realization of the defining relations for $\mathcal{U}_t$ $(\mathcal{SL}(2))$ [30] while the limit $t \to 0$ gives a representation of $\mathcal{SL}(2)$ equivalent to the metaplectic representation of the harmonic oscillator.

b. If we look for "classical" realizations (with the Poisson brackets) of the commutation relations defining QUE algebras, in order for these realizations to be significant on the associative algebra level we need to know whether the functions of the generators that appear can be written as star-functions of the same type, i.e. we need a **strong invariance condition** of the type

$$f(*tH/\hbar) = \alpha(t)\; f(\beta(t)H/\hbar) \tag{26}$$

for some analytic functions $f(x) = \sum_0^\infty a_n\, x^n$, α and β (determined by f via (26)), where in the left hand side we take star-powers of H instead of usual powers. Moreover, if we want the quantum structure to be a deformation of the classical one, we need $a_0 = 0$ in f.

The result [30] is somewhat surprising (one relation determines 3 unknown functions) :

Theorem. *For the Moyal product the only functions satisfying (26) are:* $f(x) = a_0 + a_1 x$ (*trivial case*) or $a_1 x + a_3 x^3$ ($a_1 a_3 \neq 0$, *truncated sine*), $a_0 + a_2 x^2$ ($a_0 a_2 \neq 0$, *truncated cosine*), *or a linear combination of* $a_0 \cos(cx)$ *and* $a_1 \sin(cx)$ (*with* $a_0 c \neq 0$ *and/or* $a_1 c \neq 0$).

Thus only the sine and truncated sine satisfy the strong invariance and the deformation conditions. The sine is what was found ; it is not clear whether the truncated sine corresponds to a Hopf algebra.

c. Remarks. Note that this is true only for Moyal product ; obviously (26) does not go through an equivalence. However [30] for a family of closed star-products interpolating between standard and antistandard orderings (through Moyal), for which the symmetry is the oscillator group (generated by $p, q, 1$ and pq), the exponential is still a solution of (26), the only one except for Moyal.

For higher rank algebras, the star-product realization is indispensable for classical realizations of QUE algebras because of the cubic Serre relations that appear. Explicit examples have been given for the A_n series [30], and the method generalizes to all classical algebras.

3.3. A universal rigid model for quantum groups [16]

a. The model. Let us start with $G = SU(2)$. Denote by (π_n, V_n) its $2n+1$-dimensional representation ($2n$ an integer) by matrices in $\mathcal{L}(V_n)$. The direct product $A = \Pi_{n=0}^{\infty} \mathcal{L}(V_n)$, endowed with the product topology and algebra law, becomes a Fréchet algebra into which both G and the enveloping algebra $\mathcal{U}(\mathcal{G})$ can be imbedded by $x \mapsto (\pi_n(x))$. These imbeddings are total because A is the bicommutant of the direct sum representation $\pi = \sum \pi_n$ on $V = \sum V_n$, and every π_n extends by continuity to a representation of A.

b. Universality. *If $t \notin 2\pi\mathbf{Q}$, the algebras $\mathcal{U}_q(\mathcal{G})$ can be imbedded into a dense subalgebra A_t of A ($q = e^{it}$) ; π_n are still a complete set of representations of A_t, and the Hopf structures (coproduct Δ_t, counit ϵ_t, and antipode J_t) on $\mathcal{U}_q(\mathcal{G})$ have unique extensions to topological Hopf structures (with equivalent coproducts) on A, when $A \otimes A$ is endowed with the completed projective tensor topology so that*

$$A \otimes A \approx \Pi_{n,p}\, \mathcal{L}(V_n \otimes V_p). \tag{27}$$

The classical limit is $\mathcal{U}_1(\mathcal{G}) \approx \mathcal{U}(\mathcal{G}) \otimes \theta$ where $\theta^2 = 1$ (a parity), with coproduct Δ_0 equivalent to Δ_t : there exists $P(t) \in A \otimes A$ such that $\Delta_t = P(t)\, \Delta_0\, P(t)^{-1}$. The R-matrix is $R(t) = P(-t)\, P(t)^{-1}$ and can be chosen so that A_t is a quasitriangular Hopf algebra.

c. The star-product. The topological dual A^* is isomorphic to $\oplus \mathcal{L}(V_n) = H$, endowed with the inductive topology, and can thus be viewed as polynomial functions on the complex extension $G_{\mathbf{C}}$ of G ; as Hopf algebra,

$$H \approx \mathbf{C}[a, b, c, d]/(ad - bc = 1)$$

where the product is that of functions and the coproduct is defined by the product in G. Each co-associative coproduct Δ_t induces an associative star-product on H such that $(a+d)^{*k} = (a+d)^k$, $\forall k \in \mathbf{N}$, and conversely if this is true for a $*$-product compatible with the usual coproduct Δ_0, this $*$ is induced by a coproduct $\Delta_t = P_t \Delta_0 P_t^{-1}$ on A : *we recover Manin's description of functions of non-commutative arguments.*

d. Rigidity as bialgebra. Let B be a bialgebra, an associative algebra B with coproduct $\Delta : B \to B \otimes B$. We then define [31] in a natural way the notion of isomorphism of bialgebras (associative algebra isomorphism with equivalence of coproducts), and of deformations of such structures, where we denote by $H(B, \Delta)$ the relevant 2-cohomology.

It can be shown [16] that if (for Hochschild cohomologies)

$$\tilde{H}^2(B) = 0 = \tilde{H}^1(B, B \otimes B)$$

then $H(B, \Delta) = 0$. Whitehead's lemmas therefore prove that (A, Δ_t) **is rigid for all** $t \notin 2\pi\mathbf{Q}$.

4. STAR-PRODUCTS AND CYCLIC COHOMOLOGY

4.1. Cyclic cohomology [14]

a. Hochschild cohomology. We shall consider it, to fix ideas, in our context. We take the associative algebra $N = C^\infty(W)$, where W is a symplectic manifold of dimension $2l$ with 2-form ω, and its extension $\mathcal{A} = N[[\nu]]$ to formal series. If M is a N-module, one can define the *Hochschild cohomology* with cochains $C^p(N, M)$ and the coboundary operation b (we take $f_0, ..., f_p \in N$) :

$$b\,C(f_0, ..., f_p) = f_0 C(f_1, ..., f_p) - C(f_0 f_1, f_2, ..., f_p) + ...$$
$$+(-1)^p\, C(f_0, ..., f_{p-1} f_p) + (-1)^{p+1} C(f_0, ..., f_{p-1}) f_p. \qquad (28)$$

Now to every $f \in \mathcal{A}$ one can associate an element $\tilde{f}$ of its dual $\mathcal{A}^*$ defined by $\tilde{f} : g \mapsto \int fg\,\omega^l$. This gives by composition a map

$$C^p(\mathcal{A}, \mathcal{A}) \to C^p(\mathcal{A}, \mathcal{A}^*)$$

where $\mathcal{A}$ acts on $\mathcal{A}^*$ by $(x\varphi y)(a) = \varphi(yax)$, where $a, x, y \in \mathcal{A}$ and $\varphi \in \mathcal{A}^*$. We shall denote by $\tilde{C}_r$ the image in $C^p(\mathcal{A}, \mathcal{A}^*)$ of $C_r \in C^p(\mathcal{A}, \mathcal{A})$. Obviously one has $b\tilde{C}_r = \widetilde{bC_r}$.

b. The bicomplex. We define a bicomplex by

$$C^{n,m} = C^{n-m}(\mathcal{A}, \mathcal{A}^*)$$

the space of $(n-m+1)$ linear forms ψ on $\mathcal{A}$, defined as $\{0\}$ for $n < m$. In this case the Hochschild coboundary b is of degree 1 and takes the form $(a_i \in \mathcal{A})$:

$$b\,\psi(a_0, ..., a_{n+1}) = \sum_{j=0}^{n}(-1)^j\ \psi(a_0, ..., a_j a_{j+1}, ..., a_{n+1}) \qquad (29)$$

$$+(-1)^{n+1}\ \psi(a_{n+1}a_0, ..., a_n)$$

In this bicomplex we consider another operation B of degree -1, defined as $B = A_S B_0$ where

$$(B_0\psi)(a_0, ..., a_{n-1}) = \psi(1, a_0, ..., a_{n-1}) + (-1)^n\ \psi(a_0, ..., a_{n-1}, 1) \qquad (30)$$

$$(A_S\psi)(a_0, ..., a_{n-1}) = \sum_{j=0}^{n-1}(-1)^{(n-1)j}\psi(a_j, a_{j+1}, ..., a_{n-1}, a_0, ..., a_{j-1}). \quad (31)$$

A_S is the cyclic antisymmetrization. In particular

$$B\,\tilde{C}(f, g) = C(1, f, g) - \tilde{C}(1, g, f)$$

when $\tilde{C} \in C^2(\mathcal{A}, \mathcal{A}^*)$ is a normalized 2-cochain.

As expected, one has $Bb = -bB$ and $B^2 = 0 = b^2$.

c. The cyclic cohomology $HC^n(\mathcal{A})$ of $\mathcal{A}$ is the cohomology of the complex (C^n_λ, b) where $C^n_\lambda(\mathcal{A}) \subset C^n(\mathcal{A}, \mathcal{A}^*)$ is composed of the cochains φ satisfying the cyclicity condition

$$\varphi(a_1, ..., a_n)(a_0) = (-1)^n\ \varphi(a_2, ..., a_n, a_0)(a_1).$$

A basic property of cyclic cohomology is that, at each level n :

$$HC^n(\mathcal{A}) = (ker\ b \cap ker\ B)/b(ker\ B). \qquad (32)$$

4.2. Closed star-products

a. Definition [9]. A star-product $*$ defined by (1) is **closed** if (whenever defined)

$$\int_W C_r(u, v)\omega^l = \int_W C_r(v, u)\omega^l \quad \forall r \leq l\ ;\ u, v \in N. \qquad (33)$$

Equivalently, if $u = \sum \nu^k u_k \in \mathcal{A}$, we can consider the coefficient $a_l(u, v) = \sum_{r+j+k=l} C_r(u_k, v_j)$ of ν^l in $u * v$. Then (33) can be written :

$$\int a_l(u, v)\omega^l = \int a_l(v, u)\omega^l \qquad (34)$$

If (33) is true for all r, we get (8) and the star-product is said **strongly closed.**

b. Motivation. In the notations of section (2.2), for Moyal ordering, one has :

$$Tr(\Omega_1(u)) = (2\pi\hbar)^{-l} \int_{\mathbb{R}^{2l}} u(x)dx \equiv Tr_M(u) \tag{35}$$

for $u \in N$ or $\mathcal{A}$. Thus we want to be able to define a *trace* on the algebra $\mathcal{A}$, in the general case, by *integration* over W ; or a least that this will make sense in the classical limit $\hbar \to 0$, i.e. that integration over W of the coefficient of ν^l will have the properties of a trace : this is what (34) expresses.

c. Existence. As we saw in section (2.1.d), on any symplectic manifold *there exists* a *strongly closed* star-product, obtained by "gluing" together locally defined Moyal products. A necessary condition for a star-product to be closed is that (33) will be true for $r = 2$ (it is sufficient when $l = 2$, i.e. $dim\ W = 4$), i.e. that $B\tilde{C}_2 = 0$. Now if we have a given Hochschild 2-cocycle C_1 (e.g. $C_1 = P$) we get from standard deformation theory [9] a 3-cocycle E_2, which has to be the 3-coboundary of the 2-cochain C_2. Since $bE_2 = 0$ we get $b\tilde{E}_2 = 0$, thus $B\tilde{E}_2 = 0$, and $b\tilde{C}_2 = \tilde{E}_2 \in ker\ b \cap ker\ B$. Obviously adding an element of $ker\ B$ to $\tilde{C}_2$ will not affect the results and from (32) we get that *at order 2, the obstructions to existence of closed star-products are classified by* $HC^3(\mathcal{A})$. In the same way [9] the same holds *at each order* of deformation.

d. Equivalence. If we look for a series (3) intertwining between two closed star-products $*$ and $*'$, we get that at each order t one must have $b\,T_t = C'_t - C_t + G_t \equiv H_t$ where C_t (resp. C'_t) define $*$ (resp. $*'$) and G_t is defined by C'_r, C_r and T_r with $r < t$. Equivalence to order $t-1$ assures that $bH_t = 0$. We therefore see that $H^2(N)$ classifies the obstructions to equivalence at each order. Similarly, $b\tilde{H}_t = 0 = B\tilde{H}_2$, thus $b\tilde{T}_t \in ker\ b \cap ker\ B$ and again, due to (32), the *obstructions to equivalence of closed star-products are classified, at each order, by* $HC^2(\mathcal{A})$.

e. Example of a non-closed star-product. A cochain C_2 defined by $< C, f_0\, df_1 \wedge df_2 >$, where C is a current with $dC \neq 0$, will give $B\tilde{C}_2 \neq 0$, thus a non-closed star-product.

4.3. CHARACTER OF A CLOSED STAR-PRODUCT.

a. Definition. Generalizing (35), if $f = \sum_{r=0}^{\infty} \nu^r f_r \in \mathcal{A}$, we can define a generalized **trace** by :

$$\tau(f) = \int_W f_l \, \omega^l. \tag{36}$$

The deviation from algebra homomorphism of the identity map $\mathcal{A}_0 \to \mathcal{A}_\nu$ between $\mathcal{A}$ endowed with commutative product and $\mathcal{A}$ endowed with $*_\nu$ product is expressed by

$$\theta(f_1, f_2) = f_1 * f_2 - f_1 \, f_2. \tag{37}$$

As for operator algebras [14] we can thus define the components φ_{2k} of a cocycle φ in the (B, b) bicomplex ($\varphi_{2k} = 0$ except for $l \leq 2k \leq 2l$) by :

$$\varphi_{2k}(f_0, f_1, ..., f_{2k}) = \tau(f_0 * \theta(f_1, f_2) * ... * \theta(f_{2k-1}, f_{2k})). \tag{38}$$

Definition. The **character** of the closed star-product $*$ is the cohomology class of the cocycle φ.

b. Properties. For the highest order term φ_{2l} and a closed star-product we therefore get

$$\varphi_{2l}(f_0, f_1, ..., f_{2l}) = \int f_0 \, P(f_1, f_2) ... P(f_{2l-1}, f_{2l}) \omega^l \tag{39}$$

$$= \int f_0 \, df_1 \wedge df_2 \wedge ... \wedge df_{2l-1} \wedge df_{2l}.$$

c. If $dim\, W = 4$, elementary calculations show that, in addition to φ_4 given by (39) one has $\varphi_0 = 0$ and $\varphi_2 = \tilde{C}_2$. Thus one gets [9] : $b\, \varphi_2 = -\frac{1}{2} B\, \varphi_4$. Since $HC^2(N) = Z_2(W, \mathbf{C}) \oplus \mathbf{C}$ (where Z_2 stands for de Rham closed 2-currents on W), we see that modification of C_2 with a closed 2-current will affect the corresponding class in HC^4 and therefore the class of the cocycle φ, which has thus no reason to be an integer (as required in the geometric quantization approach).

4.4. PSEUDODIFFERENTIAL CALCULUS. APPLICATIONS

a. Closedness and character. Let $W = T^*M$ where M is a compact Riemannian manifold of dimension l. If D is a compactly supported pseudodifferential operator on $C^\infty(\mathbb{R}^l)$, its symbol can be defined by [32]

$$\tau_D(q_0, p) = D\, e^{i(q-q_0)-p}|_{q=q_0} \tag{40}$$

and the symbol of a product of two pseudodifferential operators is given by

$$\tau_{DE}(q,p) = \sum_{r=0}^{\infty} \frac{i^{-r}}{r!} \frac{\partial^r \tau_D}{\partial p^r} \frac{\partial^r \tau_E}{\partial q^r} \equiv \tau_D * \tau_E \tag{41}$$

which is the star-product of standard ordering, equivalent to Moyal by an equivalence that preserves closedness. Widom [32] extended both definitions globally to any T^*M, and the closedness of this star-product can be checked directly [9]. If we denote by $\Omega_S(u)$ a pseudodifferential operator (on $C^\infty(W)$) with symbol u, formula (35) becomes :

$$Tr(\Omega_S(u)) = (2\pi\hbar)^{-l} \int_W u\omega^l + O(\hbar^{1-l}). \tag{42}$$

From (36) and (38) we therefore get $(l \leq 2k \leq 2l)$

$$\varphi_{2k}(f_0, f_1, ..., f_{2k}) = Tr(\Omega_S(f_0)\Theta(f_1, f_2)...\Theta(f_{2k-1}, f_{2k})) \tag{43}$$

where $\Theta(f_{2j-1}, f_{2j}) = \Omega_S(f_{2j-1})\Omega_S(f_{2j}) - \Omega_S(f_{2j-1}f_{2j}) \quad (j = 1, ..., k)$.

b. Applications. Combining this with index theorems [14] one gets [9] :

Theorem. *The star-product of symbols given by the pseudodifferential calculus on $W = T^*M$ (M compact Riemannian) is closed and its character φ coincides with the character defined by the trace on the operator level. Thus $\varphi \in HC^{ev}(T^*M)$ and it is given by $Td(T^*M)$ as a current over T^*M.*

As long as a closed star-product has an associated Weyl mapping that gives, for suitable functions, the compact operators, the trace is necessarily proportional to the unique trace on these operators. Thus the cyclic cocycle φ is proportional to an integral one. It is however clear from the definition of φ that this integrality condition need not be satisfied. In this way one can get examples where there are obstructions to the traditional (Bohr-Sommerfeld) quantization, but where the more general framework of deformation-quantization (with a closed star-product) can be applied. And these can be related to generalized index theorems, by finding an explicit formula for the cyclic cohomology class of the character of a closed star-product in terms of Poisson geometry. This opens the door to a "real" (not necessarily integral) index.

In particular, since in the case of Moyal, integration over W corresponds exactly (cf. (35)) to the trace, not asymptotically as in the pseudodifferential calculus (cf. (42)), a more careful study of this "symmetric pseudodifferential" calculus seems appropriate. And since in the infinite-dimensional case (field theory) normal ordering, or orderings "close to it", are the relevant ones if one wants to avoid divergences, the corresponding "complex pseudodifferential" calculus should be the appropriate framework in the case of infinite-dimensional symplectic manifolds.

References

[1] H. Weyl, Z. Phys. **46** (1927), 1 ; Gruppentheorie und Quantenmechanik, Hirzel Verlag, Leipzig (1928) (and Dover, NY (1931)).
[2] E. Wigner, Phys. Rev. **40** (1932), 749.
[3] J.E. Moyal, Proc. Cambridge Phil. Soc. **45** (1949), 99.
[4] H.J. Groenewold, Physica **12** (1946), 405.
[5] M. Flato, A. Lichnerowicz and D. Sternheimer, Compositio Mathematica **31** (1975) 47.
[6] J. Vey, Comment. Math. Helvet. **50** (1975), 421.
[7] F. Bayen, M. Flato, C. Fronsdal, A. Lichnerowicz and D. Sternheimer, Ann. Phys. **111** (1978), 61, 111 ; Lett. Math. Phys. **1** (1977), 521.
[8] A.P. Calderon and A. Zygmund, Amer. J. Math. **79** (1957) , 901.
[9] A. Connes, M. Flato and D. Sternheimer, Lett. Math. Phys. **24** (1992), 1.
[10] P.P. Kulish and N. Yu. Reshetikhin, Zap. Nauchn. Sem. LOMI **101** (1981), 101.
[11] V.G. Drinfeld, Proc. Int. Congress Math. Berkeley (1986), **1**, 798.
[12] M. Jimbo, Lett. Math. Phys. **10** (1985), 63.
[13] S.L. Woronowicz, Commun. Math. Phys. **111** (1987), 613.
[14] A. Connes, Publ. Math. IHES **62** (1985), 41 ; "Géométrie non commutative", Interéditions, Paris (1990) (English expanded version in 1993).
[15] M. Flato and D. Sternheimer, Lett. Math. Phys. **22** (1991), 155.
[16] P. Bonneau, M. Flato and G. Pinczon, Lett. Math. Phys. **25** (1992), 75.
[17] M. Gerstenhaber, Ann. Math. **79** (1964), 59 ; M. Gerstenhaber and S.D. Schack. Proc. Nat. Acad. Sci. USA **87** (1990), 478.
[18] H. Omori, Y. Maeda and A. Yoshioka, Lett. Math. Phys. **26** (1992), 285; Adv. in Math. **85** (1991), 224.
[19] M. De Wilde and P. Lecomte, Note Mat. Vol. X (1992) ; Lett. Math. Phys. **7** (1983), 487.
[20] G.S. Agarwal and E. Wolf, Phys. Rev. D **2** (1970), 2161.
[21] M. Cahen, M. Flato, S. Gutt and D. Sternheimer, J. Geom. Phys. **2** (1985), 36.
[22] H. Basart and A. Lichnerowicz, C. R. Acad. Sci. Paris **293** I (1981), 347.
[23] H. Basart, M. Flato, A. Lichnerowicz and D. Sternheimer, Lett. Math. Phys. **8** (1984), 483.
[24] D. Arnal, J.C. Cortet, P. Molin and G. Pinczon, J. Math. Phys. **24** (1983), 276.
[25] D. Sternheimer, AMS Lectures in Applied Math. **21** (1985), 255. Sem. Math. Sup. Montréal **102** (1986), 260.
[26] J. Dito, Lett. Math. Phys. **20** (1990), 125 and **27** (1993) ; J. Math. Phys. **33** (1992), 791.
[27] P. Sharan, Phys. Rev. D **20** (1979), 414. C. Cadavid and M. Nakashima. Lett. Math. Phys. **23** (1991), 111.
[28] C. Moreno and L. Valero, J. Geom. Phys. **9** (1992), 369.
[29] L. Takhtajan, Introduction to Quantum Groups in : Springer Lecture Notes in Physics **370** (1990), 3. Lectures on Quantum Groups in : Nankai Lectures on Math. Phys. (M. Ge and B. Zhao eds.), World Scientific (1990), 69.
[30] M. Flato, Z.C. Lu and D. Sternheimer, Foundations of Physics (1993). M. Flato and Z.C. Lu. Lett. Math. Phys. **21** (1991), 85. Z.C. Lu. J. Math. Phys. **33** (1992), 446.
[31] P. Bonneau, Lett. Math. Phys. **26** (1992), 277.
[32] H. Widom, Bull. Sci. Math. (2), **104** (1980), 19.

Quantum Group Symmetries in Conformal Field Theory

KRZYSZTOF GAWEDZKI
CNRS, IHES, 91440 Bures-sur-Yvette, France

Abstract. Quantum groups play the role of hidden symmetries of some two-dimensional field theories. We discuss how they appear in this role in the Wess-Zumino-Witten model of conformal field theory.

1. WZW model

Wess-Zumino-Witten (WZW) model [1] occupies an important place in the conformal field theory: it is a model of two-dimensional quantum massless fields exactly soluble for energy spectrum and correlation functions; it is also a generating theory for a rich family of other conformal field theories which may be obtained from the WZW one by various versions of the so called coset construction [2][3]. It is a $1+1$-dimensional analogue of $0+1$-dimensional particle on the group G (for simplicity, we shall assume G to be a compact matrix group). The motion of the particle on G is described by the classical equation

$$\partial_t(g\partial_t g^{-1}) = 0. \tag{1}$$

On the quantum level, the system may be solved by harmonic analysis on G. The quantum space of states is $L^2(G)$ and carries the representation of $G \times G$.

$$L^2(G) = \bigoplus_\lambda V_\lambda \otimes \overline{V}_\lambda \tag{2}$$

where V_λ denotes the irreducible highest weight (HW) λ representation of G. The Hamiltonian is proportional to the quadratic Casimir operator. The multiplication by matrix elements of $g \in G$ may be expressed in the realization (2) of $L^2(G)$ as a bilinear combination of Clebsch-Gordan coefficients for the tensor product with the fundamental representation, i.e. as a bilinear expression in intertwiners of representations of G.

The WZW model may be viewed as describing the particle on the loop group LG. The classical equation of motion is

$$\partial_-(g\partial_+ g^{-1}) = 0 \tag{3}$$

H. Araki et al. (eds.), Quantum and Non-Commutative Analysis, 239–251.

where $\partial_\pm \equiv \partial/\partial(x^1 \pm x^0)$ and we shall consider the cylindrical geometry with x^1 taken mod 2π. The quantum theory is solved by harmonic analysis on the Kac-Moody group $\widehat{LG}$, the central extension of the loop group. The space of states is

$$\bigoplus_{\lambda \text{ int.}} V_{k,\lambda} \otimes \overline{V}_{k,\lambda} \tag{4}$$

where the level k, a fixed positive integer, is the value of the central charge of $\widehat{LG}$ and λ runs through the finite (k-dependent) set of the so called integrable highest weights. The symmetry of the model is $\widehat{LG} \times \widehat{LG}$ where the factors act on, respectively, left- and right-moving degrees of freedom. The spaces $V_{k,\lambda}$ and $\overline{V}_{k,\lambda}$ combine diagonally in eq. (4) due ot the coupling between the left- and right-movers which share a finite number of degrees of freedom. The field operators of the WZW theory are bilinear combinations of intertwiners of representations of $\widehat{LG}$.

There are several ways to see manifestations of the hidden quantum group symmetry in the WZW model[1]. The labeling of the representations $V_{k,\lambda}$ is similar to that of the "good" representations [12] of the Drinfeld-Jimbo quantum group [13][14] $\mathcal{U}_q(\mathcal{G})$ for $q = \mathrm{e}^{\pi i/(k+h^\vee)}$ ($h^\vee$ is the dual Coxeter number of $\mathcal{G}$, the Lie algebra of G). The exchange algebra of the intertwining operators is given by the $6j$-symbols of $\mathcal{U}_q(\mathcal{G})$ [15]-[17]. In the free field description of the WZW theory [18][19], the quantum group describes the homology of the contours of screening charge integrals [20]-[23]. Let $j(x)$ be the current giving the infinitesimal version of a representation π of $\widehat{LG}$ by

$$\pi(X) = \int_0^{2\pi} \operatorname{tr} j(s)\, X(s)\, ds \tag{5}$$

for $X \in L\mathcal{G}$. In representation of central charge k, current j satisfies the commutation relations

$$[j(x)_1, j(y)_2] = (j(x)_1 C - C j(y)_2)\, \delta(x-y) - (ik/2\pi)\, C\, \delta'(x-y) \tag{6}$$

where $j_1 \equiv j \otimes Id$, $j_2 \equiv Id \otimes j$ and $C = \sum t^a \otimes t^a \in \mathcal{G} \otimes \mathcal{G}$, $\operatorname{tr} t^a t^b = \delta^{ab}$. Eq. (6) may be viewed as the defining relation of the Kac-Moody algebra $\widehat{L\mathcal{G}}$. Various aspects of the $\mathcal{U}_q(\mathcal{G})$ symmetry in the WZW model may be formally explained by postulating that the monodromy operators

$$M(x) = P\, \mathrm{e}^{\int_x^{x+2\pi} j(s)\, ds} \tag{7}$$

[1] See also [4]-[6] for the related work concerning the minimal or Liouville theories and [7]-[9] for an attempt at a more general, less model dependent approach relating the quantum-group-like symmetries to the theory of superselection sectors and field statistics in the spirit of [10][11]

satisfy the $\mathcal{U}_q(\mathcal{G})$ relations

$$M_1 R^- M_2 (R^+)^{-1} = R^- M_2 (R^+)^{-1} M_1 \tag{8}$$

where $R^\pm$ is a pair of solutions of the quantum Yang-Baxter equation

$$R^\pm_{12} R^\pm_{13} R^\pm_{23} = R^\pm_{23} R^\pm_{13} R^\pm_{12}. \tag{9}$$

Operators (7) appeared (implicitly) in [24] and were explicitly discussed as generators of $\mathcal{U}_q(\mathcal{G})$ in [25]. The problem is that, as written above, they are very singular objects due to the short distance singularity of currents $j(x)$.

2. Lattice Kac-Moody algebra

A possible way out from the above difficulty has been proposed in a series of papers [26]-[29] which introduced the lattice version of the Kac-Moody algebra[2]. This allows to build the regularized version of the WZW model preserving essentially all the symmetries of the continuum WZW theory and at the same time making explicit the quantum group symmetry hidden in the continuum version of the theory.

View $\mathbf{Z}_N$ as an N-point lattice in S^1. The lattice Kac-Moody algebra $\mathcal{K}_N$ proposed in [26][27] is given by matrix generators $J(n)$, $n \in \mathbf{Z}_N$, satisfying the quadratic relations

$$\begin{aligned} J(n)_1 J(n)_2 &= R^+ J(n)_2 J(n)_1 R^-, \\ J(n)_1 R^- J(n-1)_2 &= J(n-1)_2 J(n)_1, \\ J(n)_1 J(m)_2 &= J(m)_2 J(n)_1 \quad \text{for } |n-m|>1. \end{aligned}$$

The other relations are the deformations of the $det = 1$ conditions for $SL(n)$ etc. For $SL(2)$ they read $J_{11}(n)J_{22}(n) - q^{-1}J_{21}(n)J_{12}(n) = q^{1/2}$ (here the subscripts "ij" refer to the matrix element of $J(n)$). The $SL(2)$ R-matrices are

$$R^+ = q^{1/2}\begin{pmatrix} q^{-1} & 0 & 0 & 0 \\ 0 & 1 & q^{-1}-q & 0 \\ 0 & 0 & 1 & 0 \\ 0 & 0 & 0 & q^{-1} \end{pmatrix}, \quad R^- = q^{-1/2}\begin{pmatrix} q & 0 & 0 & 0 \\ 0 & 1 & 0 & 0 \\ 0 & q-q^{-1} & 1 & 0 \\ 0 & 0 & 0 & q \end{pmatrix}.$$

Operators $J(n)$ should be thought of as regularized versions of parallel transport operators $P\,\exp[(k+h^\vee)^{-1}\int_{2\pi n/N}^{2\pi(n+1)/N} j(s)\,ds]$ [26].

Algebra $\mathcal{K}_N$ contains quantum group $\mathcal{U}_q(\mathcal{G})$ as the monodromy which is now, unlike in eq. (7), defined by a regular expression:

$$M(n) = J(n+N-1)J(n+N-2)\cdots J(n). \tag{10}$$

[2] See [30]-[33] for related proposals concerning the Liouville and Toda theories

$M(n)$ satisfy relations (8). For the $SL(2)$ case, we may write

$$M = q^{3/2}\begin{pmatrix} q^{-S^3} & (q^{-1}-q)S^+ \\ 0 & q^{S^3}\end{pmatrix}\begin{pmatrix} q^{S^3} & 0 \\ (q-q^{-1})S^- & q^{-S^3}\end{pmatrix} \tag{11}$$

where

$$\begin{aligned}[S^3, S^\pm] &= \pm S^\pm, \\ [S^+, S^-] &= (q^{2S^3} - q^{-2S^3})/(q-q^{-1}).\end{aligned}$$

Algebras $\mathcal{K}_N$ may be also viewed as interpolating between quantum group $\mathcal{U}_q(\mathcal{G})$ and the enveloping algebra $\mathcal{U}(\widehat{L\mathcal{G}})$ when N goes from 1 to ∞.

3. Conformal invariance on lattice

The algebraic manifestation of the conformal invariance underlying the continuum Kac-Moody algebras is the action of the orientation-preserving diffeomorphisms of the circle $D \in \mathrm{Diff}_+S^1$ on the currents

$$j(x) \mapsto \tfrac{dD(x)}{dx}\, j(D(x))$$

which induces automorphisms of $\widehat{L\mathcal{G}}$. The latter are unitarily implementable in the HW representations:

$$\tfrac{dD(x)}{dx}\, j(D(x)) = U_D\, j(x)\, U_D{}^{-1}.$$

$D \mapsto U_D$ is a projective representation of Diff_+S^1 giving, on the infinitesimal level, a representation of the Virasoro algebra.

This structure essentially descends to the lattice. Let

$$D : \mathbf{Z} \to \mathbf{Z}$$

be an increasing map such that $D(n+N') = D(x) + N$ $(N' \le N)$. Thus D describes blocking of intervales of lattice $\mathbf{Z}_N$ into those of $\mathbf{Z}_{N'}$. D induces a "block spin" homomorphism of the lattice Kac-Moody algebras $\mathcal{D} : \mathcal{K}_{N'} \to \mathcal{K}_N$

$$\mathcal{D}(J(n)) = J(D(n+1)-1)\cdots J(D(n)+1)J(D(n)). \tag{12}$$

On the lattice, the conformal transformations are represented by a (local) renormalization group!

If we relabel algebras $\mathcal{K}_N$ by arbitrary triangulations T of S^1 (i.e. splittings of S^1 into intervals l_n) renaming the generators as $J(l_n)$ then the block spin homomorphisms give rise to homomorphisms $\iota_{T',T} : \mathcal{K}_T \to \mathcal{K}_{T'}$

for T' finer than T. One may define the continuum limit algebra[3] $\mathcal{K}_\infty$ as the inductive limit of algebras $\mathcal{K}_T$. Any $D \in \mathrm{Diff}_+S^1$ defines an isomorphism $\mathcal{D} : \mathcal{K}_T \to \mathcal{K}_{D(T)}$ which descends to $\mathcal{K}_\infty$. As a result, Diff_+S^1 acts by automorphisms on $\mathcal{K}_\infty$ showing that the block spin homomorphisms indeed encode the conformal invariance.

In continuum, the generators of the Virasoro algebra implementing conformal invariance in the HW representations of the Kac-Moody algebra are given in terms of current $j(x)$ by the Sugawara construction. On the lattice, the block spin homomorphisms $\mathcal{D}$ may be also implemented in the class of representations that we shall study below. If D is a rigid rotation of $\mathbf{Z}_N$ then we may expect that the implementing maps are expressible in terms of generators of $\mathcal{K}_N$. Such lattice Sugawara expressions for the (Minkowski time) transfer matrix are not known yet except for the $U(1)$ case, see below.

4. Free field representations of the Kac-Moody algebras

In continuum, there are various ways to approach the construction of the HW representations $V_{k,\lambda}$ of the Kac-Moody algebras.

1. In the algebraic approach [34] one starts from the concept of Verma modules of $\widehat{L\mathcal{G}}$ and analyzes their reducibility studying the structure of singular vectors with the Kac-Kazhdan determinant formula, constructing the BGG resolution etc.
2. In the geometric Borel-Weil type approach [35] one constructs $V_{k,\lambda}$ as the space of holomorphic sections of a line bundle over LG/T (T is the Cartan subgroup of G).
3. Finally, more recently, another algebraic approach to the HW representations of $\widehat{L\mathcal{G}}$ has been obtained [18][19][36][37] by representing current $j(x)$ by free fields and constructing $V_{k,\lambda}$ as a cohomology of a complex of Fock space (Wakimoto) modules of the Kac-Moody algebra.

From those three approaches at least the last one carries over to the lattice so let us sketch some of its essential points. We shall stick to the $SL(2)$ case. The free fields one uses to represent $j(x)$ are chiral scalar field $\phi(x)$ satisfying

$$[\phi(x), \phi(y)] = \tfrac{\pi i}{2(k+2)} \operatorname{sgn}(x-y) \tag{13}$$

or the corresponding $u(1)$ current $\partial\phi$ ($\phi(x+2\pi) = \phi(x) + \pi r/(k+2)$ where r is the momentum) and the $\beta\gamma$ system $\beta(x)$, $\gamma(x)$ (periodic in x) s.t.

$$[\beta(x), \gamma(y)] = \tfrac{2\pi i}{k+2}\, \delta(x-y) \tag{14}$$

[3] Its generators are $P\exp[\int_x^y (k+h^\vee)^{-1} j(s)\,ds]$ which are singular in the standard continuum Kac-Moody algebra

(all other commutators vanish). These fields act in the standard Fock spaces $\mathcal{F}_r$ (labeled by the eigenvalues of r).

$$j = \frac{k+2}{2\pi}\begin{pmatrix} i\partial\phi + :\beta\gamma: & \partial\beta - 2i\beta\partial\phi - :\beta^2\gamma: \\ \gamma & -i\partial\phi - :\beta\gamma: \end{pmatrix} \tag{15}$$

gives the $L\widehat{sl}(2)$ currents and turns the Fock spaces into the (Wakimoto) $L\widehat{sl}(2)$-modules. For $r = 1, \ldots, k+1$, using the (regularized) powers of the screening charge integral

$$Q(x) = \int_x^{x+2\pi} \gamma(s) : \mathrm{e}^{2i\phi(s)} : ds\,, \tag{16}$$

one obtains the following (x-independent) complex of Fock space modules ($p \equiv k+2$)

$$\cdots \overset{Q^r}{\rightarrow} \mathcal{F}_{-r+2p} \overset{Q^{p-r}}{\longrightarrow} \mathcal{F}_r \overset{Q^r}{\rightarrow} \mathcal{F}_{-r} \overset{Q^{p-r}}{\longrightarrow} \mathcal{F}_{r-2p} \overset{Q^r}{\rightarrow} \cdots \tag{17}$$

whose cohomology is concentrated at $\mathcal{F}_r$ and provides the HW representation of $L\widehat{sl}(2)$ of level k and spin $\frac{1}{2}(r-1)$ [19]. The intertwining chiral fields are

$$h = \begin{pmatrix} \beta : \mathrm{e}^{-i\phi} : Q & \beta : \mathrm{e}^{-i\phi} : \\ : \mathrm{e}^{-i\phi} : Q & : \mathrm{e}^{-i\phi} : \end{pmatrix}. \tag{18}$$

Each raw of the matrix defines an operator between the complexes (17) which projected onto their cohomology gives the component with magnetic number $\pm 1/2$ of the spin $1/2$ primary field of the WZW model [38]. The real fields of the WZW theory are bilinear combinations of fields h and their right-moving partners.

5. Lattice free fields

One may build the representations of the lattice Kac-Moody algebra $\mathcal{K}_N$ using (deformed) free fields. Below, $q^{1/2} \equiv \mathrm{e}^{\pi i/(2(k+2))}$ so that q is a primitive $2p$-th root of unity. We shall need lattice versions of the $u(1)$ current and of the $\beta\gamma$ system.

5.1. $u(1)$ CURRENT

The lattice $u(1)$ algebra $\mathcal{U}_N$ is generated by invertible elements $\Theta(n)$, $n \in \mathbf{Z}_N$, with the relations

$$\Theta(n)\,\Theta(n+1) = q^{1/2}\,\Theta(n+1)\,\Theta(n) \tag{19}$$

(all other generators commute). Think about $\Theta(n)$ as of a regularized version of $\exp[i\int_{2\pi n/N}^{2\pi(n+1)/N}\partial\phi(s)\,ds]$. $\mathcal{U}_N$ becomes a $*$-algebra if we put $\Theta(n)^* = \Theta(n)^{-1}$.

Consider, for N odd, the representations of $\mathcal{U}_N$ acting in the space spanned by orthonormal vectors $|\underline{\alpha}\rangle$ where $\underline{\alpha} \equiv (\alpha_0, \alpha_2, \ldots, \alpha_{N-1})$, $\alpha_{2n} \in \mathbf{Z}_{4p}$, $\sum \alpha_{2n} = 0$:

$$\begin{aligned}
\Theta(2n)\,|\underline{\alpha}\rangle &= z_{2n}\, q^{\alpha_{2n}/2}\,|\underline{\alpha}\rangle \quad \text{for } 2n < N-1,\\
\Theta(2n+1)\,|\underline{\alpha}\rangle &= z_{2n+1}\,|(\alpha_0, \ldots, \alpha_{2n}+1, \alpha_{2n+2}-1, \ldots \alpha_{N-1})\rangle,\\
\Theta(N-1)\,|\underline{\alpha}\rangle &= z_{N-1}\, q^{\alpha_{N-1}/2+(N+1)/4}\,|(\alpha_0 - 1, \alpha_2, ..\\
&\qquad\qquad \ldots, \alpha_{N-3}, \alpha_{N-1}+1)\rangle
\end{aligned}$$

where $|z_n| = 1$.

Proposition. [39] *The above formulae give irreducible $*$-representations of $\mathcal{U}_N$. They are equivalent iff they correspond to the same eigenvalues of the central elements $\Theta(n)^{4p} = z(n)^{4p}$ and $\Pi \equiv q^{-1/2}\Theta(N-1)\cdots\Theta(1)\Theta(0) = \prod z_n$. Every irreducible $*$-representation of $\mathcal{U}_N$ is equivalent to one of the above.*

Below, we shall study uniquely representations with $\Theta(n)^{4p} = 1$ (we could have included this condition into the defining relations of $\mathcal{U}_N$). In those representations, the rigid rotation of $\mathbf{Z}_N$ by two units may be implemented by elements $U \in \mathcal{U}_N$:

$$U\,\Theta(n+1) = \Theta(n-1)\,U$$

where

$$U = (4p)^{-N/2} \sum_{\alpha_n \in \mathbf{Z}_{4p}} q^{-\frac{1}{2}\alpha_0\alpha_{N-1}}\,\Theta(N-1)^{\alpha_{N-1}} \cdots \Theta(1)^{\alpha_1}\,\Theta(0)^{\alpha_0}. \qquad (20)$$

Notice that, except for the boundary term, U is a product of local expressions. Eq. (20) gives the lattice version of the Sugawara construction of the WZW Hamiltonian for the abelian group.

The case of N even is similar.

5.2. $\beta\gamma$ SYSTEM

Consider algebra $\mathcal{B}$ with generators B, Γ and relation

$$q\,\mathrm{B}\,\Gamma - q^{-1}\,\Gamma\,\mathrm{B} = q - q^{-1}. \qquad (21)$$

B^p and Γ^p generate the center of $\mathcal{B}$ and we have different classes of representations depending on the eigenvalues of those elements.

1. If in an irreducible representation $\Gamma^p \neq 0$ then Γ is invertible and we may introduce[4] $\mathrm{D} = \Gamma^{-1} - \mathrm{B}$ satisfying

$$\Gamma\,\mathrm{D} = q^2\,\mathrm{D}\,\Gamma.$$

If $\mathrm{D} = 0$ we get the 1-dimensional representation[5] of $\mathcal{B}$. If $\mathrm{D} \neq 0$, the representations of $\mathcal{B}$ are periodic: we may find a basis $|s\rangle$, $s \in \mathbf{Z}_p$, s.t.

$$\Gamma\,|s\rangle = \zeta_1\,|s+1\rangle, \quad \mathrm{D}\,|s\rangle = \zeta_2\,q^{-2s}\,|s\rangle.$$

The periodic representations are characterized by the eigenvalues of $\Gamma^p = \zeta_1^p$ and of $\mathrm{B}^p = \zeta_1^{-p} + \zeta_2^p$.

2. The case $\Gamma^p = 0$ but $\mathrm{B}^p \neq 0$ may be treated similarly.

3. Finally, if $\Gamma^p = 0$ and $\mathrm{B}^p = 0$, we either have a trivial representation or a HW representation in p-dimensional space spanned by states $|s\rangle$, $s = 0, 1, \ldots, p-1$, with the action

$$\Gamma\,|s\rangle = \begin{cases} |s+1\rangle & \text{if } s < p-1, \\ 0 & \text{if } s = p-1, \end{cases}$$

$$\mathrm{B}\,|s\rangle = \begin{cases} (1 - q^{-2s})\,|s-1\rangle & \text{if } s > 0, \\ 0 & \text{if } s = 0. \end{cases}$$

The lattice $\beta\gamma$ system is obtained by taking a copy of algebra $\mathcal{B}$ for each lattice site:

$$\mathcal{B}_N = \bigotimes_{n \in \mathbf{Z}_N} \mathcal{B}.$$

We shall denote its generators as $\mathrm{B}(n)$ and $\Gamma(n)$. $\mathcal{B}_N$ may be represented in the space spanned by vectors $|\underline{s}\rangle$, $\mathrm{B}(n)$, $\Gamma(n)$ acting on the n-th component of $\underline{s} \equiv (s_0, s_1, \ldots, s_{N-1})$. Below, we shall only consider the HW representations of $\mathcal{B}_N$ so that we could include $\mathrm{B}(\mathrm{n})^p = \Gamma(n)^p = 0$ into the defining relations of $\mathcal{B}_N$. The periodic representations might be also of interest but we shall not study them here.

In the HW representation, the rigid rotation of $\mathbf{Z}_N$ by one unit is implemented by element $U \in \mathcal{B}_N$,

$$\begin{aligned} U = \; & p^{-N} \sum_{\alpha_n, \beta_n \in \mathbf{Z}_p} \prod_{n=0}^{N-1} (1 - q^{-2\beta_n})^{-1} (1 - q^{-2(\beta_n - 1)})^{-1} \cdots (1 - q^{-2})^{-1} \\ & \cdot \Gamma(N-1)^{\beta_0} (1 - \Gamma(N-1)\mathrm{B}(N-1))^{\alpha_{N-1}} \mathrm{B}(N-1)^{\beta_{N-1}} \cdots\cdots \\ & \cdots \Gamma(1)^{\beta_2} (1 - \Gamma(1)\mathrm{B}(1))^{\alpha_1} \mathrm{B}(1)^{\beta_1} \Gamma(0)^{\beta_1} (1 - \Gamma(0)\mathrm{B}(0))^{\alpha_0} \mathrm{B}(0)^{\beta_0}, \end{aligned}$$

---

[4] I owe this observation to L. Faddeev and A. Volkov

[5] This possibility was pointed to me by R. Kashaev

such that

$$\begin{aligned} U\,\mathrm{B}(n+1) &= \mathrm{B}(n)\,U\,, \\ U\,\Gamma(n+1) &= \Gamma(n)\,U\,. \end{aligned}$$

5.3. Lattice Wakimoto representation

It will be convenient to put together the lattice $u(1)$ fields and the $\beta\gamma$ system in a somewhat twisted way defining their action on vectors $|\underline{\alpha},\underline{s}\rangle \equiv |\underline{\alpha}\rangle \otimes |\underline{s}\rangle$ by

$$\begin{aligned} \Theta(n)\,|\underline{\alpha},\underline{s}\rangle &= q^{s_{n+1}-s_n}\,(\Theta(n)\,|\underline{\alpha}\rangle)\otimes|\underline{s}\rangle\,, \\ \mathrm{B}(n)\,|\underline{\alpha},\underline{s}\rangle &= |\underline{\alpha}\rangle\otimes(\mathrm{B}(n)\,|\underline{s}\rangle)\,, \\ \Gamma(n)\,|\underline{\alpha},\underline{s}\rangle &= |\underline{\alpha}\rangle\otimes(\Gamma(n)\,|\underline{s}\rangle)\,. \end{aligned}$$

Taking different representations of $\mathcal{U}_N$ and the HW representation of $\mathcal{B}_N$, we obtain this way irreducible representations (labelled by the eigenvalue z of Π) of algebra $\mathcal{U}_N\tilde{\otimes}\mathcal{B}_N$, a twisted tensor product of $\mathcal{U}_N$ and $\mathcal{B}_N$. We shall denote by $\mathcal{H}_z$ the corresponding representation space.

The lattice version of the Wakimoto realization (15) of the Kac-Moody currents by free fields is given by the formulae

$$\begin{aligned} J_{11}(n) &= \Theta(n) + q^{-1/2}\Theta(n)^{-1}\mathrm{B}(n+1)\Gamma(n)\,, \\ J_{12}(n) &= -\Theta(n)\mathrm{B}(n) + q^{-1/2}\Theta(n)^{-1}\mathrm{B}(n+1)(1-\Gamma(n)\mathrm{B}(n))\,, \\ J_{21}(n) &= q^{1/2}\Theta(n)^{-1}\Gamma(n)\,, \\ J_{22}(n) &= q^{1/2}\Theta(n)^{-1}(1-\Gamma(n)\mathrm{B}(n)) \end{aligned} \tag{22}$$

which define a homomorphism from the lattice Kac-Moody algebra $\mathcal{K}_N$ to $\mathcal{U}_N\tilde{\otimes}\mathcal{B}_N$ and turn each representation space $\mathcal{H}_z$ of the latter into a (Wakimoto) $\mathcal{K}_N$-module. These modules are irreducible if $z^{2p} \neq 1$. For $z = q^r$, $r = -p+1,\ldots,p-1,p$, their reducibility may be studied by adapting to the lattice the cohomological constructions of [19].

6. Bernard-Felder cohomology

We shall have to adjoin to the $u(1)$ algebra $\mathcal{U}_N$ the zero mode $\Psi(0)$ s.t.

$$\begin{aligned} \Theta(0)\,\Psi(0) &= q^{-1/2}\,\Psi(0)\,\Theta(0)\,, \\ \Theta(N-1)\,\Psi(0) &= q^{-1/2}\,\Psi(0)\,\Theta(N-1) \end{aligned}$$

and all other commutators are trivial. $\Psi(0)$, which divides the eigenvalue of Π by q, may be implemented in the sum of representation spaces of $\mathcal{U}_N$ with $\Pi = q^r$. We shall also let it act in $\oplus\mathcal{H}_{q^r}$ by

$$\Psi(0)\,|\underline{\alpha},\underline{s}\rangle = q^{s_0}\,(\Psi(0)\,|\underline{\alpha}\rangle)\otimes|\underline{s}\rangle\,.$$

More symmetrically, we may construct the lattice $u(1)$ vertex operator ($\sim e^{i\phi(2\pi n/N)}$)

$$\begin{aligned} \Psi(n) &= \Theta(n-1)\ldots\Theta(1)\,\Theta(0)\,\Psi(0) \qquad \text{for } n>0\,,\\ \Psi(n) &= \Theta(n)^{-1}\ldots\Theta(-2)^{-1}\Theta(-1)^{-1}\Psi(0) \quad \text{for } n<0\,. \end{aligned} \tag{23}$$

The screening charge integral is now defined by

$$Q(n) = \Pi^{-1}\sum_{m=n}^{n+N-1}\Gamma(m)\Psi(m)^2\,, \tag{24}$$

compare eq. (16). It is related to the lower left matrix element of the monodromy matrix (10) in the Wakimoto realization (22):

$$Q(n) = q\,M_{21}(n)\,\Psi(n)^2$$

and thus (see eq. (11)) to the $\mathcal{U}_q(sl(2)$ lowering operator. The relation of the screening charge integrals to quantum groups has been observed in [20] and was developed in [21]-[23] into a theory of topological realizations of quantum groups.

$Q(n)$ acts as an operator from $\mathcal{H}_z$ to $\mathcal{H}_{z'}$, $z'=q^{-2}z$, in a nilpotent way: $Q(n)^p=0$. Besides, for $r=0,1,\ldots,p$, powers of $Q(n)$ define complexes of $\mathcal{K}_N$-modules

$$0 \;\rightarrow\; \mathcal{H}_{q^{-r}} \overset{Q^{p-r}}{\rightarrow} \mathcal{H}_{q^r} \overset{Q^r}{\rightarrow} \mathcal{H}_{q^{-r}} \;\rightarrow\; 0 \quad ,$$

$$0 \;\rightarrow\; \mathcal{H}_{q^r} \overset{Q^r}{\rightarrow} \mathcal{H}_{q^{-r}} \overset{Q^{p-r}}{\rightarrow} \mathcal{H}_{q^r} \;\rightarrow\; 0 \quad .$$

In other words, the powers of $Q(n)$ above (independent, in fact, of n) commute with the action of $\mathcal{K}_N$ as given by eqs. (22). One may show that the above complexes are exact in the middle. We conjecture that the remaining cohomology

$$\mathcal{H}_{q^r} \supset \ker Q^r \equiv \mathcal{H}'_{q^r} \;\cong\; \mathcal{H}_{q^{-r}}/\operatorname{im} Q^r \equiv \mathcal{H}'_{q^{-r}} \tag{25}$$

and

$$\mathcal{H}_{q^{-r}} \supset \ker Q^{p-r} \equiv \mathcal{H}''_{q^{-r}} \;\cong\; \mathcal{H}_{q^r}/\operatorname{im} Q^{p-r} \equiv \mathcal{H}''_{q^r} \tag{26}$$

gives irreducible representations of $\mathcal{K}_N$. This is true, for example, for $N=1$ when $\mathcal{K}_N$ reduces to $\mathcal{U}_q(sl(2))$.

7. WZW theory on lattice

The chiral intertwining operators on the lattice are

$$h(n) = \begin{pmatrix} (\Pi - \Pi^{-1})\,\Psi(n) + \mathrm{B}(n)\,\Psi(n)^{-1}\,Q(n) & \mathrm{B}(n)\,\Psi(n)^{-1} \\ \Psi(n)^{-1}\,Q(n) & \Psi(n)^{-1} \end{pmatrix}. \tag{27}$$

In comparison with the expression (18), notice additional term in the uper left corner of (27) killed by Wick ordering renormalization in the continuum. Intertwiners (27) map space $\oplus\mathcal{H}_{q^r}$ into itself and descend to cohomology (25), (26).

The above constructions have been adapted to the left-moving sector of the WZW theory. In the ones for the right-moving sector (distinguished below by the bar), we should replace q by $\bar{q} = q^{-1}$. The space of states of the lattice WZW theory may be taken as

$$\mathcal{H}' = \bigoplus_{r=1}^{p-1} \mathcal{H}'_{q^r} \otimes \bar{\mathcal{H}}'_{q^r}$$

or as $\mathcal{H}''$ using the doubly-primed spaces. The real fields of the lattice WZW model are then

$$g(n,m) = h(n)\,\bar{h}(m)^{-1} \tag{28}$$

acting in spaces $\mathcal{H}'$ or $\mathcal{H}''$, $g(n,m) = g(n+N, m+M)$.

It should be stressed that the resulting theory lives on Minkowski lattice (n, m in eq. (28) are integer valued light-cone variables). The model has essentially all the symmetries of the continuum WZW theory (in deformed form), including the conformal covariance. Its relation to quantum groups is explicit. Let us conclude by listing some open problems.

1. As mentioned above, the lattice counterpart of the Sugawara construction of energy-momentum is not known apart from the abelian case.
2. The continuum limit is not easy to understand even for the $u(1)$ case. One should expect it to take place in a weak form, for vacuum expectation values or traces of products of operators. Studying the first ones would require a choice of vacuum on the lattice (among many states invariant under lattice translations). The traces (e.g. the characters of the lattice Kac-Moody algebra) seem more accessible.
3. In the above constructions we did not study the unitarity properties of representations of $\mathcal{K}_N$ or of the intertwining fields. One knows that in the continuum this is difficult within the free field approach, even when the cohomology of the Fock space modules gives unitary HW representations of the Kac-Moody algebra. It remains to be seen if there exists a (deformed?) version of unitarity properties on the lattice.

4. There is a close relation between the two-dimensional (chiral) WZW model and the three dimensional Chern-Simons theory [40]. For example, the space of the fixed-time Chern-Simons states on the disc is the basic representation of the Kac-Moody algebra [41]. The Chern-Simons theory seems to possess lattice versions given by constructions [42][43] of 3-manifold invariants from quantum groups. The lattice WZW model described above should be related to the latter. It seems that the clue to understanding this relation should be the fusion of lattice Kac-Moody algebras generalizing the coproduct for quantum groups and allowing to glue spaces of Chern-Simons states for more complicated topologies.

References

[1] E. Witten, Non-Abelian Bosonization in Two Dimensions, *Commun. Math. Phys.*, **92**, 455-472 (1984).
[2] P. Goddard, A. Kent, D. Olive, Unitary Representations of the Virasoro and Super-Virasoro Algebras, *Commun. Math. Phys.*, **103**, 105-119 (1986).
[3] V. Drinfeld, V. Sokolov, Lie Algebras and Equations of Korteweg-de Vries type, *Journ. Sov. Math.*, **30**, 1975-2036 (1985).
[4] O. Babelon, Extended Conformal Algebra and Yang-Baxter Equation, *Phys. Lett.*, **B 215**, 523-529 (1988).
[5] G. Felder, J. Fröhlich, G. Keller, Braid Matrices and Structure Constants for Minimal Conformal Models, *Commun. Math. Phys.*, **124**, 647-664 (1989).
[6] J.-L. Gervais, The Quantum Group Structure of 2d Gravity and Minimal Models I, *Commun. Math. Phys.*, **130**, 257-283 (1990).
[7] K. Fredenhagen, K. Rehren, B. Schroer, Superselection Sectors with Braid Group Statistics and Exchange Algebras, *Commun. Math. Phys.*, **125**, 201-226 (1989)
[8] J. Fröhlich, F. Gambbiani, Braid Statistics in Local Quantum Theory, ETH-TH/90-6 preprint.
[9] G. Mack, V. Schomerus, Conformal Field Algebras with Quantum Symmetry from the Theory of Superselection Sectors, *Commun. Math. Phys.*, **134**, 139-196 (1990)
[10] S. Doplicher, R. Haag, J. Roberts, Local Observables and Particle Statistics I and II, *Commun. Math. Phys.*, **23**, 199-230 (1971) and **35**, 49-85 (1974).
[11] S. Doplicher, J. Roberts, Why There is a Field Algebra with a Compact Gauge Group Describing the Superselection Structure in Particle Physics, *Commun. Math. Phys.*, **131**, 51-108 (1990)
[12] V. Pasquier, H. Saleur, Common structures between finite systems and conformal field theories through quantum groups, *Nucl. Phys.*, **B 330**, 523-556 (1990).
[13] V. Drinfeld, Quantum Groups, in Proceedings of the International Congress of Mathematicians, Berkeley 1986, pp. 798-820.
[14] M. Jimbo, A q-difference Analogue of $U(g)$ and Yang-Baxter Equation, Lett. Math. Phys., **10**, 63-69 (1985).
[15] T. Kohno, Monodromy representations of braid groups and Yang-Baxter equations, *Ann. Inst. Fourier (Grenoble)*, **37(4)**, 139-160 (1987).
[16] G. Moore, N. Seiberg, Classical and Quantum Conformal Field Theory, *Commun. Math. Phys.*, **123**, 177-254 (1989).
[17] L. Alvarez-Gaumé, C. Gómez, G. Sierra, Duality and Quantum Groups, *Nucl. Phys.*, **B 330**, 347-398 (1990)
[18] M. Wakimoto, Fock Representations of the Affine Lie Algebra $A_1^{(1)}$, *Commun. Math. Phys.*, **104** 605-609 (1986).
[19] D. Bernard, G. Felder, Fock Representations and BRST Cohomology in $SL(2)$ Current Algebra, *Commun. Math. Phys.*, **127**, 145-168 (1990).

[20] C. Gómez, G. Sierra, Quantum Group Meaning of the Coulomb Gas, *Phys. Lett.*, **B 240**, 149-157 (1990).
[21] V. Schechtman, A. Varchenko, Arrangements of Hyperplanes and Lie Algebra Homology, *Invent. Math.*, to appear.
[22] R. Lawrence, The Topological Approach to Representations of Iwahori-Hecke Algebra, *Int. J. Mod. Phys.*, **A 5**, 3213-3219 (1990).
[23] G. Felder, C. Wieczerkowski, Topological Representations of th Quantum Group $U_q(sl_2)$, *Commun. Math. Phys.*, **138**, 583-605 (1991).
[24] E. Verlinde, Fusion Rules and Modular Transformations in 2d Conformal Field Theory, *Nucl. Phys.*, **B 300** [FS 22], 360-376 (1988).
[25] G. Moore, N. Reshetikhin, A Comment on Quantum Group Symmetry in Conformal Field Theory, *Nucl. Phys.*, **B 328**, 557-574 (1989).
[26] A. Alekseev, L. Faddeev and M. Semenov-Tian-Shansky, Hidden quantum groups inside Kac-Moody algebras, in Quantum Groups, ed. P. Kulish, Lect. Notes in Math. Vol. 1510,, Berlin, Springer 1992, pp. 148-158.
[27] A. Alekseev, L. Faddeev, M. Semenov-Tian-Shansky and A. Volkov, The unraveling of the quantum group structure in the WZNW theory, CERN preprint 1991.
[28] L. Freidel, J.M. Maillet, On Classical and Quantum Integrable Field Theories Associated to Kac-Moody Current Algebras, *Phys. Lett.*, **B 263**, 403-410 (1991).
[29] L. Faddeev, Quantum Symmetry in Conformal Field Theory by Hamiltonian Methods, Helsinki preprint 1992.
[30] A. Volkov, Miura Transformation on the Lattice, *Theor. Math. Phys.*, **74**, 135-139 (1988).
[31] O. Babelon, Exchange Formula and Lattice Deformation of the Virasoro Algebra, *Phys. Lett.*, **B 238**, 234-238 (1990).
[32] O. Babelon, L. Bonora, Quantum Toda Theory, *Phys. Lett.*, **B 253**, 365-372 (1991).
[33] L. Bonora, V. Bonservizi, Quantum sl_n Toda field theories, SISSA-ISAS preprint 1992.
[34] V. Kac, Infinite Dimensional Lie Algebras, Cambridge, Cambridge University Press 1985.
[35] A. Pressley, G. Segal, Loop Groups, Oxford, Clarendon Press 1986.
[36] B. Feigin, E. Frenkel, Representations of Affine Lie Kac-Moody algebras and bosonization, in V.G. Knizhnik Memorial Volume, Singapore, World Scientific 1989.
[37] P. Bouwknegt, J. McCarthy, K. Pilch, Quantum Group Structure in the Fock Space Resolutions of $\hat{sl}(n)$ Representations, *Commun. Math. Phys.*, **131**, 125-155 (1990).
[38] A. Tsuchiya, Y. Kanie, Vertex Operators in the Conformal Field Theory on P1 and Monodromy Representations of the Braid Group, *Adv. Stud. Pure Math.*, **16**, 297-372 (1988).
[39] K. Gawędzki, On Lattice Kac-Moody Algebras, in preparation.
[40] E. Witten, Quantum Field Theory and Jones Polynomial, *Commun. Math. Phys.*, **121**, 351-399 (1989).
[41] S. Elitzur, G. Moore, A. Schwimmer, N. Seiberg, Remarks on the Canonical Quantization of the Chern-Simons-Witten Theory, *Nucl. Phys.*, **B 326**, 104-134 (1989).
[42] N. Reshetikhin, V. Turaev, Invariants of 3-Manifolds via Link Polynomials and Quantum Groups, *Inven. Math.*, **103**, 547-597 (1991).
[43] V. Turaev, O. Viro, State sum of 3-manifolds and quantum 6j-symbols, LOMI preprint 1990.

Spectrum of an operator appears in the quantum SU(1,1) group

TOMOYUKI KAKEHI, TETSUYA MASUDA
Tsukuba University, Tsukuba 305, Japan

and

KIMIO UENO
Waseda University, Tokyo 160, Japan

Abstract. In this note, we explain how we reached to an explicitly given second order difference operator in the study of the "representation theory" of the quantum $SU(1,1)$ group of "non-compact type". We set up a situation of functional analysis in which the operator is claimed to be self-adjoint, and the spectral analysis of the operator is disccussed. The "eigenfunctions" are explicitly given in terms of the basic hypergeometric functions. We then also discuss about an explicit spectral expansion theorem which corresponds to the Fok-Mehler formula in the "classical situation ".

Key words: spectral analysis, representation theory of quantum groups

1. Introduction

In the modern mathematical research, classical Lie groups and Lie algebras are naturally related not only to the problems of algebras but also to that of geometry and then to that of analysis. These materials are quite basic, and play quite important roles in many respects. However when we turn our attention to the quantum groups, the situation is completely different. Since the quantum groups are rather new, we have been mainly concerned with the finite dimensional irreducible representations and then their applications. It is even the case that when we say quantum groups, it is usually meant to be the quantum universal enveloping algebra $U_q(\mathcal{G})$ of some semisimple Lie algebra $\mathcal{G}$. Like in the case of the classical Lie groups and Lie algebras, the finite dimensionality of the irreducible representations correspond to the compactness of the "underlying group manifold", and these finite dimensional representations would be studied purely algebraically. Now, it is quite easily imagined that these finite dimensional representations correspond to the "unitary" representations of the "compact real form", even though we have no good understanding for the "quantized compact group manifold".

If we believe that the quantum groups are natural mathematical exis-

H. Araki et al. (eds.), Quantum and Non-Commutative Analysis, 253–261.

tence, we should also be able to deal with the infinite dimensional irreducible representations which correspond to the "unitary representations"of the "non-compact real forms". Up to this point, very few people discussed about the unitarity of the infinite dimensional representations which would be expected to give us some important informations about the global "geometrical" nature of the corresponding "quantized group manifold".

In the above discussions, we put some of the words inside the double quotation marks due to the fact that these objects have not yet been very carefully and rigorously established mathematical terminologies. Among these things, many of them are still sort of virtual objects like in the case of "quantum group manifold". This is mainly due to the fact that the geometrical understanding for these quantized objects are almost perfectly missing at this moment.

As we have already mentioned above, what we have done concerning the quantum groups are mainly related to the finite dimensional representations. Then the basic tools for the research are usually the algebraic ones as we have been remarked. But if we turn to the infinite dimensional representations, we are more or less obliged to pass to some techniques of functional analysis. Here we have to claim that the unitary representation theory of the quantum groups of "non-compact type" which we can attack at the moment are very few. Among these things, $SU_q(1,1)$ is certainly one of the most appropriate examples to attack. This is mainly due to the fact that the matrix elements of the possible candidates of the unitary representations are already computed following the analogy of the "compact real form" $SU_q(2)$. (See [5].) Then the important point is that the matrix elements are explicitly expressed in terms of the basic hypergeometric functions. These matrix elements are determined by the second order difference equations of "regular-singular type", and the operator which gives us these equations comes from the "quantum Casimir element" in the quantum universal enveloping algebra $U_q(sl(2))$.

The main point is that the discussions concerning the representation theory of $SU_q(1,1)$ are only at the algebraic level at the moment. Then the main purpose of this note is to explain how we are naturally led to the concrete problem of functional analysis out of the study of the unitary representations of the quantum group $SU_q(1,1)$ of "non-compact type". We then also explain that the rigorous solution to that problem gives us some important information about the classifications of the "irreducible unitary representations"of the quantum group $SU_q(1,1)$. Here we have to point out that the problem which we disccuss in this note is just the beginning part of the establishment of the "unitary representation theory" of the quantum group $SU_q(1,1)$ including the "global informations". These things will be fully discussed in our future publications.

2. Some backgrounds on quantum groups

In the case that the Lie algebra $\mathcal{G}$ is $sl(2)$, the quantum universal enveloping algebra $U_q(sl(2))$ is defined to be the unital algebra generated by the four elements e, f, k and k^{-1} with the relations:

$$kek^{-1} = qe\,,\ kfk^{-1} = q^{-1}f\,,\ ef - fe = \frac{k^2 - k^{-2}}{q - q^{-1}}\,,\ kk^{-1} = k^{-1}k = 1$$

for a non-zero complex parameter q satisfying appropriate conditions. This algebra $U_q(sl(2))$ is naturally equipped with a Hopf algebra structure. (See [2] for the details.) Furthermore, we have the natural two-dimensional vector representation of $U_q(sl(2))$ which gives us the homomorphism $\rho_{1/2}$ from $U_q(sl(2))$ to the two by two matrix algebra $End(C^2) \cong Mat(2,C)$ over the complex number field C determined on the generators by

$$\rho_{1/2}(e) = \begin{pmatrix} 0 & 1 \\ 0 & 0 \end{pmatrix}\,,\ \rho_{1/2}(f) = \begin{pmatrix} 0 & 0 \\ 1 & 0 \end{pmatrix} \text{ and } \rho_{1/2}(k) = \begin{pmatrix} q^{1/2} & 0 \\ 0 & q^{-1/2} \end{pmatrix}.$$

Now, the elements x, u, v and y in $Hom_C(U_q(sl(2)),C)$ determined by

$$\rho_{1/2}(a) = \begin{pmatrix} x(a) & u(a) \\ v(a) & y(a) \end{pmatrix}\,,\quad \text{for } a \in U_q(sl(2))$$

generate the another Hopf algebra $A(SL_q(2))$ which we call the coordinate ring of the "quantum algebraic group" $SL_q(2)$. Through the duality argument of Hopf algebras, it is seen that the defining relations of the generators of the coordinate ring $A(SL_q(2))$ are given by:

$$\begin{aligned} &ux = qxu\,,\ vx = qxv\,,\ \ yu = quy\,,\ yv = qvy\,,\ uv = vu\,, \\ &xy - q^{-1}uv = 1\,, \qquad\qquad yx - quv = 1 \end{aligned}$$

with the same complex parameter q which we used above. The Hopf algebra structure of $A(SL_q(2))$ is also quite explicitly described. (See [4] for example.) It is seen that these two Hopf algebras $U_q(sl(2))$ and $A(SL_q(2))$ are "dual" to each other.

In this note, one of the main points is to deal with the unitarity of the representations. Since quantum groups are not groups in the ordinary sense, we need some new notions and terminologies to discuss about the unitarity of the representations. For this reason, we discuss about the real forms on the Hopf algebras to work with. The real form which we use in this note is meant to be the involutive Hopf algebra structure. That is:
(1)The Hopf algebra A is defined over the complex number field C.
(2)The Hopf algebra A is equipped with an involution $A \ni a \mapsto a^* \in A$ with the extra conditions that the coproduct $\Delta : A \to A \otimes A$ and the counit $\epsilon : A \to C$ preserve the involutive structure.

(3)The following condition (called the Woronowicz condition) holds:

$$S(S(a^*)^*) = a \text{ for any } a \in A .$$

It is also remarked that if a Hopf algebra is equipped with an involutive structure, its dual is also in the same category in a quite natural manner.

Now we pass to our basic example. As we have already pointed out, the Hopf algebras $U_q(sl(2))$ and $A(SL_q(2))$ are naturally dual to each other. We take in account of these structures and then we essentially have the following three different cases concerning the real forms.

quantum groups	$SU_q(2)$	$SU_q(1,1)$	$SL_q(2,\mathcal{R})$
involutive structure on $U_q(sl(2))$	$e^* = f$ $f^* = e$ $k^* = k$	$e^* = -f$ $f^* = -e$ $k^* = k$	$e^* = e$ $f^* = f$ $k^* = k$
conditions for the quantization parameter q	q is a non-zero real number.		$\lvert q \rvert = 1$
involutive structure on $A(SL_q(2))$	$x^* = y$ $u^* = -q^{-1}v$ $v^* = -qu$ $y^* = x$	$x^* = y$ $u^* = q^{-1}v$ $v^* = qu$ $y^* = x$	$x^* = x$ $u^* = u$ $v^* = v$ $y^* = y$
involutive algebra	$A(SU_q(2))$	$A(SU_q(1,1))$	$A(SL_q(2,\mathcal{R}))$

3. Unitary representations

For the quantum groups, the rigorous definition of the infinite dimensional unitary representation is not yet very well established as we have already pointed out. At the moment, one of the reasonable definitions of the unitary representations is to think about the unital homomorphisms from the quantum universal enveloping algebra to the matrix algebra of appropriate size preserving the involutive structures. Therefore, if we are obliged to discuss about the infinite dimensional unitary representations, there comes a subtle problem of choosing a suitable matrix algebra of infinite size. Once these non-trivial points are clarified, the above unitary representation corresponds to the unitary matrix with its entries in the "carefully chosen" coordinate ring of the quantum group and the size of this matrix is equal to the dimension of the representation. This is just a rough sketch of how we understand the unitary representations of the quantum groups neglecting some subtle technical points. Here we do not intend to go into the subtlety of the problems. However, this is closely related to the reason why we are forced to use functional analysis to deal with the quantum groups of "non-compact type". Indeed, the choice of the coordinate ring of $SU_q(1,1)$ which works for our purpose is given by the non-commutative C^*−algebra which we denote by $C_0(SU_q(1,1))$ combined with a q-analogue of the Schwartz space defined over the quantum group $SU_q(1,1)$. This Schwartz space is denoted by $\mathcal{S}(SU_q(1,1))$ and known to be a nuclear topological vector space. (See [7].)

In our situation, the homomorphisms which we use are given by π_ℓ : $U_q(sl(2)) \to Mat(I,\mathcal{C})$ where I is a suitable subset of $\mathcal{Z}$ or $\mathcal{Z} + \frac{1}{2}$. (These distinction corresponds to the even and the odd spin representations, respectively.) Here, the target of the homomorphism $Mat(I,\mathcal{C})$ is an algebra of (infinite) matrices defined over the index set I. The homomorphisms π_ℓ have a complex parameter $\ell \in \mathcal{C}$ (called the complex spin) and defined on the generators by

$$\begin{aligned}
\pi_\ell(e) &= \sum_{j\in I}[\ell - j + 1]_q E_{j-1,j}\,, \\
\pi_\ell(f) &= \sum_{j\in I}[\ell + j + 1]_q E_{j+1,j}\,, \\
\pi_\ell(k) &= \sum_{j\in I} q^{-j} E_{j,j}\,,
\end{aligned}$$

where E_{ij} are the matrix units of $Mat(I,\mathcal{C})$ and the symbol $[n]_q$ is called by the q-integer of $n \in \mathcal{Z}$ defined by

$$[n]_q := \frac{q^n - q^{-n}}{q - q^{-1}} \quad \text{for } n \in \mathcal{Z} \ .$$

Then, following the classical ideas of Bargmann [1], we are led to look at the action of the "quantum Casimir element":

$$C := ef + \frac{q^{-1}k^2 + qk^{-2} - 2}{(q - q^{-1})^2} \in U_q(sl(2))$$

which generates the center of $U_q(sl(2))$. In the classical situation, the Casimir element gives rise to a self-adjoint operator on the Hilbert space of all square integrable functions on the Lie group $SU(1,1)$ with respect to the Haar measure. In our quantized context, we also have the Hilbert space $L^2(SU_q(1,1))$ obtained by the copmletion of the quantum Schwartz space $\mathcal{S}(SU_q(1,1))$ by a suitably defined inner product which comes from the "quantum Haar measure". Then the above quantum Casimir operator C is acting on this Hilbert space.

Like in the classical situation, we have the maximal torus S^1 also embedded "inside" the quantum group $SU_q(1,1)$. This torus subgroup corresponds to one of the generators k (or k^{-1}, equivalently) of the quantum universal enveloping algebra $U_q(sl(2))$. Now we are allowed to look at the weight decomposition of the Hilbert space $L^2(SU_q(1,1))$ with respect to the action of this maximal torus from the right and from the left. Then we have the "zonal" part on which the operator is acting as the second order (ordinary) difference operator of "regular-singular type" which we will explain in the next section.

4. The second order difference operator

In this section, we pass to the concrete problem of functional analysis. The materials which we deal with in this note are considered to be defined over the "quantized objects". To make the situation clear, we consider the discrete subset

$$(-\infty, 0]_{q^2} := \{-q^{2n} : n \in \mathcal{Z}\}$$

of the half open interval $(-\infty, 0]$, where the parameter q is a real number satisfying $0 < q < 1$. We regard this set to be a q-analogue of the half open interval $(-\infty, 0]$. Then the differentiation and the integration in this context are given by

$$[D_{q^2}\varphi](\zeta) := \frac{\varphi(\zeta) - \varphi(q^2\zeta)}{(1-q^2)\zeta} \quad , \quad \zeta \in (-\infty, 0]_{q^2}$$

and

$$\int_{-\infty}^{0} \varphi(\zeta)\, d_{q^2}\zeta := \sum_{n\in\mathcal{Z}} \varphi(-q^{2n})q^{2n}(1-q^2)$$

respectively, where φ is a "function" defined over the "quantum interval" $(-\infty, 0]_{q^2}$. This integration is known as the Jackson integral. It is then natural to consider the Hilbert space $L^2 := L^2(-\infty, 0]_{q^2}$ of all "functions" on $(-\infty, 0]_{q^2}$ which are square integrable in terms of the Jackson integral. This Hilbert space L^2 corresponds to the zonal part of the Hilbert space $L^2(SU_q(1,1))$ defined over the quantum group $SU_q(1,1)$.

Now, the operator on L^2 which we obtain from the action of the "quantum Casimir element" is quite explicitly computed from the expression of C and then given by

$$P = T_{q^2}^{-1} D_{q^2}\zeta(1-q^2\zeta)D_{q^2}$$

where $T_{q^2}^{-1}$ is the inverse of the q-shifted operator defined by

$$[T_{q^2}\varphi](\zeta) := \varphi(q^2\zeta) \quad , \quad \zeta \in (-\infty, 0]_{q^2}$$

and ζ is the multiplication operator.

Here we remark that the operator P reduces to the second order ordinary differential operator of regular-singular type when we take the formal limit $q \to 1$.

As we have already mentioned in the previous section, we introduced in [7] a q-analogue of the Schwartz space $\mathcal{S}(SU_q(1,1))$ defined over the quantum group $SU_q(1,1)$. Then the zonal part is denoted by $\mathcal{S} := \mathcal{S}(-\infty, 0]_{q^2}$ and determined by the family of seminorms:

$$P_{mn}(\varphi) := \sup_{\zeta \in (-\infty,0]_{q^2}} |\zeta^m [D_{q^2}^n \varphi](\zeta)| \text{ for } m,n \in \mathcal{N}\,.$$

Theorem 1. *The operator P defined on $\mathcal{S}$ is essentially self-adjoint. Then the domain $\mathcal{D}$ of the operator obtained by the self-adjoint extension of the operator P is described by*

$$\mathcal{D} = \{\varphi \in L^2 : P\varphi \in L^2 \;,\; \lim_{\zeta \to 0} \sqrt{-\zeta}[D_{q^2}\varphi](\zeta) = 0\}\,.$$

We remark that the condition which describes the domain of the self-adjoint operator P is considered to be an analogue of the Neumann boundary condition at the origin.

5. The spectrum of the operator

In the classical situation, the Casimir operator acting on the zonal Hilbert space has no point spectrum. Then the spectrum set consists of absolutely continuous spectrum which corresponds to the principal series representations of the Lie group $SU(1,1)$. However if we turn to our quantized situation, we have both continuous and point spectrum. We put

$$\lambda(\alpha) := \frac{q^2}{(1-q^2)^2}(1 + q^2 - (\alpha + \alpha^{-1})q)$$

for a non-zero complex parameter $\alpha \in \mathcal{C}$. Then the spectrum of the operator P is described as follows.

Theorem 2. *Let $\sigma_c(P)$ and $\sigma_p(P)$ be the continuous and the point spectrum of the operator P, respectively. Then the spectrum $\sigma(P)$ of the operator P is given by $\sigma(P) = \sigma_c(P) \sqcup \sigma_p(P)$, where*

$$\sigma_c(P) = \{\lambda(e^{i\theta}) \,:\, \theta \in [0,\pi]\} = \left[\frac{q^2}{(1+q)^2}, \frac{q^2}{(1-q)^2}\right],$$
$$\sigma_p(P) = \{\lambda(-q^{2n+1}) \,:\, n = 0,1,\ldots\}\,.$$

It should be remarked that the point spectrum set lies in the "high energy region" whereas the continuous spectrum set is bounded. In other words, the unboundedness of the operator P is due to the existence of the point spectrum. This type of phenomena are very rare in the case of classical differential operators. It is also seen that when we take the formal "classical limit" $q \to 1$, the end point of the continuous spectrum set $\frac{q^2}{(1-q)^2}$ diverges to the infinity and then the point spectrum set goes away to the infinity. As a result, the point spectrum set $\sigma_p(P)$ vanishes and the spectrum set $\sigma(P) = \sigma_c(P) \sqcup \sigma_p(P)$ "approaches" to the continuous half open interval $[1/4, \infty)$.

We next pass to the description of the eigenfunctions of the operator P. For suitable parameters α, β and γ, the analytic function ${}_2\varphi_1$ of ζ is defined by

$$
{}_2\varphi_1\begin{pmatrix} \alpha \ , \ \beta \\ \gamma \end{pmatrix} : q : \zeta \Big) := \sum_{n=0}^{\infty} \frac{(\alpha : q)_n (\beta : q)_n}{(\gamma : q)_n (q : q)_n} \zeta^n
$$

where we used the notation $(\alpha : q)_n = (1-\alpha)(1-\alpha q)\cdots(1-\alpha q^{n-1})$ for $n \geq 1$ and $(\alpha : q)_n = 1$ for $n = 0$. We then put

$$
\varphi(\alpha; \zeta) := {}_2\varphi_1\left(\begin{matrix} q\alpha \ , \ q\alpha^{-1} \\ q^2 \end{matrix} : q^2 : q^2\zeta \right) .
$$

Then it is seen that the equality $(P - \lambda(\alpha))\varphi(\alpha; \zeta) = 0$ for $|\zeta| \leq 1$ and hence, by regarding $\varphi(\alpha; \zeta)$ as a sequence defined over $\mathcal{Z}$, the "function" $\varphi(\alpha, \zeta)$ is extended for the whole $\zeta \in (-\infty, 0]_{q^2}$ as an eigenfunction of the operator P belonging to the same eigenvalue $\lambda(\alpha)$.

Theorem 3.
(1)The eigenfunction associated with the eigenvalue $\lambda(-q^{2n+1})$, $n = 0, 1, \ldots$ *is given by* $\varphi(-q^{2n+1}; \zeta) = \varphi_n(\zeta)$.
(2)The generalized eigenfunction associated with the eigenvalue $\lambda(e^{i\theta})$, $\theta \in [0, \pi]$ *is given by* $\varphi(e^{i\theta}; \zeta) = \varphi_\theta(\zeta)$.

By making use of these eigenfunctions, we have the following spectral expansion theorem.

Theorem 4. *For* $\varphi \in L^2(-\infty, 0]_{q^2}$,

$$
\varphi(\zeta) = \int_0^\pi c(\theta)\varphi_\theta(\zeta) \left\{ \int_{-\infty}^0 \varphi_\theta(\eta)\varphi(\eta) d_{q^2}\eta \right\} d\theta
$$

$$+\sum_{n=0}^{\infty}[2n+1]_q\varphi_n(\zeta)\left\{\int_{-\infty}^{0}\varphi_n(\eta)\varphi(\eta)d_{q^2}\eta\right\}$$

where $c(\theta)$ is a q-analogue of the c-function given by

$$c(\theta)=\frac{2}{\pi}\frac{q^2}{1-q^2}\left[\frac{(q^4:q^4)_\infty}{(q^2:q^4)_\infty}\right]^2\frac{(e^{2i\theta}:q^4)_\infty(e^{-2i\theta}:q^4)_\infty}{(e^{2i\theta}q^2:q^4)_\infty(e^{-2i\theta}q^2:q^4)_\infty}.$$

The proofs of these theorems are obtained by the explicitly described Green function of the operator P. Then the essential part of the existence theorem of the Green function is due to the "connection formula" (see [6]) for the basic hypergeometric function ${}_2\varphi_1$.

The details on the spectral theory of the operator P is given in [3].

References

[1] Bargmann, V., Irreducible unitary representations of the Lorentz group, *Ann. Math.*, **48**, 568-640 (1947).

[2] Jimbo, M., A q-difference analogue of $U_q(\mathcal{G})$ and the Yang-Baxter equation, *Lett. Math. Phys.*, **10**, 63-69 (1985).

[3] Kakehi, T., Masuda, T. and Ueno, K., Spectral analysis of a q-difference operator which arises from the quantum $SU(1,1)$ group, *Preprint.*

[4] Masuda, T., Mimachi, K., Nakagami, Y., Noumi, M. and Ueno, K., Representations of the quantum group $SU_q(2)$ and the little q-Jacobi polynomials, *J. Funct. Anal.*, **99**, 357-386 (1991).

[5] Masuda, T., Mimachi, K., Nakagami, Y., Noumi, M., Saburi, Y. and Ueno, K., Unitary representations of quantum group $SU_q(1,1)$ I:Structure of the dual space of $U_q(sl(2))$, II:Matrix elements of unitary representations and the hypergeometric functions, *Lett. Math. Phys.*, **19**, 187-204 (1990).

[6] Mimachi, K., Connection problem in harmonic q-difference system associated with a Jackson integral of Jordan-Pochhammer type, *Nagoya Math. J.*, **116**, 149-161 (1989).

[7] Masuda, T. et Watanabé, J., Sur les espaces vectoriels topologiques associés aux groupes quantiques $SU_q(2)$ et $SU_q(1,1)$, *C. R. Acad. Sci. Paris*, **312**, 827-830 (1991).

Takesaki Duality for the Crossed Product by Quantum Groups

YOSHIOMI NAKAGAMI
Department of Mathematics, Yokohama City University

Abstract. The concepts of a coaction of a quantum group on a von Neumann algebra and their crossed product by means of the coaction are introduced and a Takesaki duality is proved for the quantum group $SU_q(n)$, $q \in (0,1)$. The AFD II_∞-factor is then expressed by a crossed product of the AFD III_λ-factor by $SU_q(n)$ with $\lambda = q^2$.

Key words: quantum group, von Neumann algebra, Hopf *-algebra, Woronowicz algebra, coaction, crossed product

1. Introduction

It is more or less known by specialists that the fixed point subalgebra of the Powers factor R_λ of type III_λ by the product type action of the quantum group $SU_q(n)$ is an AFD (approximately finite dimensional) factor of type II_1 generated by Jones projections and that the Markov trace gives us Jones polynomials with one or two variables, [1,7,8]. When we study such actions of quantum groups on operator algebras, it is desirable to formulate the quantum groups in termes of the operator algebras. The quantum group considered is then described by an operator algebra containing the coordinate ring of the quantum group as its dense *-subalgebra and also by the group structure comming from its Hopf *-algebra. In Definition 2.1 we will introduce two kinds of definitions for a quantum group by using a C^*-algebra or a von Neumann algebra, which is called a Hopf C^*-algebra or a Hopf W^*-algebra, respectively. The former is suitable for describing the topological aspects of quantum groups and the latter for describing the measure theoretical aspects. It is easy to verify that the concept of the Hopf C^*-algebra is coherent with the definition of compact matrix quantum groups due to Woronowicz. As for the difinition of the Hopf W^*-algebra, we have used extensively the idea of the Kac algebra, [4]. In order to confirm the validity of this definition we have exibited the duality in Theorem 3.1 similarly as the case of locally compact groups. The existence of the Hopf W^*-algebra $L_q^\infty(SU(n))$ corresponding to the Lie group $SU(n)$ is assured in Proposition 3.2.

The quantum universal enveloping Lie algebra is known to be the dual

H. Araki et al. (eds.), Quantum and Non-Commutative Analysis, 263–281.

object of quantum groups and the representation on a Hilbert apace is usually unbounded. Therefore the treatment by means of the operator algebra would have many problems to be solved. However, it is shown in Proposition 3.3 that if the representation of the quantum universal enveloping Lie algebra $U_q(su(2))$ is regular, then it preserves the self-adjointness and generates an O^*-algebra in the sense of Schmüdgen.

In Sections 4 and 5 we will discuss the coaction of the Hopf W^*-algebra $L_q^\infty(SU(n))$, $0 < q < 1$ on a von Neumann algebra. Using the infinite tensor product of the fundamental unitary representation of the quantum group $SU_q(n)$, we define, in Proposition 4.2, a coaction δ of $L_q^\infty(SU(n))$ on the Powers (Araki-Words) factor R_λ of type III$_\lambda$, where λ is given by q^2. The fixed point subalgebra of R_λ by this coaction turns out to be an AFD factor of type II$_1$, since the restriction of the Powers state to the fixed point subalgebra becomes the canonical trace.

In Section 5 we define the crossed product of a von Neumann algebra by a Hopf W^*-algebra $L_q^\infty(SU(n))$ coacting on it. By using the dual Hopf W^*-algebra, we may construct the double crossed product. Then a Takesaki duality for the double crossed product with respect to the quantum group $SU_q(n)$ holds as in Theorem 5.5. As a result of this theorem an Araki-Connes-Takesaki type structure theorem is obtained as follows: an AFD factor of type II$_\infty$ or III$_\lambda$ is obtained by the crossed product of an AFD factor of type III$_\lambda$ or II$_\infty$ by the Hopf W^*-algebra $L_q^\infty(SU(n))$ or its dual Hopf W^*-algebra, respectively. The former assertion gives us a geometric picture for an AFD factor of type II$_\infty$ by using the quantum group $SU_q(n)$.

2. Quantum groups in operator algebras

We designate the *-algebra, the C^*-algebra and the von Neumann algebra describing the quantum group G_q corresponding to the group G by $A_q(G)$, $C_q(G)$ and $L_q^\infty(G)$, respectively. The accompanying group structure is endowed with the Hopf *-algebra, the Hopf C^*-algebra and the Hopf W^*-algebra, respectively, [11,12].

Definition 2.1. (i) The family $(A,\delta,\varepsilon,\kappa)$ is called a *Hopf *-algebra* if it is a Hopf algebra such that A ia a *-algebra on $\mathbf{C}$, the comultiplication δ and the counit ε are *-homomorphisms and the antipode κ satisfies the Woronowicz condition $\kappa^{\pm 1}(a^*) = \kappa^{\mp 1}(a)^*$ for $a \in A$.

(ii) The family $(\mathcal{A},\delta_\mathcal{A},\varepsilon_\mathcal{A},\kappa_\mathcal{A})$ is called a *Hopf C^*-algebra* if $\mathcal{A}$ is a C^*-algebra, $\mathcal{A}\hat{\otimes}\mathcal{A}$ is the completion of $\mathcal{A} \otimes \mathcal{A}$ with respect to the extended Haagerup norm due to Effros and Ruan [2], the comultiplication $\delta_\mathcal{A}$ is a coassociative *-isomorphism of $\mathcal{A}$ to $\mathcal{A}\hat{\otimes}\mathcal{A}$, the counit $\varepsilon_\mathcal{A}$ is a *-homomorphism of $\mathcal{A}$ to $\mathbf{C}$ with the counit condition $(\mathrm{id} \otimes \varepsilon_\mathcal{A}) \circ \delta_\mathcal{A}=\mathrm{id}=(\varepsilon_\mathcal{A} \otimes \mathrm{id}) \circ \delta_\mathcal{A}$ and the antipode $\kappa_\mathcal{A}$ is a *-antiautomorphism on a dense *-subalgebra A of $\mathcal{A}$ with $(A, \delta_\mathcal{A}\lceil_A, \varepsilon_\mathcal{A}\lceil_A, \kappa_\mathcal{A})$ a Hopf *-algebra.

(iii) The family $\mathbf{W}=(M,\delta_{\mathbf{W}},R,\{\tau_t\},h)$ is called a *Hopf W^*-algebra* or *Woronowicz algebra* if M is a von Neumann algebra, the comultiplication $\delta_{\mathbf{W}}$ is a coassociative normal *-isomorphism of M to $M\overline{\otimes}M$, the *unitary antipode* R is a *-antiautomorphism with

$$\sigma\circ(R\otimes R)\circ\delta_{\mathbf{W}}=\delta_{\mathbf{W}}\circ R$$

(σ is the flip), the *deformation automorphism* $\{\tau_t\}$ is a one parameter *-automorphism group with

$$(\tau_t\otimes\tau_t)\circ\delta_{\mathbf{W}}=\delta_{\mathbf{W}}\circ\tau_t \quad\text{and}\quad R\circ\tau_t=\tau_t\circ R$$

and the *Haar weight* h is a $\{\tau_t\}$-invariant faithful normal weight on M_+ with the left invariance $(\mathrm{id}\otimes h)(\delta_{\mathbf{W}}(a))=h(a)1$ for $a\in M_+$, the symmetry

$$(R\circ\tau_{-i/2}\otimes h)(\delta_{\mathbf{W}}(b^*)(1\otimes a))=(\mathrm{id}\otimes h)((1\otimes b^*)(\delta_{\mathbf{W}}(a)))$$

for entire analytic elements a,b in $\boldsymbol{n}_h$ with respect to $\{\tau_t\}$ and the commutativity of h and $h\circ R$, where $\boldsymbol{n}_h$ is the left ideal $\{a\in M : h(a^*a)<\infty\}$ of M.

Examples. a) Let $U=(x_{i,j})$ be the quantum matrix for $SU_q(n)$, $0<q<1$, i.e. $TU_1U_2=U_1U_2T$, $\det_q U=1$ and $U^*U=UU^*=1$, where T is the R-matrix of type A_{n-1}:

$$T=\sum_{i,j=1}^{n}q^{\delta_{i,j}}e_{i,j}\otimes e_{j,i}+(q-q^{-1})\sum_{i<j}e_{i,i}\otimes e_{j,j}.$$

The Hopf *-algebra $A_q(SU(n))$ is the *-algebra generated by $\{x_{i,j} : i,j=1,\cdots,n\}$ whose comultiplicaton δ, counit ε and antipode κ are given by matrices

$$(\delta(x_{i,j}))=U\dot{\otimes}U,\quad(\varepsilon(x_{i,j}))=(\delta_{i,j}),\quad(\kappa(x_{i,j}))U=U(\kappa(x_{i,j}))=(\delta_{i,j}1).$$

Let R and τ_t be a *-antiautomorphism and a *-automorphism of $A_q(SU(n))$ defined by

$$R(x_{l,k})=(-1)^{l-k}\det_q U_{k,l},\qquad \tau_t(x_{l,k})=q^{2it(l-k)}x_{l,k},\qquad t\in\mathbf{R},$$

where $U_{k,l}$ is the $(n-1)\times(n-1)$ matrix obtained from U by deleting the k-th row and the l-th colum. Then the antipode κ is written in the form $R\circ\tau_{-i/2}$.

b) Let $C_q(SU(n))$ be the envelopping C^*-algebra of $A_q(SU(n))$ and δ, ε their natural extensions to $C_q(SU(n))$. Since $A_q(SU(n))$ is identified with a dense *-subalgebra of $C_q(SU(n))$, the antipode is the same as that of $A_q(SU(n))$. Thus $C_q(SU(n))$ is a Hopf C^*-algebra.

3. Duality for Woronowicz algebras

Let $\mathbf{W}$ be a Woronowicz algebra and W the Kac-Takesaki operator on $L^2(M,h) \otimes L^2(M,h)$ defined by

$$W(a \otimes b) = \delta_{\mathbf{W}}(b)(a \otimes 1), \;\; a, b \in \boldsymbol{n}_h,$$

where $L^2(M,h)$ is the completion of the left ideal $\boldsymbol{n}_h$ with respect to the inner product defined by $(a|b) = h(b^*a)$. Then the Kac-Takesaki operator W satisfies the pentagonal relation:

$$W_{1,2}W_{2,3} \;=\; W_{2,3}W_{1,3}W_{1,2} \tag{1}$$

on $L^2(M,h) \otimes L^2(M,h) \otimes L^2(M,h)$ and implements the comultiplication: $\delta_{\mathbf{W}}(a) = W(1 \otimes a)W^*$. An element in the intersection of the predual M_* and the dual space of $\boldsymbol{n}_h$ with respect to the pairing $\langle a, \varphi \rangle = \varphi(a)$, $a \in M$, $\varphi \in M_*$ is said to be L^2-bounded. The set of all L^2-bounded elements is identified with a dense subspace of $L^2(M,h)$. If $\psi \in M_*$ is L^2-bounded, so is $\varphi * \psi$ for each $\varphi \in M_*$, where $\varphi * \psi = (\varphi \otimes \psi) \circ \delta_{\mathbf{W}}$. Let $\widehat{M}$ be the von Neumann algebra generated by operators $\widehat{\pi}(\varphi)$, $\varphi \in M_*$ defined by $\widehat{\pi}(\varphi)\psi = \varphi * \psi$. Then $\delta_{\widehat{\mathbf{W}}}(a) = \widehat{W}(1 \otimes a)\widehat{W}^*$ is a coassociative comultiplication on $\widehat{M}$, where $\widehat{W}$ is $\sigma(W^*)$ and σ is the flit map. Let $\widehat{R}(\widehat{\pi}(\varphi)) = \widehat{\pi}(\varphi \circ R)$, $\hat{\tau}_t(\widehat{\pi}(\varphi)) = \widehat{\pi}(\varphi \circ \tau_t)$ and $\hat{h}(\widehat{\pi}(\psi)^* \widehat{\pi}(\varphi)) = (\varphi|\psi)$ for L^2-bounded φ, ψ. Then $\widehat{\mathbf{W}} = (\widehat{M}, \delta_{\widehat{\mathbf{W}}}, \widehat{R}, \{\hat{\tau}_t\}, \hat{h})$ is a Woronowicz algebra, which is called the *dual* Woronowicz algebra and sometimes written by $\mathbf{W}\hat{}$.

Theorem 3.1. ([11,12]) *If* $\mathbf{W}$ *is a Woronowicz algebra, then the bidual Woronowicz algebra* $(\mathbf{W}\hat{})\hat{}$ *is *-isomorphic to* $\mathbf{W}$.

Let $\mathbf{W}' = (M', \delta_{\mathbf{W}'}, R', \{\tau'_t\}, h')$ and $\mathbf{W}^o = (M^o, \delta_{\mathbf{W}^o}, R^o, \{\tau^o_t\}, h^o)$ be Woronowicz algebras called the commutant Woronowicz algebra and the opposite Woronowicz algebra defined respectively by

$$M' = JMJ, \;\; \delta_{\mathbf{W}'} = \mathrm{Ad}_{J \otimes J} \circ \delta_{\mathbf{W}} \circ \mathrm{Ad}_J, \;\; \tau'_t = \mathrm{Ad}_J \circ \tau_{-t} \circ \mathrm{Ad}_J, \;\; h' = h \circ \mathrm{Ad}_J$$

and

$$M^o = M, \;\; \delta_{\mathbf{W}^o} = \sigma \circ \delta_{\mathbf{W}}, \;\; R^o = R, \;\; \tau^o_t = \tau_{-t}, \;\; h^o = h \circ R,$$

where M is identified with the von Neumann algebra acting on $L^2(M,h)$ by the left multiplication and J is the modular conjugation operator. In the proof of the above theorem we utilized the fact that $(\mathbf{W}^o)\hat{} = (\widehat{\mathbf{W}})'$ and $(\mathbf{W}')\hat{} = (\widehat{\mathbf{W}})^o$. In what follows $(\widehat{\mathbf{W}})'$ is denoted by $\widehat{\mathbf{W}}'$. Therefore $((\widehat{\mathbf{W}}')\hat{})' = (\mathbf{W}^o)'$ and the *-isomorphism of $\mathbf{W}$ to $(\mathbf{W}^o)'$ is given by $\mathrm{Ad}_{J\hat{J}}$. These facts will be used in Section 5.

There exists a unique normalized Haar measure h on the C^*-algebra $\mathcal{A}$ of the Hopf C^*-algebra $C_q(SU(n))$ by [16]. Let $L^\infty_q(SU(n))$ be the weak closure of $\pi_h(\mathcal{A})$ together with other structures in Definition 2.1(iii), where $\{\pi_h, L^2(M,h)\}$ is the GNS representation defined by $h(c^*ab) = (\pi_h(a)b|c)$.

Proposition 3.2. ([11,12]) *$L_q^\infty(SU(n))$ is a Woronowicz algebra.*

Let $U_q(su(n))$ be the Hopf *-algebra obtained as the quantum universal enveloping algebra for a Lie algebra $su(n)$. For example, $U_q(su(2))$ is a Hopf *-algebra generated by four elements $k^\pm$, e, f with relations

$$kk^- = k^-k = 1, \;\; kek^- = qe, \;\; kfk^- = q^{-1}f, \;\; ef - fe = \frac{k^2 - k^{-2}}{q - q^{-1}}$$

and with the involution

$$(k^\pm)^\sharp = k^\pm, \;\; e^\sharp = f, \;\; f^\sharp = e \quad \left(\text{or } (k^\pm)^\flat = k^\pm, \;\; e^\flat = q^2 f, \;\; f^\flat = q^{-2}e\right).$$

The comultiplication δ, the counit ε and the antipode κ are given by

$$\delta(k^\pm) = k^\pm \otimes k^\pm, \;\; \delta(e) = e \otimes k^- + k \otimes e, \;\; \delta(f) = f \otimes k^- + k \otimes f$$

and

$$\varepsilon(\kappa^\pm) = 1, \;\; \varepsilon(e) = \varepsilon(f) = 0; \quad \kappa(k^\pm) = k^\mp, \;\; \kappa(e) = -qe, \;\; \kappa(f) = -q^{-1}f.$$

The pair of Hopf *-algebras $A_q(SU(2))$ and $U_q(su(2))$ is a dual pair with respect to the pairing given by the relations

$$(k^\pm(x_{i,j})) = \begin{pmatrix} q^{\pm 1/2} & 0 \\ 0 & q^{\mp 1/2} \end{pmatrix}, \;\; (e(x_{i,j})) = \begin{pmatrix} 0 & 1 \\ 0 & 0 \end{pmatrix}, \;\; (f(x_{i,j})) = \begin{pmatrix} 0 & 0 \\ 1 & 0 \end{pmatrix}.$$

The left (resp. right) invariant action of $U_q(su(n))$ on $A_q(SU(n))$ is given by

$$\lambda(\varphi)a = (\mathrm{id} \otimes \varphi)(\delta(a)), \quad (\text{resp.}\, \rho(\varphi)a = (\varphi \otimes \mathrm{id})(\delta(a))),$$

for $\varphi \in U_q(su(n))$. These actions $\lambda(\varphi)$ and $\rho(\varphi)$ are identified with unbounded operators in $L^2(M,h)$.

Proposition 3.3. ([9]) (i) *The unbounded operator algebra $\lambda(U_q(su(2))$ (resp. $\rho(U_q(su(2))))$ is an O^*-algebra and the representation preserves the self-adjointness in the sense that*

$$\lambda(k^\pm)^- = \lambda(k^\pm)^*, \quad \lambda(e^\sharp)^- = \lambda(e)^*, \quad \lambda(f^\sharp)^- = \lambda(f)^*$$

$$(\text{resp. } \rho(k^\pm)^- = \rho(k^\pm)^*, \quad \rho(e^\flat)^- = \rho(e)^*, \quad \rho(f^\flat)^- = \rho(f)^*),$$

where a^- denotes the closure of the operator a.

(ii) *The bicommutant of $\lambda(U_q(su(2)))$ or $\rho(U_q(su(2)))$ is *-isomorphic to the von Neumann algebra in the dual Woronowicz algebra $L_q^\infty(SU(2))\widehat{}$ or its commutant, respectively.*

4. Coactions of $L_q^\infty(SU(n))$, $0<q<1$ on von Neumann algebras

Let's consider an action of a quantum group on a von Neumann algebra.

Definition 4.1. Let $\mathbf{W}=(M,\delta_{\mathbf{W}},R,\{\tau_t\},h)$ be a Woronowicz algebra. A normal *-homomorphism δ of a von Neumann algebra $\mathcal{M}$ to $\mathcal{M}\overline{\otimes}M$ (resp. $M\overline{\otimes}\mathcal{M}$) is called a *right* (resp. *left*) *coaction* of the Woronowicz algebra $\mathbf{W}$ on $\mathcal{M}$ if it satisfies

$$(\delta\otimes \mathrm{id})\circ\delta=(\mathrm{id}\otimes\delta_{\mathbf{W}})\circ\delta \quad (\text{resp.} \quad (\mathrm{id}\otimes\delta)\circ\delta=(\delta_{\mathbf{W}}\otimes \mathrm{id})\circ\delta).$$

Let $U=(x_{i,j})$ be the quantum matrix for $SU_q(n)$, which is the fundamental unitary representation of $SU_q(n)$ on an n-dimensional vector space $\mathbf{C}^n$. Denote by $U^{(m)}$ the Kronecker tensor product of m copies of U:

$$U^{(m)}=\sum_{i_1,\ldots,i_m,j_1,\ldots,j_m=1}^{n} e_{i_1,j_1}\otimes\cdots\otimes e_{i_m,j_m}\otimes x_{i_1,j_1}\cdots x_{i_m,j_m}.$$

Then the matrix $U^{(m)}$ is a unitary representation of $SU_q(n)$ on an nm-dimensional vector space $(\mathbf{C}^n)^{\otimes m}$. Then the *-isomorphism $\delta^{(m)}:\mathcal{L}(\mathbf{C}^n)^{\otimes m}\longmapsto \mathcal{L}(\mathbf{C}^n)^{\otimes m}\overline{\otimes}L_q^\infty(G)$ defined by

$$\delta^{(m)}(x)=U^{(m)}(x\otimes 1)U^{(m)*},\qquad x\in\mathcal{L}(\mathbf{C}^n)^{\otimes m}$$

is a right coaction of $L_q^\infty(SU(n))$ on $\mathcal{L}(\mathbf{C}^n)^{\otimes m}$. Indeed, since $U^{(m)}$ is an element in $\mathcal{L}(\mathbf{C}^n)^{\otimes m}\overline{\otimes}L_q^\infty(G)$ and $(\mathrm{id}\otimes\delta_{\mathbf{W}})(U^{(m)})=U_{1,2}^{(m)}U_{1,3}^{(m)}$, we see that

$$\begin{aligned}(\delta^{(m)}\otimes \mathrm{id})(\delta^{(m)}(x))&=U_{1,2}^{(m)}U_{1,3}^{(m)}(x\otimes 1\otimes 1)U_{1,3}^{(m)*}U_{1,2}^{(m)*}\\ &=(\mathrm{id}\otimes\delta_{\mathbf{W}})(U^{(m)})(x\otimes 1\otimes 1)(\mathrm{id}\otimes\delta_{\mathbf{W}})(U^{(m)})^*\\ &=(\mathrm{id}\otimes\delta_{\mathbf{W}})(U^{(m)}(x\otimes 1)U^{(m)*})=(\mathrm{id}\otimes\delta_{\mathbf{W}})(\delta^{(m)}(x)).\end{aligned}$$

Let φ be a state on $\mathcal{L}(\mathbf{C}^n)$ defined by $\varphi(x)=\mathrm{Tr}(Hx)$, $x\in\mathcal{L}(\mathbf{C}^n)$, where $H=(q-q^{-1})(q^n-q^{-n})^{-1}\sum_{j=1}^n q^{2j-n-1}e_{j,j}$. Let ${}^tU$ denote the transpose of U. Since ${}^tU(H\otimes 1)\,{}^tU^*=H\otimes 1$, i.e. $\sum_{k=1}^n q^{2k}x_{k,i}x_{k,j}^*=q^{2i}\delta_{i,j}$, it follows that

$$\begin{aligned}\sum_{i,j,k,l=1}^{n} e_{l,k}He_{i,j}\otimes x_{i,j}x_{k,l}^* &=\sum_{i,j,l=1}^{n}\frac{q^{2i-2}}{[n]_{q^2}}e_{l,j}\otimes x_{i,j}x_{i,l}^*\\ &=\sum_{j,l=1}^{n}e_{l,j}\otimes\left(\sum_{i=1}^{n}\frac{q^{2i-2}}{[n]_{q^2}}x_{i,j}x_{i,l}^*\right)\\ &=\sum_{j,l=1}^{n}e_{l,j}\otimes\frac{q^{2j-2}}{[n]_{q^2}}\delta_{j,l}1=H\otimes 1,\end{aligned}\tag{2}$$

where $[n]_{q^2} = (1-q^{2n})(1-q^2)^{-1}$. Hence, for any $x \in \mathcal{L}(\mathbf{C}^n)^{\otimes m}$, we have

$$\begin{aligned}(\varphi^{\otimes m} \otimes \mathrm{id})(\delta^{(m)}(x)) &= (\mathrm{Tr} \otimes \mathrm{id})((H^{\otimes m} \otimes 1)U^{(m)}(x \otimes 1)U^{(m)*}) \\ &= \sum \mathrm{Tr}\,(H^{\otimes m}(e_{i_1,j_1} \otimes \cdots \otimes e_{i_m,j_m})x(e_{l_1,k_1} \otimes \cdots \otimes e_{l_m,k_m})) \\ &\qquad \cdot x_{i_1,j_1} \cdots x_{i_m,j_m} x^*_{k_m,l_m} \cdots x^*_{k_1,l_1} \\ &= \sum \mathrm{Tr}\,((e_{l_1,k_1} \otimes \cdots \otimes e_{l_m,k_m})H^{\otimes m}(e_{i_1,j_1} \otimes \cdots \otimes e_{i_m,j_m})x) \\ &\qquad \cdot x_{i_1,j_1} \cdots x_{i_m,j_m} x^*_{k_m,l_m} \cdots x^*_{k_1,l_1}.\end{aligned}$$

By (2) we find that

$$(\varphi^{\otimes m} \otimes \mathrm{id})(\delta^{(m)}(x)) = \mathrm{Tr}((H^{\otimes m})x)1 = \varphi^{\otimes m}(x)1.$$

Thus the product state $\varphi^{\otimes m}$ on $\mathcal{L}(\mathbf{C}^n)^{\otimes m}$ satisfies the $SU_q(n)$-invariance, i.e. $(\varphi^{\otimes m} \otimes \mathrm{id})(\delta^{(m)}(x)) = (\varphi^{\otimes m})(x)1$ for all $x \in \mathcal{L}(\mathbf{C}^n)^{\otimes m}$.

Let $\lambda = q^2$ $(0 < q < 1)$ and R_λ the AFD factor of type III_λ:

$$(R_\lambda, \varphi_\lambda) = \bigotimes_{m=1}^{\infty} (\mathcal{L}(\mathbf{C}^n), \varphi_m), \qquad \varphi_m = \varphi.$$

Then φ_λ is a Powers (Araki-Woods) state on R_λ. In what follows, we may assume that $\{R_\lambda, \mathcal{H}\}$ is standard. Let π_m be the canonical embedding of $\mathcal{L}(\mathbf{C}^n)^{\otimes m}$ to R_λ such that $\pi_m(x) = x \otimes 1 \otimes 1 \otimes \cdots$ and $\mathcal{M}_m = \pi_m(\mathcal{L}(\mathbf{C}^n)^{\otimes m})$. Then $\varphi^{\otimes m} = \varphi_\lambda \circ \pi_m$. Let T_k denote the element $1^{\otimes(k-1)} \otimes T \otimes 1 \otimes \cdots$ in R_λ, where T is the R-matrix of type A_{n-1}. The set $\{T_k : k \in \mathbf{N}\}$ satisfies the relation of generators of Heck algebra.

Proposition 4.2. *There exists a product type right coaction δ of the Woronowicz algebra $L^\infty_q(SU(n))$ on the AFD factor R_λ of type* III_λ *such that*

(i) *δ is the extension to R_λ of $\delta^{(m)}$ for all m;*

(ii) *φ_λ is $SU_q(n)$-invariant, i.e. $(\varphi_\lambda \otimes \mathrm{id})(\delta(x)) = \varphi_\lambda(x)1$ for all $x \in R_\lambda$;*

(iii) *the fixed point subalgebra $(R_\lambda)^\delta$ by the coaction δ is an AFD factor of type* II_1 *generated by $\{T_k : k \in \mathbf{N}\}$; and*

(iv) *the restriction of φ_λ to the fixed point subalgebra $(\mathcal{M}_m)^\delta$ by δ is a Markov trace of modulus β for the inclusion $(\mathcal{M}_{m-1})^\delta \subset (\mathcal{M}_m)^\delta$, where $\beta = [2]_{q^2}[n]_{q^2}([n]_{q^2} - 1)^{-1}$.*

Proof. (i) By the $SU_q(n)$-invariance of the product state $\varphi^{\otimes m}$ under the coaction $\delta^{(m)}$, it holds that

$$(\varphi_\lambda \otimes \mathrm{id})(\delta^{(\infty)}(x)) = \varphi_\lambda(x)1, \quad x \in \mathcal{M}_\infty, \tag{3}$$

where $\mathcal{M}_\infty = \bigcup \mathcal{M}_m$ and $\delta^{(\infty)}$ is the natural extension of $\delta^{(m)}$ to $\mathcal{M}_\infty$. Since $\{R_\lambda, \mathcal{H}\}$ is standard, there exists a cyclic and separating vector ξ_λ for R_λ with $\varphi_\lambda(x) = (x\xi_\lambda|\xi_\lambda)$ for $x \in R_\lambda$. Let ξ_h be the cyclic and separating vector for M with $h(a) = (a\xi_h|\xi_h)$ for $a \in M$. Denote by p' the projection onto the closure of $\delta^{(\infty)}(\mathcal{M}_\infty)(\xi_\lambda \otimes \xi_h)$ which is a closed subspace of $\mathcal{H} \otimes L^2(M,h)$. The weak closure $\mathcal{N}$ of $\delta^{(\infty)}(\mathcal{M}_\infty)$ is a von Neumann subalgebra of $R_\lambda \overline{\otimes} M$ and the projection p' is an element in $\mathcal{N}'$. It is immediate from equation (3) that the linear mapping

$$x\xi_\lambda \to \delta^{(\infty)}(x)(\xi_\lambda \otimes \xi_h), \quad x \in \mathcal{M}_\infty$$

is isometric with respect to the norms in Hilbert spaces $\mathcal{H}$ and $\mathcal{H} \otimes L^2(M,h)$. There exists an isometry u from $\mathcal{H}$ onto $p'(\mathcal{H} \otimes L^2(M,h))$ with $u(x\xi_\lambda) = \delta^{(\infty)}(x)(\xi_\lambda \otimes \xi_h)$ for $x \in \mathcal{M}_\infty$. Then $\delta^{(\infty)}(x)p' = uxu^{-1}$ for $x \in \mathcal{M}_\infty$, and hence the *-isomorphism $\delta^{(\infty)}$ can be extended to a *-isomorphism from R_λ to $\mathcal{N}_{p'}$, where $\mathcal{N}_{p'}$ is the restriction of $\mathcal{N}$ to the range of p'. On the other hand, the vector $\xi_\lambda \otimes \xi_h$ is separating for $R_\lambda \overline{\otimes} M$ as well as $\mathcal{N}$. Hence $\mathcal{N}_{p'}$ is *-isomorphic to $\mathcal{N}$. Here we denote the composite of the above two *-isomorphisms $R_\lambda \to \mathcal{N}_{p'}$ and $\mathcal{N}_{p'} \to \mathcal{N}$ by δ. Then δ is an extension of $\delta^{(\infty)}$, which is a normal *-isomorphism of R_λ to $R_\lambda \overline{\otimes} M$. Since $\delta^{(\infty)}$ satisfies the equation $(\delta^{(\infty)} \otimes \mathrm{id}) \circ \delta^{(\infty)} = (\mathrm{id} \otimes \delta_{\mathbf{W}}) \circ \delta^{(\infty)}$ on $\mathcal{M}_\infty$, the extension δ satisfies the same equation. Thus δ is a right coaction of $\mathbf{W}$ on R_λ.

(ii) The $SU_q(n)$-invariance of the state φ_λ under the coaction δ is clear from (3).

(iii) Suppose that x is an element in the fixed point subalgebra $(R_\lambda)^\delta$. It is apporoximated by a net $\{x_i\}$ in $\mathcal{M}_\infty$ in the σ-weak topology. Put $y_i = (\mathrm{id} \otimes h)(\delta(x_i))$. Since the Haar measure is left invariant, we have

$$\begin{aligned} \delta(y_i) &= (\mathrm{id} \otimes \mathrm{id} \otimes h)((\delta \otimes \mathrm{id})(\delta(x_i))) \\ &= (\mathrm{id} \otimes \mathrm{id} \otimes h)((\mathrm{id} \otimes \delta_{\mathbf{W}})(\delta(x_i))) = y_i \otimes 1 \end{aligned}$$

and hence $y_i \in (R_\lambda)^\delta$. Each x_i belongs to $\mathcal{M}_m$ for some $m \in \mathbf{N}$. As $\delta(x_i) = \delta^{(m)}(x_i)$ belongs to $\mathcal{M}_m \overline{\otimes} M$, the element y_i belongs to $(\mathcal{M}_m)^\delta$. Further, $\delta(y_i)$ converges σ-weakly to $(\mathrm{id} \otimes \mathrm{id} \otimes h)((\delta \otimes \mathrm{id})(\delta(x))) = \delta(x)$ and hence y_i converges σ-weakly to x. This means that $(\mathcal{M}_\infty)^\delta$ is σ-weakly dense in $(R_\lambda)^\delta$. It is known that $(\mathcal{M}_m)^\delta$ is generated by $T_1, \cdots, T_{m-1}$. Hence $(R_\lambda)^\delta$ is generated by $\{T_j : j \in \mathbf{N}\}$. On the other hand, the density matrix $H \otimes H$ of the product state $\varphi \otimes \varphi$ commutes with T. Hence $\varphi_\lambda(T_jT_k) = \varphi_\lambda(T_kT_j)$ for $j, k \in \mathbf{N}$. The restriction of the state φ_λ to the fixed point subalgrbra $(R_\lambda)^\delta$ is then a normalized faithful normal trace on it. Consequently, the von Neumann algebra $(R_\lambda)^\delta$ is finite.

Let ρ be the unilateral shift on $(R_\lambda, \varphi_\lambda) = \bigotimes_{m=1}^\infty(\mathcal{L}(\mathbf{C}^n), \varphi_m)$, namely, the unital injective normal *-endomorphism on R_λ with

$$\rho(x_1 \otimes x_2 \otimes \cdots) = 1 \otimes x_1 \otimes x_2 \otimes \cdots.$$

Then $(\rho \otimes \mathrm{id}) \circ \delta = \delta \circ \rho$ holds on the fixed point subalgebra $(R_\lambda)^\delta$. Using the conditions for the generators of a Hecke algebra, we have

$$\mathrm{Ad}_{T_1T_2\cdots T_m}(T_j) = T_{j+1} \quad \text{for} \quad j < m,$$

where $\mathrm{Ad}_a(b) = aba^{-1}$. Let ρ_∞ be the inductive limit of the action ρ_m defined by $\mathrm{Ad}_{T_1 \cdots T_m}$ on $(\mathcal{M}_\infty)^\delta$. Then it is a $*$-endomorphism with $\rho_\infty = \rho$ on $(\mathcal{M}_\infty)^\delta$. Let τ be the above trace on $(R_\lambda)^\delta$. Suppose that z is an element in the center of the fixed point subalgebra $(R_\lambda)^\delta$. For any $\varepsilon > 0$ there exists an element y in $(\mathcal{M}_\infty)^\delta$ with $\|y - z\|_2 < \varepsilon$, where $\|a\|_2 = \tau(a^*a)^{1/2}$. Fix an integer $m_0 \in \mathbf{N}$ with $y \in (\mathcal{M}_{m_0})^\delta$. If $x \in \mathcal{M}_\infty$, then

$$\tau(zx) = \tau(\rho_m^k(zx)) = \tau(z\rho_m^k(x)) = \tau(z\rho_\infty^k(x)) = \tau(z\rho^k(x))$$

for any $m, k \in \mathbf{N}$ and $\tau(y\rho^k(x)) = \tau(y)\tau(x)$ for $k \geq m_0$. Therefore

$$\begin{aligned} |\tau(zx) - \tau(z)\tau(x)| &= |\tau(z\rho^k(x)) - \tau(z)\tau(x)| \\ &\leq |\tau((z-y)\rho^k(x))| + |\tau(y-z)||\tau(x)| \leq 2\|z-y\|_2\|x\| < 2\varepsilon\|x\|, \end{aligned}$$

for $k \geq m_0$. Hence $z = \tau(z)1$. Consequently, the von Neumann algebra $(R_\lambda)^\delta$ is a factor of type II_1.

(iv) The R-matrix T of type A_{n-1} has the spectral decomposition $q(1 - e) - q^{-1}e$, where e is the projection

$$\sum_{i<j} \frac{1}{q+q^{-1}} \left(q^{-1} e_{i,i} \otimes e_{j,j} - e_{i,j} \otimes e_{j,i} - e_{j,i} \otimes e_{i,j} + q e_{j,j} \otimes e_{i,i} \right).$$

Put $e_j = (q+q^{-1})^{-1}(q1 - T_j)$ for $j \in \mathbf{N}$. Then $\{e_j : j \in \mathbf{N}\}$ is a subset of projections in $(R_\lambda)^\delta$ which satisfies

$$\begin{cases} e_i e_j e_i - \frac{1}{(q+q^{-1})^2} e_i = e_j e_i e_j - \frac{1}{(q+q^{-1})^2} e_j & (|i-j| = 1) \\ e_i e_j = e_j e_i & (|i-j| \neq 1). \end{cases}$$

It is straight forward to show that

a) $\varphi_\lambda(e_j) = \beta^{-1}$ for all $j \in \mathbf{N}$;
b) $\varphi_\lambda(e_i e_j) = \varphi_\lambda(e_i)\varphi_\lambda(e_j)$ if $i \neq j$; and
c) $\varphi_\lambda(x e_m) = \beta^{-1}\varphi_\lambda(x)$ for $x \in (\mathcal{M}_m)^\delta$.

Hence φ_λ is a Markov trace on $(R_\lambda)^\delta$, [5]. We begin by showing assertion a) as follows:

$$\begin{aligned} &\left(\frac{q^n - q^{-n}}{q - q^{-1}}\right)^2 (\varphi \otimes \varphi)(e) \\ &= \frac{q^{-1}}{q+q^{-1}} \sum_{i<j} q^{(2i-n-1)+(2j-n-1)} + \frac{q}{q+q^{-1}} \sum_{i<j} q^{(2j-n-1)+(2i-n-1)} \\ &= \frac{1}{2}\left\{ \left(\sum_{i=1}^n q^{2i-n-1} \right)^2 - \sum_{i=1}^n q^{2(2i-n-1)} \right\} \\ &= \frac{1}{2}\left\{ \left(\frac{q^n - q^{-n}}{q - q^{-1}} \right)^2 - \frac{q^{2n} - q^{-2n}}{q^2 - q^{-2}} \right\}, \end{aligned}$$

hence $(\varphi \otimes \varphi)(e) = \beta^{-1}$ and so $\varphi_\lambda(e_j) = \beta^{-1}$.

If $|i-j| \geq 2$, then $\varphi_\lambda(e_i e_j) = \varphi_\lambda(e_i)\varphi_\lambda(e_j)$. It remains to show the case where $|i-j| = 2$. For this it suffices to show that $\varphi^{\otimes 3}((e \otimes 1)(1 \otimes e)) = \beta^{-2}$. Since the value of the state $\varphi^{\otimes 3}$ on the off diagonal part of $(e \otimes 1)(1 \otimes e)$ vanishes, we have only to consider the diagonal part, which is given by

$$\frac{1}{(q+q^{-1})^2}\left(\sum_{i<j<l} q^{-2} e_{i,i} \otimes e_{j,j} \otimes e_{l,l} + \sum_{i<j,k<j} e_{i,i} \otimes e_{j,j} \otimes e_{k,k} \right.$$
$$\left. + \sum_{i<j,i<l} e_{j,j} \otimes e_{i,i} \otimes e_{l,l} + \sum_{k<i<j} q^2 e_{j,j} \otimes e_{i,i} \otimes e_{k,k}\right).$$

Therefore

$$\left(\frac{q^n - q^{-n}}{q-q^{-1}}\right)^3 \varphi^{\otimes 3}((e \otimes 1)(1 \otimes e))$$
$$= \frac{q^{-3n-3}}{(q+q^{-1})^2}\left(\sum_{i<j<l} q^{-2} q^{2(i+j+l)} + \sum_{i<j,l<j} q^{2(i+j+l)} \right.$$
$$\left. + \sum_{i<j,i<k} q^{2(i+j+k)} + \sum_{k<i<j} q^2 q^{2(i+j+k)}\right)$$
$$= \frac{q^{-3n-3}}{(q+q^{-1})^2}\left(\sum_{1<j<l} \frac{q^2 - q^{2j}}{1-q^2} q^{2(j+l)-2} + \sum_{l<j} \frac{q^2 - q^{2j}}{1-q^2} q^{2(j+l)} \right.$$
$$\left. + \sum_{i<k} \frac{q^{2i+2} - q^{2n+2}}{1-q^2} q^{2(i+k)} + \sum_{k<i<n} \frac{q^{2i+2} - q^{2n+2}}{1-q^2} q^{2(i+k)+2}\right).$$

Summing over l and k, we have

$$(q^{3n+3}(q+q^{-1})^2)\left(\frac{q^n - q^{-n}}{q-q^{-1}}\right)^3 \varphi^{\otimes 3}((e \otimes 1)(1 \otimes e))$$
$$= \frac{q^2 - q^{2n}}{1-q^2}\left\{\sum_{j=2}^{n} \frac{(q^2 - q^{2j})q^{2j}}{1-q^2} + \sum_{i=1}^{n-1} \frac{(q^{2i+2} - q^{2n+2})q^{2i+2}}{1-q^2}\right\}$$
$$= \frac{q^2 - q^{2n}}{1-q^2} \frac{(q^4 - q^{2n+4})(q^2 - q^{2n})}{(1-q^2)^2}.$$

Consequently, we have $\varphi^{\otimes 3}((e \otimes 1)(1 \otimes e)) = \beta^{-2}$. Thus assertion (b) is proved. Assertion (c) is an immediate consequence from assertions (a) and (b).

Q.E.D.

When $n = 2$, $\{T_k : k \in \mathrm{N}\}$ generates the Temperly-Lieb algebra and the Markov trace gives the Jones polynomials. When $n > 2$, the Markov trace gives the two variable Jones polynomials, [1,6].

5. Takeseki duality for quantum group $SU_q(n)$

The following Definition 5.1, Lemmas 5.2 and 5.3, and the commutativity of the diagram in Theorem 5.5(ii) are essentially the same as the assertions for the Kac algebras obtained by Enock [3]. See also [13].

Definition 5.1. Let δ be the right coaction of a Woronowicz algebra $\mathbf{W}$ on a von Neumann algebra $\mathcal{M}$. The crossed product $\mathcal{M} \times_\delta \mathbf{W}$ of $\mathcal{M}$ by $\mathbf{W}$ with respect to δ is the von Neumann algebra generated by $\delta(\mathcal{M})$ and $\mathbf{C} \otimes \widehat{M}'$, where $\widehat{M}'$ is the von Neumann algrebra of the commutant Woronowicz algebra $\widehat{\mathbf{W}}'$ of the dual Woronowicz algebra $\widehat{\mathbf{W}}$.

Lemma 5.2. *Let $\widehat{\delta}$ be the mapping defined by*

$$\widehat{\delta}(x) = \mathrm{Ad}_{1\otimes((J\hat{J}\otimes J\hat{J})W^*(J\hat{J}\otimes J\hat{J}))}(x \otimes 1), \ \ x \in \mathcal{M} \times_\delta \mathbf{W}.$$

Then $\widehat{\delta}$ is a right coaction of $\widehat{\mathbf{W}}'$ on the crossed product $\mathcal{M} \times_\delta \mathbf{W}$.

Proof. Let $\widehat{R}$ be the unitary antipode of $\widehat{\mathbf{W}}$. Since $\widehat{R}(b) = Jb^*J$ by [12], it follows from $\delta_{\widehat{\mathbf{W}}} = \sigma \circ (\widehat{R} \otimes \widehat{R}) \circ \delta_{\widehat{\mathbf{W}}} \circ \widehat{R}$ that

$$\delta_{\widehat{\mathbf{W}}}(b) = \mathrm{Ad}_{(J\otimes J)W^*(J\otimes J)}(b \otimes 1), \quad b \in \widehat{M}.$$

Since $\delta_{\widehat{\mathbf{W}}'} \circ \mathrm{Ad}_{\hat{J}} = \mathrm{Ad}_{\hat{J}\otimes\hat{J}} \circ \delta_{\widehat{\mathbf{W}}}$, it follows that

$$\delta_{\widehat{\mathbf{W}}'}(b') = \mathrm{Ad}_{(J\hat{J}\otimes J\hat{J})W^*(J\hat{J}\otimes J\hat{J})}(b' \otimes 1), \ b' \in \widehat{M}'.$$

Therefore $\widehat{\delta}$ agrees with $\mathrm{id} \otimes \delta_{\widehat{\mathbf{W}}'}$ on $\mathbf{C} \otimes \widehat{M}'$. Since $W \in M \overline{\otimes} \widehat{M}$ and $R(a) = \hat{J}a^*\hat{J}$ by [12], we see that $\widehat{\delta}(\delta(x)) = \mathrm{Ad}_{1\otimes((J\hat{J}\otimes J\hat{J})W^*(J\hat{J}\otimes J\hat{J}))}(\delta(x) \otimes 1) = \delta(x) \otimes 1$ for $x \in \mathcal{M}$. Thus $\widehat{\delta}$ maps $\mathcal{M} \times_\delta \mathbf{W}$ to $(\mathcal{M} \times_\delta \mathbf{W})\overline{\otimes}\widehat{M}'$ and satisfies $(\widehat{\delta} \otimes \mathrm{id}) \circ \widehat{\delta} = (\mathrm{id} \otimes \delta_{\widehat{\mathbf{W}}'}) \circ \widehat{\delta}$. Hence $\widehat{\delta}$ is the right coaction of $\widehat{\mathbf{W}}'$ on $\mathcal{M} \times_\delta \mathbf{W}$.

Q.E.D.

The mapping $\widehat{\delta}$ is called the *dual* coaction of δ. Since $((\widehat{\mathbf{W}}')^\wedge)' = (\mathbf{W}^o)'$, the bidual coaction $\widehat{\widehat{\delta}}$ is defined to be a right coaction of $(\mathbf{W}^o)'$ on $\mathcal{N} \times_{\widehat{\delta}} \widehat{\mathbf{W}}'$ with $\mathcal{N} = \mathcal{M} \times_\delta \mathbf{W}$:

$$\widehat{\widehat{\delta}}(y) = \mathrm{Ad}_{1_{\mathcal{N}}\otimes((J\hat{J}\otimes J\hat{J})(\widehat{W}')^*(J\hat{J}\otimes J\hat{J}))}(y \otimes 1). \tag{4}$$

Lemma 5.3. *Let δ be a right coaction of $\mathbf{W}$ on $\mathcal{M}$. The double crossed product $(\mathcal{M} \times_\delta \mathbf{W}) \times_{\widehat{\delta}} \widehat{\mathbf{W}}'$ is $*$-isomorphic to the von Neumann algebra $\delta(\mathcal{M}) \vee (\mathbf{C} \otimes \mathcal{L}(L^2(M,h)))$.*

Proof. We will show that the composite mapping $\pi = \mathrm{Ad}_{1\otimes\widehat{W}'} \circ (\delta \otimes \mathrm{id})$ gives a *-isomorphism of $\delta(\mathcal{M}) \vee (\mathbf{C} \otimes \mathcal{L}(L^2(M,h)))$ onto the double crossed product $(\mathcal{M} \times_\delta \mathbf{W}) \times_{\widehat{\delta}} \widehat{\mathbf{W}}'$, where $\widehat{W}'$ is the Kac-Takesaki operator for $\widehat{\mathbf{W}}'$. Since $R(a) = \hat{J}a^*\hat{J}$ for $a \in M$, it follows that

$$\begin{aligned}\delta_{\mathbf{W}}(a) &= \sigma \circ (R \otimes R) \circ \delta_{\mathbf{W}} \circ R(a) \\ &= \mathrm{Ad}_{(\hat{J}\otimes\hat{J})\sigma(W)(\hat{J}\otimes\hat{J})}(a \otimes 1) = \mathrm{Ad}_{(\widehat{W}')^*}(a \otimes 1). \end{aligned} \tag{5}$$

Therefore, if $x \in \mathcal{M}$, then

$$\pi(\delta(x)) = \mathrm{Ad}_{1\otimes\widehat{W}'}((\delta \otimes \mathrm{id})(\delta(x)) = \mathrm{Ad}_{1\otimes\widehat{W}'}((\mathrm{id} \otimes \delta_{\mathbf{W}})(\delta(x))) = \delta(x) \otimes 1.$$

On the other hand, $\widehat{\delta}(\delta(x)) = \delta(x) \otimes 1$ as in the proof of Lemma 5.2. Hence $\pi(\delta(x)) = \widehat{\delta}(\delta(x))$ for $x \in \mathcal{M}$. If $b' \in \widehat{M}'$ and $a' \in M'$ $((M^o)' = M')$, then

$$\pi(1 \otimes b') = \mathrm{Ad}_{1\otimes\widehat{W}'}(1 \otimes 1 \otimes b') = \widehat{\delta}(1 \otimes b')$$

and

$$\pi(1 \otimes a') = \mathrm{Ad}_{1\otimes\widehat{W}'}(1 \otimes 1 \otimes a') = 1 \otimes 1 \otimes a'.$$

It is known that $M' \vee \widehat{M}' = \mathcal{L}(L^2(M,h))$, for $M \cap \widehat{M} = \mathbf{C}$ by [12]. Consequently, the mapping π is the desired *-isomorphism.

Q.E.D.

Lemma 5.4. *Let* $\mathbf{W}$ *be the Woronowicz algebra* $L_q^\infty(SU(n))$.

(i) *There exists a counit* $\tilde{\varepsilon}$ *on* $\mathbf{W}$ *which is a Hahn-Banach extension of the counit of* $C_q(SU(n))$ *(or* $A_q(SU(n))$*).*

(ii) *There exists a counit on the dual Woronowicz algebra* $\widehat{\mathbf{W}}$.

Proof. (i) Let $\tilde{\varepsilon}$ be a Hahn-Banach extension to M of the counit ε of $C_q(SU(n))$ (or $A_q(SU(n))$), where M is the von Neumann algebra of $\mathbf{W}$. Then $\tilde{\varepsilon}(1) = 1 = \| \tilde{\varepsilon} \|$, which implies that $\tilde{\varepsilon}$ is a state on M. It suffices to show the counit condition: $(\tilde{\varepsilon} \otimes \mathrm{id}) \circ \delta_{\mathbf{W}} = \mathrm{id} = (\mathrm{id} \otimes \tilde{\varepsilon}) \circ \delta_{\mathbf{W}}$. Let $W^{(\gamma)} = (w_{i,j}^{(\gamma)})_{i,j\in I_\gamma}$ be an irreducible unitary representation of the quantum group $SU_q(n)$ associated with $\gamma \in \widehat{SU(n)}$ such that $w_{i,j}^{(\gamma)} \in A_q(SU(n))$ for all $i,j \in I_\gamma$. Denote $SU(n)$ by G. Then the Kac-Takesaki operator $W \in M \overline{\otimes} \widehat{M}$ is expressed in the form

$$W = \sum_{\gamma\in\widehat{G}} W^{(\gamma)} \otimes 1_\gamma = \sum_{\gamma\in\widehat{G}} \sum_{i,j\in I_\gamma} \frac{\| \xi_j \|}{\| \xi_i \|} w_{i,j}^{(\gamma)} \otimes (e_{j,i}^{(\gamma)} \otimes 1_\gamma), \tag{6}$$

where $1_\gamma = \sum_{k\in I_\gamma} e_{k,k}^{(\gamma)}$, $\sum_{\gamma\in\widehat{G}} \sum_{i\in I_\gamma} e_{i,i}^{(\gamma)} \otimes 1_\gamma = 1$, (ξ_j) is the first colum of $W^{(\gamma)}$ and $\|\xi\| = h(\xi^*\xi)^{1/2}$. Since each $W^{(\gamma)}$ is irreducible and unitary, $\|\xi_i\| \neq 0$ for all $i \in I_\gamma$. The comultiplication $\delta_{\mathbf{W}}$ is of the form

$$\delta_{\mathbf{W}}(a) = \sum_{\gamma,\gamma'\in\widehat{G}} \sum_{i,j\in I_\gamma} \sum_{k,l\in I_{\gamma'}} \frac{\| \xi_j \| \| \xi_l \|}{\| \xi_i \| \| \xi_k \|} w_{i,j}^{(\gamma)} w_{k,l}^{(\gamma')*} \otimes (e_{j,i}^{(\gamma)} \otimes 1_\gamma) a (e_{k,l}^{(\gamma')} \otimes 1_{\gamma'})$$

the unitary v belongs to $\mathbf{C} \otimes M' \overline{\otimes} \widehat{M}'$. Further, $\widehat{\delta}(y) = v(y \otimes 1)v^*$ for $y \in \mathcal{N}$ by Lemm 5.2.

Let $\widehat{\varepsilon}$ be the counit on $\widehat{\mathbf{W}}$ obtained in Lemma 5.4(ii). Then $\widehat{\varepsilon}'$ defined by $\widehat{\varepsilon} \circ \mathrm{Ad}_{\hat{J}}$ is the counit of $\widehat{\mathbf{W}}'$. Thus

$$(\mathrm{id} \otimes \widehat{\varepsilon}') \circ \delta_{\widehat{\mathbf{W}}'} = \mathrm{id} = (\widehat{\varepsilon}' \otimes \mathrm{id}) \circ \delta_{\widehat{\mathbf{W}}'},$$

where $\delta_{\widehat{\mathbf{W}}'}$ is the comultiplication of $\widehat{\mathbf{W}}'$ and $\delta_{\widehat{\mathbf{W}}'}(a) = \widehat{W}'(1 \otimes a)\widehat{W}'^*$ for $a \in \widehat{M}'$.

According to the same discussion as the first step in the proof of (i), we have an equality corresponding to (9):

$$\widehat{\delta}(\mathcal{N}) = \Big(\mathcal{N} \overline{\otimes} \mathcal{L}(L^2(M,h))\Big) \cap \Big(v(\mathcal{L}(\mathcal{H} \otimes L^2(M,h)) \otimes \mathbf{C})v^*\Big)$$

by using the property of the counit $\widehat{\varepsilon}'$ on $\widehat{M}'$. We also have an equality corresponding to (10):

$$\widehat{\delta}(\mathcal{N}) \;=\; \Big(\mathcal{N} \overline{\otimes} \mathcal{L}(L^2(M,h))\Big) \cap (v(\mathbf{C}1_{\mathcal{N}} \otimes M)v^*)' \cap (\mathbf{C}1_{\mathcal{N}} \otimes \widehat{M})' \qquad (12)$$

by using the fact that $M' \cap \widehat{M}' = \mathbf{C}$. It remains to show that if $z \in \mathcal{N} \times_{\widehat{\delta}} \widehat{\mathbf{W}}'$ satisfies $\widehat{\widehat{\delta}}(z) = z \otimes 1$, then z belongs to the right hand side of (12). By the same reason as the proof of (i), we have only to show that z commutes with both $v(\mathbf{C}1_{\mathcal{N}} \otimes M)v^*$ and $\mathbf{C}1_{\mathcal{N}} \otimes \widehat{M}$.

If $a \in M$, then

$$\mathrm{Ad}_v(1_{\mathcal{N}} \otimes a) = \mathrm{Ad}_{J_{\mathcal{M}} \otimes J \otimes \hat{J}}(\mathrm{Ad}_{1_{\mathcal{M}} \otimes W}(1_{\mathcal{N}} \otimes R(a^*))),$$

where R is the unitary antipode of $\mathbf{W}$ and is written in the form $R(a) = \hat{J}a^*\hat{J}$. Hence, by $\delta_{\mathbf{W}} \circ R = (R \otimes R) \circ \sigma \circ \delta_{\mathbf{W}}$,

$$\mathrm{Ad}_v(1_{\mathcal{N}} \otimes a) = \mathrm{Ad}_{1_{\mathcal{M}} \otimes J\hat{J} \otimes 1}(1_{\mathcal{M}} \otimes (\sigma \circ \delta_{\mathbf{W}})(a^*)),$$

which commutes with $\mathbf{C}1_{\mathcal{N}} \otimes M'$. It is clear that it commutes with $\widehat{\delta}(\mathcal{N})$. Therefore the element z in $\mathcal{N} \times_{\widehat{\delta}} \widehat{\mathbf{W}}'$ commutes with $v(\mathbf{C}1_{\mathcal{N}} \otimes M)v^*$.

The bidual coaction $\widehat{\widehat{\delta}}$ is implemented by

$$1_{\mathcal{N}} \otimes (\hat{J} \otimes J)\widehat{W}'(\hat{J} \otimes J) = (1_{\mathcal{N}} \otimes 1 \otimes J\hat{J})(1_{\mathcal{N}} \otimes \sigma(W^*))(1_{\mathcal{N}} \otimes 1 \otimes J\hat{J}).$$

If $\widehat{\widehat{\delta}}(z) = z \otimes 1$, then z belongs to $\mathcal{L}(\mathcal{H} \otimes L^2(M,h)) \overline{\otimes} \widehat{M}'$. Thus z commutes with $\mathbf{C}1_{\mathcal{N}} \otimes \widehat{M}$.

Q.E.D.

This theorem gives us an Araki-Connes-Takesaki type decomposition for AFD factors of type III_λ , i.e. an AFD factor of type III_λ is expressed by a crossed product of an AFD factor of type II_∞ by $L_q^\infty(SU(n))^{\frown\prime}$ as follows:

Theorem 5.6. *Let δ be the right coaction of $L_q^\infty(SU(n))$ on R_λ in Proposition 4.2.*

(i) $R_\lambda \times_\delta L_q^\infty(SU(n)) \cong (R_\lambda)^\delta \overline{\otimes} \mathcal{L}(L^2(M,h))$.

(ii) $R_\lambda \cong \left(R_\lambda \times_\delta L_q^\infty(SU(n))\right) \times_{\widehat{\delta}} L_q^\infty(SU(n))^{\frown\prime}$.

Proof. (i) Let W be the Kac-Takesaki operator for $\mathbf{W}$. Then by (6) it is expressed in the form

$$W = \sum_{\gamma\in\widehat{G}} \sum_{i,j\in I_\gamma} \frac{\|\xi_j\|}{\|\xi_i\|} w_{i,j}^{(\gamma)} \otimes (e_{j,i}^{(\gamma)} \otimes 1_\gamma)$$

with $w_{i,j}^{(\gamma)} \in A_q(G)$, where $G = SU(n)$. Due to the Weyl's reciprocity theorem for the product representation, there exists a unitary representation $(x_{i,j}^{(\gamma)})$ such that $x_{i,j}^{(\gamma)} \in R_\lambda$ and $\delta(x_{i,j}^{(\gamma)}) = \sum_{k\in I_\gamma} x_{i,k}^{(\gamma)} \otimes w_{k,j}^{(\gamma)*}$. Here we set

$$u(\gamma) = \sum_{i,j\in I_\gamma} \frac{\|\xi_i\|}{\|\xi_j\|} x_{i,j}^{(\gamma)} \otimes (e_{i,j}^{(\gamma)} \otimes 1_\gamma), \qquad \gamma \in \widehat{G}.$$

Since $e_{i,j}^{(\gamma)} \otimes 1_\gamma \in \widehat{M}$, the unitary $u(\gamma)$ belongs to the von Neumann algebra $R_\lambda \overline{\otimes} \widehat{M}$ and satisfies

$$\begin{aligned}
\overline{\delta}(u(\gamma)) &= \sum_{i,j,k\in I_\gamma} \frac{\|\xi_i\|}{\|\xi_j\|} x_{i,k}^{(\gamma)} \otimes (e_{i,j}^{(\gamma)} \otimes 1_\gamma) \otimes w_{k,j}^{(\gamma)*} \\
&= \left(\sum_{i,k\in I_\gamma} \frac{\|\xi_i\|}{\|\xi_k\|} x_{i,k}^{(\gamma)} \otimes (e_{i,k}^{(\gamma)} \otimes 1_\gamma) \otimes 1\right) \\
&\qquad \cdot \left(\sum_{\gamma'\in\widehat{G}} \sum_{l,j\in I_{\gamma'}} \frac{\|\xi_l\|}{\|\xi_j\|} (1 \otimes (e_{l,j}^{(\gamma')} \otimes 1_{\gamma'}) \otimes w_{l,j}^{(\gamma')*})\right) \\
&= (u(\gamma) \otimes 1)(1 \otimes \sigma(W^*)),
\end{aligned}$$

where $\overline{\delta} = (\mathrm{id} \otimes \sigma) \circ (\delta \otimes \mathrm{id})$ is the right coaction of $\mathbf{W}$ on $R_\lambda \overline{\otimes} \mathcal{L}(L^2(M,h))$. Let $u = \sum_{\gamma\in\widehat{G}} u(\gamma)$. Then u is a unitary in $R_\lambda \overline{\otimes} \widehat{M}$ satisfying $\overline{\delta}(u) = (u \otimes 1)(1 \otimes \widehat{W})$, for $\widehat{W} = \sigma(W^*)$. Consequently, the right coaction $\tilde{\delta}$ is conjugate to $\overline{\delta}$ on $R_\lambda \overline{\otimes} \mathcal{L}(L^2(M,h))$.

(ii) It is clear from Theorem 5.5.

Q.E.D.

References

[1] Deguchi, T., M. Wadati and Y. Akutsu, Knot theory based on solvable models at criticality, "Integrable Systems in Quantum Field Theory and Statistical Mechanics", *Adv. Studies in Pure Math.*, **19**, 193-285, 1989.

[2] Effros, E. G. and Z. Ruan, Operator convolution algebras: an approach to quantum groups, preprint, 1991.

[3] Enock, M., Produit croisé d'une algèbre de von Neumann par une algèbre de Kac, *J. Functional Analysis,* **27**(1977), 16-47.

[4] Enock, M. and J. M. Schwartz, Une dualité dans les algèbre de von Neumann, *Bull. Soc. Math. France Supple. Mem.*, **44**(1975), 1-144.

[5] Goodmann, F. M., P. de la Harpe and V. F. R. Jones, *Coxeter Graphs and Towers of Algebras,* MSRI Publ. **14**, Springer-Verlag, 1989.

[6] Jimbo, M., A q-difference analogue of $U(\mathrm{g})$ and Yang-Baxter equation, *Letter in Math. Phys.,* **10**(1985), 63-69.

[7] Jones, V. R. F., *Subfactors and Knots,* Regional Conference Ser. **80**, Amer. Math. Soc., 1991.

[8] Konishi, Y., A note on actions of compact matrix quantum groups on von Neumann algebras, preprint of Niigata Univ., 1991.

[9] Kurose, H. and Y. Nakagami, Representation of Quantum $U_q(su(n))$ in unbounded operator algebras, in preparation.

[10] Masuda, T., K. Mimachi, Y. Nakagami, M. Noumi and K. Ueno, Representation of quantum groups, *Mapping of Operater Algebras,* Proc. Japan-U.S. Joint Seminar, Ed. by H. Araki and R. V. Kadison, 119-128, Birkhäuser, 1990.

[11] Masuda, T. and Y. Nakagami, An operator algebraic framework for the duality of quantum groups, *Mathematical Physics X,* Proc. IAMP-91, Ed. by K. Schmüdgen, Springer-Verlag, 1992.

[12] Masuda, T. and Y. Nakagami, in preparation.

[13] Nakagami, Y. and M. Takesaki, *Duality of the Crossed Product of von Neumann Algebras,* Lecture Notes in Math., **731**, Springer-Verlag, 1979.

[14] Noumi, M., H. Yamada and K. Mimachi, Finite dimentional representations of the quantum group $GL_q(n; \mathbf{C})$ the zonal spherical function on $U_q(n-1)\backslash U_q(n)$, preprint of Nagoya Univ.

[15] Reshetikhin, N. Yu., L. A. Takhitadzhyan and L. D. Faddeev, Quantization of Lie groups and Lie algebras, *Leningrad Math.J.,* **1**(1990), 193-225.

[16] Woronowicz, S. L., Tannaka-Krein duality for compact matrix pseudogroups. Twisted $SU(\mathrm{N})$ groups, *Invent. Math.,* **93**(1989), 193-285.

[17] Wassermann, A., Coactions and Yang-Baxter equations for ergodic actions and subfactors, *Operator Algebras and Applications,* Ed. by D. Evans and M. Takesaki, London Math. Soc. Lecture Note Ser. **136**, 203-236, Cambridge Univ. Press, 1988.

4. Subfactors and Index Theory

A New Role of Graph Projections in Index Theory

T. NATSUME*
State University of New York at Buffalo
Buffalo, NY 14214, U. S. A.

Abstract. The orthogonal projection corresponding to the graph of a closed operator is a useful tool in Operator Theory. We will show that it is also useful in noncommutative geometry.

Key words: Cyclic cocycle, graph projection, K-theory.

1. Background

A densely defined closed operator T from a Hilbert space H_1 to a second Hilbert space H_2 gives rise to various bounded operators. For instance, the partial isometry part of T in its polar decomposition is such an operator. It is a common method to use partial isometry parts in Operator Theory. This method is also used in noncommutative geometry to produce Fredholm modules [2] .

The graph of a closed operator is, by definition, a closed subspace of $H_1 \oplus H_2$. Denote by e_T the corresponding orthogonal projection. The operator e_T is another bounded operator associated to T. It is an old idea to use those projections in the study of closed operators. This projection was introduced by M. H. Stone and J. von Neumann [9], [10] . They called e_T the characteristic projection of T . Since the word "characteristic" seems to be overused over many years, we call e_T the graph projection of T, following the terminology of [5]. The analysis of graph projections has been quite successful. For instance, in the beautiful paper [8] of A. E. Nussbaum, it is applied to establishing a complete reduction theory for closed operators.

Recently, graph projections have found its new role in K-theory of C^*-algebras.

2. K-theory

In [5] graph projections are used to produce a continuous field of projections. It is shown there that the Bott periodicity theorem for C^*-algebras can be deduced from a description of a particular field of C^*-algebras over $[0,1]$.

* Partially supported by NSF Grant

H. Araki et al. (eds.), Quantum and Non-Commutative Analysis, 285–290.

The fibres of this field, labelled by h, are $C_0(\mathbf{R}^2)$ when $h = 0$, and the C^*-algebra $\mathcal{K}$ of all compact operators on $L^2(\mathbf{R})$ when $h \neq 0$. The idea of the proof is to reduce the calculation of the K-groups of $A \otimes C_0(\mathbf{R}^2)$, where A is a given C^*-algebra, to the calculation of $A \otimes \mathcal{K}$. It is further reduced to showing that the Bott projection in the fibre $M_2(C_0(\mathbf{R}^2)^\sim)$ extends to a projection in each fibre $M_2(\mathcal{K}^\sim)$, which together with the unit of $\mathcal{K}^\sim$ generates $K_0(\mathcal{K}^\sim)$. This is done by explicitly constructing a continuous field of projections.

When one identifies $\mathbf{R}^2$ with the complex plane $\mathbf{C}$, the Bott projection b is given by

$$b = \begin{bmatrix} \frac{1}{1+|z|^2} & \frac{\overline{z}}{1+|z|^2} \\ \frac{z}{1+|z|^2} & \frac{|z|^2}{1+|z|^2} \end{bmatrix},$$

and the difference $[b] - \left[\begin{pmatrix} 0 & 0 \\ 1 & 1 \end{pmatrix}\right]$ is the generator of $K_0(C_0(\mathbf{R}^2)) \cong \mathbf{Z}$. The projection b amounts to be the graph projection of the multiplication operator T by the function z . The function z is the symbol of the differential operator

$$D = x + \frac{d}{dx}$$

on $\mathbf{R}$. If you let $D_h = x + h\frac{d}{dx}$, $0 < h \leq 1$, the graph projections e_h of D_h belong to $\mathcal{K}^\sim$. Moreover $\{e_h\}$ is an extension of the Bott projection b for the field over $[0,1]$. Of course, the point is continuity at $h = 0$.

In this way, not only a simple proof of the Bott periodicity theorem, but also a simple proof of the Connes isomorphism theorem for one-parameter crossed products is obtained in [5].

3. Index Theorem on $\mathbf{R}^n$

As explained above, a proof of the Bott periodicity is deduced from comparing $K_0(C_0(\mathbf{R}^2))$ and $K_0(\mathcal{K})$, by exhibiting a continuous field of projections. There is a second way to do so, by evaluating a certain field (ω_h) of cyclic 2-cocycles on any sufficiently smooth field of projections extending the Bott projection. Combining these two methods, one gets that

$$\left\langle \omega_0, [b] - \left[\begin{pmatrix} 0 & 0 \\ 0 & 1 \end{pmatrix}\right]\right\rangle = \left\langle \omega_1, [e_0] - \left[\begin{pmatrix} 0 & 0 \\ 0 & 1 \end{pmatrix}\right]\right\rangle$$

by continuity of both fields of projections and of cocycles.

The cocycle ω_1 is cohomologous to the image of the canonical trace on $\mathcal{K}$ under the periodicity operator, and the right-hand side of the equality above is precisely the index of the operator $D = x + d/dx$. The left-hand side is a formula given in terms of the symbol $\sigma(D)$ of D. Thus the equality above may be interpreted as an index formula for the differential operator D.

This can be generalized to obtain an index theorem for a certain class of pseudo-differential operators on $\mathbf{R}^n$ [6].

4. Index Formula for Operators on foliations

If P is a longitudinal elliptic operator on a foliated manifold (X, F), then its index $ind(P)$ is a class in the K_0-group of the foliation C^*-algebra $C^*(X,F)$ [1], [4]. When the foliation has a transverse invariant measure ν, it induces a trace on $C^*(X,F)$, which in turn defines an additive map $\tau : K_0(C^*(X,F)) \to \mathbf{R}$. Evaluating τ on $ind(P)$, we obtain a numerical invariant $ind_\nu(P) = \tau(ind(P))$. The index theorem of A. Connes [1] describes $ind_\nu(P)$ in terms of the symbol of the operator and the foliation cycle corresponding to ν. Those foliations that have nontrivial transverse invariant measure are rather "tame" ones, and many interesting foliations, e.g. Anosov foliations, have no nontrivial transverse invariant measures. Thus, in order to obtain a numerical invariant for operators on such foliations, we need substitutes of traces. A natural candidate is the pairing between K_0-group and even-dimensional cyclic cocycles. In [7] we obtained a result in this direction.

We focus on a particular class of foliations. Let Γ be a discrete group, acting freely on a manifold $\tilde{M}$ so that $M = \tilde{M}/\Gamma$ is a closed orientable manifold. Given a (right) Γ-action on the circle S^1 by orientation preserving diffeomorphisms, one can construct a bundle $X \to M$ with fibre S^1. This is the associated bundle

$$X = \tilde{M} \times_\Gamma S^1 \to M = \tilde{M}/\Gamma,$$

where the Γ-action on $\tilde{M} \times S^1$ is diagonal. A product foliation on $\tilde{M} \times S^1$, whose leaves are $\tilde{M} \times \{t\}, t \in S^1$, descends to a foliation F on X. The S^1-bundle X with the foliation F is called a foliated S^1-bundle. Each leaf of F is a covering space of M.

Example. Let $\sum$ be a closed Riemann surface of genus ≥ 2, furnished with the metric of constant curvature -1. The geodesic flow on the unit tangent bundle $T_1 \sum$ gives rise to a foliation F_A called an Anosov foliation. The S^1-bundle $T_1 \sum$ with F_A is a foliated bundle, where $\Gamma = \pi_1(\sum)$ acts on S^1 by linear fractional transformations.

When the Γ-action on S^1 satisfies a certain condition (Condition 2.1 of [7]), the foliation C^*-algebra can be described in a simple way as follows.

Let $\mathcal{A}$ be the space of all C^∞-functions $k : \tilde{M} \times \tilde{M} \times S^1 \to \mathbf{C}$ such that

i. $k(mg, ng, xg) = k(m, n, x), (m, n, x) \in \tilde{M} \times \tilde{M} \times S^1, g \in \Gamma$,

ii. k has Γ-compact support, i.e. the image of supp(k)is compact in $\left(\tilde{M} \times \tilde{M} \times S^1\right)/\Gamma$.

Let ω_0 be a Γ-invariant volume form on $\tilde{M}$. The collection $\mathcal{H} = \left(L^2\left(\tilde{M} \times \{x\}\right)\right)_{x \in S^1}$ is in a natural way a continuous field of Hilbert spaces. The Γ-action on $\tilde{M} \times S^1$ induces a Γ-action on $\mathcal{H}$. The space $End_\Gamma(\mathcal{H})$ of Γ-equivariant bounded measurable fields of operators is a C^*-algebra.

For $f \in \mathcal{A}$, an operator $\rho(f) \in End_\Gamma(\mathcal{H})$ is defined by

$$[\rho(f)\xi](m,x) = \int_{\tilde{M}} f(m,n,x)\,\xi(n,x)\,\omega_0$$

for $\xi \in L^2\left(\tilde{M} \times \{x\}\right)$, and $\rho(\mathcal{A})$ is a $*$-subalgebra of $End_\Gamma(\mathcal{H})$. The norm closure of $\rho(\mathcal{A})$ in $End_\Gamma(\mathcal{H})$ is, by definition, the C^*-algebra $C^*(X,F)$ of the foliated bundle (X,F).

Define a linear functional μ on $\mathcal{A}$ by

$$\mu(f) = \int_{\mathcal{D} \times S^1} f(m,m,x)\,\omega_0 \wedge dx,$$

where $\mathcal{D}$ is the fundamental domain of the covering space $\tilde{M} \to M$, and dx is the standard volume form on S^1. choose a Γ-invariant volume form ω_1 on $\tilde{M} \times S^1$ so that there exists a positive smooth function ψ on $\tilde{M} \times S^1$ satisfying

$$\omega_0 \wedge dx = \psi\omega_1.$$

Set $\phi = \log\psi$ and denote by $\dot{\phi}$ the partial derivative of ϕ in the direction of S^1. A cyclic 2-cocycle τ on $\mathcal{A}$ is defined by

$$\tau\left(f^0, f^1, f^2\right) = \mu\left(f^0\left[\phi, f^1\right]\left[\dot{\phi}, f^2\right] - f^0\left[\dot{\phi}, f^1\right]\left[\phi, f^2\right]\right).$$

The cocycle τ extends to a cyclic 2-cocycle on a dense Banach subalgebra B of $C^*(X,F)$ with the property

$$K_0(B) \cong K_0(C^*(X,F)).$$

Thus τ induces an additive map from $K_0(C^*(X,F))$.

The definition of $C^*(X,F)$ is generalized to that of foliation C^*-algebras with coefficient $C^*(X,F,E)$, where E is a Hermitian vector bundle over X, and construction of τ is also carried out to yield a cyclic cocycle on $C^*(X,F,E)$. It may be worth pointing out why we need foliation C^*-algebras with coefficient. In the definition of $ind(P)$, if P is a pseudo-differential operator $C^\infty(E_1) \to C^\infty(E_2)$, then $ind(P)$ is the difference of classes of idempotents

$$[p] - \left[\begin{pmatrix} 0 & 0 \\ 0 & 1 \end{pmatrix}\right],$$

and the difference $p - \begin{pmatrix} 0 & 0 \\ 0 & 1 \end{pmatrix}$, in fact, belongs to $C^*(X, F, E_1 \oplus E_2)$. As $C^*(X, F, E)$ and $C^*(X, F)$ are strongly Morita equivalent, their K_0-groups are isomorphic, but not in a canonical way. Thus, as an abstract K_0-class, we can regard $ind(P)$ as a class in $K_0(C^*(X, F))$. However, when you evaluate a cyclic cocycle on $ind(P)$, you need to know to which algebra the element actually belongs.

Suppose that $\tilde{M}$ is endowed with a Γ-invariant spin structure. So, $M = \tilde{M}/\Gamma$ is also a spin manifold. Let D^+ be the positive Dirac operator on $\tilde{M}$. Denote by D^+_x the Dirac operator on $\tilde{M} \times \{X\}, X \in S^1$. Then the family $(D^+_x)_{x \in S^1}$ is a Γ-equivariant family of elliptic operators parameterized by S^1, and defines a longitudinal elliptic operator on (X, F), which we shall denote again by

$$D^+ : C^\infty(S^+) \to C^\infty(S^-).$$

The graph projection e_x of D^+_x is a bounded operator on $L^2\left(\tilde{M} \times \{x\}, S^+ \oplus S^-\right)$. Denote by e the family $(e_x)_{x \in S^1}$. Then $e \in End_\Gamma(\mathcal{H})$, and

$$e - \begin{pmatrix} 0 & 0 \\ 0 & 1 \end{pmatrix} \in C^*(X, F, S^+ \oplus S^-).$$

Therefore $[e] - \left[\begin{pmatrix} 0 & 0 \\ 0 & 1 \end{pmatrix}\right] \in K_0(C^*(x, F, S^+ \oplus S^-))$. Moreover, we have [7] that

$$ind(D^+) = [e] - \left[\begin{pmatrix} 0 & 0 \\ 0 & 1 \end{pmatrix}\right].$$

The foliation F on X is of codimension one and transversely oriented. To such a foliation is assigned a secondary characteristic class $gv(F) \in H^3(X, \mathbf{R})$. The main result of [7] is the following.

THEOREM *(Theorem 8.11 of [7]). We have*

$$\langle [\tau], ind(D^+) \rangle = C\left(gv(F) \cup \hat{A}(M)\right)[X],$$

where C is a universal constant dependent only on the dimension of M, and $\hat{A}(M)$ is the lift of the $\hat{A}$-class of M to X.

As explained above, the index $ind(P)$ of P is defined to be the difference of classes of idempotents [3], [4]. The idempotent p is described in terms of a parametrix of P . Then, it is automatic that the difference $p - \begin{pmatrix} 0 & 0 \\ 0 & 1 \end{pmatrix}$ belongs to the domain of τ. However it is not an easy task to carry out the calculation of the pairing, partly because there is no cononical way to choose a parametrix. In other words, the idempotent is not canonically chosen. If we use the graph projection, the regularity is sacrificed and it has to be checked that the difference $\hat{e} = e - \begin{pmatrix} 0 & 0 \\ 0 & 1 \end{pmatrix}$ belongs to the domain of the cyclic cocycle. When $\dim M = 2$, the element $\hat{e}$ belongs to the domain of an extension of τ. When $\dim M = 2n, n \geq 2$, the element $\hat{e}$ belongs to the domain of a cyclic $2n$-cocycle which is an extension of $S^{n-1}\tau$ (S is the periodicity operator in cyclic cohomology theory). This situation is quite similar to that for various Schatten classes of compact operators. The canonical trace Tr is defined only on $\mathcal{L}^1(H)$. So, in order to get a cyclic cocycle on $\mathcal{L}^p(H), p > 1$, we must apply the operator S to Tr sufficiently many times. Once it is shown that $\hat{e}$ belongs to the domain, by a straightforward computation, we get the pairing.

References

[1] Connes, A.: Sur la théorie noncommutative de l'integration, Algébres d' Opérateurs, Lecture Notes in Math. 725, Springer-Verlag, 1979, 19-143.

[2] ______: Non-commutative differential geometry, Parts I and II, Publ. Math. IHES, 62 (1986), 257-360.

[3] ______, Moscovici, H.: Cyclic cohomology, the Novikov conjecture and hyperbolic groups, Topology 29, (1990), 345-388

[4] ______, Skandalis, G.: The longitudinal index theorem for foliations, Publ. Research Inst. Math. Sci., Kyoto Univ., 20 (1984), 1139-1183.

[5] Elliott, G. A., Natsume, T., Nest, R.: The Heisenberg group and K-theory, preprint, 1992.

[6] ______________: The Atiyah-Singer index theorem as passage to the classical limit in quantum mechanics, in preparation.

[7] Moriyoshi, H., Natsume, T.: The Godbillon-Vey cyclic cocycle and longitudinal Dirac operators, preprint, 1992.

[8] Nussbaum, A. E. : Reduction theory for unbounded closed operators in Hilbert space, Duke Math. J., 31 (1964), 33-44.

[9] Stone, M. H.: On unbounded operators in Hilbert space, J. Indian Math. Soc., 15 (1951), 155-192.

[10] von Neumann, J.: "Functional Operators", vol.2: The Geometry of Orthogonal Spaces, Ann.Math. Studies No. 22, Princeton Univ. Press, Princeton, 1950.

Endomorphisms and Automorphisms for Factor Inclusions

MARIE CHODA
Department of mathematics, Osaka Kyoiku University, Tennoji, Osaka, 543, Japan

Abstract. We investigate some kinds of *-endomorphisms and automorphisms for inclusions of type II_1 factors in connection with Jones index theory.

Key words: subfactor, index

1. Introduction

Let M be a finite von Neumann algebra with a faithful normal trace τ, $\tau(1) = 1$ and $End(M, \tau)$ be the set of *-endomorphisms σ of M with $\tau(\sigma(x)) = \tau(x)$ for all $x \in M$. Then $\sigma \in End(M, \tau)$ preserves the identity 1 of M and $\sigma(M)$ is a von Neumann subalgebra of M. If there exists a $\theta \in Aut(M)$ with $\sigma_2 \cdot \theta = \theta \cdot \sigma_1$ for $\sigma_i \in End(M, \tau)(i = 1, 2)$ then σ_1 and σ_2 are said to be *conjugate* and if there exist a $\theta \in Aut(M)$ and a unitary $u \in M$ such that $Adu \cdot \sigma_2 \cdot \theta = \theta \cdot \sigma_1$, then σ_1 and σ_2 are *outer conjugate.* The von Neumann subalgebra $\cap_n \sigma^n(M)$ is an outer conjugacy invariants so that conjugacy invariant for $\sigma \in End(M, \tau)$. The most interesting class of $\sigma \in End(M, \tau)$ is those satisfying the ergodic property ;

$$\bigcap_k \sigma^k(M) = C1.$$

A $\sigma \in End(M, \tau)$ satisfying the ergodic property is called a *shift* ([31]).

In the case of factor M, as the second invarient for $End(M, \tau)$, we have the Jones index $[M : \sigma(M)]$ of the subfactor $\sigma(M)$ in M. The index Indσ of $\sigma \in End(M, \tau)$ is defined by $[M : \sigma(M)]$. Jones [16] obtained the indices $I(R)$ for subfactors of the hyperfinite II_1 factor R. Since each subfactor N of R with finite index is also hyperfinite, we have an $\sigma \in End(R, \tau)$ with $\sigma(R) = N$. Those implies

$$\{Ind(\sigma); \sigma \in End(R, \tau)\} = I(R) = \{4cos^2(\pi/n); n = 3, 4, ...\} \cup [4, \infty).$$

By the classification for subfactors of R, some numbers of indices are completely invariant for subfactors of R (for example, $4con^2(\pi/n)$ is completely invariant for all odd number $n \geq 3$ ([16], [17], [22], [28])). Back to

H. Araki et al. (eds.), Quantum and Non-Commutative Analysis, 291–304.

the conjugacy classes of *-endomorphisms, we have the following fact as a generalization of [31] for shifts of R with index 2:

For each $r \in I(R)$, there exist at least countably infinite shifts of R with index r which are not outer conjugate ([3]).

This note is a report of some results in [5], [7] and [8] on *-endomorphisms which are based on above those facts.

2. The inclusion $B(\Xi) \supset C(\Xi)$

In this section we review the method in [11] to give an inclusion $B(\Xi) \supset C(\Xi)$ of factors from a Coxeter graph Ξ. Let $C_0 \subset B_0$ be a pair of finite dimensional von Neumann algebras. Let $[C_0 \to B_0]$ be the inclusion matrix for $C_0 \subset B_0$. Let $\beta = \|[B_0 \to C_0]\|^2$. Assume $2 \leq \beta \leq 4$. Then the Bratteli diagram for $C_0 \subset B_0$ is one of the Coxeter graphs Ξ of types $A_n, D_n, A_n^{(1)}, D_n^{(1)}$, and $E_m, E_m^{(1)}$ for $m = 6, 7, 8$. Let q be a number with $\beta = 2 + q + q^{-1}$. Then normalized Markov trace τ of modulus β for $C_0 \subset B_0$ has an extension τ to the fundamental extension algebra $B_1 =< B_0, e_0 >$, where e_0 is the projection from $L^2(B_0, \tau)$ onto $L^2(C_0, \tau)$. By the induction, we have the projection e_j from $L^2(B_{j-1}, \tau)$ onto $L^2(B_{j-2}, \tau)$, $B_j =< B_{j-1}, e_{j-1} >$ and the normalized Markov trace τ with modulus β on B_j for all j. Let $v_j = (q+1)e_j - 1$ and $C_1 = v_0 B_0 v_0^{-1}$. Let C_j be the algebra generated by $\{C_{j-1}, e_{j-1}\}$. Let π be the GNS representation of $\bigcup_j B_j$ with respect to τ and put

$$B(\Xi) = \pi(\bigcup_j B_j)'' \text{ and } \mathrm{C}(\Xi) = \pi(\cup_j \mathrm{C}_j)''$$

Then $C(\Xi)$ is a subfactor of the hyperfinite II_1 factor $B(\Xi)$ with $[B(\Xi) : C(\Xi)] = \beta$.

The principal graph for an inclusion $N \subset M$ of II_1 factors is defined in [11] as the Bratteli diagram ([2]) for the inclusion of the tower of relative commutant algebras of N in the tower of basic constructions. It is a bipartite graph with the weights on the vertices, which are given by the restriction of the Markov trace of the basic extension algebras. If $[M : N] < 4$, then the principal graph is one of the Coxeter graphs of type A_n, D_{2n}, E_6, E_8 ([16], [17], [13], [15], [22] [23], [28]).

Proposition. *Let Ξ be one of the Coxeter graphs A_n, D_{2m}, E_6, E_8, or $E_n^{(1)}$ for $n = 6, 7, 8$. Then the principal graph of $B(\Xi) \supset C(\Xi)$ is Ξ.*

Proof. This fact was first announced for E_6 in [1] by the differnt method from the method given in [7] for A_n, D_{2m}, $E_n (n = 6, 8)$. For the graphs $E_n^{(1)} (n = 6, 7, 8)$, we can gove a proof by the same method with one in [7].

3. Endomorphisms and Jones tower

Let $M \supset N$ be a pair of II_1 factors with $[M:N] =< \infty$. By iterating the Jones' basic construction [16], we obtain a whole tower of factors with Jones projections $\{e_i\}_i$:

$$N = M_{-1} \subset M = M_0 \subset M_1 \subset \cdots \subset M_j =< M_{j-1}, e_j > \subset \cdots.$$

Let

$$A_k = M' \cap M_k \qquad and \qquad B_k = M_1' \cap M_k \qquad (k \geq 0).$$

The canonical trace τ of M is extended to the Markov trace on each M_j, which we denote by the same notation τ. The τ satisfies $\tau(xe_j) = \lambda\tau(x)$ for $x \in M_{j-1}$ for all j. Here $\lambda = [M:N]^{-1}$. The canonical conjugation J_k on the Hilbert space $L^2(M_k, \tau)$ is defined by $J_k(x) = x^*$ for all $x \in M_k$. We denote by M_∞ the factor $\pi(\cup_j M_j)''$, where π is the GNS representation with respect to τ. Let

$$A = M' \cap M_\infty \qquad and \qquad B = M_1' \cap M_\infty.$$

3.1. CANONICAL SHIFT

In a series of studies on index theory for pairs $N \subset M$ of infinite factors, Longo's canonical endomorphism γ for $N \subset M$ is used to investigate the relation of the inclusion $N \subset M$ ([20], [21]). The endomorphism γ is literally canonical for the incluson $N \subset M$ but the range of it has the relation $M \supset N \supset \gamma(M)$.

The finite case version Γ of the canonical endomorphism γ is introduced by Ocneanu [22] to classify subfactors of the hyperfinite II_1 factor with index less than 4.

The *canonical shift* Γ is an endomorphism on the von Neumann algebra $A = M' \cap M_\infty$ which is defined by $\Gamma(x) = J_{n+1}J_nxJ_nJ_{n+1}$ for $n \geq k$ for all $x \in M' \cap M_{2k}$. The definition does not depend on n ([22], see also [6]) and Γ is τ-preserving ([5], [6]). Also the followings hold.

(1) The extension Γ to A is a shift of A ([5],ch) and

(2)

$$\Gamma(M_l' \cap M_k) = M_{l+2}' \cap M_{k+2}, \quad for\ all \quad l \leq k \ ([22],[6]).$$

3.2. BASIC *-ENDOMORPHISM

We give some definition for $\rho \in End(M,\tau)$ of a factor M. A $\rho \in \mathrm{End}(M,\tau)$ is irreducible if $\rho(M)' \cap M = \mathbf{C}1$. Let E_ρ be the conditional expectation $E_{\rho(M)}$ of M onto $\rho(M)$. Assume that ρ have finite index. If there exists a projection $e \in M$ which satisfies that

$$E_\rho(e) = (Ind\rho)^{-1}1, \quad \rho^2(M) = \{e\}' \cap \rho(M)$$

then ρ is said to be *basic* and such a projection e is called a *basic projection* for ρ. If $M = \{\bigcup_j(\rho^j(M)' \cap M)\}''$ then ρ is said to have the *generating property*. If ρ is basic and satisfy the generating property, ρ is said to be *standard*. The ρ has finite depth if

$$\sup_k\{\text{dimension of } Z(\rho^k(M)' \cap M)\} < +\infty,$$

where $Z(L)$ is the center of the algebra L. The *principal graph* (resp. *dual principal graph*) for ρ is the principal graph (resp. dual principal graph) for the pair $M \supset \rho(M)$.

Proposition. ([5]) *Let* $\rho \in End(M, \tau)$. *Then*

(1) ρ *is basic if and only if* M *is the basic extension for* $\rho(M) \supset \rho^2(M)$.

(2) Assume that ρ *is basic. Then there exists a* $\rho_\infty \in Aut(M_\infty)$ *which satisfies for all* k

$$\rho_\infty(e_k) = e_{k-1}, \qquad \rho_\infty(M_k) = M_{k-1} \qquad \text{and} \qquad \rho_\infty|_M = \rho.$$

Hence the principal graph of ρ *is the same graph with the dual principal graph of* ρ.

3.3. Example of a basic *-endomorphism

Let Ξ be one of the Coxeter graphs of types $A_n, D_n, A_n^{(1)}$, and $D_n^{(1)}, E_n^{(1)}$. Let us consider the pair $B(\Xi) \supset C(\Xi)$ and the unitary v_j in §2. The *-endomorphism Φ is defined for all $x \in B(\Xi)$ by

$$\Phi(x) = \lim_{k\to\infty} v_0v_1...v_kxv_k^*...v_1^*v_0^*.$$

The subfactor $\Phi(B(\Xi))$ of $B(\Xi)$ coinsides with $C(\Xi)$.

The $\Phi \in End(B(\Xi), \tau)$ is basic but does not satisfy the generating property if dim $C_0 \geq 2$.

3.4. Example of a standard *-endomorphism

The typical example of standard endomorphisms is the canonical shift Γ of the factor $A = M' \cap M_\infty$ for the pair $M \supset N$ of type II_1 factors with $[M : N] < \infty$ and finite depth. In fact, Γ is basic with a basic projection

$$f = [M : N]e_3e_2e_4e_3.$$

Since $\Gamma(A_j) = A_{j+2}$, it satisfies the generating property by the following :

$$(\cup_j((\Gamma^j(M' \cap M_\infty))' \cap M' \cap M_\infty))'' \supset (\cup_j(M_{2j} \cap M'))''.$$

If $M \supset N$ has finite depth, then Γ has the generating property . Hence Γ is standard.

3.5. SELF CONJUGATE *-ENDOMORPHISMS

Under the condition that A is a factor, we study *-endomorphisms of A for the inclusion $N \subset M$ with respect to relations of basic endomorphisms. For a $\sigma \in \mathrm{End}(A,\tau)$, let

$$Aut(A,\sigma) = \{\theta \in Aut(A) : \theta(\sigma^n(A)) = \sigma^n(A) \; for \; all \; n\}.$$

The following three statements are equivalent for $\rho, \sigma \in End(A,\tau)$;
(1) there exists a $\theta \in Aut(A,\sigma)$ with $\rho = \sigma \cdot \theta$,
(2) there exists a $\theta' \in Aut(\sigma(A),\sigma)$ with $\rho = \theta' \cdot \sigma$,
(3) $\sigma^n(A) = \rho^n(A)$ for all n.

Two $\rho, \sigma \in End(A,\tau)$ are *equivalent* if they satisfies one of the above equivalent conditions. The conjugate $\bar{\rho}$ of ρ with the property $\rho(A) = B$ is defined by $\bar{\rho} = \rho^{-1}\Gamma$. A $\rho \in \mathrm{End}(A,\tau)$ with $\rho(A) = B$ is *self conjugate* if ρ is equivalent to $\bar{\rho}$.

A von Neumann subalgebra D of a von Neumann algebra C is called *normal* in C if $(D' \cap C)' \cap C = D$. If $[M : N] \leq 4$, then by Skau's lemma [11], $\{e_i; i \geq 2\}' \cap M_\infty = M$. Hence M is normal in M_∞. If $M = N \rtimes_\alpha G$ for an outer action α of G of a factor N, then M is also normal in M_∞ (cf. see [32]).

Theorem. ([8]). *(1) If $\rho \in End(M,\tau)$ is basic and $\rho(M) = N$, then the restriction of ρ_∞^{-1} to A is self-conjugate and satisfies*

$$\rho_\infty^{-1}(M_k' \cap M_l) = M_{k+1}' \cap M_{l+1} \; for \; all \; k \leq l.$$

(2) If M is normal in M_∞ and a $\rho \in End(A,\tau)$ is self-conjugate, then we have

$$\rho(M_i' \cap M_j) = M_{i+1}' \cap M_{j+1}, \; (j \geq i).$$

Remark. If the principal graph and the dual principal graph for an inclusion $N \subset M$ are different, then there does not exist any self conjugate $\rho \in \mathrm{End}(A,\tau)$. For example, consider $M = N \rtimes_\alpha G$ by an outer action α of a non abelian finite group G. Then M is normal in M_∞. Since G is non abelian, the dual principal graph for $N \subset M$ is different from the principal graph. Hence any $\rho \in End(A,\tau)$ is not self-conjugate by the above Theorem

4. Automorphisms and Bipartite Graphs

A finite bipartite graph is considered as a Bratteli diagram for an inclusion $A \supset B$ of finite dimensional C^* algebras. If an automorphism θ of the graph preserves the inclusion relation, then θ can be considered as an automorphism of A with $\theta(B) = B$. Since the crossed products $C =< \theta > \triangleright A \supset D =< \theta > \triangleright B$ is also an inclusion of finite dimensional C^*- algebras, we have a new Bratteli diagram for $C \supset D$. Here we obtain the relation between two graphs. Let Ξ be a graph, two vertices of which are connected by at most one edge. Let $V(\Xi)$ be the set of vertices of Ξ and $E(\Xi)$ the set of edges. The edge combining v and w in $V(\Xi)$ is denoted by (v, w). A graph Ξ with bicolored vertices means that two vertices joined by an edge are colored in such a way that one is black and the other is white. We denote by $B(\Xi)$ the set of vertices colored by black and by $W(\Xi)$ the set of vertices with white color. An automorphism θ of Ξ is a mapping from $V(\Xi)$ onto $V(\Xi)$ which satisfies that if $(v, w) \in E(\Xi)$ then $(\theta(v), \theta(w)) \in E(\Xi)$ for all pair of $v, w \in V(\Xi)$. When $\theta(v)$ has the same color as v for all $v \in V(\Xi)$, θ is said to be *color-preserving.*

We treate only color-preserving automorphisms of a finite graph with bicolored vertices. The period $p = p(\theta)$ of θ is the smallest number satisfying that θ^p is the identity on Ξ. For a $v \in V(\Xi)$, we put

$$q(v) = min\{k; \theta^k(v) = v\}, \quad r(v) = \frac{p}{q(v)}, \quad \hat{v} = \{\theta^k(v); 1 \leq k \leq q(v)\}.$$

We discuss only the automorphisms θ with $p(\theta) = q(v)$ for all $v \in V(\Xi)$. Let consider $r(v)$ abstruct elements $\hat{v}(0), \hat{v}(1), ..., \hat{v}(r(v) - 1)$ and identify $\hat{v}(0)$ with $\hat{v}$.

4.1. Dual graph

The dual graph $\hat{\Xi} = \hat{\Xi}(\theta)$ of Ξ by θ has the set

$$V(\hat{\Xi}) = \{\hat{v}(i); v \in \Xi, 0 \leq i \leq r(v) - 1\}$$

as vertices. Two vertices $\hat{v}(i)$ and $\hat{w}(j)$ in $V(\hat{\Xi})$ is combined by an edge in $E(\hat{\Xi})$ if there exist $v' \in \hat{v}$ and $w' \in \hat{w}$ such that $(v', w') \in E(\Xi)$, and

$$\{iq(v) + k; 0 \leq k \leq q(v) - 1\} \cap \{jq(w) + l; 0 \leq l \leq (w) - 1\} \neq \emptyset.$$

The $\hat{v}(i)$ is given the same color with $v \in V(\Xi)$ for all i.

4.2. Dual weighted graph

A *weight*(v) for a $v \in V(\Xi)$ is given by the method in [11]. Let $B(\Xi) = \{b_1, b_2, \cdots, b_m\}$ and $W(\Xi) = \{w_1, w_2, \cdots, w_n\}$. The m-by-n matrix $mat(\Xi) = [x_{ij}]_{ij}$ is defined from Ξ by $x_{ij} = 1$, if $(b_i, w_j) \in E(\Xi)$ and $x_{i,j} = 0$, if $(b_i, w_j) \notin E(\Xi)$.

Assume that Ξ is connected. Then there exists a Perron-Frobenius eigen (column) vector u for $(mat(\Xi))(mat\ (\Xi))^t$. Let $s = (mat(\Xi))^t u$ and $t = mat(\Xi)s$. We give a weight for $b_i \in B(\Xi)$ by $weight(b_i) = t(i),\ for\ t = (t(i))_i$ and for $w_k \in W(\Xi)$ by $weight(w_k) = s(k),\ for\ s = (s(k))_k$. We identify weighted vectors with s and t.

Let Ξ be an weighted finite graph with bicolored vertices and θ a color and weight preserving automorphism of Ξ. Let $x \in V(\hat{\Xi})$ be $x = \hat{v}(i)$ for some $v \in V(\Xi)$ and $i(0 \leq i \leq r(v)-1)$. Then the $weight(x)$ of x is given by $weight(x) = q(v)weight(v)$.

Let Ξ be one of the Coxeter graphs A_{2n+1}, D_{2n} and E_6. Then the symmetry θ of Ξ in the figure induces an outer automorphism θ of $B(\Xi)$ such that $\theta(C_j) = C_j$ and $\theta(B_j) = B_j$ for all j.

By the definition of dual weighted graphs and the result of the principal graph for $B(\Xi) \supset C(\Xi)$, we have :

Proposition. ([7]) *Let $B(\Xi) \supset C(\Xi)$ be the pair obtained from Ξ and θ the automorphism of $B(\Xi) \supset C(\Xi)$ induced by the symmetry of Ξ.*

(1) If Ξ is A_{4n-3} (resp. D_{2n}), then the principal graph of the crossed product $Z_2 \rtimes_\theta B(\Gamma) \supset Z_2 \rtimes_\theta C(\Xi)$ is D_{2n} (resp. A_{4n-3}).

(2) If Ξ is E_6, then $B(\Xi) \supset C(\Xi)$ is conjugate to $Z_2 \rtimes_\theta B(\Xi) \supset Z_2 \rtimes_\theta C(\Xi)$ and those principal graph is E_6 .

5. square roots of Γ

Let $M \supset N$ be a pair of II_1 factors with finite index. We call a $\sigma \in End(M,\tau)$ with $\sigma(M) = N$ a *square root* of Γ if $\sigma^2 = \Gamma$ for the canonical shift Γ of the inclusion $M \supset N$.

If A is a factor, then a square root of Γ is self-conjugate.

We obtain square roots of the canonical shift for the pair of hyperfinite type II_1 factors with index smaller than 4 and $\{e_i : i \geq 1\}" \supset \{e_i : i \geq 2\}"$.

5.1. CASE OF $B(\Xi) \supset C(\Xi)$

Let Ξ be one of the Coxeter garphs and the extended Coxeter garphs. Let $M = B(\Xi)$ and N = C(Ξ). For convenience sake, we denote the projection e_j implimenting the conditional expectation of B_j onto B_{j-1} by e_{-j}. Also we denote by $e_n (n \geq 1)$ the projection for the pair $M_{n-1} \supset M_{n-2}$ and $v_j = qe_j - (1-e_j)$ for all $j \in \mathbf{Z}$. Let Φ be the basic *-endomorphism of M

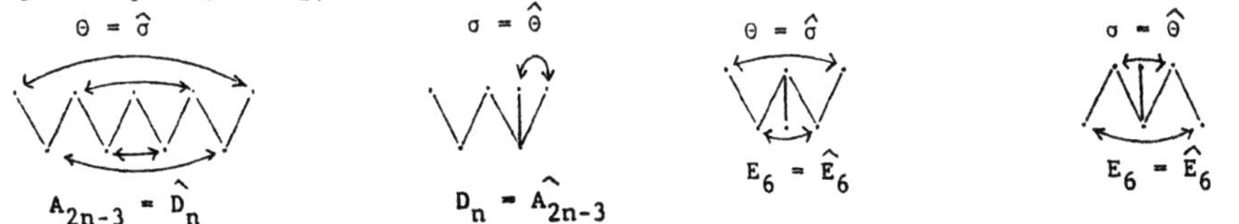

Fig. 1. Examples of the dual pairs

onto N. The Φ has the extension (which we denote by the same notation Φ) to M_∞. The restriction of Φ^{-1} to A has the form

$$\Phi^{-1}(x) = \lim_{k\to\infty} v_k v_{k-1}...v_1 x v_1 ... v_{k-1} v_k, (x \in A).$$

Proposition. ([8]) *Assume that Ξ is neither of type A_3 nor D_4. Then the restriction of Φ^{-1} to $A = M' \cap M_\infty$ for $M = B(\Xi) \supset N = C(\Xi)$ is a square root of Γ.*

5.2. Case of crossed products

Next we study *-endomorphisms for the pair given by the crossed products, in order to obtain a square root of Γ for the pair $M \supset N$ with the principal graph of A_3 or D_4.

Let N be a finite factor, G a finite abelian group and α an outer action of G on N. Let $M = N \lhd_\alpha G$. We denote the canonical unitary in M which corresponds $g \in G$ by $u_0(g)$. The action $\hat{\alpha}$ is defined by $\hat{\alpha}_\gamma(u_0(g)a) =< \gamma, g > u_0(g)a$, $(\gamma \in \hat{G}, g \in G, a \in N)$. Then the automorphism $\hat{\alpha}_\gamma$ induces an unitary $u_1(\gamma)$ on $L^2(M, \tau)$. The factor generated by M and $\{u_1(\gamma); \gamma \in \hat{G}\}$ is the basic extension algebra M_1. Iterating this method, we have the sequence $\{u_{2i-1}(\gamma); \gamma \in \hat{G}\}_{i\geq 1}$ and $\{u_{2i}(g); g \in G\}_{i\geq 1}$ of unitaries, which satisfy

$$M_{2i} = \{M_{2i-1}, u_{2i}(G)\}'' \qquad and \qquad M_{2i+1} = \{M_{2i}, u_{2i+1}(\hat{G})\}''.$$

Proposition. ([8]) *Let $M = G \rhd_\alpha N$ and ρ be an isomorphism of A onto B. Then $\rho^2 = \Gamma$ if and only if there exist an isomorphism ψ_ρ from G onto $\hat{G}$ and a $\chi_\rho \in \hat{G}$ which satisfy that*

$$\begin{aligned} \rho(u_{2i}(g)) &= <\chi, g> u_{2i+1}(\psi(g)), \ for\ all\ g \in G, \\ \rho(u_{2i+1}(\psi(g))) &= <\chi, g>^{-1} u_{2i+2}(g), \ for\ all\ g \in G. \end{aligned}$$

As a consequence of Lemma, Proposition and the results of classification of subfactors, we have :

Theorem. ([8]). *Let $M \supset N$ be one of the following pairs : (1) $M \supset N$ is the Jones pair $R = \{e_i : i \geq 1\}'' \supset R_\lambda = \{e_i : i \geq 2\}''$ for all $\lambda \in I(R)$. (2) $M = N \lhd_\alpha G$ of II_1 factor N by an outer action α of an abelian finite group G. (3) M is hyperfinite and $[M : N] < 4$. (4) M is hyperfinite and the principal graph of $M \supset N$ is one of $E_n^{(1)}$ for $n = 6, 7, 8$. Then there exists a square root σ of the canonical shift Γ.*

5.3. How many square roots of Γ

Here we give a characterization for a $\rho \in End(A, \tau)$ to be a square root of Γ. By the fact that $e_i \in M'_{i-2} \cap M_i$ for all $i \geq 2$, we have the following commuting square coming from $M \supset N$:

$$\begin{array}{ccc} A = M' \cap M_\infty & \supset & B = M_1' \cap M_\infty \\ \cup & & \cup \\ R = \{e_i : i \geq 2\}'' & \supset & R_\lambda = \{e_i : i \geq 3\}''. \end{array}$$

Here $\lambda = [M : N]^{-1}$. We denote the tower of basic constructions for $R \subset A$ as follows:

$$R \subset A \subset A_1(R) = < A, e_R > \subset A_2(R) \subset \cdots.$$

Proposition. *If the principal graph of $M \supset N$ is one of Coxeter graphs $A_n (n \geq 4), D_{2n} (n \geq 3)$, $E_6, E_8, E_i^{(1)}$ $(i = 6, 7, 8)$ and A_∞, then a square root σ of Γ must satisfy $\sigma(e_i) = e_{i+1}$ for all $i \geq 2$.*

This proposition implies that the restriction of any square root to R in the commuting square is unique.

Assume that $2 \neq [M : N] \leq 4$ and $\dim(M' \cap M_2) = 2$. Then the number of square roots of Γ is smaller than the cardinal number of projections $p \in A' \cap A_2(R)$ with $\tau(p)[A : R] = 1$. Here τ is the Markov trace of modulud $[A : R]^{-1}$ for the pair $R \subset A$.

Conclusion. If $M \supset N$ is of type A_3 or D_4, then all square roots of Γ are known in 5.2 proposition. The other cases are obtained using the daul principal graph for the inclusion $R \subset A$.

1. *If the principal graph for $M \supset N$ is one of the Coxeter graphs of type $A_n (n > 3)$, then the square root of Γ is unique.*
2. *If $M \supset N$ is a pair of hyperfinite II_1 factors with the principal graph E_6, then there exists the unique square root of Γ.*
3. *If $M \supset N$ has the principal graph $D_{2n} (n \geq 3)$, then there exist precisely two square roots of Γ by 4.2 and 5.2.*

6. Entropy for *-endomorphisms

The notion of the entropy for *-automorphisms of finite von Neumann algebras is introduced by Connes and Størmer [9] and generalized to completely positive linear maps on C^*algebras [10]. For the sake of completeness, we describe the definition of the entropy for trace preserving *-endomorphisms of a finite von Neumann algebra.

In this section, M is a finite von neumann algebra with a faithful normal trace τ, $\tau(1) = 1$. The letter η designates the function on $[0, \infty)$ defined by $\eta(t) = -t \log t$. For each k, we let S_k be the set of all families $(x_{i_1,i_2,\ldots,i_k})_{i_j \in \mathrm{N}}$

of positive elements of M, zero except for a finite number of indices and satisfying

$$\sum_{i_1,\ldots,i_j,\ldots i_k} x_{i_1,\ldots,i_k} = 1.$$

For $x \in S_k$, $j \in 1,2,\ldots,k$ and $i_j \in \mathrm{N}$, put

$$x^j_{i_j} = \sum_{i_1,\ldots,i_{j-1},i_{j+1},\ldots,i_k} x_{i_1,i_2,\ldots,i_k}.$$

Let $N_1, N_2, \ldots, N_k$ be finite dimensional von Neumann subalgebras of M. Then

$$H(N_1,\ldots,N_k) = Sup_{x\in S_k}[\sum_{i_1,\ldots,i_k} \eta\tau(x_{i_1,\ldots i_k}) - \sum_j \sum_{i_j} \tau\eta E_{N_j}(x^j_{i_j})].$$

For $\sigma \in End(M,\tau)$ and a finite dimensional von Neumann subalgebra N of M,

$$H(N,\sigma) = \lim_{k\to\infty} \frac{1}{k} H(N,\sigma(N),\ldots,\sigma^{k-1}(N))$$

exists. The entropy $H(\sigma)$ for σ is defined as the supremum of $H(N,\sigma)$ for all finite dimensional subalgebras N of M.

If there exists an increasing sequence $(N_j)_j$ of finite dimensional subalgebras which generates M, then

$$H(\sigma) = \lim_{j\to\infty} H(N_j,\sigma).$$

The relative entropy $H(M \mid N)$ for a von Neumann subalgebra N of M is defined ([25]) as an extension form of one ([9]) by

$$H(M \mid N) = Sup_{x\in S_1} \sum_i [\tau\eta(x_i) - \tau\eta E_N(x_i)].$$

6.1. n-SHIFTS ON TOWERS OF ALGEBRAS

Assume that n is a natural number, A is a finite von Neumann algebra with a faithful normal trace $\tau, \tau(1) = 1$ and $(A_j)_j$ is an increasing sequence of finite dimensional von subalgebra of A which generate A. A $\sigma \in End(A,\tau)$ is an *n-shift* if the following conditions are satisfied:

(1) For all j and m, the von Neumann algebra $\{A_j, \sigma(A_j), \ldots, \sigma^m(A_j)\}''$ generated by $\{\sigma^j(A_j); j = 0,\ldots,m\}$ is contained in A_{j+nm}.

(2) There exists a sequence $(k_j)_{j\in\mathrm{N}}$ of integers with the properties:

$$\lim_{j\to\infty} \frac{nk_j - j}{j} = 0$$

and

$$x\sigma^m(y) = \sigma^m(y)x, \quad \tau(z\sigma^{lk_j}(x)) = \tau(z)\tau(x)$$

for all $l \in \mathbf{N}$, $x, y \in A_j$, $m \in k_j\mathrm{N}$ and $z \in \{A_j, \sigma^{k_j}(A_j), ..., \sigma^{(l-1)k_j}(A_j)\}''$.
(3) For an $j \geq n$

$$E_{A_j} E_{\sigma(A_j)} \;=\; E_{\sigma(A_{j-n})}$$

(4) For each j, there exists a τ-preserving *-automorphism β of A_{nj+n} such that $\sigma(A_{nj}) = \beta(A_{nj})$.

Theorem.([5]) *If a τ-preserving *-endomorphism σ of A satisfies the condition (1) and (2) in Definition 1 for the tower $(A_j)_j$ of A, then*

$$H(\sigma) = \lim_{k\to\infty} \frac{H(A_{nk})}{k}.$$

6.2. EXAMPLES OF n-SHIFTS

(1) Let S be the *-endomorphism corresponding to the translation by 1 in the infinite tensor product $R = \bigotimes_{i\in N}(M_i, tr_i)$of the algebra M_i of $m \times m$ matrices with the normalized trace tr_i on M_i for each $i \in N$. For each j, let $A_j = \bigotimes_{i=1}^{j}(M_i, tr_i)$. Then for all n, S^n is an n-shift on the tower $(A_j)_j$ for R.

(2) Let $(e_j)_j$ be the sequence of projections with the following properties for some natural number k and $\lambda \in (0, 1/4] \cup \{1/(4cos^2(\pi/n); n \geq 3\}$:

(a) $e_i e_j e_i = \lambda e_i$ if $|i - j| = k$

(b) $e_i e_j = e_j e_i$ if $|i - j| \neq k$

(c) $(e_j)_j$ generates the hyperfinite type II_1-factor R

(d) $\tau(we_i) = \lambda\tau(w)$ for the trace τ of R and a reduced word w on $\{1, e_1, ..., e_{i-1}\}$.

Let A_j be the von Neumann algebra generated by $\{e_1, ..., e_j\}$. Then, by [16], A_j is finite dimensional. Let σ be the *-endomorphism of R such that $\sigma(e_i) = e_{i+1}$. Then σ^n is an n-shift on the tower $(A_j)_j$ of R for all n.

(3) Let $M \supset N$ be an inclusion of factors with finite index and finite depth. Then the canonical shift Γ is a 2-shift on the tower $(A_j)_j$ of relative commutants for $N \subset M$. And a square root of Γ is a 1-shift on $(A_j)_j$.

6.3. Entropy of *-endomorphisms and relative entropy of subalgebras

If M is a finite dimensional von Neumann algebra and τ a fixed faithful normal trace of M with $\tau(1) = 1$, then M is decomposed into the direct summands:

$$M = \sum_{l \in K} \bigoplus M_l$$

where M_l is the algebra of $d(l) \times d(l)$ matrices and $K = K_M$ is a finite set. Then the vector $d_M = d = (d(l))_{l \in K}$ is called the *dimension vector* of M.

Let $(A_j)_j$ be an increasing sequence of finite dimensional von Neumann algebras. Let $A_j = \sum_{k \in K_j} \oplus A_j(k)$ and d_j the dimension vector of A_j. Then we say that $(A_j)_j$ satisfies *the bounded growth conditions* if the following two conditions are satisfied :

(i)

$$\sup_j \frac{|(K_j)|}{j} < +\infty$$

and

(ii) For some m, $A_{j+1}(l)$ contains at most $d_j(k)$ $A_j(k)$- components for all $j \geq m$, where $|(K_j)|$ is the cardinal number of K_j.

For examples, let us consider the towers of finite dimensional von Neumann algebras which are treated in 6.2 of n-shifts. They all satisfy the bounded growth conditions .

Let $\sigma \in End(A, \tau)$ for a finite von Neumann algebra generated by an increasing sequence $(A_j)_j$ of finite dimensional subalgeras, then the followings hold [5].

1. *If σ is an n-shift on the tower $(A_j)_j$ for A, then*

$$H(A \mid \sigma(A)) \leq 2H(\sigma).$$

2. *If the bounded growth conditions are satisfied, for the tower $(A_{nj})_j$*

$$H(A \mid \sigma(A)) = 2H(\sigma).$$

3. *If σ satisfies the conditions (1), (3) and (4) in Definition for n, then*

$$H(A \mid \sigma(A)) = \lim_{j \to \infty} H(A_{nj+n} \mid A_{nj}).$$

4. *For all integer n,*

$$H(A \mid \sigma^n(A)) = nH(A \mid \sigma(A)).$$

The shifts S in Example and σ_λ for $\lambda > (1/4)$ for Jones sequence of projections satisfy the following relation thanks of the bounded growth condition

$$H(M \mid \sigma(R)) = 2H(\sigma) = \log[R : \sigma(R)].$$

However, if $\lambda \leq (1/4)$, then the shifts σ_λ in Example have the following relation:

$$H(R \mid \sigma(R)) = 2H(\sigma) < \log[R : \sigma(R)].$$

6.4. Entropy of square roots of Γ

The canonical shift Γ is a 2-shift on the tower of the relative commutnt algebras and a square root ρ of Γ is a 1-shift on the same tower, and so we can compute the entropy of those *-endomorphisms. Popa([30]) called an inclusion $N \subset M$ is *extremal* if it satisfies $E_{M'\cap M_1}(e_1) = [M : N]^{-1}1$. Irreducible pair is of course irreducible and if $M \supset N$ has finite depth, then $M \supset N$ is extremal([28]).

Assume that $\sigma \in End(M, \tau)$ is basic for a II_1-factor M and has finite index. Then we have

1. $$H(\sigma) = \lim_{k\to\infty} \frac{H(M \cap \sigma^k(M)')}{k}$$
2. $$2H(\sigma) \leq \log Ind(\sigma),$$
   and the equality holds if σ has finite depth.
3. If $M \supset N$ is extremal, then a self-conjugate ρ is standard and
   $$2H(\rho) = H(\Gamma).$$

Combining these results, we have :

Theorem .[8] *If $N \subset M$ has finite depth (for example $[M : N] < 4$ ([28])), then a self conjugate ρ (hence a square root of Γ) satisfies*

$$2H(\rho) = H(\Gamma) = \log[M : N].$$

References

[1] J. Bion-Nadal, Subfactors of the hyperfinite II_1 factor with Coxeter graph E_6 as invariant, *Current topics in Operator algebras, World Scientific*, (1991).

[2] O. Bratteli, Inductive limits of finite dimensional C*-algebras, *Trans. amer. Math. Soc.*, **171**, 195-234 (1972).

[3] M. Choda, Shifts on the hyperfinite II_1-factor, *J. Operator Theory*, **17**, 223-235 (1987).

[4] M. Choda, Entropy for *-endomorphisms and relative entropy for subalgebras, to appear in *J. Operator Theory*.

[5] M. Choda, Entropy for Canonical shifts, to appear in *Trans. Amer. Math. Soc.*.

[6] M. Choda and F. Hiai, Entropy for canonical shifts II, *Publ., RIMS, Kyoto Univ.*, **27**, 461-489 (1991).

[7] M. Choda, Duality for finite bipartite graphs (with an application to II_1 factors), to appear in *Pacific J. Math.*.

[8] M. Choda, Square roots of the canonical shifts, Preprint.

[9] A. Connes and E. Støormer, Entropy for automorphism of II_1 von Neumann algebras, *Acta Math.*,**134**, 288-306 (1975).

[10] A. Connes, H.Narnhofer and W. Thirrng, Dynamical entropy of C^*-algebras and von Neumann algebras, *Commun. Math. Phys.*, **112**, 691-719 (1987).

[11] F.Goodman, P.de la Harpe and V. Jones, Coxter graphs and towers of algebras, (Springer, 1989).

[12] U. Haagerup : in preparation.

[13] M. Izumi: Application of fusion rules to classification of subfactors, *Publ., RIMS, Kyoto Univ.*, **24**, 953-994 (1988).

[14] M. Izumi and Y. Kawahigashi: Classification of subfactors with the principal graph $D_n^{(1)}$ Preprint.

[15] M. Izumi: On flatness of the Coxeter graph E_8, Preprint.

[16] V. F. R. Jones, Index for subfactors, *Invent. Math.*, **72**, 1-25 (1983).

[17] Y. Kawahigashi : On flatness of Ocneanu's connections on the Dynkyn diagrams and classification of subfactors, Preprint.

[18] Y. Kawahigashi : Exactly solvable orbifold models and subfactors, to appear in Functional analysis conference at Kyoto, (1991).

[19] R. Longo, Simple injective subfactors, *Adv. in Math.*, **63**, 152-171 (1987).

[20] R. Longo: Index of subfactors and statistics of quantum fields I, *Commun. Math. Phys.*, **126**, 217-247 (1989).

[21] R. Longo: Index of subfactors and statistics of quantum fields II, *Commun. Math. Phys.*, **130**, 285-309(1990).

[22] A. Ocneanu : Quantized groups, string algebras and Galois theory, Preprint. Quantized groups, string algebras and Galios theory in Operator algebras and applications, vol. II, *Cambridge Univ. Press*, (1988).

[23] A. Ocneanu : Quantum symmetry, differential geometry of finite graphs and classification of subfactors, Notes by Kawahigashi.

[24] S. Okamoto : Invariants for subfactors arizing from Coxeter graphs, Current topics in Operator algebras, *World Scientific*, (1991).

[25] M. Pimsner and S. Popa, Entropy and index for subfactors, *Ann. Sci. Ec. Norm. Sup.*, **19**, 57-106 (1986).

[26] M. Pimsner and S. Popa, Iterating the basic construction, *Trans. A. M. S.*, **310**, 127-134(1988).

[27] S. Popa, Maximal injective subalgebras iin factors associated with free groups, *Adv. in Math.*, **50**, 27-48(1983).

[28] S. Popa, Classification of subfactors : the Reduction to commuting squares, *Invent. Math.*, **101**, 19-43(1990).

[29] S. Popa, Sur classification des sousfacteurs d'indice fini du facteur hyperfini, *C. R. Acad. Sc. Paris*, **311** , 95-100(1990).

[30] S. Popa, Classification of amenable subfactors of type II_1, Preprint.

[31] R. T. Powers, An index theory for semigroups of *-endomorphisms of $B(H)$ and type II_1 factors, *Canad. J. Math.*, **40**,86-114 (1988).

[32] A. J. Wassermann, Coaction and Yang-Baxter equations for ergodic actions and subfactors, Operator algebras and applications, *Cambridge Univ. Press*, II, (1988).

Automorphisms in the Irreducible Decomposion of Sectors

HIDEKI KOSAKI
Department of Mathematics
College of General Education
Kyushu University
Fukuoka, 810, Japan

Abstract. Automorphisms for an inclusion of factors with finite index are considered. For such automorphisms, we introduce the notion of strong outerness and obtain a characterization of the strong outerness based on the sector technique.

Key words: inclusion of factors, strongly outer automorphism, sector, canonical endomorphism, Frobenius reciprocity

1. Introduction

Let $N \subset M$ be an inclusion of factors, and $E : M \longrightarrow N$ be a normal conditional expectation with $Ind\ E < \infty$ ([11,15,22]). Assume that an (outer) action $\theta : G \longrightarrow Aut(M)$ of a finite group G is given and that each θ_g leaves N invariant (i.e., $\theta_g \in Aut(M, N)$). Then we obtain the new pair $B = N \times_\theta G \subset A = M \times_\theta G$ of crossed product factors. For example, when the inclusin $N \subset M$ is irreducible, $\theta_g \circ E = E \circ \theta_g$ by the uniqueness of a normal conditional expectation, and hence E naturally induces the normal conditional expectation $\hat{E} : A \longrightarrow B$. Although $Ind\ E = Ind\ \hat{E}$ (see for example [18]), the new inclusion $B \subset A$ might be quite different from the original inclusion $N \subset M$. In fact, they may have different towers of the higher relative commutants ([5,11]). This fact was pointed out in several papers (see for example [2,12]).

It is well-known (see for example [3,18]) that the Jones tower of $B \subset A$ is $\{M_i \times_\theta G\}_i$. Here,

$$N(= M_{-1}) \subset M(= M_0) \subset M_1 \subset M_2 \subset \cdots$$

is the Jones tower with the Jones projections $\{e_j\}$, and θ was uniquely extended to the tower subject to the condition $\theta(e_j) = e_j$. To compute $(M_i \times_\theta G) \cap B'$, it is convenient to begin by determining $(M_i \times_\theta G) \cap N'$. By looking at the "Fourier coefficients" of an element in this larger relative commutant, we immediately encounter the following commutation relation:

H. Araki et al. (eds.), Quantum and Non-Commutative Analysis, 305–316.

$$x \in M_i \text{ satisfies } nx = x\theta_g(n) \text{ for all } n \in N$$

In certain cases, this relation automatically implies $x = 0$ (for all i). In that case, we say that θ_g is *strongly outer*. If $\theta_g, g \neq e$, are strongly outer, then the computation of the tower is very easy as was shown in [3].

The purpose of this article is to obtain a characterization of the strong outerness. Our basic tool to investigate this problem is the sector technique ([23], see also [8,9]), or equivalently, the bimodule technique ([25,27] and see also [31,32] where a very readable account is given). Especially, various versions of the Frobenius reciprocity for sectors will play important roles. When $N = \rho(M)$ with an endomorphism ρ (and factors are properly infinite), we can perform the irreducible decomposition of $(\rho\bar{\rho})^n$ (as sectors). We will show that non-strongly outer automorphisms are characterized as those appearing in this irreducible decomposition. Some consequences of our characterization are explained. For example, it follows immediately from our characterization that non-strongly outer automorphisms form a group.

A subject closely related to the one in this article is studied in [13], where the effect of an automorphism on (relative) central sequences is carefully analyzed and some results in the AFD II_1 factor case are obtained.

Acknowledgement

The author profited from fruitful discussions with M. Izumi and R. Longo on sectors and bimodules, and he is also grateful to M. Choda and Y. Kawahigashi for various communications on automorphisms for a factor-subfactor pair.

2. Sectors and the Frobenius Reciprocity

For simplicity we now assume that $N \subset M$ are isomorphic properly infinite factors. In particular, we can find $\rho \in End(M)$, the unital normal *-endomorphisms of M, satisfying $N = \rho(M)$. In what follows we will deal with an endomorphism $\sigma \in End(M)$ such that $M \supset \sigma(M)$ has finite index, and $d\sigma$, the statistical dimension, means the square root of the minimal index of the above inclusion ([6,7,22], see also [18,24] for basic properties). Let $Sect(M) = End(M)/Int(M)$, the sectors. Hence, σ and $Adu \circ \sigma \quad (u \in \mathcal{U}(\mathcal{M}))$ are identified as sectors, and the class of σ is denoted by $[\sigma]$. However, in most cases below there is no possibility of confusion, and we will often write σ instead of $[\sigma]$.

Let $\overline{[\sigma]}$ be the conjugate sector defined by $\overline{[\sigma]} = [\bar{\sigma}]$, where $\bar{\sigma} = \sigma^{-1} \circ \gamma$ and γ is the canonical endomorphism associated with the inclusion $M \supset \sigma(M)$. (see [21]) The sum and the product (i.e., the composition) of sectors can be also defined in the natural way (see [8,23] for details). Considering sectors is the same thing as considering bimodules, and the product of two sectors corresponds to the relative tensor product of the relevant bimodules

([9,29]). Also taking the conjugate sector means to look at the contragredient bimodule.

A sector σ is called irreducible when $\sigma(M)' \cap M$ is one dimensional. A non-irreducible sector σ can be (uniquely) decomposed as the sum of irreducible sectors in the following way: Choose a (finite......since $d\sigma < \infty$) partition $\{p_i\}_{i=i,2,\cdots,n}$ of the unity consisting of minimal projections in $\sigma(M)' \cap M$, and choose isometries $\{v_i\}_{i=1,2,\cdots,n}$ in M satisfying $v_i v_i^* = p_i$. Then we have

$$\sigma = v_1^* \sigma(\cdot) v_1 \oplus v_2^* \sigma(\cdot) v_2 \oplus \cdots \oplus v_n^* \sigma(\cdot) v_n.$$

Notice that each $v_i^* \sigma(\cdot) v_i$ is irreducible from the construction. In what follows, we will use the notations $\prec$ and $\succ$ as in the representation theory.

One of the most important results in [23] (see also [8]) is that, for irreducible sectors ρ_1, ρ_2, the following three statements are equivalent: (i) $\rho_1 = \overline{\rho_2}$ (as sectors), (ii) $id \prec \rho_1\rho_2$, and (iii) $id \prec \rho_2\rho_1$. Here, id means the identity sector. Based on this fact, one easily obtains the next resut.

PROPOSITION 1 (Frobenius reciprocity). *Let ζ, ξ, ρ be sectors.*
(i) When ζ, ξ, ρ are irreducible, we have

$$\zeta\rho \succ \xi \iff \xi\overline{\rho} \succ \zeta.$$

(ii) When ρ is irreducible, we have

$$\zeta \succ \rho \iff \zeta\overline{\rho} \succ id \iff \overline{\rho}\zeta \succ id.$$

(iii) When ζ, ξ are irreducible, we have

$$\rho\zeta\overline{\rho} \succ \xi \iff \overline{\rho}\xi\rho \succ \zeta.$$

These might be generalized to several directions ([31,32] for example), but the above versions are sufficient for our later purpose. The above formulations were noticed during conversations with R.Longo.

3. Main Results

Let $M \supset N = \rho(M)$ as in §2, and for simplicity we will assume that ρ is irreducible in the rest of the article. Let θ be an automorphism of M leaving N invariant (i.e., $\theta \in Aut(M, N)$). Let $E : M \longrightarrow N$ be the minimal expectation ($d\rho = (Ind\ E)^{1/2}$), see [6,7,22]. Then the uniqueness of a minimal expectation guarantees $\theta \circ E = E \circ \theta$. Hence, as in §1, θ has the canonical extension to the Jones tower

$$N(= M_{-1}) \subset M(= M_0) \subset M_1 \subset M_2 \subset \cdots.$$

Definition([3]) An automorphism θ ($\in Aut(M, N)$) is called strongly outer when the following condition is satisfied for each i:

if $x \in M_i$ satisfies $nx = x\theta(n)$ for all $n \in N$, then we must have $x = 0$.

The properly infiniteness guarantees that $M \supset N$ can act on a Hilbert space $\mathcal{H}$ with a common cyclic and separating vector. We thus have the modular operators J_M and J_N on the same Hilbert space $\mathcal{H}$. The Jones tower $\{M_i\}$ can be easily identified with

$$M_{2i-1} = \gamma^{-i}(N) \text{ and } M_{2i} = \gamma^{-i}(M),$$

where γ ($= AdJ_N J_M$) is the canonical endomorphism ([21]). (An advantage to deal with properly infinite factors is that one can (standardly) represent all the basic (and downward basic) extensions on the same Hilbert space.) For the sake of completeness, we prove the following lemma:

LEMMA 2. *Sectors σ and τ admit a non-zero intertwiner (i.e., $x\sigma(m) = \tau(m)x$, $m \in M$, for some non-zero $x \in M$) if and only if there exists an irreducible sector ζ such that $\zeta \prec \sigma$ and $\zeta \prec \tau$.*

Proof. Assume the existence of a non-zero intertwiner x as above, and let $x = u|x|$ be the polar decomposition. Then the initial and final projections of u belong to the finite dimensional algebras $\sigma(M)' \cap M$ and $\tau(M)' \cap M$ respectively. Let p be a spectral projection ($\in \sigma(M)' \cap M$) of $|x|$. Then we have

$$\tau(m)u|x|p = u|x|\sigma(m)p = u|x|p\sigma(m),$$
$$\text{and} \quad pu^*\tau(m)u|x|p = pu^*u|x|p\sigma(m) = p|x|p\sigma(m).$$

Since $|x|p = \lambda p$ ($\lambda > 0$), we conclude

$$pu^*\tau(m)up = p\sigma(m).$$

Notice $pu^* = pu^*(upu^*)$. Being a spectral projecton of $u|x|u^* = |x^*|$, upu^* belongs to $\tau(M)' \cap M$. Choose an isometry $v \in M$ with $vv^* = p$, and set $w = uv$. We compute

$$ww* = uvv^*u^* = upu^*,$$
$$(1 \geq)w^*w = v^*u^*uv \geq v^*pv = 1.$$

Hence, w is an isometry with the final projection $upu^* \in \tau(M)' \cap M$. The above equation thus implies

$$v^*\sigma(m)v = v^*p\sigma(m)v = v^*pu^*\tau(m)upv = w^*(upu^*)\tau(m)w = w^*\tau(m)w.$$

Obviously an irreducible component ζ of $v^*\sigma(\cdot)v = w^*\tau(\cdot)w$ does the job.

Conversely assume that an irreducible sector ζ is included in σ and τ. One finds isometries $u, v \in M$ satisfying $\zeta = u^*\sigma(\cdot)u = v^*\tau(\cdot)v, uu^* \in \sigma(M)' \cap M, vv^* \in \tau(M)' \cap M$. Then, we have $\zeta(\cdot)u^* = u^*\sigma(\cdot)$ and $\zeta(\cdot)v^* = v^*\tau(\cdot)$. Therefore, $x = vu^*(\neq 0)$ does the job. □

Given an automorphism in $Aut(M, N)$, the next result tells whether it is strongly outer and where the strong outerness breaks (if not).

THEOREM 3. *For an automorphism $\theta \in Aut(M,N)$, the strong outerness breaks at the n th extension M_n if and only if $\theta \prec (\rho\bar{\rho})^{n+1}$.*

Proof. (see Theorem 2 in [3]) We just prove the result when $n(=2i)$ is even. Assume that the strong outerness breaks at M_n, that is,

$$nx = x\theta(n), n \in N,$$

for some $x \neq 0 \in M_n$. Applying γ^i $(=(\rho\bar{\rho})^i$ on $M)$ to the both sides, we get

$$(\rho\bar{\rho})^i(n)\gamma^i(x) = \gamma^i(x)(\rho\bar{\rho})^i(\theta(n)) = \gamma^i(x)\theta(\rho\bar{\rho})^i(n)$$

for all $n \in N$. Here, $\rho\bar{\rho}$ and θ commute up to an inner automorphism ([1]), and we let $\gamma^i(x)$ $(\in M)$ absorb the relevant unitary. Therefore, $(\rho\bar{\rho})^i\rho$ and $\theta(\rho\bar{\rho})^i\rho$ admit a non-zero intertwiner, and Lemma 2 guarantees that these two sectors contain a common irreducible sector. Therefore, we conclude

$$id \prec ((\rho\bar{\rho})^i\rho)\overline{(\theta(\rho\bar{\rho})^i\rho)} = (\rho\bar{\rho})^{n+1}\bar{\theta}.$$

The automorphism θ being irreducible as a sector, Proposition 1,(ii) implies

$$\theta \prec (\rho\bar{\rho})^{n+1}.$$

Conversely, assume that $\theta \prec (\rho\bar{\rho})^{n+1}$. Then $id \prec (\rho\bar{\rho})^{n+1}\bar{\theta}$ and as before $(\rho\bar{\rho})^i\rho$ and $\theta(\rho\bar{\rho})^i\rho$ contain a common irreducible sector. Hence, by Lemma 2 again, there exists a non-zero intertwiner between them, and the strong outerness breaks at M_n. □

A few remarks are in order.
(i) In the first part of the above proof, γ and θ actually commute up to an inner automorphism of N. Also notice that, if $x \in M_{2i-1}$, then $\gamma^i(x)$ falls into N. The odd case will be proved based on these facts at the end of this section.
(ii) Let us now require the commutation relation in the definition of the strong outerness for all $m \in M$ (instead of for all $n \in N$). Then, one can find a non-zero intertwiner in M_n if and only if $\theta \prec (\rho\bar{\rho})^n$.
(iii) We assumed that $N \subset M$ are isomorphic properly infinite factors. But, in the applications below, one can actually drop this assumption by the standard "tensoring trick" (see [24]). But, it is also possible to repeat similar arguments (for the above proof and the applications in the rest of the article) by using bimodules.

Assume that an automorphism θ $(\in Aut(M))$ satisfies $\theta \prec (\rho\bar{\rho})^n$. This may or may not be adjusted to an automorphism in $Aut(M,N)$ (i.e., we may or may not find a unitary $u \in \mathcal{U}(\mathcal{M})$ satisfying $Adu \circ \theta \in Aut(M,N)$). The next result completely clarifies this phenomenon. Therefore, Theorem 3 and the next result together give us the complete description of non-strongly outer automorphisms. (see [13] for related results)

PROPOSITION 4. *An automorphism θ ($\in Aut(M)$) can be adjusted to an automorphism in $Aut(M,N)$ if and only if there exists an automorphism α satisfying $\theta\rho = \rho\alpha$ (as sectors). Furthermore, if $\theta \prec (\rho\bar{\rho})^n$, then we have $\alpha \prec (\bar{\rho}\rho)^n$.*

Proof. When θ belongs to $Aut(M,N)$, we obviously have

$$\theta\rho(M) = \theta(N) = N.$$

Therefore, one can find a (uniquely determined by θ) automorphism $\alpha \in Aut(M)$ satisfying $\theta\rho = \rho\alpha$. Conversely, when $\theta\rho = \rho\alpha$ as sectors, one can find a unitary $u \in M$ satisfying $Adu \circ \theta\rho(m) = \rho\alpha(m), m \in M$. Hence, the adjusted $Adu \circ \theta$ obviously leaves N invariant. (In the bimolule version, the role of the above α is played by the irreducible $N-N$ bimodule associated with the restriction $\theta\mid_N \in Aut(N)$.)

We now assume $\theta\rho = \rho\alpha$. The assumption $\theta \prec (\rho\overline{\rho})^n$ implies $id \prec (\rho\overline{\rho})^n\overline{\theta}$, and we compute

$$(\rho\overline{\rho})^n\overline{\theta} = ((\rho\overline{\rho})^{n-1}\rho)\overline{(\theta\rho)}$$

(when $n \geq 1$). Notice that $\theta\rho$ is irreducible because

$$\theta\rho(M)' \cap M = \theta(\rho(M)' \cap M)$$

is one dimensional. Therefore, Proposition 1, (ii) implies

$$\theta\rho \prec (\rho\overline{\rho})^{n-1}\rho.$$

We thus conclude

$$\alpha \prec \overline{\rho}\rho\alpha = \overline{\rho}\theta\rho \prec \overline{\rho}(\rho\overline{\rho})^{n-1}\rho = (\overline{\rho}\rho)^n.$$

When $n = 1$, $\theta \prec \rho\overline{\rho}$ implies $\rho = \theta\rho$. Therefore, we get $\rho = \rho\alpha$ and $\alpha \prec \overline{\rho}\rho\alpha = \overline{\rho}\rho$. □

Even if ρ is not irreducible, $\theta \prec (\rho\overline{\rho})^n$ implies $\alpha \prec (\overline{\rho}\rho)^{n+1}$ because of

$$\alpha \prec \overline{\rho}\rho\alpha = \overline{\rho}\theta\rho.$$

The irreducible sectors appearing in $(\rho\bar{\rho})^n, (\rho\bar{\rho})^n\rho, (\bar{\rho}\rho)^n$, and $(\bar{\rho}\rho)^n\bar{\rho}$ ($n = 1,2,\cdots$) are corresponding to the irreducible $M-M, N-M, N-N$, and $M-N$ bimodules respectively considered in [25]. Also, for a sector σ in $(\rho\bar{\rho})^n$ for example (i.e., an $M-M$ bimodule), the composition $\sigma\rho$ corresponds to the restricted $N-M$ bimodule. Therefore, in terms of bimodules the equation

$$\theta\rho = \rho\alpha, \text{ that is, } \overline{\theta\rho} = \bar{\alpha}\bar{\rho}$$

means the following: Look at the irreducible $M-M$ bimodule corresponding to the automorphism θ. There shoud exists an irreducible $N-N$ bimodule (at the same depth, but in the other graph) corresponding to an automorphism of N such that the restricted $N-M$ bimodule of the former is exactly the contragredient bimodule of the induced $M-N$ bimodule of the latter.

Notice that $\theta \prec \rho\bar{\rho}$ if and only if $\theta\rho = \rho$ (since ρ is irreducible). Therefore, we have:

COROLLARY 5. *Assume that an automorphism θ ($\in Aut(M)$) satisfies $\theta \prec (\rho\bar{\rho})^n$. Assume further that the identity is the only irreducible sector in $(\bar{\rho}\rho)^k, k = 1, 2, \cdots$, with statistical dimension 1. Then we have:*
(i) If $\theta \prec \rho\bar{\rho}$, then θ can be adjusted in such a way that $\theta\,|_N = Id_N$.
(ii) If θ does not appear in $\rho\bar{\rho}$, then θ cannot be adjusted to an automorphism in $Aut(M, N)$.

The Haagerup subfactor with index $(5 + \sqrt{13})/2$ satisfies the extra assumption in Corollary 5. The "dual" principal graph is Y-shaped, and the three end-points correspond to irreducible sectors (in $(\rho\bar{\rho})^3$) with statistical dimension 1. One of them is of course the identity. The other two do not appear in $\rho\bar{\rho}$ so that they cannot be adjusted to leave N invariant. (M.Izumi and Y.Kawahigashi informed the author of this example.)

Proof of THEOREM 3 (when $n = 2i+1$ is odd). Assume that x ($\neq 0$) $\in M_n$ satisfies

$$nx = x\theta(n), n \in N.$$

As before we get

$$(\rho\bar{\rho})^{i+1}(n)\gamma^{i+1}(x) = \gamma^{i+1}(x)\theta(\rho\bar{\rho})^{i+1}(n), n \in N.$$

Recall that $\gamma^{i+1}(x) \neq 0$ falls into N (,i.e., $(\rho\bar{\rho})^{i+1}\rho$ and $\theta(\rho\bar{\rho})^{i+1}\rho$ admit a non-zero intertwiner as $M-N$ sectors in the sense of [9]). Therefore, $\rho^{-1}(\rho\bar{\rho})^{i+1}\rho$ and $\rho^{-1}\theta(\rho\bar{\rho})^{i+1}\rho$ admit a non-zero intertwiner (in M). On the other hand, because of $\theta \in Aut(M, N)$, we can find an automorphism α satisfying $\theta\rho = \rho\alpha$ (Proposition 4). Notice that

$$\rho^{-1}(\rho\bar{\rho})^{i+1}\rho = (\bar{\rho}\rho)^{i+1},$$

$$\rho^{-1}\theta(\rho\bar{\rho})^{i+1}\rho = \rho^{-1}(\rho\bar{\rho})^{i+1}\theta\rho = \rho^{-1}(\rho\bar{\rho})^{i+1}\rho\alpha = (\bar{\rho}\rho)^{i+1}\alpha.$$

We thus get

$$id \prec (\bar{\rho}\rho)^{n+1}\alpha \text{ and } \bar{\alpha} \prec (\bar{\rho}\rho)^{n+1}.$$

Since $\bar{\alpha}\,\bar{\rho} = \bar{\rho}\bar{\theta}$, Proposition 4 (applied to $\bar{\alpha}$ and $\bar{\rho}$) implies

$$\bar{\theta} \prec (\rho\bar{\rho})^{n+1}, \text{ that is, } \theta \prec (\rho\bar{\rho})^{n+1}.$$

We conversely assume $\theta \prec (\rho\overline{\rho})^{n+1}$. Proposition 4 guarantees the existence of an automorphism α satisfying

$$\theta\rho = \rho\alpha \text{ and } \alpha \prec (\overline{\rho}\rho)^{n+1}.$$

Since $id \prec (\overline{\alpha}(\overline{\rho}\rho)^{i+1})(\overline{\rho}\rho)^{i+1}$, $(\overline{\rho}\rho)^{i+1}$ and $(\overline{\rho}\rho)^{i+1}\alpha$ admit a non-zero intertwiner, that is, there exists a non-zero $z \in M$ such that

$$(\overline{\rho}\rho)^{i+1}(y)z = z(\overline{\rho}\rho)^{i+1}\alpha(y)$$

$$= z(\overline{\rho}\rho)^{i}\overline{\rho}\rho\alpha(y) = z(\overline{\rho}\rho)^{i}\overline{\rho}\theta\rho(y)$$

for all $y \in M$. This means that

$$(\overline{\rho}\rho)^{i}\overline{\rho}(n)z = z(\overline{\rho}\rho)^{i}\overline{\rho}\theta(n), n \in N.$$

Applying ρ to the both sides and setting $x = \rho(z) \neq 0 \in N$, we get

$$\rho(\overline{\rho}\rho)^{i}\overline{\rho}(n)x = x\rho(\overline{\rho}\rho)^{i}\overline{\rho}\theta(n), n \in N.$$

Since $\rho(\overline{\rho}\rho)^{i}\overline{\rho} = (\rho\overline{\rho})^{i+1}$, we are done. □

4. Consequences

In this section we point out some consequences and applications of our characterization.

COROLLARY 6. *Non-strongly outer automorphisms form a group.*

The result follows from the fact that sectors ε with $d\varepsilon = 1$ appearing in $(\rho\overline{\rho})^k, k = 1, 2, \cdots$, form a group.

The quotient of this group modulo the unitary normalizers of N in M can be easily determined in many examples based on Theorem 3 and Proposition 4.

COROLLARY 7. *Let $M \supset N, A \supset B$ be factor-subfactor pairs of finite index, and α, β be automorphisms in $Aut(M, N)$ and $Aut(A, B)$ respectively. Then the tensor product $\alpha \otimes \beta (\in Aut(M \otimes A, N \otimes B))$ is strongly outer if and only if at least one of α, β is strongly outer.*

Proof. When $N = \rho_1(M)$ and $B = \rho_2(A)$, we have $N \otimes B = \rho(M \otimes A)$ with $\rho = \rho_1 \otimes \rho_2$ (in the obvious sense). It is elementary to see

$$(\rho\overline{\rho})^n = (\rho_1\overline{\rho_1})^n \otimes (\rho_2\overline{\rho_2})^n.$$

Let

$$(\rho_1\overline{\rho_1})^n = \alpha_1 \oplus \alpha_2 \oplus \cdots \oplus \alpha_m,$$

$$(\rho_2\overline{\rho_2})^n = \beta_1 \oplus \beta_2 \oplus \cdots \oplus \beta_k$$

be the irreducible decompositions. Then we get

$$(\rho\overline{\rho})^n = \sum_{i,j} \alpha_i \otimes \beta_j,$$

and each $\alpha_i \otimes \beta_j$ is irreducible. (Observe that $d(\alpha_i \otimes \beta_j) = d\alpha_i d\beta_j$.) If $\alpha \otimes \beta = \alpha_i \otimes \beta_j$ as sectors, then α_i, β_j are automorphisms (because of $d\alpha_i d\beta_j = 1$) and $\alpha^{-1}\alpha_i \otimes \beta^{-1}\beta$ is an inner automorphism of $M \otimes A$. The second statement means that both of $\alpha^{-1}\alpha_i, \beta^{-1}\beta_j$ are inner automorphisms (of M and A respectively). Therefore, $\alpha\otimes\beta$ is non-strongly outer if and only if both of α, β are non-strongly outer. □

The next three corollaries were pointed out in [3].

COROLLARY 8. *Assume that the identity is the only irreducible sector appearing in $(\rho\bar{\rho})^n, n = 1, 2, \cdots$, with statistical dimension 1. If an outer automorphism of M leaves N invariant, then it is strongly outer.*

The assumption of this corollary is satisfied for an inclusion whose principal graph is one of the Coxeter graphs $A_{2n}, D_{2n}(n \geq 3), E_8, A_\infty$, etc. Thanks to the characterization given in [26], the unitary normalizers of N in M are exactly the unitary group $\mathcal{U}(N)$ in this case. Therefore, an automorphism in $Aut(M, N)$ is either strongly outer or of the form $Adu, u \in \mathcal{U}(N)$.

COROLLARY 9. *If an inclusion $M \supset N$ has finite depth, then any aperiodic automorphism of M leaving N invariant is strongly outer.*

Let $M \supset N$ be an inclusion of type III_1 factors with finite depth. Then, a modular automorphism σ_t $(t \neq 0)$ in $Aut(M, N)$ is aperiodic and hence strongly outer. (This fact is closely related to the main result in [10])

COROLLARY 10. *Let $M = N \times_\alpha G$ with an outer action $\alpha : G \longrightarrow Aut(N)$ of a finite group G, and θ be an automorphism in $Aut(M, N)$. If $\alpha_g \circ \theta \mid_N$ is an outer automorphism of N for each $g \in G$, then θ is strongly outer.*

This result follows from the description of irreducible $M - M$ bimodules obtained in [19,25].

Let $M \supset N$ be an inclusion whose principal graph is given by the Coxeter graph A_{2n+1}. As was shown in [8], a period 2 automorphism $\theta \in Aut(M, N)$ appears in $(\rho\bar{\rho})^n$ and plays an extremely important role. For this θ the strong outerness breaks at the $(n-1)$ st extension M_{n-1}, i.e., a non-zero intertwiner can be found here. Also recall that the Loi invariant ([20]) of θ is trivial. In fact, the tower $\{M_i \cap N'\}$ of the higher relative commutants is generated by the Jones projections.

Assume $2n+1 = 4m-1$. The crossed product pair $M \times_\theta \mathcal{Z}_2 \supset N \times_\theta \mathcal{Z}_2$ has the same principal graph as the original inclusion $M \supset N$ ($N \times_\theta \mathcal{Z}_2$ is not a factor when $n = 1$) since the both inclusions have the index $4\cos^2(\frac{\pi}{4m})$. Therefore, we conclude:

COROLLARY 11. *Assume* $2n+1 = 4m-1$, $m = 2, 3, \cdots$. *For the above automorphism* $\theta \in Aut(M, N)$, *there exists a non-zero intertwiner in* M_{n-1}, *but this cannot be chosen from the fixed point subalgebra* $(M_{n-1})^\theta$.

Let us now assume that M, N are AFD II_1 factors and $2n+1 = 4m-3$. In this case, the above intertwiner can be indeed chosen from the fixed point subalgebra $(M_{n-1})^\theta$. (This fact was communicated to the auther by Y. Kawahigashi.) Therefore, the principal graph of $M \times_\theta \mathcal{Z}_2 \supset N \times_\theta \mathcal{Z}_2$ is strictly larger than that of $M \supset N$, and hence this inclusion has to the unique one with the Coxeter graph D_{2m} (see [12,25,28]). The dual action $\hat{\theta} \in Aut(M \times_\theta \mathcal{Z}_2, N \times_\theta \mathcal{Z}_2)$ is strongly outer as was pointed out above. One can show that this (period 2) automorphism is the one flipping the two end-points of D_{2m} (see [2, 3, 18, 20]).

The above automorphisms appearing in the Coxeter graphs A_{4m-3} also play important roles in analysis of subfactors in the type III setting. If factors are of type III_1 and the graph has finite depth, a modular automorphism cannot appear in $(\rho\bar{\rho})^n, n = 1, 2, \cdots$, as was pointed out above. But, when factors are of type $III_\lambda, 0 < \lambda < 1$, instead, it is indeed possible (since a modular automorphism could have a finite period), and this is responsible for the difference of the "type III and type II principal graphs" ([10], see also [20] for typical examples). The same phenomenon also occurs in the type III_0 setting in a more complicated fashion. In [16,17,18] many non-conjugate inclusions $M \supset N$ of AFD type III_0 factors were constructed with the following properties: (i) M and N have the same flow of weights ([4]), (ii) the "type III principal graph" is A_{4m-3}, and (iii) the "type II principal graph" is D_{2m}. To distinguish non-conjugacy classes of subfactors, we constructed a certain two-to-one ergodic extension of the flow of weights of M (or equivalently, that of N) from the inclusion data. (see [16,17] for details) To do so, we looked at the dual action (in the sense of the structure theorem of type III factors) on the tower of the higher relative commutants of the associated inclusion of the type II_∞- algebras. In these examples the automorphism θ appearing at the end of the graph A_{4m-3} is actually shown to be an extended modular automorphism ([4]) of period 2. The dual action $\hat{\theta}$ (of $\mathcal{Z}_2$) has a non-trivial module ([4]) and all the other invariants in the sense of [14,30] are trivial. The above two-to-one ergodic extension is given by the module of $\hat{\theta}$ together with the flow of weights of the factor on which $\hat{\theta}$ acts.

References

[1] Araki,H., Some properties of modular conjugation operator of von Neumann algebras and a non-commutative Radon-Nikodym theorem with a chain rule, *Pacific J. Math.*, 50(1974), 309-359.
[2] Choda,M., Duality for finite bipartite graphs (with applications to II_1-factors), to appear in *Pacific J. Math.*
[3] Choda,M. and Kosaki,H., Strongly outer actions for an inclusion of factors, preprint.
[4] Connes,A. and Takesaki,M., The flow of weights on factors of type III, *Tôhoku J. Math.*, 29(1977), 473-575.
[5] Goodman,F., de la Harpe,P. and Jones,V.F.R., Coxeter Dynkin Diagrams and Towers of Algebras, Springer, 1989.
[6] Havet,J.F., Espérance conditionnelle minimale, *J. Operator Theory*, 24(1990), 33-55.
[7] Hiai,F. Minimizing indices of conditional expectations onto a subfactor, *Publ., RIMS, Kyoto Univ.*, 24(1988), 673-678.
[8] Izumi,M., Application of fusion rules to classification of subfactors, *Publ., RIMS, Kyoto Univ.*, 27(1991), 953-994.
[9] Izumi,M., Goldman's type theorem for index 3, preprint.
[10] Izumi,M., On type III and type II principal graphs for subfactors, in preparation.
[11] Jones,V.F.R., Index of subfactors, *Invent. Math.*, 66(1983), 1-25.
[12] Kawahigashi,Y., On flatness of Ocneanu's connections on the Dynkin diagrams and classification of subfactors, preprint.
[13] Kawahigashi,Y., Centrally trivial automorphisms and an analogue of Connes' $\chi(M)$ for subfactors, preprint(preliminary version).
[14] Kawahigashi,Y., Sutherland,C. and Takesaki,M., The structure of automorphism group of an injective factor and the cocycle conjugacy of discrete abelian group actions, to appear in *Acta Math.*
[15] Kosaki,H., Extension of Jones' theory on index to arbitrary factors, *J. Funct. Anal.*, 66(1986), 123-140.
[16] Kosaki,H., AFD factor of type III_0 with many isomorphic index 3 subfactors, to appear in *J. Operator Theory.*
[17] Kosaki,H., in preparation.
[18] Kosaki,H. and Longo,R., A remark on the minimal index of subfactors, to appear in *J. Funct. Anal.*
[19] Kosaki,H. and Yamagami,S., Irreducible bimodules associated with crossed product algebras, to appear in *Internat. J. Math.*
[20] Loi,P.H., On automorphisms of subfactors, preprint.
[21] Longo,R., Simple injective subfactors, *Adv. in Math.*, 63(1987), 152-172.
[22] Longo,R., Index of subfactors and statistics of quantum fields I, *Commun. Math. Phys.*, 126(1989), 217-247.
[23] Longo,R., Index of subfactors and statistics of quantum fields II, *Commun. Math. Phys.*, 130(1990), 285-309.
[24] Longo,R., Minimal index and braided subfactors, to appear in *J. Funct. Anal.*
[25] Ocneanu,A., Quantum Symmetry, Differential Geometry of Finite Graphs, and Classification of Subfactors, University of Tokyo Seminary Notes, 1991.
[26] Pimsner,A. and Popa,S., Entropy and index of subfactors, *Ann. Sci. École Norm. Sup.*, 19(1986), 57-106.
[27] Popa,S., Correspondences, preprint.
[28] Popa,S., Classification of subfactors: reduction to commuting squares, *Invent. Math.*, 101(1990), 19-43.
[29] Sauvageot,J.L., Sur le produit tensoriel relatif d'espaces de hilbert, *J. Operator Theory*, 9(1983), 237-252.
[30] Sutherland,C. and Takesaki,M., Actions of discrete amenable groups on injective factors of type $III_\lambda, \lambda \neq 1$, *Pacific J. Math.*, 137(1989), 405-444.
[31] Yamagami,S., A report on Ocneanu's lectures on Jones' index theory, preprint.
[32] Yamagami,S., Modular theory in bimodules, preprint.

Notes:

After completion of the present work the author received "Classification of actions of discrete amenable groups on amenable subfactors of type II" by S. Popa, where the subject similar to the one in the present article is studied.

Fusion Rules and Classification of Subfactors

MASAKI IZUMI
Research Institute for Mathematical Sciences, Kyoto University, Kyoto 606, Japan

Abstract. We survey two applications of sector theory to classification of subfactors.
1. A proof is given to A. Ocneanu's announcement on the non-existence of the paragroups corresponding to the Coxeter graphs D_{odd} and E_7.
2. Goldman's type theorem for subfactors with the principal graph A_5 is obtained.

Key words: sectors, fusion rules

1. Introduction

The notion of sectors for infinite factors was introduced by R. Longo in analogy with the case of quantum field theory [14]. He proved basic properties of sectors such as characterization of conjugate sectors, and showed how they are related to index theory, which was initiated by V. Jones for II_1 factors [8] and extended by H. Kosaki for arbitrary factors [10]. Recently, several people used sector theory as a fundamental tool for analysis of subfactors [1, 5, 6, 7, 12].

The aim of this article is to present two applications of sector theory to classification of subfactors. One can find basics of index theory in [2, 10], and those of sector theory in [6, 14]. There is a brief summary for sector theory in H. Kosaki's article in this book.

2. Fusion Rules of Sectors and Principal Graphs

Let $M \supset N$ be a pair of properly infinite factors with finite index and finite depth. For simplicity we assume $M \cong N$. (This assumption is not essential because in the case of $M \not\cong N$, we can modify the definition of $Sect(M)$ to obtain the same result [6].) Due to the assumption we can find $\rho \in End(M)$ satisfying $N = \rho(M)$. Then the downward tower for $M \supset N$ is as follows [14]:

$$M \supset \rho(M) \supset \rho\bar{\rho}(M) \supset \rho\bar{\rho}\rho(M) \supset \cdots.$$

Let $\mathcal{G}$ be the dual principal graph of the inclusion [2]. It is known that $\mathcal{G}$ is a bipartite graph with a distinguished point $*$, which is an even vertex. We denotes by $\mathcal{G}^{even}$ and $\mathcal{G}^{odd}$ the sets of even and odd vertices. The structure of the higher relative commutants of the above tower is determined by $\mathcal{G}$, and

H. Araki et al. (eds.), Quantum and Non-Commutative Analysis, 317–320.

every vertex of the corresponding Bratteli diagram is labeled by a vertex in $\mathcal{G}^0 = \mathcal{G}^{even} \cup \mathcal{G}^{odd}$. On the other hand, in general every minimal projection in $M \cap \sigma(M)'$ $\sigma \in End(M)$ corresponds to an irreducible component of σ. Therefore, by decomposing $\rho, \rho\overline{\rho}, \rho\overline{\rho}\rho \cdots$ and taking the irreducible components, we get irreducible sectors labeled by the vertices in $\mathcal{G}^0$, say $\{\rho_v\}_{v \in \mathcal{G}^0}$. Note that $\rho_* = id$. Let Δ be the incidence matrix of $\mathcal{G}$. Then the following hold.

$$\rho_v \rho = \bigoplus_w \Delta_{v,w} \rho_w, \quad v \in \mathcal{G}^{even}, \tag{2.1}$$

$$\rho_v \overline{\rho} = \bigoplus_w \Delta_{v,w} \rho_w, \quad v \in \mathcal{G}^{odd}.$$

Taking the statistical dimension of the both sides, we have the following equation.

$$d(\rho)d(\rho_v) = \sum_w \Delta v, wd(\rho_w),$$

where we use the additivity and the multiplicativity of the statistical dimension [12, 15] and $d(\overline{\rho}) = d(\rho)$. This means that $d(\rho)$ is the Perron-Frobenius eigenvalue of Δ and $(d(\rho_v))_{v \in \mathcal{G}^0}$ is the corresponding eigenvector, normalized by the condition $d(\rho_*) = 1$. Recalling that $d(\rho)$ is the square root of the minimum index $[M : \rho(M)]_0$, (for the minimum index, see [4]), we have the following theorem.

Theorem 2.1. [5, Theorem 6.1] *Let $\mathcal{G}$ be a principal graph of an inclusion of factors with finite depth. Then the following hold.*

1. *The distinguished point $*$ must be the point with the minimal component of the Perron-Frobenius eigenvector.*
2. *If we normalize the Perron-Frobenius eigenvector such that the minimal component is one, then all the components must belong to the set $\{2\cos\frac{\pi}{N}; N = 3, 4, 5, \ldots\} \cup [2, \infty)$.*

3. Applications

3.1. A proof of Ocneanu's announcement

The following theorem was announced by A. Ocneanu [16] without proof. Y. Kawahigashi [K] and V.S. Sunder, A.K. Vijayarajan [18] also independently obtained it in different methods.

Theorem 3.1. [I, Corollary 3.9] *The principal graph of the inclusion of the factors is neither D_{odd} nor E_7.*

In the case of E_7, by direct computation one can see that E_7 does not satisfy the criterion in Theorem 2.1. For the proof of the D_{odd} case, first we suppose the existence of a family $\{\rho_v\}$ satisfying (2.1), and compute decomposition rules of two endomorphisms in the family, using associativity, distributivity and the Frobenius reciprocity of sectors. (For the Frobenius reciprocity of sectors, see Kosaki's article.) Then we arrive at contradiction. We can refer to this fact as "*There is no fusion rule on* D_{odd}".

3.2. GOLDMAN'S TYPE THEOREM FOR INDEX 3

Let $M \supset N$ be a pair of factor with index 2. Then there is an outer action of $\mathbf{Z}_2$ on N such that M is written as $M = N \times \mathbf{Z}_2$. This fact has been known as Goldman's theorem [3] in the case of II_1 factors, and is one of Jones' motivations of initiating the index theory. It is natural to consider the same type of expression for the index 3 case. However one can no longer expect the simple result obtained by replacing Z_2 with Z_3. Let $M \supset N$ be a pair of factors with index 3. Then it is known that the principal graph of this pair is D_4 or A_5 [2]. In the first case, M is expressed as the crossed product of N by an outer action of $\mathbf{Z}_3$ [11, 17]. Theorem 3.2 below determines the structure of the pair in the latter case. Since we do not use classification results of subfactors and group actions, we need no assumption of AFD and the types of factors involved.

Theorem 3.2. **[6, Theorem 3.1]** *Let $M \supset N$ be a pair of factors with the principal graph A_5. The there are a subfactor $L \subset N$ and an outer action of the symmetric group S_3 on N satisfying*

$$(M \supset N) \cong (L \times S_3 \supset L \times S_2).$$

To prove this theorem we take three steps.

1. First we find a natural subfactor $L \subset N$ with index 2, or equivalently (by Goldman's theorem) a natural $\mathbf{Z}_2$ action on N.
2. Next we show that the depth of $M \supset L$ is two.
3. Finally we prove $M = L \times S_3$.

For simplicity we assume that M and N are properly infinite. (In the case of II_1 factors, we can follow essentially the same procedure, treating bimodules instead of sectors [6].) Let $\gamma : M \longrightarrow N$ be the canonical endomorphism [13]. Since the Perron-Frobenius eigenvector of A_5 has two minimal components, by decomposing $\gamma^2|_N$ we obtain a suitable $\mathbf{Z}_2$ action on N. To perform the second step, the basic theorem of R. Longo [13] on the relation between canonical endomorphisms and conditional expectations is essential. For an inclusions of factors with depth 2, it is known that the larger factor is expressed by the crossed product of the smaller factor by a Kac algebra [19] (cf. Longo's contribution), (although we did not use this fact in [5, 6]). For the final step, we implicitly proved in [5, Lemma 5.3] that there is no non-trivial 6 dimensional Kac algebra.

References

[1] Choda, M. and Kosaki, H., Strongly outer actions for inclusion of factors, preprint.

[2] Goodman, F., de la Harpe, P. and Jones, V., *Coxeter Graph and tower of algebras* , MSRI Publictions 14, Springer Verlag, (1989).

[3] Goldman, M., On subfactors of type II_1, *Mich. Math. J.*, **7**, 167-172, (1960).

[4] Hiai, F., Minimizing indices of conditional expectations on a subfactors, *Publ. RIMS, Kyoto Univ.* **24**, 673-678, (1988).

[5] Izumi, M., Application of fusion rules to classification of subfactors, *Publ. RIMS, Kyoto Univ.,* **27**, 953-994, (1991).

[6] Izumi, M., Goldman's type theorem for index 3, to appear in *Publ. RIMS, Kyoto Univ.*

[7] Izumi, M., On type II and type III principal graph of subfactors, preprint.

[8] Jones, V., Index for subfactors, *Invent. Math.*, **72**, 1-25, (1983).

[9] Kawahigashi, Y. On flatness of Ocneanu's connection on Dynkin diagrams and classification of subfactors, preprint.

[10] Kosaki, H., Extension of Jones theory on index to arbitrary factors, *J. Funct. Anal.*, **66**, 123-140, (1986).

[11] Kosaki, H., Characterization of crossed product (properly infinite case), *Pacific. J. Math.*, **137**, 159-167, (1989).

[12] Kosaki, H. and Longo, R., A remark on the minimal index of subfactors, *J. Funct. Anal.*, **107**, 458-470, (1992).

[13] Longo, R., Index of subfactors and statistics of quantum field. I, *Commun. Math. Phys.*, **126**, 217-247, (1989).

[14] Longo, R., Index of subfactors and statistics of quantum field. II, *Commun. Math. Phys.*, **130**, 285-309, (1990).

[15] Longo, R., Minimal index and braided subfactors, to appear in *J. Funct. Anal.*

[16] Ocneanu, A., *Quantized group, string algebra and Galoi theory for algebra,* in Operator algebras and applications, **2** (Warwick, 1987), *London Math. Snc. Lect. Note Series* **136**, Cambridge University Press, 119-172, (1988).

[17] Pimsner, M. and Popa, S., Finite dimensinal approximation of algebras and obstruction of the index, *J. Funct. Anal.*, **98**, 270-291, (1991).

[18] Sunder, V.S. and Vijayarajan, A.K., On the non-occurrence of the Coxeter graphs E_7 and D_{2n+1} as the principal graph of inclusion of II_1 factors, to appear in *Pacific J. Math.*

[19] Szymanski, W., Finite index subfactors and Hopf algebra crossed products, preprint.

Vector Bundles and Bimodules

YAMAGAMI SHIGERU
Department of Mathematics
College of General Education
Tohoku University
Sendai, 980, Japan

Abstract. A vector bundle construction for bimodules is discussed with an application to finite dimensional Kac algebras.

Key words: vector bundle, bimodule, Kac algebra

Introduction

In our previous paper [8], we have introduced a certain tensor category of vector bundles over a group and constructed a categorical equivalence of this with a full subcategory of bimodules generalizing the crossed product construction in operator algebras.

In this note, we shall review some of the main results in [8] and supplements their proofs (we assumed there the finiteness of actiong von Neumann algebras for technical reasons) as well as an application to finite dimensional Kac algebras.

Notation.

For a von Neumann algebra N, the canonical standard space of N is denoted by $L^2(N)$ (see [5] or [17] for the canonical construction of standard spaces). For a positive normal linear functional φ of N, we denote by $\varphi^{1/2}$ the associated vector in the natural positive cone of $L^2(N)$.

1. Bimodules

Let A, B, C be factors (not necessarily finite). An A-B bimodule X is, by definition, a Hilbert space X with a pair of normal representations of A and B° (= the opposite algebra of B) which commute with each other. Such simultaneous representations give rise to a left action of A and a right action of B, in which sense, we shall write ${}_AX_B$ to express an A-B bimodule.

If we denote by B' the commutant of (the right action of) B, giving an A-B bimodule structure on X is nothing but considering an inclusion relation

H. Araki et al. (eds.), Quantum and Non-Commutative Analysis, 321–329.

of von Neumann algebras $A \subset B'$ realized on X. In this sense, the study of bimodules could be paralleled to that of inclusion relations of von Neumann algebras. The bimodule-point of view has some advantages, however, and one of key points is in its categorical structure.

For two A-B bimodules X and Y, we denote by $Hom(X,Y)$ the set of intertwiners from X into Y; $Hom(X,Y)$ is the vector space of bounded linear maps, say T, from X into Y which commutes with A-B actions. When $X = Y$, we also use the notation $End(X)$ in place of $Hom(X,X)$. Recall that a bimodule X is called irreducible if $End(X) = \mathbf{C}1_X$. The direct sum $X \oplus Y$ is defined in the obvious way and it is easy (or rather trivial) to see that these operations give an abelian category for each fixed pair of A and B. Bimodules also provide another kind of categorical operations adjoint and tensor product:

(i) To each A-B bimodule X, the adjoint Hilbert space X^* (this is the Hilbert space of bounded linear functional on X and is identified with the conjugate Hilbert space of X) admits a B-A bimodule structure by

$$b\xi^* a = (a^*\xi b^*)^*.$$

Here ξ^* denotes the linear functional associated with $\xi \in X$ through the inner product. The bimodule ${}_BX^*_A$ is called the adjoint of X.

(ii) Given an A-B bimodule X and a B-C bimodule Y, we construct an A-C bimodule $X \otimes^B Y$, where $X \otimes^B Y$ denotes the relative tensor product of X_B and ${}_BY$ and an A-C action is defined by the multiplication from the outside ([14], [17]).

All these operations are canonical with respect to the categorical structure of bimodules and provides a 'categorical *-algebra', i.e., a tensor category with adjoint. For example, if $S \in Hom(X_1, X_2)$ and $T \in Hom(Y_1, Y_2)$, then $S \otimes T \in Hom(X_1 \otimes X_2, Y_1 \otimes Y_2)$ is associated and satisfies the expected properties.

For a bimodule $X = {}_AX_B$, its dimension is defined to be the square root of the minimal index for the inclusion relation $A \subset B'$ ([3], [4], [9], and also cf. [17]). Here are basic properties of dimension (see [6], [7], and [10] besides the references cited above):

(i) $\dim X \in [1, +\infty]$ and $\dim X = 1$ iff $A = B'$.
(ii) $\dim X^* = \dim X$.
(iii) $\dim(X \oplus Y) = \dim X + \dim Y$.
(iv) $\dim(X \otimes Y) = \dim X \dim Y$.
(v) The dimension function is a conjugacy invariant: Let X_i be an A_i-B_i bimodule for $i = 1, 2$ and suppose that there are isomorphisms of von Neumann algebras $\alpha : A_1 \to A_2$, $\beta : B_1 \to B_2$, and a unitary map $u : X_1 \to X_2$ such that $u(a\xi b) = \alpha(a)u(\xi)\beta(b)$ for $a \in A_1$, $b \in B_1$, and $\xi \in X_1$. Then $\dim X_1 = \dim X_2$.

The existence of such a function is very useful. For example, if $\dim X < +\infty$, the properties (i) and (iii) imply that X is decomposed into a finitely many direct sum of irreducible submodules. In particular, $End(X)$ is finite dimensional.

According to Ocneanu, we associate two graphs $\mathcal{L}(X)$ and $\mathcal{R}(X)$ to a bimodule $X = {}_AX_B$ of finite index: The left graph $\mathcal{L}(X)$ is defined to be an unoriented bipartite graph $\mathcal{G}$ with vertex set ${}_A\mathcal{V}_B \sqcup {}_B\mathcal{V}_B$. Here, ${}_A\mathcal{V}_B$ (resp. ${}_B\mathcal{V}_B$) denotes the set of equivalence classes of irreducible bimodules appearing as components in X, $X \otimes_B X^* \otimes_A X, \cdots$ (resp. $X^* \otimes_A X$, $X^* \otimes_A X \otimes_B X^* \otimes_A X$, $\cdots$), and ${}_AY_B \in {}_A\mathcal{V}_B$, ${}_BZ_B \in {}_B\mathcal{V}_B$ are joined by m edges iff Y appears in $X \otimes_B Z$ with multiplicity m.

Similarly, switching the acting algebras on the right hand side, we obtain the right graph $\mathcal{R}(X)$ of X.

Remark. The left graph $\mathcal{L}(X)$ (resp. the right graph $\mathcal{R}(X)$) coincides with the dual principal (resp. principal) graph of the inclusion relation $A \subset B'$ (cf. [18]).

2. Vector Bundles

Take and fix a (not necessarily finite and even non-countable) group Γ and consider finite subgroups G, H, and K in Γ.

Definition. *A vector bundle V with base space Γ is called an H-K bundle if*
(i) *fibres of V are Hilbert spaces,*
(ii) *V admits an H-K (left H- and right K-) action which extends unitarily right and left translations on the base space Γ.*

Take another H-K bundle W. A vector bundle map $T : V \to W$ is called a morphism if (i) $T(V_g) \subset W_g$ for $g \in \Gamma$, (ii) $T(hvk) = hT(v)k$ for $h \in H$, $k \in K$, $v \in V$, and (iii) $\sup_{g\in\Gamma} \|T|_{V_g}\| < +\infty$.

The set of such morphisms is denoted by $Hom(V, W)$. With the obvious definition of direct sum for H-K bundles, we obtain an abelian category of H-K bundles. We can also define categorical operations analogous to adjoint and relative tensor product: The adjoint V^* (which is a K-H bundle) of an H-K bundle V is defined so that $(V^*)_g$ is the dual space of $V_{g^{-1}}$ with the obvious actions of K and H. The tensor product bundle $V \otimes_H W$ (which is a G-K bundle) is defined for a G-H bundle V and an H-K bundle W in such a way that

$$(V \otimes_H W)_g = \bigoplus_{g_1 g_2 = g} (V_{g_1} \otimes W_{g_2}) / \sim,$$

where the quotient (of vector spaces) in the right hand side is done with respect to the equivalence relation

$$vh \otimes w \sim v \otimes hw, \qquad h \in H.$$

In this way, we obtain a categorical *-algebra $\mathcal{C}(\Gamma)$.

3. Crossed Products

Let $\alpha : \Gamma \to Aut(N)$ be an outer action on a factor N. We denote by $N \lhd G$ the crossed product factor by restricting α to a finite subgroup G, i.e., $N \lhd G$ is a universal von Neumann algebra generated by N and G with the relation

$$g^* = g^{-1}, \qquad gxg^{-1} = \alpha_g(x), \quad \text{for } g \in G,\ x \in N.$$

Now we shall construct a categorical equivalence between $\mathcal{C}(\Gamma)$ and a certain class of bimodules.

For an H-K bundle V, we associate an $N \lhd H$-$N \lhd K$ bimodule $\hat{V}$ in such a way that

$$\hat{V} = \bigoplus_{g \in \Gamma} L^2(N) \otimes V_g$$

as a Hilbert space with left and right actions given by

$$\begin{cases} a(\xi \otimes v) = (a\xi) \otimes v \\ h(\xi \otimes v) = u_h\xi \otimes hv \end{cases}, \qquad \begin{cases} (\xi \otimes v_g)a = \xi\alpha_g(a) \otimes v_g \\ (\xi \otimes v)k = \xi \otimes vk. \end{cases}$$

Here $a \in N$, $\xi \in L^2(N)$, $v \in V$, $h \in H$, $k \in K$, and $u_h \in \mathcal{B}(L^2(N))$ is the canonical unitary implementation of α_h in the standard space.

As to the correspondence of morphisms, we asociate $\hat{\Phi} \in Hom(\hat{V}, \hat{W})$ to each $\Phi \in Hom(V, W)$ by

$$\hat{\Phi}(\xi \otimes v) = \xi \otimes \Phi(v).$$

Theorem. *The correspondence $\mathcal{C}(\Gamma) \ni V \mapsto \hat{V}$ gives a categorical isomorphism which preserves *-algebraic operations:*

(i) $\hat{V}^* = \widehat{V^*}$.

(ii) $\widehat{V \oplus W} = \hat{V} \oplus \hat{W}$.

(iii) $\widehat{V \otimes W} = \hat{V} \otimes \hat{W}$.

(iv) $Hom(V, W) \cong Hom(\hat{V}, \hat{W})$.

Proof. The proof is given in [8] but, in the proof of (ii), we assumed N to be a finite factor there. Here we shall deal with the general case. To this end, it is useful to give another realization of the Hilbert space $\hat{V}$: Through the isomorphism

$$\bigoplus_{g\in\Gamma} L^2(N)\otimes V_g \ni \xi\otimes v \mapsto v\otimes u_g^{-1}\xi \in \bigoplus_{g\in\Gamma} V_g\otimes L^2(N),$$

$\hat{V}$ is identified with $\oplus_g V_g \otimes L^2(N)$. Similarly for $\hat{W}$.

Take a representative set $S \subset \Gamma$ for the quotient Γ/H. Then we get the following isomorphism of Hilbert spaces:

$$\bigoplus_{s\in S} V_s\otimes L^2(N \triangleleft H) \ni \bigoplus_{s\in S} v_s\otimes(h\xi) \mapsto \bigoplus_{s\in S, h\in H} v_s h\otimes\xi \in \bigoplus_{g\in\Gamma} V_g\otimes L^2(N).$$

Combining this with the above realization of $\hat{V}$, we get an isomorphism

$$\hat{V} \cong \bigoplus_{s\in S} V_s\otimes L^2(N\triangleleft H).$$

In other words, $\hat{V}$ is realized as a free right $N\triangleleft H$-module. By the absorbing property $L^2(M)\otimes^M L^2(M) = L^2(M)$ of relative tensor product, we have an isomorphism

$$\hat{V}\otimes^{N\triangleleft H}\hat{W} \cong \bigoplus_{s\in S} V_s\otimes\hat{W}.$$

Now the desired isomorphism $V \widehat{\otimes_H} W \to \hat{V}\otimes^{N\triangleleft H}\hat{W}$ is given by

$$(v\otimes_H w)\otimes\xi \mapsto v\otimes(w\otimes\xi).$$

Here $v \in V_s$ with $s \in S$, $w \in W$, and $\xi \in L^2(N)$. Although we have taken a special realization of the isomorphism, this turns out to be independent of the choice of the representative S and behave well under the categorical operations. All of these can be checked step by step computations and the details are left to the readers. ■

Remark. If one uses the formalism in [17], then the isomorphism in (ii) takes the form

$$\varphi^{1/2}\otimes(v\otimes_H w) \mapsto (\varphi^{1/2}\otimes v)\otimes_{N\triangleleft H}\psi^{-1/2}\otimes_{N\triangleleft H}(\varphi^{1/2}\otimes w)$$

Here, given a normal state $\varphi \in N_*$, ψ denotes the induced state of the crossed product algebra $N\triangleleft H$, i.e.,

$$\psi(xh) = \begin{cases} \varphi(x) & \text{if } h \neq e \\ 0 & \text{otherwise} \end{cases}$$

for $x \in N$ and $h \in H$.

By the realization of this form, we can give a much natural proof for the isomorphism.

4. Index Formula

Let V be an H-K bundle. The following index (or dimension) formula for the associated bimodule $\hat{V}$ is given in [8] with a proof for finite factors:

$$\dim \hat{V} = \frac{|V|}{\sqrt{|H||K|}} \quad \text{with } |V| = \sum_{g \in \Gamma} \dim V_g.$$

Here we shall give consider the general case.

From the discussion in [KY, §2], we have

$$_N\hat{V}_N = {}_NL^2(N \lhd H) \otimes^{N \lhd H} \hat{V} \otimes^{N \lhd K} L^2(N \lhd K)_N,$$

which implies that

$$\dim({}_N\hat{V}_N) = \dim({}_NL^2(N \lhd H)_{N \lhd H}) \dim(\hat{V}) \dim({}_{N \lhd K}L^2(N \lhd K)_N).$$

From the definition, $_N\hat{V}_N$ is decomposed as N-N bimodule by

$$\bigoplus_{g \in \Gamma} {}_N(L^2(N) \otimes V_g)_N.$$

Moreover if we take orthonormal bases $\{v_{g,i}\}_{i \geq 1}$ for each V_g, then the bimodule $_N(L^2(N) \otimes V_g)_N$ is further decomposed as

$$\bigoplus_{i \geq 1} {}_N(L^2(N) \otimes v_{g,i})_N.$$

Thus

$$\dim({}_N\hat{V}_N) = \sum_{g,i} \dim {}_N(L^2(N) \otimes v_{g,i})_N.$$

Since $_N(L^2(N) \otimes v_{g,i})_N$ is isomorphic to $_NL^2(N)g_N$ by the map $\xi \otimes v_{g,i} \mapsto \xi g$ and since left and right actions in $_NL^2(N)g_N$ generates the commutants of each other, we have $\dim {}_N(L^2(N) \otimes v_{g,i})_N = 1$ for all g and i. In this way we obtaine a formula

$$\dim({}_N\hat{V}_N) = |V|.$$

Similarly (or as a special case of this) we have $\dim {}_NL^2(N \lhd G)_N = |G|$ and the identity

$$_NL^2(N \lhd G)_N = {}_NL^2(N \lhd G) \otimes^{N \lhd G} L^2(N \lhd G)_N$$

gives

$$\dim({}_N L^2(N \triangleleft G)_{N \triangleleft G}) = |G|^{1/2}.$$

Combining all these, we get the formula.

Remark. $(\dim \hat{V})^2 \in \{1, 2, 3, \cdots\}$.

5. Irreducibile Bimodules

As described in [8], we can give a criterion of irreducibility for vector bundles: An H-K bundle V is irreducible, i.e., $\hat{V}$ is irreducible if and only if

(i) V is supported by a single H-K orbit in Γ, and
(ii) for $g \in \Gamma$ in the support of V, the H-K stabilizer at g acts irreducibly on the fibre V_g
([KY, §1]).

Rewriting this, we obtain a parametrization of irreducible H-K bundles by

$$\bigsqcup_{\dot{g} \in H \backslash \Gamma / K} \widehat{H \cap gKg^{-1}}$$

Since tensor products for vector bundles are described, in some sense, by a combination of induction and restriction of the representations at stabilizers, their fusion rules can be read off from representation theory of finite groups. In particular, their graph invariants as well as connections are described by means of Mackey method. Consult [8] for the details.

6. Kac Algebras

A bipartite graph $\mathcal{G}$ with a distinguished point $*$ is called to have depth n if graphical distance from $*$ to any vertex is not greater than n and there is a vertex with distance exactly n.

The following is well-known and can be checked without difficulty in the present formalism.

Lemma. *For an irreducible bimodule X of finite index, the following conditions are equivalent.*
(i) *$End(X \otimes X^* \otimes X)$ is a factor.*
(ii) *$End(X^* \otimes X \otimes X^*)$ is a factor.*
(iii) *$\mathcal{L}(X)$ is of depth 2.*
(iv) *$\mathcal{R}(X)$ is of depth 2.*

We remark here that Frobenius reciprocity for bimodules (cf. [17]) gives rise to a canonical pairing between $End(X \otimes X^*)$ and $End(X^* \otimes X)$ without assumptions on the depth of X ([18]).

For a bimodule of depth 2, a finite dimensional Kac algebra is associated through this pairing

$$End(X \otimes X^*) \times End(X^* \otimes X) \to \mathbf{C},$$

i.e., comultiplications defined by this pairing are algebra-homomorphisms and the antipode is given by transposed map ([15]). Recall that, for $a \in Hom(X \otimes X^*)$, its transposed ${}^t a$ is defined by

$${}^t a(\xi) = (a^*(\xi^*))^*.$$

Here $*$ for $\xi \in X \otimes X^* = L^2(End(X_B))$ refers to the modular involution in the standard space.

Proposition. *Let H, K be finite subgroups of Γ such that*
(i) *HK is a subgroup,*
(ii) *$H \cap K = \{e\}$.*
Then the trivial H-K vector bundle supported by HK is of depth 2 and hence gives rise to a finite dimensional Kac algebra.

This, with the association of Kac algebras to depth 2 bimodules, gives a bimodule-theoretic interpretation of Majid's bicrossed product construction ([11]). The details will be discussed elsewhere.

Remark. We can also define a canonical Kac-algebra action of $End(X \otimes X^*)$ on $End(X_B)$ so that $End(X \otimes X_A^*)$ is (canonically) isomorphic to the crossed product $End(X_B) \triangleleft End(X \otimes X^*)$ (cf. [15]).

Since there are a lot of examples of above pairs H, K, we have many examples of finite dimensional Kac algebras with outer actions on factors. Among such examples, the following yields the 24-dimensional Kac algebra discussed in [1, Appendix].

Example. In S_4, $H = S_3$ and $K = \langle(1234)\rangle$ satisfy the above conditions. The associated graphs are given by

K-K

Left Graph

H-K

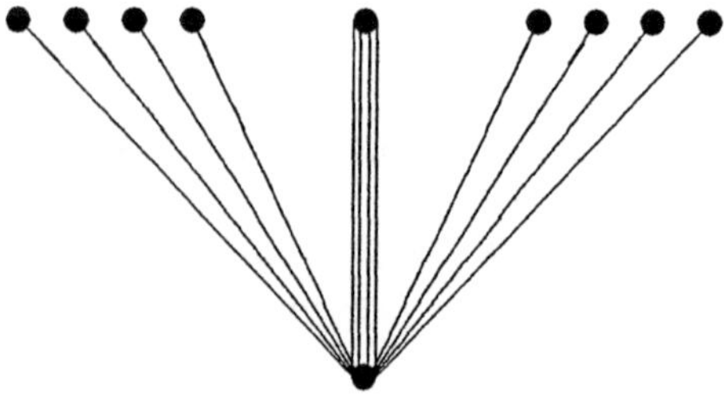

H-H

Right Graph

H-K

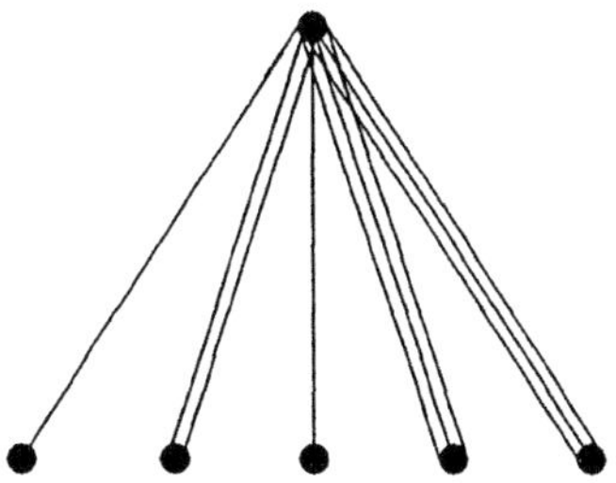

References

[1] S. Baaj and G. Skandalis, *Unitaires multiplicatifs et dualité pour les produits croisés de C^*-algébres:deux nouveaux appendices*, preprint.

[2] F. M. Goodman, P. de la Harpe, and V. Jones, *Coxeter-Dynkin Diagrams and Towers of Algebras*, Springer, (1989).

[3] J. F. Havet, *Espérance conditionnelle minimale*, J. Operator Theory, 24(1990), 33–55.

[4] F. Hiai, *Minimizing indices of conditional expectations onto a subfactor*, Publ. RIMS, Kyoto Univ, 24(1988), 673–678.

[5] H. Kosaki, *Canonical L^p-spaces associated with an arbitrary abstract von Neumann algebra*, Thesis at UCLA, (1980).

[6] ______, *Extension of Jones' theory on index to arbitrary factors*, J. Funct. Anal., 66(1986), 123–140.

[7] H. Kosaki and R. Longo, *A remark on the minimal index of subfactors*, J. Funct. Analysis, 107(1992), 458–470.

[8] H. Kosaki and S. Yamagami, *Irreducible bimodules associated with crossed product algebras*, to appear in International. J. Math.

[9] R. Longo, *Index of subfactors and statistics of quantum fields. II*, Commun. Math. Phys., 130(1990), 285–309.

[10] ______, *Minimal index and braided subfactors*, preprint.

[11] S. Majid, *Physics for algebraists: Non-commutative and non-cocommutative Hopf algebras by a bicrossproduct construction*, J. Algebra, 130(1990), 17–64.

[12] A. Ocneanu, *Quantized groups, string algebras, and Galois theory for algebras*, Operator algebras and applications, vol. 2, Warwick(1987), Cambridge University Press, (1988).

[13] ________, *Quantum Symmetry, Differential Geometry of Finite Graphs and Classification of Subfactors*, University of Tokyo Seminary Notes, (1991).

[14] J.-L. Sauvageot, *Sur le prodiut tensoriel relatif d'espaces de hilbert*, J. Operator Theory, 9(1983), 237–252.

[15] W. Szymanski, *Finite index subfactors and Hopf algebra crossed products*, preprint.

[16] Yamagami S., *Algebraic aspects in modular theory*, to appear in Publ. RIMS.

[17] __________, *Modular theory for bimodules*, preprint(1992).

[18] __________, *A note on Ocneanu's approach to Jones' index theory*, preprint(1992).

Lattice Structure of Intermediate Subfactors

YASUO WATATANI
Department of Mathematics
Hokkaido University
Sapporo, 060, Japan

Abstract. Any irreducible subfactor of finite index has only finitely many intermediate subfactors. Any finte lattice with at most five elements can be realized as an intermediate subfactor lattice.

Key words: subfactor, intermediate subfactor, lattice, modular identity

1. Lattices of von Neumann subalgebras

Let H be a Hilbert space and $B(H)$ the algebra of bounded linear operators on H. Let M and N be von Neumann subalgebras of $B(H)$. Define

$$M \vee N = (M \cup N)'' \quad \text{and} \quad M \wedge N = M \cap N.$$

Then the set $\mathcal{L}(B(H))$ of all von Neuman subalgebras is a lattice by the above operations $\wedge$ and $\vee$. The lattice $\mathcal{L}(B(H))$ was considered by Murray and von Neumann [7] in 1936.

In algebraic quantum field theory ([4], [5]), lattice structure of bounded open regions $\mathcal{O}$ of a space-time and lattice structure of algebras $A(\mathcal{O})$ of local observables have been investigated.

In 1963, H. Araki [1] established a lattice isomorphism from a lattice of subspaces of a Hilbert space and a lattice of von Neumann subalgebras associated with the quantum theory of a free Bose field.

More generally, the structure of von Neumann subalgebras were investigated by Skau [12] and Christensen [3].

But these lattices of von Neumann subalgebras are too large to describe. An atractive feature in our study of intermediate subfactor lattices is that they are in fact finite lattices and easily seen by their Hasse diagrams.

2. Intermediate subfactor lattices

Let M be a factor of type II_1 and N a subfactor of M. Jones [6] introduced Jones index $[M : N]$ which is a fundamental invariant for subfactors. Ocneanu [10] and Popa [11] show that the analysis of higher relative commutants is essential for the classification of subfactors.

H. Araki et al. (eds.), Quantum and Non-Commutative Analysis, 331–333.

In this note we advertize that the lattice structure of intermediate subfactors is also an interesting object for the study of subfactors [13].

Let M be a factor and N a subfactor of M such that $N' \cap M = \boldsymbol{C}$. Then the set $Lat(N \subset M)$ of all intermediate subfactors for $N \subset M$ is a lattice. For an intermediate subfactor K, consider the Jones projection $e_K^M : L^2(M) \to L^2(K)$. Then we have $e_K^M \in N' \cap \langle M, e_N^M \rangle$. A nice characterization of intermediate subfactors K in terms of projections in $N' \cap \langle M, e_N^M \rangle$ has been obtained by Bisch [2] and Ocneanu:

THEOREM 1 (Bisch and Ocneanu). *Let M be a II_1 factor and N a subfactor of M such that $[M : N] < \infty$ and $N' \cap M = \boldsymbol{C}$. Put $L = \langle M, e_N^M \rangle$. Choose a subfactor $P \subset N$ such that $M = \langle N, e_P^N \rangle$. Let $IS(N, M)$ be the set of projections q such that $q \geq e_N^M$, $E_M^L(q) \in \boldsymbol{C}$ and $E_M^L(q e_p^N e_N^M)$ is a scalar multiple of a projection. Then $IS(N, M)$ is precisely the set of Jones projections coming from intermediate subfactors for $N \subset M$.*

We regard intermediate subfactor lattice as a quantization of continuous geometry [9] by von Neumann. Then the operation of taking Jones projections is something like a forgetful functor for the quantization.
We consider the following question:

PROBLEM 1. *What kind of lattices appear as intermediate subfactor lattices ?*

The following theorem shows that we can draw a picture of an intermediate subfactor by its Hasse diagram.

THEOREM 2. *Let M be a factor of type II_1 and N a subfactor of M. If $N' \cap M = \boldsymbol{C}$ and $[M : N] < \infty$, then the lattice $Lat(N \subset M)$ of the intermediate subfactors is a finite lattice.*

Any subgroup lattice $\mathcal{L}(G)$ of a finite group G and its dual lattice can be realized as intermediate subfactor lattices by the Galois correspondence for factors of type II_1 by Nakamura-Takeda [8]. But there exists a finite lattice $\mathcal{L}$ such that $\mathcal{L}$ is not isomorphic to $\mathcal{L}(G)$ or its dual lattice for any finite group G but $\mathcal{L}$ can be realized as an intermediate subfactor lattice.

PROPOSITION 3. *Any finite lattice with at most five elements can be realized as an intermediate subfactor lattice.*

We can put weights for edges of a lattice by Jones index. Then Problem 1 can be refined as follows:

PROBLEM 2. *What kind of weighted lattices appear as intermediate subfactor lattices ?*

3. Modular identity

A lattice $\mathcal{L}$ is called a modular lattice if $\mathcal{L}$ satisfies the following modular identity.

If $x \leq z$ then $x \vee (y \wedge z) = (x \vee y) \wedge z$. For example, the set of normal subgroups of a group is a modular lattice. In particular subgroup lattices of abelian groups are modular lattices. Commuting square conditions and co-commuting square conditions are closely related with the modular identity in $Lat(N \subset M)$, see [13]. In particular we have the following

PROPOSITION 4. *Let M be a factor of type II_1 and N a subfactor of M with $[M : N] < \infty$ and $N' \cap M = \boldsymbol{C}$. Let $M_1 = \langle M, e_N \rangle$ and $M_2 = \langle M_1, e_M \rangle$. If $N' \cap M_1$ and $M' \cap M_2$ are both commutative, then the intermediate subfactor lattice* Lat$(N \subset M)$ *is a modular lattice.*

References

[1] H. Araki, *A lattice of von Neumann algebras associated with the quantum theory of a free Bose field,* J. Math. Phys. 4(1963) 1343-1362.

[2] D. Bisch, *A note on intermediate subfactors,* preprint, 1992.

[3] E. Christensen, *Subalgebras of a finite algebra,* Math. Ann. 243(1979), 17-29.

[4] S. Doplicher, R. Haag and J.E. Roberts, *Fields, observables and gauge transformations I,* Comm. Math. Phys. 13(1969), 1-23.

[5] R. Haag and D. Kastler, *An algebraic approach to quantum field theory,* Comm. Math. Phys. 5(1964), 848-864.

[6] V. Jones, *Index for subfactors,* Invent. Math. 72(1983), 1-15.

[7] F.J. Murray and J.von Neumann, *On rings of operators,* Ann. Math. 37(1936), 116-229.

[8] M. Nakamura and Z. Takeda, *A Golois theory for finite factors,* Proc. Japan Acad. 36(1960), 258-260.

[9] J. von Neumann, *Continuous Geometry,* Princeton, 1960.

[10] A. Ocneanu, *Quantized group, string algebras and Galois theory for algebras,* in "Operator Algebras and Applications, vol.2" London Math. Soc. Lect. Note Series, Vol.136, (1988), 119-172.

[11] S. Popa, *Sur la classification des sousfacteurs d'indice fini du facteur hyperfini,* C. R. Acad. Sci. Paris, 311(1990), 95-100.

[12] C. Skau, *Finite subalgebras of a von Neumann algebra,* J. Funct. Anal. 25(1977), 211-235.

[13] Y. Watatani, *Lattices of intermediate subfactors,* preprint.

Minimal Index and Unimodular Sectors

ROBERTO LONGO
Dipartimento di Matematica
Università di Roma "Tor Vergata"
via della Ricerca Scientifica
I-00133 Roma, Italy

1. Introduction

We give a somewhat more direct argument for the multiplicativity of the minimal index [3], in the case of a chain of inclusions in the Jones tower; the general case is a consequence of this result [4]. The proof still relies on an argument in [8], now extended to the type III setting.

In section 3, we discuss a characterization of the sectors such that the minimal index coincides with the Jones index in terms of the unimodularity of the subsectors.

We refer to the original papers for motivations, applications and further references.

2. A proof of the equality $\mathrm{Ind}(N, M_1) = \mathrm{Ind}(N, M)^2$

Let $N \subset M$ be an inclusion of factors and $M \subset M_1$ its extension [6]; we assume $\mathrm{Ind}(N, M) < \infty$ where $\mathrm{Ind}(N, M)$ denotes the minimal index of N in M, see [2].

Let $\varphi \in N_*^+$ a faithful state. For any (faithful, normal) expectation $E \in C(M, N)$, let $E' \in C(M_1, M)$ be the dual expectation and

$$\tilde{\varphi} \equiv \varphi \circ E \circ E'$$

the corresponding extension of φ to M_1. The modular group $\sigma^{\tilde{\varphi}}$ leaves N and M, hence $N' \cap M_1$ and $N' \cap M$, globally invariant therefore, by Takesaki theorem, there exist normal expectations

$$\begin{aligned} E_1^{(\varphi)} &\in C(M_1, N' \cap M_1) \\ E_0^{(\varphi)} &\in C(M_1, M' \cap M_1) \end{aligned}$$

leaving $\tilde{\varphi}$ invariant, that satisfy

$$\mathcal{E}^{(\varphi)} \circ E_1^{(\varphi)} = E_0^{(\varphi)} \tag{1}$$

H. Araki et al. (eds.), Quantum and Non-Commutative Analysis, 335–340.

where $\mathcal{E}^{(\varphi)} \in C(N' \cap M_1, M' \cap M_1)$ is the $\tilde{\varphi}$–invariant expectation.

Denote by $e \in M_1$ the Jones projection (that depends on E).

Lemma 2.1. *e belongs to the centralizer of $\tilde{\varphi}$.*

Proof. Let $\xi \in \mathcal{H}$ be a cyclic vector representative of $\tilde{\varphi}|M$ so that

$$e = [N\xi] \in N' \ ;$$

if Δ is the modular operator associated with M, ξ we have

$$\Delta^{it} e\mathcal{H} = \overline{\Delta^{it} N\xi} = \overline{N\xi} = e\mathcal{H}$$

namely $\Delta^{it} e \Delta^{-it} = e$, therefore Δ^{it} implements a one-parameter group of automorphisms α_t of $M_1 = \langle M, e\rangle$, moreover

$$E(\alpha_t(x)) = \alpha_t(E(x)) \ , \quad x \in M_1 \ .$$

Since all elements of M_1 are sums of monomials xey with $x, y \in M$ and

$$\tilde{\varphi}(xey) = \tilde{\varphi}(E'(xey)) = \tilde{\varphi}(xE'(e)y) = \lambda\tilde{\varphi}(xy)$$

($\lambda^{-1} = \mathrm{Ind}_E(N, M)$ is the index of E [2]), we may easily check that α_t satisfies the KMS condition with respect to $\tilde{\varphi}$, hence e is fixed by the modular group α of M_1 with respect to $\tilde{\varphi}$.

Proposition 2.2. (cf. [7] for the type II case). *The conditional expectation $E \in C(M, N)$ is minimal if and only if*

$$E_0^{(\varphi)}(e) \in \mathbb{R} \ .$$

In this case

$$E_0^{(\varphi)}(e) = \mathrm{Ind}(N, M)^{-1}$$

and in particular $E_0^{(\varphi)}(e)$ does not depend on φ.

Proof. Let j be the modular conjugation of M and $j(\cdot) \equiv J \cdot J$; we have

$$j(N' \cap M) = M' \cap M_1 \ .$$

Let $M' \cap M_1 = \oplus_k A_k$ be the decomposition of $M' \cap M_1$ into matrix algebras and $\{e_{il}^{(k)}\}$ matrix units for A_k; we may choose the $e_{il}^{(k)}$ so that

$$\tilde{\varphi}(j(e_{il}^{(k)})) = \delta_{il}\tilde{\varphi}(j(e_{il}^{(k)})) \tag{2}$$

and with $j(e_{ii}^{(k)})$ in the centralizer of $\tilde{\varphi}$ ($\tilde{\varphi}|_{j(A_k)}$ is given by a density matrix that we diagonalize).

Since $e \in N' \cap M_1$, by (1) we have to show that E is minimal iff

$$\mathcal{E}^{(\varphi)}(e) \in \mathbb{R}$$

Now $\mathcal{E}^{(\varphi)}$ is the orthogonal projection of $L^2(N' \cap M_1)$ onto $L^2(M' \cap M_1)$ with respect to the scalar product $\langle x, y\rangle_\varphi \equiv \tilde{\varphi}(y^* x)$ and $\{e_{il}^{(k)}\}$ is an orthogonal basis of $L^2(M' \cap M_1)$ hence

$$\begin{aligned}\mathcal{E}^{(\varphi)}(e) &= \sum_{k,i,l} \frac{\langle e, e_{il}^{(k)}\rangle_\varphi e_{il}^{(k)}}{\|e_{il}^{(k)}\|_\varphi^2} \\ &= \sum \frac{\tilde{\varphi}(e_{li}^{(k)} e) e_{il}^{(k)}}{\tilde{\varphi}(e_{ll}^{(k)})} \\ &= \sum \frac{\tilde{\varphi}(e e_{li}^{(k)} e) e_{il}^{(k)}}{\tilde{\varphi}(e_{ll}^{(k)})}\end{aligned}$$

(because of lemma 2.1)

$$\begin{aligned}&= \sum \frac{\tilde{\varphi}(j(e) e_{li}^{(k)} j(e)) e_{il}^{(k)}}{\tilde{\varphi}(e_{ll}^{(k)})} \\ &= \sum \frac{\tilde{\varphi}(j(e j(e_{li}^{(k)}) e)) e_{il}^{(k)}}{\tilde{\varphi}(e_{ll}^{(k)})} \\ &= \sum \frac{\tilde{\varphi}(j(e_{li}^{(k)}))\tilde{\varphi}(e) e_{il}^{(k)}}{\tilde{\varphi}(e_{ll}^{(k)})}\end{aligned}$$

(because $exe = E(x)e$, $x \in M$, and $E|N' \cap M = \tilde{\varphi}|N' \cap M$)

$$= \sum_{i,k} \lambda \frac{\tilde{\varphi}(j(e_{ii}^{(k)}))}{\tilde{\varphi}(e_{ii}^{(k)})} e_{ii}^{(k)} \tag{3}$$

(because of (2) and $\tilde{\varphi}(e) = \tilde{\varphi}(E'(e)) = \lambda$ where $\lambda = \mathrm{Ind}_E(N, M)^{-1}$).

Now (3) is a scalar ($= \lambda$) if and only if

$$\tilde{\varphi}(j(e_{ii}^{(k)})) = \tilde{\varphi}(e_{ii}^{(k)}) \tag{4}$$

in fact $\tilde{\varphi}(j(e_{ii}^{(k)}))$ and $\tilde{\varphi}(e_{ii}^{(k)})$ must be proportional, hence equal because

$$\sum \varphi(j(e_{ii}^{(k)})) = \sum \varphi(e_{ii}^{(k)}) = 1 \; . \tag{5}$$

Therefore we have to show that (4) holds iff E is minimal. Now for any projection $p \in N' \cap M$ in the centralizer of $\tilde{\varphi}$ we have [6.2],

$$\mathrm{Ind}_{E_p}(N_p, M_p) = E(p)E'(j(p)) \, \mathrm{Ind}_E(N, M) = \tilde{\varphi}(p)\tilde{\varphi}(j(p)) \, \mathrm{Ind}_E(N, M)$$

in particular

$$\mathrm{Ind}(N_{j(e_{ii}^{(k)})}, M_{j(e_{ii}^{(k)})}) = \tilde{\varphi}(e_{ii}^{(k)})\tilde{\varphi}(j(e_{ii}^{(k)}))\,\mathrm{Ind}_E(N,M)$$

and summing up over i and k, by the additivity of the square root of the minimal index,

$$\begin{aligned}\mathrm{Ind}(N,M)^{\frac{1}{2}} &= \sum \mathrm{Ind}(N_{j(e_{ii}^{(k)})}, M_{j(e_{ii}^{(k)})})^{\frac{1}{2}} \\ &= \sum \tilde{\varphi}(e_{ii}^{(k)})^{\frac{1}{2}}\tilde{\varphi}(j(e_{ii}^{(k)}))^{\frac{1}{2}}\,\mathrm{Ind}_E(N,M)^{\frac{1}{2}}.\end{aligned}$$

Therefore E is minimal if and only if

$$\sum \tilde{\varphi}(e_{ii}^{(k)})^{\frac{1}{2}}\tilde{\varphi}(j(e_{ii}^{(k)}))^{\frac{1}{2}} = 1$$

equivalently because of (5) iff

$$\sum(\tilde{\varphi}(e_{ii}^{(k)})^{\frac{1}{2}} - \tilde{\varphi}(j(e_{ii}^{(k)}))^{\frac{1}{2}})^2 = 0$$

namely iff (4) holds. □

Corollary 2.3. *E is minimal iff $E \circ E'$ is minimal.*

Proof. Let $e_0 = e, e_1, e_2 \ldots$ be the iterated Jones projections. As in Pimsner–Popa [8] the Jones projection for $N \subset M_1$ is a multiple of $e_1e_0e_2e_1$ and one easily checks that $E \circ E'$ satisfies proposition 2.2. □

3. Unimodular sectors

In the following M is a II_∞–factor with a semifinite trace τ. We denote by Sect(M) the sectors (endomorphisms up to inner automorphisms) of M, with finite index.

A sector $\rho \in$ Sect(M) is called *minimal* if

$$d(\rho)^2 = \mathrm{Ind}_\tau(\rho)$$

where $\mathrm{Ind}_\tau(\rho) \equiv \mathrm{Ind}_\tau(\rho(M), M)$ denotes the Jones index and $d(\rho) \equiv \mathrm{Ind}(\rho(M), M)^{\frac{1}{2}}$. ρ is called *unimodular* if

$$\mathrm{mod}(\rho) = \mathrm{Ind}_\tau(\rho)^{\frac{1}{2}} \tag{6}$$

where the *modulus* mod(ρ) is defined by

$$\tau(\rho(x)) = \mathrm{mod}(\rho)\tau(x)\,, \quad x \in M^+\,. \tag{7}$$

Since $\rho\overline{\rho}$ is the canonical endomorphism $\gamma : M \to \rho(M)$, it follows that

$$\mathrm{mod}(\rho\overline{\rho}) = \mathrm{Ind}_\tau(\rho) \tag{8}$$

so that ρ is unimodular iff $\mathrm{mod}(\rho) = \mathrm{mod}(\overline{\rho})$.

Lemma 3.1. $\mathrm{mod}(\cdot)$ *is a (not involutive) homomorphism of* $\mathrm{Sect}(M)$ *into* $\mathbb{R}^+$

$$\mathrm{mod}(\rho_1\rho_2) = \mathrm{mod}(\rho_1)\,\mathrm{mod}(\rho_2) \tag{9}$$

$$\mathrm{mod}(\rho_1 \oplus \rho_2) = \mathrm{mod}(\rho_1) + \mathrm{mod}(\rho_2)\ . \tag{10}$$

If ρ *is minimal, in particular if* ρ *is irreducible, then* $d(\rho)$ *is the geometric mean of* $\mathrm{mod}(\rho)$ *and* $\mathrm{mod}(\overline{\rho})$

$$d(\rho) = \sqrt{\mathrm{mod}(\rho)\,\mathrm{mod}(\overline{\rho})}\ . \tag{11}$$

Proof. The multiplicativity of $\mathrm{mod}(\cdot)$ is immediate. To check the additivity notice that $\rho \equiv \rho_1 \oplus \rho_2 \oplus \ldots \oplus \rho_n$ is defined by

$$\rho(x) = \sum_i v_i\rho_i(x)v_i^*$$

where the $v_i \in M$ are isometries and $\sum v_i v_i^* = 1$; hence

$$\tau(\rho(x)) = \sum_i \tau(v_i\rho_i(x)v_i^*) = \sum_i \tau(\rho_i(x)) = \sum_i \mathrm{mod}(\rho_i)\tau(x)\ . \tag{12}$$

Formula (11) follows immediately from (8). □

Lemma 3.2. *Let* $\rho \in \mathrm{Sect}(M)$ *be unimodular. Then* ρ *is minimal iff all subsectors of* ρ *are unimodular.*

Proof. Let

$$\rho = \rho_1 \oplus \ldots \oplus \rho_n$$

be the decomposition of ρ into irreducible components. Since ρ and $\overline{\rho}$ are unimodular, by (8) and (10) we have

$$\begin{aligned} \mathrm{Ind}_\tau(\rho)^{1/2} &= \mathrm{mod}(\rho) = \sum \mathrm{mod}(\rho_i) \\ &= \mathrm{Ind}_\tau(\overline{\rho})^{1/2} = \mathrm{mod}(\overline{\rho}) = \sum \mathrm{mod}(\overline{\rho}_i)\ . \end{aligned} \tag{13}$$

On the other hand the ρ_i are irreducible, hence minimal, and by (11)

$$d(\rho) = \sum d(\rho_i) = \sum \sqrt{\mathrm{mod}(\rho_i)\,\mathrm{mod}(\overline{\rho}_i)} \tag{14}$$

and composing (13) and (14) we see that $d(\rho) = \mathrm{Ind}_\tau(\rho)^{1/2}$ iff

$$2\sum \sqrt{\mathrm{mod}(\rho_i)\,\mathrm{mod}(\overline{\rho}_i)} = \sum(\mathrm{mod}(\rho_i) + \mathrm{mod}(\overline{\rho}_i)) \tag{15}$$

namely iff $\sum(\operatorname{mod}(\rho_i)^{1/2} - \operatorname{mod}(\overline{\rho}_i)^{1/2})^2 = 0$, i.e. the ρ_i are unimodular. □

Proposition 3.2. *Let $\rho \in \mathrm{Sect}(M)$. Then ρ is minimal iff all subsectors of $\rho\overline{\rho}$ are unimodular.*

Proof. By corollary 2.3 ρ is minimal iff $\rho\overline{\rho}$ is minimal. Since $\rho\overline{\rho}$ is unimodular, we may apply lemma 3.2. □

Next corollary applies to arbitrary factors.

Corollary 3.3. (cf. [1,7]) *Let $N \subset M$ be an inclusion of factors with finite index. With the notations of Section 2, E is minimal iff $j : x \in N' \cap M_1 \to N' \cap M_1$ is a $\tilde{\varphi}$–invariant anti–automorphism of $N' \cap M_1$; in this case $\tilde{\varphi}|N' \cap M_1$ ($\equiv E \circ E'|N' \cap M_1$) is a j–invariant trace.*

Proof. We may assume that N and M are infinite. Now j maps minimal projections of $N' \cap M_1$ to minimal projections of $N' \cap M_1$. This map corresponds to the conjugation map $\eta \to \overline{\eta}$ on the irreducible subsectors of the canonical endomorphism $\gamma : M_1 \to M$. If e is a minimal projection of $N' \cap M_1$ and η is the corresponding subsector one has

$$\tau(e) = \frac{\operatorname{mod}(\gamma)}{\operatorname{mod}(\eta)}$$

(similarly as in the formula (12)) therefore

$$\tau(e) = \tau(j(e)) \quad \text{iff} \quad \eta \quad \text{is unimodular.}$$

By proposition 3.2, this proves the corollary in the case $N \subset M$ are isomorphic and semifinite, because property (4) characterizes the minimal expectation. In general, after tensoring by a suitable III_1–factor, we may assume that N and M are isomorphic III_1–factors [3] one obtains the general result and by Takesaki duality. □

References

[1] Hiai, F., Minimizing indices of conditional expectations onto a subfactors, *Publ. Res. Inst. Math. Sci.*, **24**, 673–678 (1988).

[2] Kosaki, H., Extension of Jones' theory on index to arbitrary factors, *J. Funct. Anal.*, **66**, 123–140 (1986).

[3] Kosaki, H., Longo, R., A remark on the minimal index of subfactors, *J. Funct. Anal.*, **107**, 458–470 (1992).

[4] Longo, R., Index of subfactors and statistics of quantum fields. I & II *Commun. Math. Phys.*, **126**, 217–247 (1989) and **130**, 285–309 (1990).

[5] Longo, R., Minimal index and braided subfactors, *J. Funct. Anal.* (to appear).

[6] Jones, V.F.R., Index for subfactors, *Invent. Math.*, **66**, 1–25 (1983).

[7] Pimsner, M., Popa, S., Entropy and index for subfactors, *Am. Sci. École Norm. Sup.*, **19**, 57–107 (1986).

[8] Pimsner, M., Popa, S., Iterating the basic construction, *Trans. Amer. Math. Soc.*, **310**, 127–133 (1988).

Subfactors and Conformal Field Theory

DAVID E. EVANS
University of Wales, Swansea, SA2 8PP, WALES

and

YASUYUKI KAWAHIGASHI
University of Tokyo, Tokyo, 113, JAPAN

Abstract. We discuss relations between the combinatorial structure of subfactors, solvable lattice models, (rational) conformal field theory, and topological quantum field theory.

Key words: conformal field theory, modular automorphism group, modular invariant, orbifold construction, statistical mechanics, subfactors, topological invariant

1. Introduction

Since the breakthrough by V. F. R. Jones [48, 49], the theory of subfactors has had unexpectedly deep connections with topology and mathematical physics. The combinatorial structure of subfactors plays a key role, and Ocneanu's paragroup theory [64] gives the most powerful machinery for understanding this at present. Ocneanu, however, has not published details of his theory, and a lack of understanding of the paragroup theory prevented detailed study of the relationship between the paragroup theory and the topology and physics — beyond noticing apparent similarities in the various fields. Recently, full details and several applications of the paragroup theory have appeared [11, 32, 33, 34, 42, 44, 45, 52, 53, 54, 66, 67, 83] and our aim here is to give a full account of the recent interactions.

A paragroup is a (discrete) quantization of a (finite) group, and it appeared as a combinatorial characterization of higher relative commutants of finite index and finite depth subfactors of the approximately finite dimensional (AFD) factor of type II_1. (See [38] for the basics of subfactor theory and [71] for analytic aspects of the classification of subfactors.) It has been realized that the combinatorial structure of paragroups has considerable similarity with several other structures in mathematics and physics. The following table shows correspondences between various notions in paragroup theory, exactly solvable lattice model theory, quantum $6j$-symbols, rational conformal field theory and topological quantum field theory. Further details on these will be presented in later sections. Some correspondences are more

H. Araki et al. (eds.), Quantum and Non-Commutative Analysis, 341–369.

exact than others — some merely express similarity.

<table>
<tr><th>Paragroups (path algebras)</th><th>Paragroups (bimodules)</th><th>Lattice models</th><th>Quantum 6j-symbols</th><th>RCFT</th><th>TQFT</th></tr>
<tr><td colspan="2">connection</td><td>Boltzmann weight</td><td>6j-symbol</td><td>braiding matrix</td><td>local data for tetrahedron</td></tr>
<tr><td colspan="5">unitarity</td><td>invariance under moves II, III</td></tr>
<tr><td>commuting square</td><td>Frobenius reciprocity</td><td>crossing symmetry</td><td>S_4-symmetry</td><td colspan="2">tetrahedral symmetry</td></tr>
<tr><td>flatness</td><td>associativity</td><td>Yang-Baxter equation</td><td>pentagon relation</td><td>braiding-fusion relation</td><td>invariance under move I</td></tr>
</table>

See [52, 53, 65, 66, 74] for the path algebra approach to paragroups, and [33, 66, 67, 84] for the bimodule approach to paragroups.

This work was partially supported by financial support from the British Council, Miller Research Fellowship, MSRI (Berkeley), SERC, and University College of Swansea. We thank Professors D. Bisch, M. Choda, U. Haagerup, M. Izumi, V. F. R. Jones, A. Kato, T. Kohno, H. Kosaki, T. Miwa, R. Nest, A. Ocneanu, M. Okado, S. Okamoto, S. Popa, N. Yu. Reshetikhin, Ph. Roche, M. Rosso, M. Wakui, Y. Watatani, H. Wenzl, F. Xu, S. Yamagami, and J.-B. Zuber for several useful comments and or conversations during this work.

2. Statistical Mechanics and Subfactors

We enter this maze of interactions involving operator algebras, conformal quantum field theory, and topological quantum field theory, with certain statistical mechanical models on a two dimensional infinite oriented lattice L (which for our purposes here will be $\mathbf{Z}^2$). The configuration space of the classical model will be (compatible) distributions of the edges of some oriented graph Γ (here taken to be finite) on the edges of L. If there is at most one edge between any pair of vertices of Γ, and if there is an edge from α to β then there is another edge from β to α, then this will be the same as distributing the vertices of Γ over L so that labels of two vertices in L are joined in Γ if the original vertices are joined in L. For the Dynkin diagram

A_3 with vertices labeled as

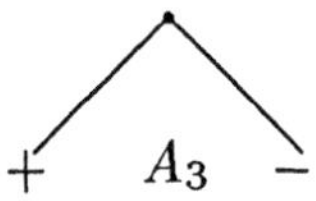

and square lattice L (with downward orientation)

$L = \mathbf{Z}^2$

one obtains two copies of the configuration space of the Ising model by placing the frozen spin · on the even or odd sublattices of L.

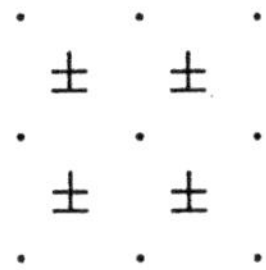

To complete the data for the model, we assign a complex number $W(\alpha, \beta, \gamma, \delta)$, the Boltzmann weight to each allowed configuration:

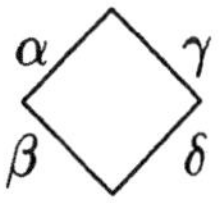

For the A_3 graph, we assign the appropriate Ising weights to the eight possible configurations:

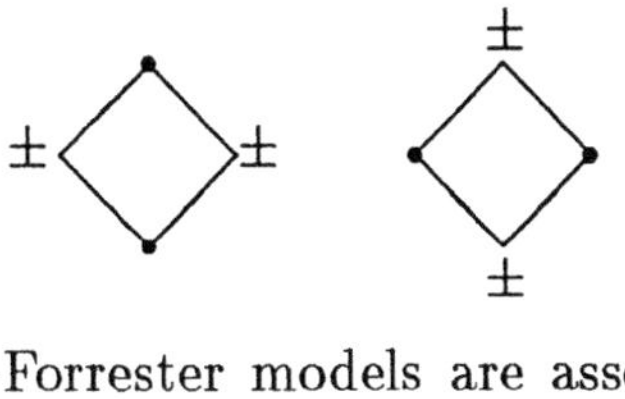

The Andrews-Baxter-Forrester models are associated with the Dynkin diagrams A_n [2]:

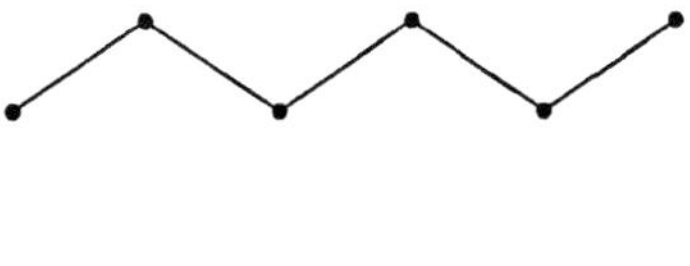

Thinking of this as an oriented graph with basic transitions, we can generalize to the Weyl alcove $\mathcal{A}^{(n)}$ of the level k-integrable representations of the Kac-Moody algebra $\mathcal{A}^{(1)}_{N-1} = \widehat{SU(N)}$. For $SU(3)$ instead of $SU(2)$, we replace

by

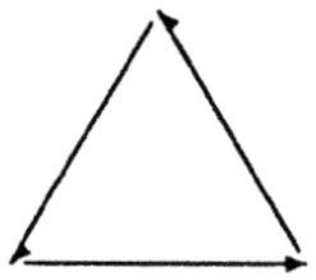

i.e., transitions generated by 3 vectors (instead of 2) which add to zero. Let

$$P^{(n)}_{++} = \{\lambda = \sum_{i=1}^{N-1} \lambda_i \Lambda_i ; \lambda_i \geq 1, \sum_{i=1}^{N-1} \lambda_i \leq n-1\},$$

where the Λ_i's are the $N-1$ weights of the fundamental representations and $n = k + N$. The vertices of the graph $\mathcal{A}^{(n)}$ are the elements of $P^{(n)}_{++}$ and its oriented edges are given by N vectors e_i defined by

$$\begin{aligned} e_1 &= \Lambda_1, \\ e_i &= \Lambda_i - \Lambda_{i-1}, \qquad i = 1, 2, \ldots, N-1, \\ e_N &= -\Lambda_{N-1}. \end{aligned}$$

These models were introduced and studied by Date, Jimbo, Miwa and Okado [21].

The notion of an orbifold of such a model, dividing out by a symmetry group G on Γ which preserves the Boltzmann weights was introduced by [36, 37, 22]. The model based on Γ is replaced by one built on an orbifold graph, denoted by Γ/G; i.e., we consider a model on the orbifold graph Γ/G equipped with some Boltzmann weights. Fendley and Ginsparg considered the $\mathbf{Z}_2$-orbifold of the A-series associated to $SU(2)$ and Di Francesco and Zuber for the $\mathbf{Z}_3$-orbifolds of the corresponding $SU(3)$ orbifolds.

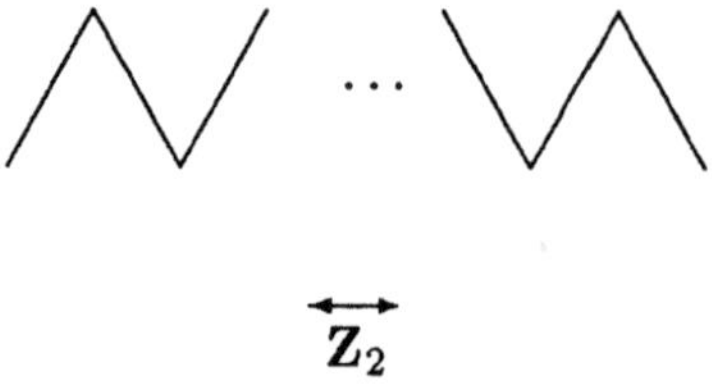

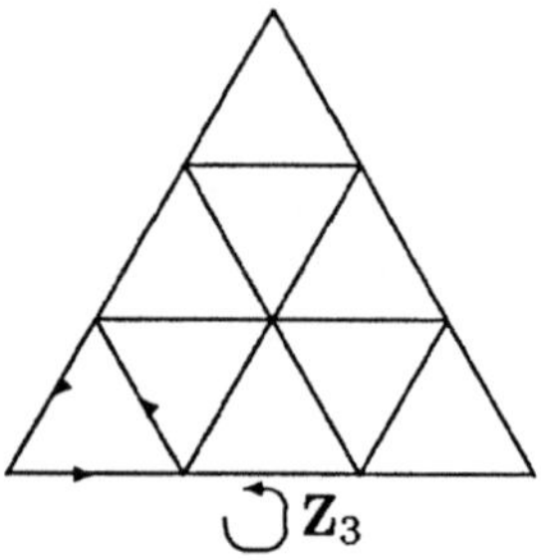

In particular, they expressed the Boltzmann weights for the original $\mathcal{A}^{(n)}$-model (for $N = 2, 3$ respectively) as

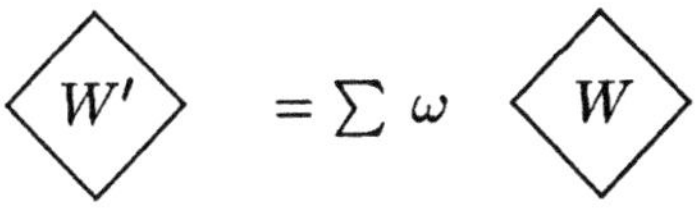

(1)

with complex coefficients $\omega^N = 1$, and could show by direct computation that the Yang-Baxter equation for the orbifold graph was inherited from the corresponding equation for the original graph. In this framework, the Yang-Baxter equation is expressed as equality of the partition functions

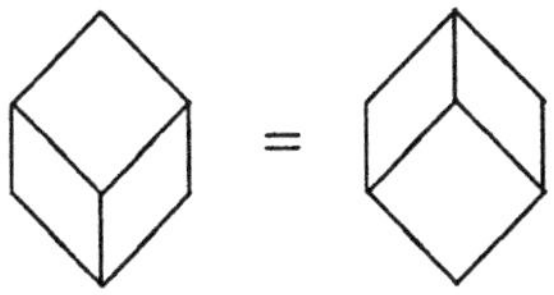

where both figures have the same configurations (edges of Γ) along the boundary and the sum (the product of 3 Boltzmann weights) over internal edges.

The orbifold procedure is best understood in the path algebras associated to graphs. The transfer matrix method transforms the two dimensional classical model to a one-dimensional quantum system [70, 60, 35, 3, 13, 58, 30]. The classical observables are $C(P)$, the commutative C^*-algebra of all continuous functions on the compact Hausdorff space of configurations, and at each inverse temperature β we may be interested in the simplex K_β of equilibrium states. The transfer matrix formalism transfers this setting to that of a one dimensional quantum model represented by a non-commutative C^*-algebra A, a path algebra on the graph Γ or the algebra generated by the local transfer matrices. For each inverse temperature β, one looks for a map from equilibrium states in K_β to linear functionals on A and a map $F \mapsto F_\beta$ from (local) classical observables in $C(P)$ to the quantum algebra A such that one recover the classical expectation or correlation functions from a knowledge of the quantum ones alone: $\mu(F) = \varphi_\mu(F_\beta)$.

The non-commutative path algebra $A(\Gamma)$ associated to the graph Γ is defined as follows. First take the space of all one-sided paths in Γ (beginning in some specified subset of vertices of Γ). Drawing the paths vertically, beginning with the initial vertices at level zero, we go down one level at a time, using the graph Γ, so that at each level we have a family of vertices. Consider the finite dimensional algebra $A[m,n]$ between the two levels m and n, generated by matrix units or partial isometries $|\xi\rangle\langle\eta| \equiv (\xi,\eta)$, where ξ, η are paths from level m to n, with $(\xi,\eta) = 0$ unless ξ, η have the same initial and same terminal vertices. Let $A(\Gamma) = \varinjlim A[m,n]$ be the completion, i.e., the

C^*-algebra generated by such pairs of paths (ξ, η), with $A[m, n]$ embedded into $A[m', n']$, $(m' \leq m \leq n \leq n')$ by

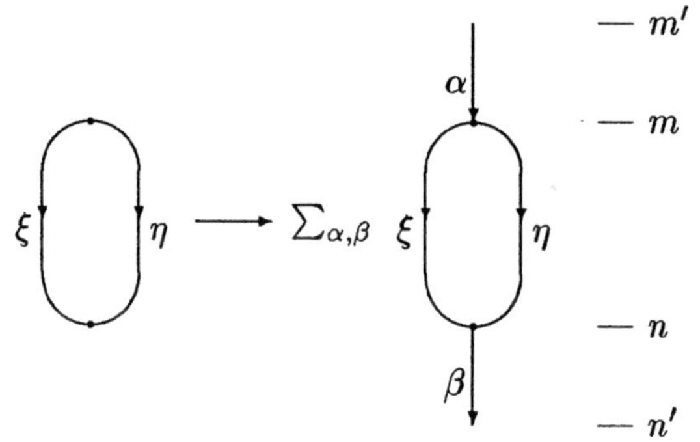

$$(\xi, \eta) \quad \rightarrow \quad \sum_{\alpha,\beta} (\alpha\xi\beta, \alpha\eta\beta),$$

where the summation is over paths α from level m' to level m, β from level n to n'.

Writing

$W = \sum$ 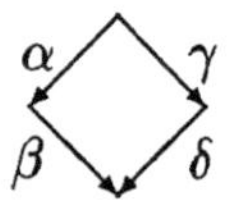 $(\alpha\beta, \gamma\delta)$

we see that being given a (translation invariant) Boltzmann weight is the same as begin given an element W in the path algebra $A[n-1, n+1]$, for some n. Then for any n, letting W be the corresponding element of the path algebra, the Yang-Baxter equation corresponds to

$$W_n W_{n+1} W_n \;=\; W_{n+1} W_n W_{n+1}. \tag{2}$$

In general, the $\{W_n; n \in \mathbf{N}\}$ will not generate the path algebra, but they will in the case of the $\mathcal{A}^{(n)}$-models of [21], when W_n gives a representation of the Hecke algebra of type A. If the group G acts on Γ, then (under reasonable hypotheses) the fixed point algebra $A(\Gamma)^G$ is isomorphic to $A(\Gamma/G)$, the path algebra of the orbifold graph; with this isomorphisms explicitly leading to (2), as a Boltzmann weight invariant under the group action gives an invariant element of the path algebra $A(\Gamma)$. This gives a natural meaning for the construction of the orbifold graph. The fixed points of the graph $\mathcal{A}^{(n)}$ are replaced by N vertices representing the orbit and, the N vertices which are moves cyclically onto each other are replaced by singleton as in the following examples:

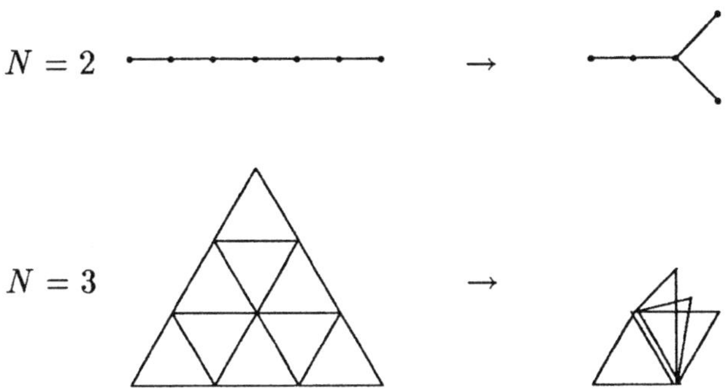

(3)

These models were first introduced by Pasquier [69] in the case of $N = 2$ by an ad hoc method of finding Boltzmann weights for the D-series which satisfies the Yang-Baxter equation, and later by [36, 37] for the case $N = 2$, $N = 3$, who used (1) to write down the Boltzmann weights and verified by hand that they satisfied the Yang-Baxter equation. Di Francesco and Zuber predicted similar models for general N would follow as a consequence of identities between theta functions. The methods of [32] shows that the best way to understand these models is via the path algebra picture outlined here. (Pasquier also used ad hoc methods to obtain a model at criticality for E_6; an embedding of A_{11} in E_6 via the Temperley-Lieb algebra.) This is a discrete version of orbifold theory in the structure theory of C^*-algebras, where the existence of fixed points in the symmetry group of some underlying structure tends to new phenomena. For example, consider the $\mathbf{Z}_2$-action of flipping the circle; $z \mapsto \bar{z}$. The crossed product of a pair of distinct points $\{z, \bar{z}\}$ moved by a symmetry $\mathbf{Z}_2$ is $C\{z, \bar{z}\} \times \mathbf{Z}_2 = M_2$, the 2×2-complex matrices generated by the diagonal $\mathbf{C}^2$, and the $\mathbf{Z}_2$-flip which interchanges the components. The crossed product of a singleton $\{z\}$ by $\mathbf{Z}_2$-action is $\mathbf{C}^2$ $(= C(\hat{\mathbf{Z}}_2))$. Gluing this together, the crossed product $C(\mathbf{T}) \times \mathbf{Z}_2$ is the M_2-valued functions on the interval (or a semicircle) diagonal at the endpoints, represented geometrically as

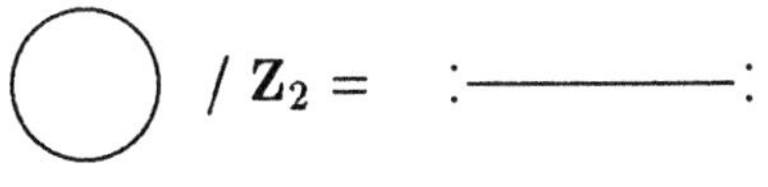

This example is the continuous analogue of (3) and is at the heart of recent progress in the structure theory of amenable C^*-algebras, where for example, one can show the existence of compact group actions on AF algebras whose

fixed point algebras are not AF [5, 59, 6, 7, 8]. The analogous $\mathbf{Z}_2$-action on the non-commutative two torus A_θ leads to an AF algebra $A_\theta \times \mathbf{Z}_2$ [9, 10] (so that the dual action has again non-AF fixed point algebra, essentially A_θ); the fixed point algebra $A_\theta^{\mathbf{Z}_2}$ is also AF, and this may help in understanding the Cantor spectra of discrete Schrödinger operators with almost periodic potentials.

Returning to our central framework, note that the graph A_3 is self-dual

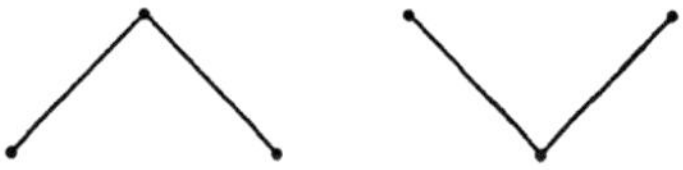

In this case, the process taking the orbifold model, is a non-trivial identification of the Ising model with itself; indeed it is the high temperature-low temperature duality

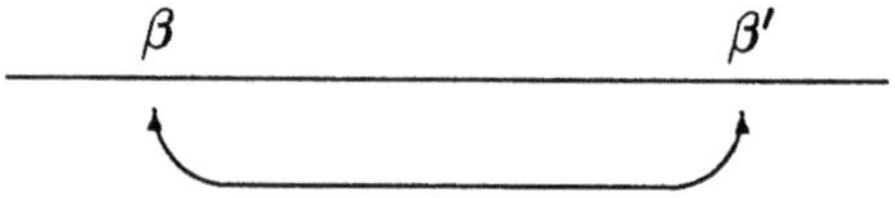

of Kramers and Wannier [19, 35]. Note that again the existence of a fixed point under this transformation is what leads to the interesting phenomena; it is the critical temperature of Onsagar β_c.

We now use the path formulation of AF algebras [28, 29, 64, 65, 75] to construct subfactors, with the main interest centered on those constructed from the solvable lattice models. Consider the following

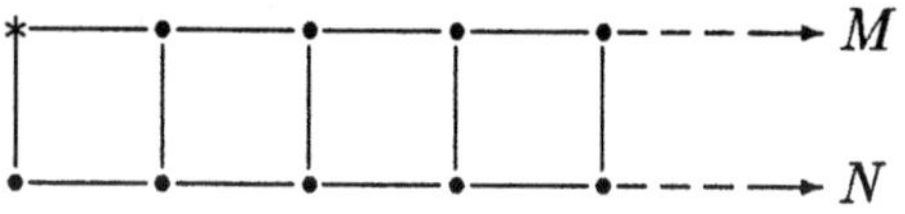

(4)

Here each line represents a graph, and each node represents a sequence of vertices say perpendicular to the plane of the paper, where at the top left hand corner we have one distinguished vertex $*$. Taking the Bratteli diagrams

and

we obtain two AF algebras, completing with respect to some traces we obtain von Neumann algebras N and M; the vertical graphs will allow us to embed N in M — for the time being we will not be precise about this. We want to extend this picture to obtain a double complex and a sequence of factors $N \subset M \subset M_1 \subset \cdots$:

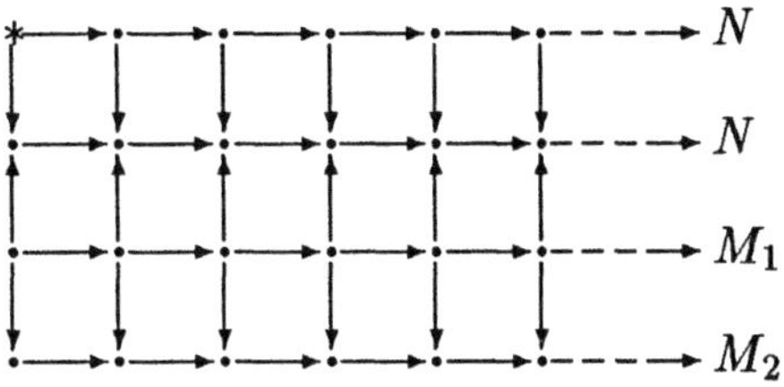

(5)

We do not wish to generate a random sequence of factors, but the Jones tower associated to the first inclusion $N \subset M$. The Jones basic extension from $N \subset M$ to $M \subset M_1$ can be thought of as a reflection of N through M; this can be achieved by reflecting the vertical graphs as indicated by alternating the orientations. This using the graphs $\mathcal{A}^{(n)}$, where the vertices are N-colorable, we obtain the following picture of colors [32]:

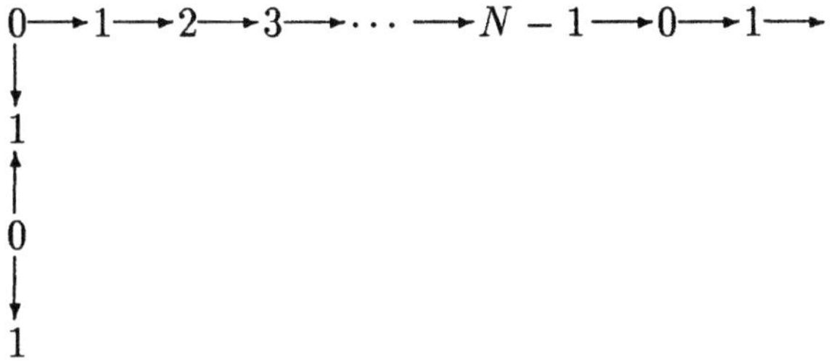

(6)

The Jones tower of relative commutants $N' \cap N \subset N' \cap M \subset N' \cap M_1 \subset N' \cap M_2 \cdots$ is an inclusion of finite dimensional C^*-algebras in the case of finite index; its Bratteli diagram being the iteration of a single graph called the principal graph Γ (which may bear no relation in general to the graphs used in (4) to construct the subfactor). The inclusion $N \subset M$ is said to be of finite depth when the principal graph is finite, in which case the norm of its incidence matrix is the (square root) of the Jones index of the subfactor. Our task is to identify the principal graph in certain examples.

To compute the relative commutant $N' \cap M_k$

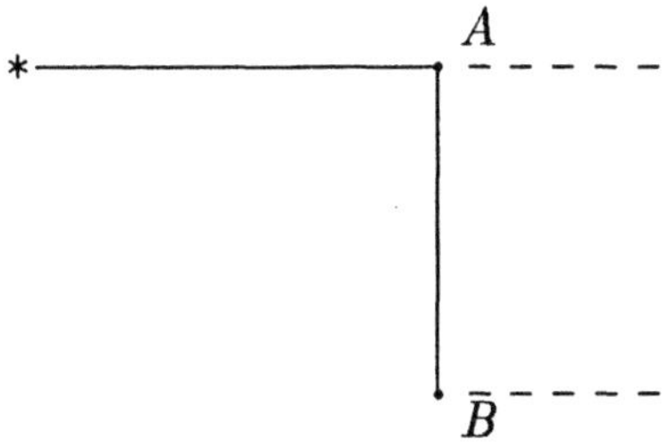

first consider at the finite dimension level the commutant $A' \cap B$. Straightening this out,

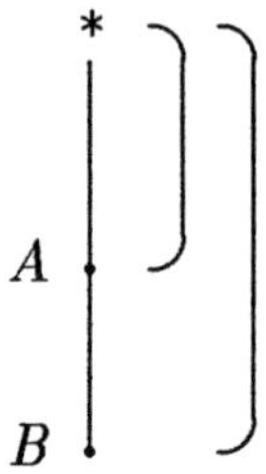

we need the commutant of a path algebra $A[0,n]$ in $A[0,m]$ which is easily seen to be $A[n,m]$, the vertical path algebra

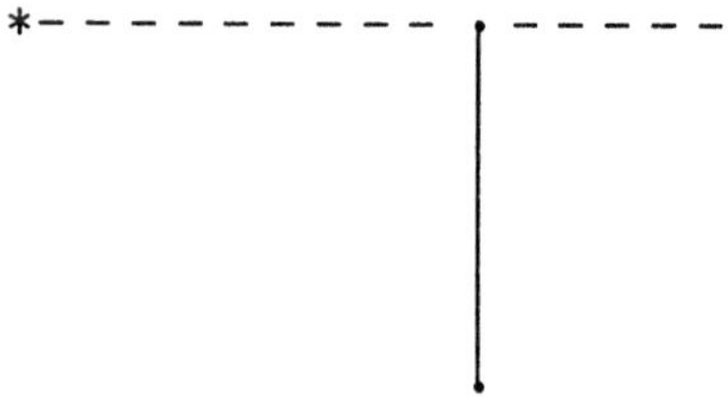

Since we need the relative commutant at the infinite stage, it is reasonable to expect (and it is the case when the four basic graphs in (4) are finite) that what we should take is that part of the vertical algebra which is transformed

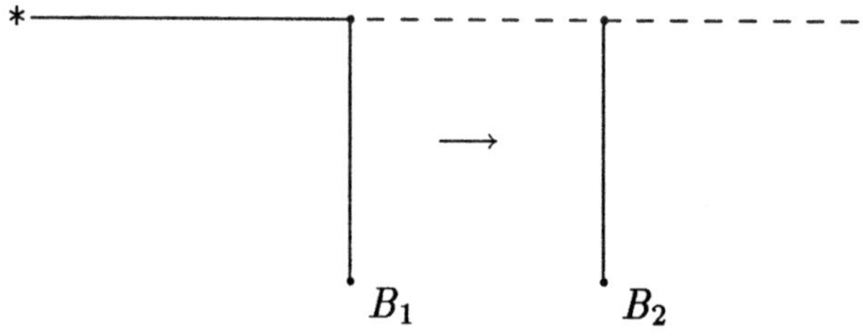

to another vertical path algebra when we embed B_1 in B_2. This is the notion of parallel transport of Ocneanu [64, 66]; the principal graph is general being the flat part (of a given complex (5)). The simplest case to describe is when everything vertical transforms in a parallel fashion to a vertical algebra; in which case the principal graph can be identified with the graph appearing in the left most vertical part of the complex (5).

So far, we have not been specific about the vertical inclusions. If we take trivial diagonal embeddings

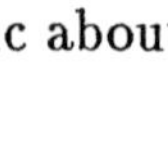

then we would be in the worst possible situation where something vertical becomes horizontal. We need to make a non-diagonal identification, and this is where the Boltzmann weights become useful. They provide us, with suitable specialization of the spectral parameter, with a unitary to identify the path algebra with  and when normalized, they give rise to a unitary to identify paths  with . This is a connection in the sense of Ocneanu [64, 65, 66]; the unitarity condition corresponds to the first inversion relation in the solvable lattice model theory, the biunitarity (unitarity of  under renormalization) or commuting square condition corresponds to the second inversion relation. It is the Yang-Baxter equation which is useful in helping to compute the flat part of a graph. By symmetry, flatness is the same as vertical parallel transport of the horizontal path algebras. Thus taking a face operator in the horizontal path algebra we have [73]:

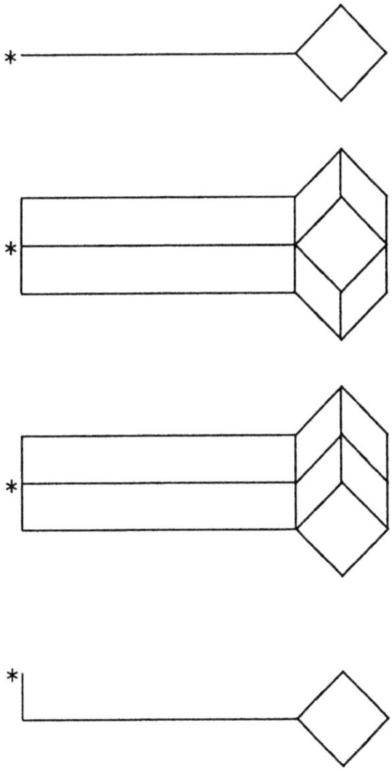

using the connection

by Yang-Baxter equation

by unitarity

Thus if face operators generate the (first) horizontal algebra in (2), then we certainly have flatness. This is the case for the $\mathcal{A}^{(n)}$ models with connection given by the Boltzmann weights [46] so that the principal graphs of these subfactors (first constructed by Wenzl of [80] who computed their index) are their 0-1 parts [32] (c.f. (3)). Similarly embedding A_{11} in E_6 (since a Temperley-Lieb algebra can be embedded in the path algebra of a bipartite graph [38]), the principal graph is the intertwining graph (computed in [69, 31]) and so we recover the result of Okamoto [68]. Note that here we are using the intertwining Yang-Baxter equation between the (critical) Boltzmann weights on A_{11} and E_6 (see [22]).

3. Modular Invariants and Subfactors

The A-D-E classification of (level k) $SU(2)$ Wess-Zumino-Witten models is as follows. The $SU(2)_k$ Kac-Moody algebras has $k+1$ integrable representations $[\varphi_j]$, $2j = 0,\ldots,k$, with $SU(2)$ isospin j and affine character $\chi_\lambda(\tau) = \mathrm{Tr}_{\mathcal{H}_j} e^{2\pi\tau i(L_0 - c/24)}$, $\lambda = 2j+1$. A one loop partition function on the torus should be expressed as $Z(\tau,\tau') = \sum N_{\lambda,\bar\lambda}\chi_\lambda(\tau)\overline{\chi_{\bar\lambda}}(\tau')$. (a continuum version of the transfer matrix formalism outlined above). The classification problems to find all such partition functions of the above form, where $N_{\lambda,\lambda'}$ are non-negative integers (arising from a decomposition of the underlying Hilbert space as $\oplus N_{\lambda,\bar\lambda}\mathcal{H}_\lambda \otimes \mathcal{H}_{\bar\lambda}$), and where Z is modular invariant (due to invariance under reparametrizations of the torus). Such modular invariant sesqui-linear forms were shown [12, 50] to be of type A-D-E.

The identification of a modular invariant with a graph is through noting that the diagonal part (writing $Z = \sum_{\lambda\in\mathcal{E}} |\chi_\lambda|^2$+remainder) of a modular invariant is described by a set ($\mathcal{E}$) of exponents of an A-D-E Dynkin diagram. Thus for $k = 16$, we have three invariants corresponding to A_{17}, D_{10}, E_7 as follows:

modular invariant	exponents	graph
$\lvert\chi_1\rvert^2 + \lvert\chi_2\rvert^2 + \cdots + \lvert\chi_{17}\rvert^2$	$1,2,3,\ldots,17$	A_{17}
$\lvert\chi_1+\chi_{17}\rvert^2+\lvert\chi_3+\chi_{15}\rvert^2+\lvert\chi_5+\chi_{13}\rvert^2 + \lvert\chi_7+\chi_{11}\rvert^2 + 2\lvert\chi_9\rvert^2$	$1,3,5,7,9,11,13,15,17$	D_{10}
$\lvert\chi_1+\chi_{17}\rvert^2+\lvert\chi_5+\chi_{13}\rvert^2+\lvert\chi_7+\chi_{11}\rvert^2 + \lvert\chi_9\rvert^2 + (\chi_3+\chi_{15})\chi_9^* + \chi_9(\chi_3+\chi_{15})^*$	$1,5,7,9,11,13,17$	E_7

In the subfactor theory, only A-D-E Dynkin diagrams, with D_{even} and E_{even} appear as principal graphs of subfactors with index less than four. In the rational conformal field theory of $SU(2)$ Kac-Moody WZW models, described by A-D-E Dynkin diagrams, there is a degeneracy so that only D_{even} and E_{even} need be counted. For example in the case of $k = 16$, the modular invariant for E_7 reduces to that of D_{10} under the simple interchange of blocks χ_9 and $\chi_3+\chi_{15}$. Note that there are two kinds of modular invariants here

$$\sum|\chi_i|^2 \qquad \text{type I}$$

$$\sum\chi_i\bar\chi_{\sigma(i)} \qquad \text{type II}$$

where σ is a permutation of the extended fusion rules. In this case, the flip on the extended blocks identifies the two modular invariants for D_{10} and E_7, and led Zuber to conjecture that the subfactor built from the E_7 connection

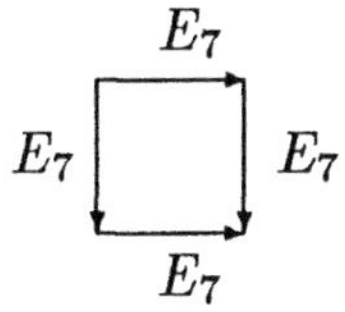

is the D_{10} subfactor (i.e., the flat part of E_7 is D_{10}). This conjecture was verified in [34]; indeed using a commuting square or connection

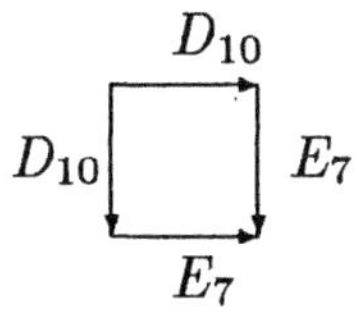

where the two D_{10}'s are glued together by a non-trivial identification corresponding exactly to the interchange blocks $\chi_9 \leftrightarrow \chi_3 + \chi_{15}$.

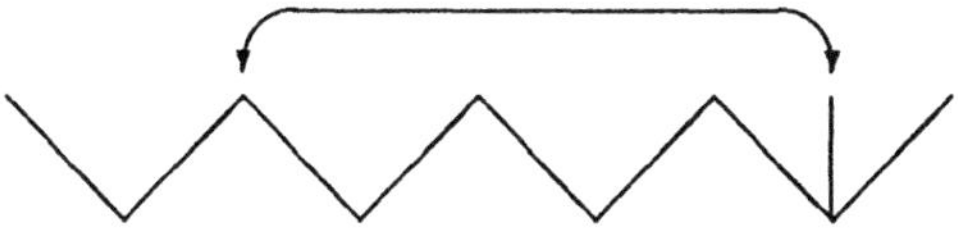

This is related to an observation in [24].

Note that the flat parts of the other Dynkin diagrams had been known; A_n, D_{2n}, E_6, E_8 are flat and D_{2n+1}, E_7 are non-flat as announced by Ocneanu [64] and proved by [4, 43, 44, 52, 76]. It is easy to see that the flat part of D_{2n+1} is A_{4n-1} and again corresponds to a degeneracy of a type II modular invariant for D_{2n+1} to a type I invariant for A_{4n-1}. See also [42] for an approach to flatness based on ideas from quantum groups.

4. Rational Conformal Field Theory (RCFT) and Subfactors

The combinatorial approach to rational conformal field theory such as [63, 82] has much in common with paragroup theory, and was explicitly worked out in [11] for the first time. We will explain this work [11] briefly and then show how F. Xu [83] combined this method and the orbifold construction of [32, 52]. The operator algebraic meaning of this construction will be clarified in Section 6.

Moore and Seiberg [63] showed that the braiding and fusion matrices and the modular matrix satisfying certain axioms contain the duality data of a conformal field theory. If we start from a connected, simply connected, compact and simple Lie group G and an integer k called a level, one gets such combinatorial data and can compute an expectation value of knotted graphs as in [82]. In [11], de Boer and Goeree showed that this kind of combinatorial data produce a paragroup from an RCFT and a choice of a primary field. The biunitary connection is directly obtained from entries of the braiding matrices, and the only non-trivial axiom we have to verify is flatness. Their observation is that if we have a partition function diagram with $*$ at all the four corners, the computation of the value is reduced to that of the expectation value of the following knotted diagram.

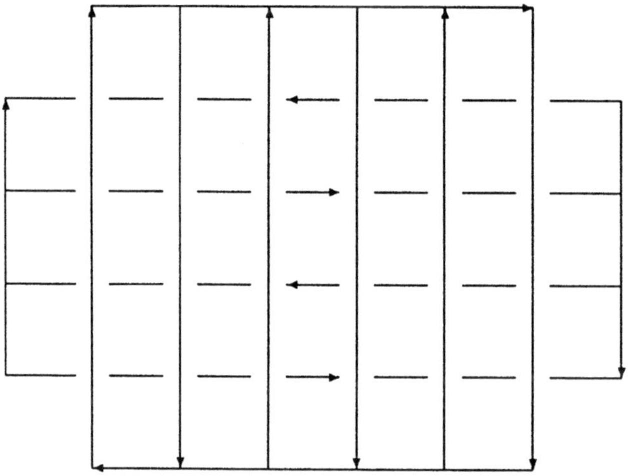

This diagram can be split into two pieces because the vertical arrows are over the horizontal arrows at all the crossings. Then it is very easy to compute the expectation value and we get flatness of the connection.

Next we explain the orbifold construction of subfactors from this viewpoint. The orbifold construction was first used in subfactor setting in [52] to prove classification of subfactors corresponding to the Dynkin diagram D_n which was announced by Ocneanu without a proof in [64], and established as a general method in [32]. The main idea is that if a paragroup has a symmetry of finite order, we can produce a candidate for a new paragroup which is the quotient of the original paragroup by the symmetry. This is interesting especially when the symmetry moves the distinguished vertex $*$ of the paragroup. In such a case, the symmetry cannot act on the higher relative commutants naturally, but we realize the subfactor in non-canonical way so that the symmetry can act. That is, the path algebra starts not just from $*$ but from all the images of $*$ under the symmetry. Then we can take fixed point algebras of the double sequence of the path algebras and get a natural biunitary connection on the double sequence of the fixed point algebras, then the only remaining problem is to verify flatness. The paragroups corresponding to the Dynkin diagrams A_{2n-3} naturally have $\mathbf{Z}_2$-symmetries, and the fixed point algebras under the $\mathbf{Z}_2$-action on the double sequence of the path algebras starting from the two endpoints of A_{2n-3} give the connection on D_n. Then the flatness problem of D_n is reduced to the following identity of the partition functions as in [52].

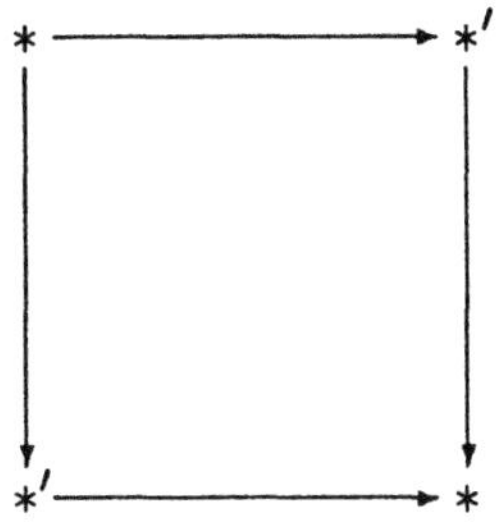

$= 1$

where $*'$ is the other endpoint than $*$, and the size of the diagram is $(2n-4)\times(2n-4)$. Then direct computation based on induction as in [52] shows that the value of the left hand side is $(-1)^n$, and this proves that the Dynkin diagrams D_{2n} are realized as paragroups (in unique way) and the diagrams D_{2n+1} are impossible. This gave the first complete proof of this statement in the literature. Impossibility of D_{2n+1} has been independently proved by [43] and [76] based on fusion rules, the original proof of realization of D_{2n} of Ocneanu is now known [53, appendix], and Izumi's method for E_8 also gives a simpler proof of realization of D_{2n}, but this orbifold method still gives the most natural conceptual understanding of the phenomena as seen below.

In [32], we first computed the higher relative commutants of the Hecke algebra subfactors of type A of Wenzl [80] based on solutions to the Yang-Baxter equation by Jimbo-Miwa-Okado [46] as explained in section 1, and then applied the orbifold construction to these series. The sequence of subfactors of Wenzl with indices converging to N^2 corresponds to $SU(N)$. We proved in [32] that if $N=3$, the orbifold construction always gives a flat connection. Thus $SU(2)$ and $SU(3)$ behave quite differently in this question of flatness. We further noted with a little bit more work that the same result holds for odd prime N, and conjectured that the parity of N causes this difference. F. Xu [83] proved that this is indeed the case. That is, if N is odd, the orbifold construction always gives a flat connection, and if N is even, the sequence of the resulting connections contains a flat connection and a non-flat connection alternately. Xu's method is more general and applicable to any subfactor arising from RCFT corresponding to G_k Wess-Zumino-Witten models as explained as above, if we work at the symmetry of the paragroup arising from the center Z of the Lie group G.

By [32] and [83], the problem of flatness is reduced to verifying the identity

$$\begin{array}{ccc} * & \xrightarrow{\ \eta\ } & \rho^j(*) \\ {\scriptstyle \xi}\big\downarrow & & \big\downarrow{\scriptstyle \rho^j(\xi)} \\ \rho^k(*) & \xrightarrow[\ \rho^k(\eta)\ ]{} & \rho^{j+k}(*) \end{array} = 1,$$

where ρ denote the automorphism of the graph arising from the symmetry. Xu further noticed that it is enough to assume $k = -j$. By the same argument as in [11], this computation is reduced to that of the expectation value of the following knotted graph.

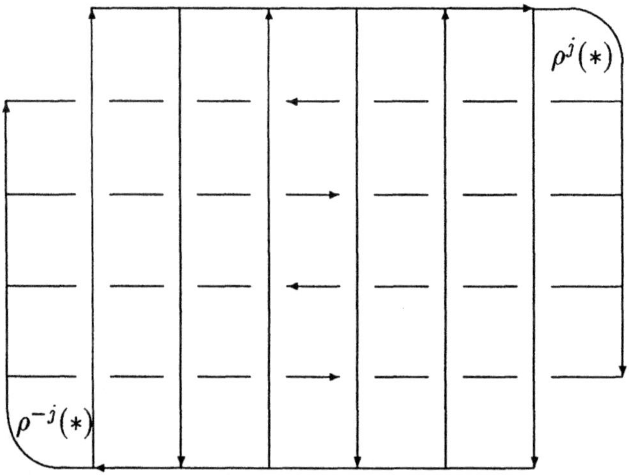

We again want to simplify the picture, but this expectation value is invariant only under regular isotopy, and this time we have a writhe for a curve at the upper right corner of the original graph. If we remove this writhe with the first Reidemeister move, we have a coefficient $\exp(\pm 2\pi i h_j)$, where h_j is the conformal dimension of $\rho^j(*)$, as in [82, page 639]. Thus we get flatness if and only if h_j's are integers. The conformal dimensions for $SU(N)$ have been known and we can decide when it is an integer.

Thus the difference between the Dynkin diagrams D_{2n} and D_{2n+1}, first observed by Ocneanu in paragroup settings, is naturally understood as a very special case of general phenomena caused by conformal dimensions and the orbifold construction.

Xu further noticed that the condition on the Lie group G and level k equivalent to flatness of the resulting connections in the orbifold construction is the same as the one equivalent to existence of Chern-Simons gauge theory based on G/Z at level k.

We conclude this section by an important remark. Although the orbifold construction has its origin in conformal field theory [26] and works well in paragroup setting as seen above, it does not produce the combinatorial data necessary for rational conformal field theory in the following sense. The simplest case is the series D_{2n} obtained as Z_2-orbifolds of the series A_{4n-3} which comes from rational conformal field theory corresponding to $SU(2)$. Although the Dynkin diagrams D_{2n} are realized as paragroups, we cannot have a corresponding rational conformal field theory as in [11] since we do not have symmetric S-matrices.

5. Topological Quantum Field Theory (TQFT) and Subfactors

Turaev-Viro [78] introduced a topological invariant of 3-dimensional closed manifolds based on triangulations using the quantum $6j$-symbols of Kirillov-Reshetikhin [56]. They define the invariant as a state sum for a triangulation; topological invariance of this state sum is based on certain algebraic relations of the $6j$-symbols. To achieve this, they proved that classical Alexandar moves [1] can be replaced simpler local moves of tetrahedra. (For a simple proof of this claim in 3 and 4 dimensions, see [39].) A. Ocneanu [67] claimed without a full proof that this type of invariants can be generalized to a subfactor setting and that algebraic axioms necessary for topological invariance are equivalent to the paragroup axioms. Although he has not published the exact statement of the theorem or a complete proof, a complete proof was given in [33] and the full machinery was worked out there. We discuss this aspect of paragroup theory here.

The basic idea of the construction in [78] is as follows. We have a set of finitely many colors with multiplication rules and each color has a certain weight. For a fixed triangulation of a given 3-dimensional compact manifold, we consider all the admissible configurations of colors assigned to edges appearing in the triangulation. (Admissibility is determined by the multiplication rule.) Each tetrahedron with six colors on six edges has a complex value called a $6j$-symbol. For a given configuration, we take the product of all the $6j$-symbol values, and then take a weighted sum of all the products over all configurations, which is called the state sum. In order to show that this complex number is really a topological invariant, we have to prove that the $6j$-symbol has a certain symmetry and that the state sum does not depend on a choice of triangulation. Turaev-Viro proved that these two properties follow from three axioms on $6j$-symbols and presented the Kirillov-Reshetikhin quantum $6j$-symbol for $U_q(sl_2)$ [56] satisfied these axioms. The symmetry of $6j$-symbols we need means that the value given to a tetrahedron does not depend on the way we project a tetrahedron on paper, and is called tetrahedral symmetry. The other two axioms in [78] are unitarity and the pentagon relation. In order to prove that this construction gives a complex number independent on triangulation, we have to compare two triangulations of the same manifold. Alexandar [1] already showed that such two triangulations can be transformed to each other by finite steps of local moves, called Alexandar moves. These moves, however, are infinitely many, and are not easy to handle, so Turaev-Viro replaced the Alexandar moves in 3-dimensions by three local moves. The unitarity and the pentagon relation are axioms equivalent to invariance of the state sum under these three moves. Dijkgraaf-Witten proposed a similar invariant arising from 3-cocycles of finite groups based on physical ideas. A rigorous presentation of this invariant based on Alexandar's theorem is given by [79].

To generalize this machinery to a general subfactor setting, we first need an appropriate $6j$-symbol. This can be done as the generalization of a connection as in [67]. In paragroup theory, we look at four bimodules, and composition of four intertwiners among the tensor products of these four bimodules and ${}_NL^2(M)_M$ and ${}_ML^2(M)_N$. Extending this setting, we look at *six* bimodules and composition of four intertwiners among their tensor products. We also have a fusion algebra consisting of four kind of bimodules: N-N, N-M, M-N, and M-M. Each bimodule has a weight given by the corresponding Jones index. In this way, we have a generalized $6j$-symbols with tetrahedral symmetry, unitarity, and flatness. The tetrahedral symmetry (after a certain normalization) comes from the Frobenius reciprocity of bimodules and intertwiners [67, 84] and unitarity is just as usual. So all we have to prove is the pentagon relation.

First write down the following identity as 3×3-flatness. Then split the square into two pieces as follows.

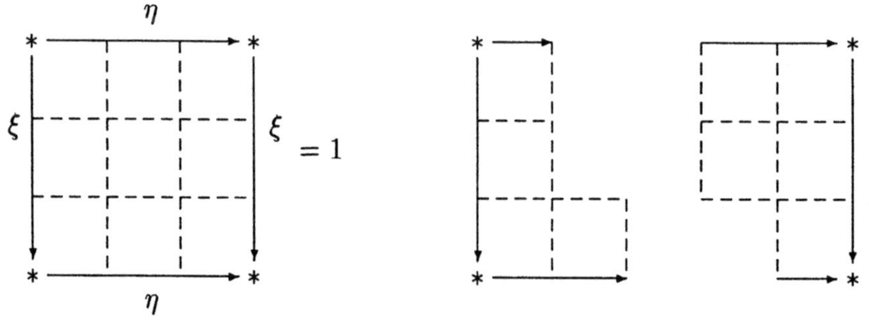

It is not difficult to see that the Cauchy-Schwarz inequality implies that these two pieces have the same partition function value. (This kind of use of the Cauchy-Schwarz inequality was first made in [32, §5].) It is easy to see that all the four corner square of the picture have value 1 up to some normalization constant. Thus we have non-trivial values for two squares for the left picture and three square for the right picture. The pentagon relation is obtained immediately by writing down this identity among five squares.

Now we turn to the problem of the construction in the converse direction, constructing a subfactor from a topological invariant. Of course, we assume that the topological invariant is of Turaev-Viro type in the sense that it is a state sum based on triangulation. We also need a finitely generated fusion algebra with positive weights on the generators. Such conditions are often called rationality and unitarity respectively. By reversing the above argument, we get a biunitary connection and its 3×3-flatness in the sense that partition functions of size 3×3 have value 1. All we need is (ordinary) flatness now. This can be done as follows. First we split the 3×3-picture in the following way.

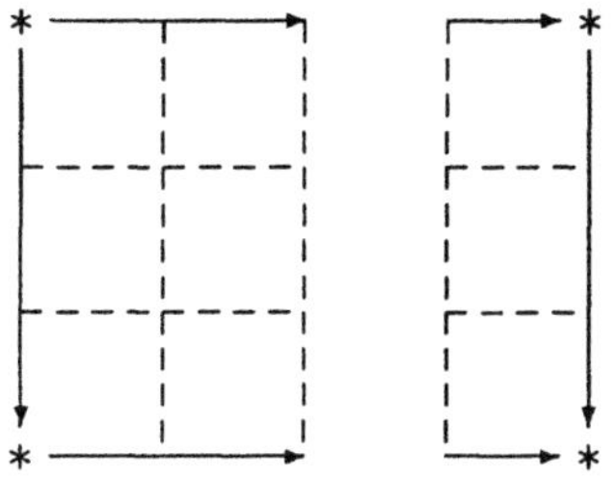

Then again by the Cauchy-Schwarz argument, we can reduce the size $3 \times n$ of a large diagram to 3×3 as follows.

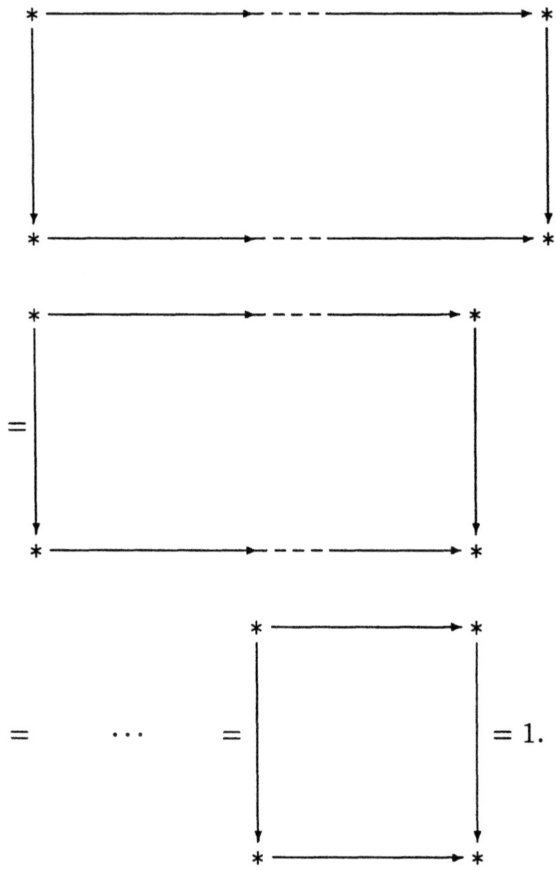

Then by the same kind of argument, we can reduce the size $m \times n$ of a large diagram to $3 \times n$. In this way, we can prove ordinary flatness from 3×3-flatness. This is an analogue of the argument of de Boer-Goeree [11] mentioned in Section 4 in the sense that we get a global identity from local identities.

In this way, we can say that flatness is equivalent to the pentagon relation which means invariance of the state sum under the following move which we call move I.

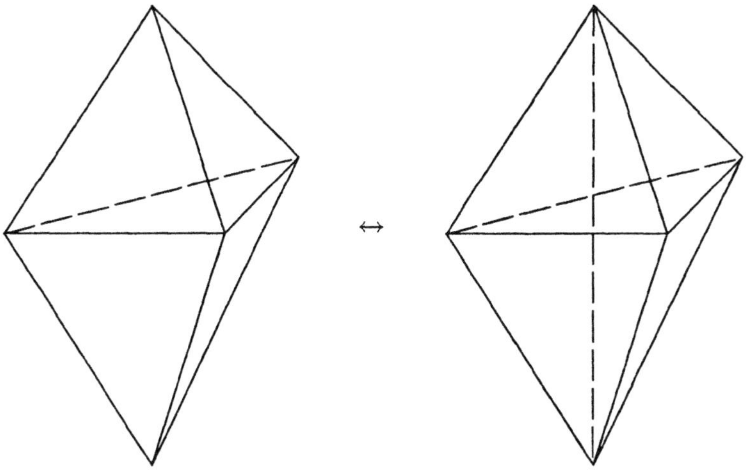

where the left picture has two tetrahedra and the right has three tetrahedra.

At the end of this section, we mention two subtle differences between the Turaev-Viro construction and topological invariants arising from subfactors. The first difference is unitarity, or positivity of the trace. If we start from a subfactor, each element of the fusion algebra has a positive weight given by the Jones index, and we need this positivity for the construction in the converse direction. Turaev-Viro machinery, however, does not require this positivity, and actually, the Kirillov-Reshetikhin quantum $6j$-symbol does not satisfy this positivity in general. At the end of [78], they compute the invariant for the case Dynkin diagram A_2 for two choices of roots of unity and get different values. One of the two choices does not satisfy positivity condition and does not come from a subfactor. In general, we have to choose a root of unity q to be $\exp(\pi i/k)$ to get this positivity. The other difference comes from $\mathbf{Z}_2$-grading of vertices of the principal graph. In the above state sum approach, each vertex has an assignment of M or N in our setting unlike in [78]. We do *not* have to take a sum over configurations of M or N on vertices since each fixed configuration on vertices gives the same value as proved in [33]. (The key argument for this is as follows. In one of the three local moves, one vertex disappears. The assignment of M or N to this vertex does not change the value we obtain.) In particular, we can use only N-N bimodules or M-M bimodules. In the Turaev-Viro case of A_n Dynkin diagrams, this means that we use only a fusion subalgebra generated by "colors" labeled by integers instead of the entire algebra generated by "colors" labeled by half-integers. This causes another difference between the Turaev-Viro invariant and the invariant coming from the Jones subfactor of type A_n.

6. Automorphisms of Subfactors

As seen above, the orbifold construction gives new series of interesting subfactors as simultaneous fixed point algebras of finite group actions on the original subfactors. This suggests that automorphisms appearing in this construction have interesting operator algebraic properties. The aim of this section is to show that this is indeed the case. Beginning with a series of papers by A. Connes, there has been extensive work on the automorphisms of operator algebras, and many results on the classification of automorphisms and group actions on injective factors have been obtained. In this work, the roles of two classes of automorphisms have been emphasized. One is a class of approximately inner automorphisms and the other is that of centrally trivial automorphisms. In the subfactor setting, Loi introduced analogues of these two classes for subfactors and we use the same names "approximately inner" and "centrally trivial" for these analogues. Let $N \subset M$ be a subfactor of type II_1. An automorphism $\alpha \in \mathrm{Aut}(M,N)$ is said to be approximately inner if it is of the form $\lim_n \mathrm{Ad}(u_n)$ with $u_n \in N$ and it is said to be centrally trivial if α acts trivially on $N^\omega \cap M'$, where ω is a free ultrafilter over $\mathbf{N}$.

In the following, let $N \subset M$ be a strongly amenable subfactor of type II_1 in the sense of S. Popa [71]. This condition is equivalent to the generating property in the sense that the higher relative commutants reproduce the subfactor for some choice of tunnel [71]. P. Loi studied the classification problem of $\alpha \in \mathrm{Aut}(M,N)$ to classify type III_λ subfactors, $0 < \lambda < 1$, via subfactor version of Connes' discrete decomposition in [61], and introduced his invariant of automorphisms of subfactors. This invariant is given by the action of (the extension of) an automorphism on the higher relative commutants, and he proved that triviality of this invariant is equivalent to approximate innerness of the automorphism. He also proved that if any non-trivial power of $\alpha \in \mathrm{Aut}(M,N)$ is outer on M and $N \subset M$ has finite depth, then no non-trivial power of this α is centrally trivial. He further noticed that if $\alpha \in \mathrm{Aut}(M,N)$ is approximately inner and no non-trivial power of α is centrally trivial, a Connes type argument as in [17] uniquely determines α up to outer conjugacy.

Consider our orbifold action α of $\mathbf{Z}_2$ on subfactors of type A_{4n-3}. Loi's invariant is always trivial on subfactors of type A_n, so all automorphisms of these subfactors are approximately inner. Because the simultaneous fixed point algebra gives a different subfactor D_{2n}, this action α is different from the "standard" action of $\mathbf{Z}_2$, which is of the form $\mathrm{id} \otimes \sigma$ on $N \bar{\otimes} R \subset M \bar{\otimes} R$, where σ is a (unique) outer action of $\mathbf{Z}_2$ on the common hyperfinite II_1 factor R and $N \bar{\otimes} R \subset M \bar{\otimes} R$ is conjugate to $N \subset M$. By applying Loi's Connes type argument to $\mathbf{Z}_2$, we conclude that this action α is centrally trivial.

This method, however, is quite indirect, and does not work in the case of A_{4n-1}, where the simultaneous fixed point algebras are of the same type.

A more direct method using partition functions gives a better proof. Let α be an automorphism appearing in the orbifold construction for subfactors arising from RCFT as in section 4. Then computations based on topological moves of knotted graphs show the following identities for any k.

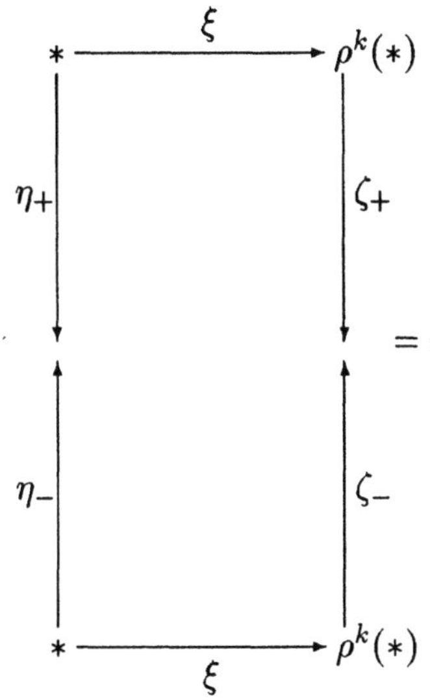

$$= \delta_{\rho^k(\eta_+),\zeta_+}\delta_{\rho^k(\eta_-),\zeta_-},$$

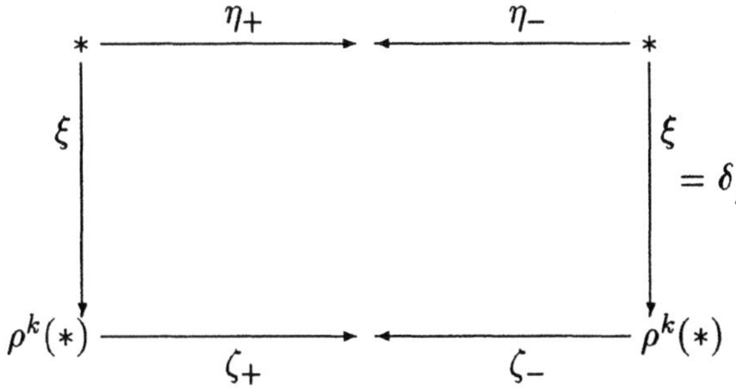

$$= \delta_{\rho^k(\eta_+),\zeta_+}\delta_{\rho^k(\eta_-),\zeta_-},$$

where ρ is the induced symmetry of the graph.

We can show that the first identity implies that α acts trivially on higher relative commutants, and the second implies that α acts trivially on central sequences. (See [32, Remark 5.9], [54, Lemma 4.3] respectively.) In this sense, approximate innerness and central triviality appear in a symmetric way, and we get that $\alpha \in \mathrm{Ct}(M,N) \cap \overline{\mathrm{Int}}(M,N)$. Note that we get this regardless of whether the orbifold construction produces a flat connection.

In a single factor case, Connes [16] introduced an invariant $\chi(M) = (\mathrm{Ct}(M) \cap \overline{\mathrm{Int}}(M))/\mathrm{Int}(M)$ and proved several interesting results. (Also see [47].) The above observation means that the orbifold construction produces non-trivial elements in the subfactor version of $\chi(M)$. We define $\chi(M,N) = (\mathrm{Ct}(M,N)\cap\overline{\mathrm{Int}}(M,N))/\mathrm{Int}(M,N)$, where $\mathrm{Ct}(M,N)$, $\overline{\mathrm{Int}}(M,N)$, $\mathrm{Int}(M,N)$ mean the class of centrally trivial automorphisms, the approximately inner automorphisms, and inner automorphisms given by unitaries in N, respectively. This is an abelian group as in the single factor case, and we have an analogue [54, Theorem 1.1] of the exact sequence in [16].

Next we aim to show that the automorphisms appearing in the orbifold construction for subfactors arising from RCFT as in section 4 actually give the entire $\chi(M,N)$. For example, in the case of Hecke algebra subfactors of type A by Wenzl [80] with indices $\sin^2(N\pi/k)/\sin^2(\pi/k)$, we get $\chi(M,N) = \mathbf{Z}_d$ with $d=(k,N)$. (In [54, Proposition 4.5], this was proved only for the case where N is prime, but the same method based on the above computation of partition functions gives this.) This claim is proved by getting an upper bound for the size of the group $\mathrm{Ct}(M,N)$ as follows.

Popa and Choda-Kosaki independently introduced the same notion of outerness of an automorphism in different situations. Popa calls an $\alpha \in \mathrm{Aut}(M,N)$ *properly outer* if the identity $ax = \alpha(x)a$ for some $a \in M_k$ and all x implies $a=0$. Choda-Kosaki calls the same condition *strongly outer*. Popa [72] proved that if a subfactor is strongly amenable, then not being properly outer is equivalent to central triviality. Choda-Kosaki [15, 57] proved that $\alpha \in \mathrm{Aut}(M,N)$ is not strongly outer if and only if α appears at even vertices of the dual principal graph in sector/bimodule approach. (If a subfactor is of type III, this condition means that α appears in irreducible decompositions of powers of Longo's canonical endomorphism.) Combining these two results, we know that the order of the group $\mathrm{Ct}(M,N)/\mathrm{Int}(M,N)$ is bounded by the number of even vertices of the dual principal graph with normalized Perron-Frobenius weights 1. In the case of subfactors arising from RCFT as in Section 4, this upper bound is attained and we get $\chi(M,N)$.

This upper bound was proved in the finite depth case first in [54] based on Ocneanu's approach with central sequence subfactors and asymptotic inclusion [64, 66]. (The statements, made without proof, by Ocneanu needed in this approach were proved in [54].) Although this approach is more complicated than combining results of Choda-Kosaki and Popa, it is expected to give more information on subfactors with finite depth. For example, Ocneanu has suggested that considerable information on TQFT of Section 5 can be obtained from central sequence subfactors and asymptotic inclusions.

Note that central triviality is easier to handle with partition functions than the property of not being properly outer is, although these two are equivalent in our situation. For example, in the orbifold construction for A_{2n-3}, it is easy to write down a non-zero $a \in M_k$ with $ax = \alpha(x)a$, $x \in N$, explicitly as in [52] if n is even, but it is not clear at all how to get such an a, if n is odd, while central triviality is easily established in both cases. With this advantage, we can prove the following: If a subfactor arises from

RCFT as in Section 4, automorphisms appearing on even vertices of the dual principal graph are the same as those appearing in the orbifold construction. That is, the orbifold construction gives a very concrete realization of these automorphisms appearing in a more abstract way in the sector/bimodule approach.

Note that a recent construction of a subfactor with finite depth and index $(5+\sqrt{13})/2$ by Haagerup [40] shows that the above upper bound is not always attained. (This is related to a problem whether one can adjust a certain automorphism M so that it fixes N globally. See [57].)

Thus we have many examples of subfactors with non-trivial $\chi(M, N)$, and it is not difficult to realize all finite abelian groups as $\chi(M, N)$ for subfactors with finite depth using tensor products. In the case of single factor, Connes [16, 47] constructed several examples with non-trivial $\chi(M)$, but these are rather unusual II_1 factors in the sense that they are non-AFD with property Γ. Subfactors arising from RCFT such as Hecke algebras subfactors of Wenzl are very natural and basic objects in subfactor theory, and they should not regarded as analogues of such exotic single factors. The right analogues of these subfactors in the single factor case turn out to be (injective) type III factors.

Loi's invariant has considerable similarity with the Connes-Takesaki module [20], and paragroups can be regarded as discrete analogues of flows of weights [20], as pointed out in [51]. For example, an automorphism of a strongly amenable subfactors is approximately inner if and only if its Loi invariant is trivial [61, Theorem 5.4], while an automorphism of an injective type III factor is approximately inner if and only if its module is trivial, as announced by Connes [18, section 3.8] and proved by [55, Theorem 1].

As to the other important class of automorphisms, centrally trivial automorphisms, Connes [18] announced that an automorphism α of an injective type III factor is centrally trivial if and only if it is an extended modular automorphism up to inner perturbation, and this was proved in [55, Theorem 1]. From this viewpoint, we come to the idea that an orbifold action arising from RCFT is a (discrete) analogue of modular automorphisms. With this understanding, we can interpret the result of Choda-Kosaki [15, 57] naturally as an analogue of a result of Haagerup-Størmer [41, Proposition 5.4].

It is easy to see that the orbifold $\mathbf{Z}_2$ actions on A_{4n-3} and the $\mathbf{Z}_2$ actions flipping the two tails of D_{2n} are in Takesaki type duality. This duality is again a discrete analogue of that between the Connes-Takesaki module [20] and the Sutherland-Takesaki modular invariant [77].

Popa [72] obtained the classification result of discrete amenable group actions on strongly amenable subfactors in the sense that properly outer such actions are classified by the Loi invariant. Because such actions are centrally free, this is again an analogue of the classification of centrally free actions of discrete amenable groups on injective type III factors by the Connes-Takesaki module.

Longo had already noticed in [62] that his canonical endomorphisms are similar to modular automorphism groups. From our viewpoints, automorphisms appearing in the powers of canonical endomorphisms have further similarity to modular automorphisms, and in this approach the analogy makes sense even in type II_1 settings.

This kind of central sequence approach is useful for the classification of group actions on subfactors. For example, there are two and only two outer actions of $\mathbf{Z}_2$ on subfactors of type A_{2n+1}, and one is given by the standard form and the other is given by the orbifold construction. There is also an analogue of modular obstruction. For more on this classification, see [54, §5].

A basic philosophy in the above type of analogy is that any phenomenon on type III (single) factors appears in type II_1 subfactor in a discrete form. A mixture of the both will appear in type III subfactors. Winsløw's simultaneous generalization of the Connes-Takesaki module and Loi invariant in [81] is the first example of such a mixture.

It is a very important problem to distinguish non-amenable subfactors with the same indices using $\chi(M,N)$ as an analogue of Connes' construction in [16] for single II_1 factors. Our exact sequence [54, Theorem 1.1] is expected to be useful for this purpose, but so far, no example has been computed in the non-amenable subfactor case.

Finally, we discuss the problem of finding non-commutative symmetries of paragroups. In Xu's construction [83], the finite group acting on subfactors arising from RCFT is given as the center of the Lie group with which we started. In particular, this finite group is always abelian. In general, there seems to be no reason why a symmetry group of a given paragroup is abelian and the problem to find an interesting non-abelian symmetry has caught attention of physicists. From the viewpoint of automorphisms, $\chi(M,N)$ is always abelian, but $\mathrm{Ct}(M,N)/\mathrm{Int}(M,N)$ is not necessary abelian. (A trivial example is a subfactor $R^G \subset R$ with G finite, non-commutative, and acting on R outerly.) We feel that "interesting" symmetries should give $\chi(M,N)$ and so are commutative by some deeper reason, but nothing is clear at present.

References

[1] J. W. Alexandar, The combinatorial theory of complexes, *Ann. Math.* **(2) 31** 294–322, (1930).

[2] G. E. Andrews, R. J. Baxter, & P. J. Forrester, Eight vertex SOS model and generalized Rogers-Ramanujan type identities, *J. Stat. Phys.* **35** 193–266, (1984).

[3] H. Araki & D. E. Evans, On a C^*-algebra approach to phase transition in the two dimensional Ising model, *Comm. Math. Phys.* **106** 241–266, (1986).

[4] J. Bion-Nadal, An example of a subfactor of the hyperfinite II_1 factor whose principal graph invariant is the Coxeter graph E_6, in "Current Topics in Operator Algebras",

World Scientific Publishing, pp. 104–113, (1991).
[5] B. Blackadar, Symmetries of the CAR algebra, *Ann. Math.* **131** 589–623, (1990).
[6] O. Bratteli, G. A. Elliott, D. E. Evans, & A. Kishimoto, Non-commutative spheres, *Inter. J. Math.* **2** 139–166, (1991).
[7] O. Bratteli, G. A. Elliott, D. E. Evans, & A. Kishimoto, Finite group actions on AF algebras obtained by folding the interval, (to appear in *K-theory*).
[8] O. Bratteli, G. A. Elliott, D. E. Evans, & A. Kishimoto, Non-commutative spheres II: rational rotations, (to appear in *J. Operator Theory*).
[9] O. Bratteli, D. E. Evans, & A. Kishimoto, Crossed products of totally disconnected spaces by $\mathbf{Z}_2 * \mathbf{Z}_2$, preprint, Swansea, (1991).
[10] O. Bratteli & A. Kishimoto, Non-commutative spheres III: irrational rotations, preprint, Trondheim, (1991).
[11] J. de Boer & J. Goeree, Markov traces and II_1 factors in conformal field theory, *Comm. Math. Phys.* **139** 267–304, (1991).
[12] A. Cappelli, C. Itzykson, & J.-B. Zuber, The *A-D-E* classification of minimal and $A_1^{(1)}$ conformal invariant theories, *Comm. Math. Phys.* **113** 1–26, (1987).
[13] A. L. Carey & D. E. Evans, The operator algebras of the two dimensional Ising model, in "Braids" eds. J. Birman, A. Libgober, *Comtemp. Math.* **78** 117–165, (1988).
[14] M. Choda, Duality for finite bipartite graphs, to appear in *Pac. J. Math.*
[15] M. Choda & H. Kosaki, Strongly outer actions for inclusion of factors, preprint, (1992).
[16] A. Connes, Sur la classification des facteurs de type II, *C. R. Acad. Sc. Paris* **281** 13–15, (1975).
[17] A. Connes, Outer conjugacy classes of automorphisms of factors, *Ann. Sci. École Norm. Sup.* **8** 383–419, (1975).
[18] A. Connes, On the classification of von Neumann algebras and their automorphisms, *Symposia Math.* **XX** 435–478, (1976).
[19] A. Connes & D. E. Evans, Embeddings of $U(1)$-current algebras in non-commutative algebras of classical statistical mechanics, *Comm. Math. Phys.* **121** 507–525, (1989).
[20] A. Connes & M. Takesaki, The flow of weights on factors of type III, *Tohoku Math. J.* **29** 473–555, (1977).
[21] E. Date, M. Jimbo, T. Miwa, & M. Okado, Solvable lattice models, in "Theta functions — Bowdoin 1987, Part 1," Proc. Sympos. Pure Math. Vol. 49, Amer. Math. Soc., Providence, R.I., pp. 295–332.
[22] P. Di Francesco & J.-B. Zuber, $SU(N)$ lattice integrable models associated with graphs, *Nucl. Phys.* **B338** 602–646, (1990).
[23] P. Di Francesco & J.-B. Zuber, $SU(N)$ lattice integrable models and modular invariance, (to appear in Proc. Trieste Conf. Recent Developments in Conformal Field Theories, Trieste, 1989).
[24] R. Dijkgraaf & E. Verlinde, Modular invariance and the fusion algebra, Proceedings of the Annecy Conference on Conformal Field Theory. *Nucl. Phys.* B(Proc. Suppl.) **5B** 87–97, (1988).
[25] R. Dijkgraaf & E. Witten, Topological gauge theories and group cohomology, *Comm. Math. Phys.* **129** 393–429, (1990).
[26] L. Dixon, J. A. Harvey, C. Vafa, & E. Witten, Strings on orbifolds, *Nucl. Phys.* **B261**, 678–686, (1985); **B274** 285–314, (1986).
[27] B. Durhuus, H. P. Jakobsen, & R. Nest, Topological quantum field theories from generalized 6j-symbols, preprint, (1991).
[28] D. E. Evans, The C^*-algebras of topological Markov chains, Tokyo Metropolitan University Lecture Notes, (1984).
[29] D. E. Evans, Quasi-product states on C^*-algebras, in "Operator algebras and their connections with topology and ergodic theory", Springer Lecture Notes in Math., **1132** 129–151, (1985).
[30] D. E. Evans, C^*-algebraic methods in statistical mechanics and field theory, *Inter. J. Mod. Phys.* **4** 1069–1118, (1990).

[31] D. E. Evans & J. D. Gould, Embeddings and dimension groups of non-commutative AF algebras associated to models in classical statistical mechanics, preprint, Swansea.
[32] D. E. Evans & Y. Kawahigashi, Orbifold subfactors from Hecke algebras, preprint, (1992).
[33] D. E. Evans & Y. Kawahigashi, From subfactors to 3-dimensional topological quantum field theories and back, preprint, (1992).
[34] D. E. Evans & Y. Kawahigashi, The E_7 commuting squares produce D_{10} as principal graph, preprint, (1992).
[35] D. E. Evans & J. T. Lewis, On a C*-algebra approach to phase transition in the two-dimensional Ising model II, *Comm. Math. Phys.* **102** 521–535, (1986).
[36] P. Fendley, New exactly solvable orbifold models, *J. Phys.* **A22** 4633–4642, (1989).
[37] P. Fendley & P. Ginsparg, Non-critical orbifolds, *Nucl. Phys.* **B324** 549–580, (1989).
[38] F. Goodman, P. de la Harpe, & V. F. R. Jones, "Coxeter graphs and towers of algebras", MSRI publications 14, Springer, (1989).
[39] M. Gross & S. Varsted, Elementary moves and ergodicity in D-dimensional simplicial quantum gravity, *Nucl. Phys.* **B378** 367–380, (1992).
[40] U. Haagerup, in preparation.
[41] U. Haagerup & E. Størmer, Pointwise inner automorphisms of von Neumann algebras (with an appendix by C. Sutherland), *J. Funct. Anal.* **92** 177–201, (1990).
[42] T. Hayashi, Quantum group symmetry of partition functions of IRF models and their application to Jones' index theory, preprint, (1992).
[43] M. Izumi, Application of fusion rules to classification of subfactors, *Publ. RIMS Kyoto Univ.* **27** 953–994, (1991).
[44] M. Izumi, On flatness of the Coxeter graph E_8, to appear in *Pac. J. Math.*
[45] M. Izumi & Y. Kawahigashi, Classification of subfactors with the principal graph $D_n^{(1)}$, to appear in *J. Funct. Anal.*
[46] M. Jimbo, T, Miwa, & M. Okado, Solvable lattice models whose states are dominant integral weights of $A_{n-1}^{(1)}$, *Lett. Math. Phys.* **14** 123–131, (1987).
[47] V. F. R. Jones. Notes on Connes' invariant $\chi(M)$, unpublished.
[48] V. F. R. Jones, Index for subfactors, Invent. Math. **72** 1–15, (1983).
[49] V. F. R. Jones, A polynomial invariant for knots via von Neumann algebras Bull. Amer. Math. Soc. **12** 103–112, (1985).
[50] A. Kato, Classification of modular invariant partition functions in two dimensions, Mod. Phys. Lett. **A 2** 585–600, (1987).
[51] Y. Kawahigashi, Automorphisms commuting with a conditional expectation onto a subfactor with finite index, to appear in *J. Operator Theory.*
[52] Y. Kawahigashi, On flatness of Ocneanu's connections on the Dynkin diagrams and classification of subfactors, University of Tokyo, preprint, (1990).
[53] Y. Kawahigashi, Exactly solvable orbifold models and subfactors, to appear in Proceedings of the Conference on Functional Analysis and Related Topics, Kyoto, 1991.
[54] Y. Kawahigashi, Centrally trivial automorphisms and an analogue of Connes' $\chi(M)$ for subfactors, preprint, (1992).
[55] Y. Kawahigashi, C. E. Sutherland, & M. Takesaki, The structure of the automorphism group of an injective factor and the cocycle conjugacy of discrete abelian group actions, *Acta Math.* **169** 105–130, (1992).
[56] A. N. Kirillov & N. Yu. Reshetikhin, Representations of the algebra $U_q(sl_2)$, q-orthogonal polynomials and invariants for links, in "Infinite dimensional Lie algebras and groups" (V. G. Kac, ed.), Adv. Ser. in Math. Phys., vol. 7, pp. 285–339, (1988).
[57] H. Kosaki, Automorphisms in irreducible decompositions of sectors, in this volume.
[58] R. Kuik, On the q-state Potts model by means of non-commutative algebras, Thesis Groningen, (1986).
[59] A. Kumjian, An involutive automorphism of the Bunce-Deddens algebra, *CR Math. Rep. Acad. Sci. Canada* **10** 217–218, (1988).
[60] J. T. Lewis, P. N. M. Sisson, A C^*-algebra of the 2-dimensional Ising model, *Comm. Math. Phys.* **44** 279–292, (1975).

[61] P. H. Loi, On automorphisms of subfactors, preprint, (1990).
[62] R. Longo, Simple injective subfactors, *Adv. Math.* **63** 152–171, (1987).
[63] G. Moore & N. Seiberg, Classical and quantum conformal field theory, *Comm. Math. Phys.* **123** 177–254, (1989).
[64] A. Ocneanu, Quantized group, string algebras and Galois theory for algebras, in "Operator algebras and applications, Vol. 2 (Warwick, 1987)," London Math. Soc. Lect. Note Series Vol. 136, Cambridge University Press, pp. 119–172, (1988).
[65] A. Ocneanu, "Graph geometry, quantized groups and nonamenable subfactors", Lake Tahoe Lectures, June–July, (1989).
[66] A. Ocneanu, "Quantum symmetry, differential geometry of finite graphs and classification of subfactors", University of Tokyo Seminary Notes 45, (Notes recorded by Y. Kawahigashi), (1991).
[67] A. Ocneanu, An invariant coupling between 3-manifolds and subfactors, with connections to topological and conformal quantum field theory, unpublished announcement, (1991).
[68] S. Okamoto, Invariants for subfactors arising from Coxeter graphs, in "Current Topics in Operator Algebras," World Scientific Publishing, pp. 84–103, (1991).
[69] V. Pasquier, Operator content of the ADE lattice models, *J. Phys. A. Math. Gen.* **20** 5701-5717, (1987).
[70] S. Pirogov, States associated with the two dimensional Ising model, *Theor. Math. Phys.* **11(3)** 614–617, (1972).
[71] S. Popa, Classification of amenable subfactors of type II, IHES preprint, (1992).
[72] S. Popa, Classification of actions of discrete amenable groups on amenable subfactors of of type II, IHES preprint, (1992).
[73] Ph. Roche, Ocneanu cell calculus and integrable lattice models, *Comm. Math. Phys.* **127** 395–424, (1990).
[74] J. Schou, Commuting squares and index for subfactors, Ph. D. Thesis, Odense University, 1990.
[75] V. S. Sunder, A model for AF-algebras and a representation of the Jones projections, *J. Operator Theory* **18** 289–301, (1987).
[76] V. S. Sunder & A. K. Vijayarajan, On the non-occurrence of the Coxeter graphs β_{2n+1}, E_7, D_{2n+1} as principal graphs of an inclusion of II_1 factors, (to appear in Pac. J. Math.).
[77] C. E. Sutherland & M. Takesaki, Actions of discrete amenable groups on injective factors of type III_λ, $\lambda \neq 1$, *Pac. J. Math.* **137** 405–444, (1989).
[78] V. G. Turaev & O. Y. Viro, State sum invariants of 3-manifolds and quantum $6j$-symbols, preprint.
[79] M. Wakui, On Dijkgraaf-Witten invariant for 3-manifolds, (to appear in Osaka J. Math.).
[80] H. Wenzl, Hecke algebras of type A_n and subfactors, *Invent. Math.* **92** 345–383, (1988).
[81] C. Winsløw, Approximately inner automorphisms on inclusions of type III_λ factors, preprint, (1992).
[82] E. Witten, Gauge theories and integrable lattice models, *Nucl. Phys.* **B322** 629–697, (1989).
[83] F. Xu, Orbifold construction in subfactors, preprint, (1992).
[84] S. Yamagami, A note on Ocneanu's approach to Jones' index theory, preprint, (1992).

5. Operator Algebras and Related Topics

A Classification of Certain Simple C*-Algebras

GEORGE A. ELLIOTT
Mathematics Institute, Universitetsparken 5, 2100 Copenhagen, Denmark
and
Department of Mathematics, University of Toronto, Toronto, Canada M5S 1A1

Abstract. Certain separable simple C*-algebras which are inductive limits of algebras of type I (and are therefore semifinite) are shown to be classified by K-theoretical and tracial invariants. The question is left open whether this holds for all such simple C*-algebras.

Key words: matrix algebras over the interval, C*-algebra inductive limits, K-theoretical invariants, classification

1. In [8] and [9], certain inductive limits of separable C*-algebras of type I were classified by K-theory (in a way similar to AF algebras in [7]). These limits were assumed to be of real rank zero (i.e., the invertible selfadjoint elements dense — see [5]).

In [2], for simple unital C*-algebras which are inductive limits of a certain kind of separable C*-algebra of type I (including those considered in [8] and [9]), having real rank zero was shown to be equivalent to having a single tracial state for each state on K_0.

This suggests that, considering only the case of simple unital inductive limits, one might try to drop the property of real rank zero (i.e., traces determined by K_0), and just adjoin the simplex of traces to the invariant. As shown recently in [10], this simplex may be an arbitrary one.

2. The purpose of the present note is to initiate this project. For simplicity, — to display what is new most clearly — , attention will be restricted to very special type I algebras, namely, finite direct sums of matrix algebras over the interval (i.e., the tensor products of finite-dimensional C*-algebras with $C([0,1])$). One may hope that it will be possible to deal with other cases by similar methods.

In the case of just a single matrix algebra over $C([0,1])$ at each stage, the classification theorem takes a particularly simple form: the invariant is a subgroup of $\mathbf{Q}$ containing 1 (the order unit K_0-group), together with a simplex. (It was already in this case that Thomsen showed in [10] that the simplex may be arbitrary.)

In the case of finite direct sums at each stage, it is necessary to keep track of the pairing between traces and K_0.

Theorem. *Let A and B be separable simple unital C*-algebras each of*

H. Araki et al. (eds.), Quantum and Non-Commutative Analysis, 373–385.

which is the inductive limit of a sequence of type I C^*-algebras of the form $C([0,1]) \otimes$ *(finite-dimensional). Let*

$$\phi_0 : \mathrm{K}_0(A) \to \mathrm{K}_0(B)$$

be an isomorphism of ordered groups preserving the class of the unit, and let

$$\phi_{\mathrm{T}} : \mathrm{T}(B) \to \mathrm{T}(A)$$

be a continuous affine isomorphism of the simplex of tracial states of B onto that of A. Suppose that ϕ_0 and ϕ_{T} are compatible, in the sense that

$$\langle \phi_0 g, \tau \rangle = \langle g, \phi_{\mathrm{T}} \tau \rangle, \qquad g \in \mathrm{K}_0(A), \quad \tau \in \mathrm{T}(B)\,.$$

It follows that there exists an isomorphism

$$\phi : A \to B$$

which gives rise to the maps ϕ_0 and ϕ_{T}.

3. **Remark.** A modification of the proof of Theorem 2 yields the analogous statement in which $\phi_0, \phi_{\mathrm{T}}$, and ϕ are just asserted to be homomorphisms (rather than isomorphisms).

4. **Proof of Theorem 2.** Roughly speaking, the proof is an intertwining argument, like that of [7], but of an approximate nature — like that of [8] and [9]. (This is not surprising, since in the case of real rank zero Theorem 2 is included in Theorem 7.1 of [8].)

What is different is that now the intertwining must take the traces into account, too, as well as the K-groups. The techniques required for this, it turns out, are just those developed by Thomsen in [10], in connection with the construction of an inductive limit algebra with specified simplex of traces. (The argument in [10] is an embryonic form of the present one, and in fact suggested the present one.)

More precisely, the existence theorem given below (Theorem 5) is essentially contained in [10]. The uniqueness theorem, following that (Theorem 6), is based on the case of unique trace dealt with in [2].

These two results are combined as follows. Let $A_1 \to A_2 \to \cdots$ and $B_1 \to B_2 \to \cdots$ be sequences of finite direct sums of the basic building blocks $\mathrm{M}_k(\mathrm{C}([0,1]))$, $k = 1, 2, \ldots$, with inductive limits A and B. Passing to subsequences as in [7] we may suppose that the isomorphism ϕ_0 lifts to a commutative diagram

$$\begin{array}{ccccccc}
\mathrm{K}_0 A_1 & \to & \mathrm{K}_0 A_2 & \to & \cdots & \to & \mathrm{K}_0 A \\
\downarrow & \nearrow & \downarrow & \nearrow & & & \downarrow\uparrow \\
\mathrm{K}_0 B_1 & \to & \mathrm{K}_0 B_2 & \to & \cdots & \to & \mathrm{K}_0 B
\end{array}$$

in which the vertical maps are positive and preserve the class of the unit. (This uses that K_0 of a basic building block is $\mathbf{Z}$.)

The main step is now to lift in a similar way the isomorphism ϕ_{T}, to a diagram

$$\begin{array}{ccccccc} \mathrm{T}A_1 & \leftarrow & \mathrm{T}A_2 & \leftarrow & \cdots & \leftarrow & \mathrm{T}A \\ \uparrow & \swarrow & \uparrow & \swarrow & & & \uparrow\downarrow \\ \mathrm{T}B_1 & \leftarrow & \mathrm{T}B_2 & \leftarrow & \cdots & \leftarrow & \mathrm{T}B \end{array}$$

(after passing to subsequences). We do not expect this diagram to be exactly commutative, but only approximately so, — in a suitable sense. This lifting can most easily be achieved (and the approximate commutativity most easily even defined) by formulating the problem in terms of the affine function spaces: We wish to have a diagram of unital positive maps

$$\begin{array}{ccccccc} \mathrm{C}(X_1) & \rightarrow & \mathrm{C}(X_2) & \rightarrow & \cdots & \rightarrow & \mathrm{Aff}\,(\mathrm{T}A) \\ \downarrow & \nearrow & \downarrow & \nearrow & & & \downarrow\uparrow \\ \mathrm{C}(Y_1) & \rightarrow & \mathrm{C}(Y_2) & \rightarrow & \cdots & \rightarrow & \mathrm{Aff}\,(\mathrm{T}B) \end{array}$$

which is approximately commutative, in a fairly obvious sense, made precise in Theorem 2.1 of [8] (and recalled below). Here, X_i is the spectrum of A_i, and Y_i that of B_i, and the horizontal sequences are the canonical ones, with the normed spaces $\mathrm{Aff}\,(\mathrm{T}A)$ and $\mathrm{Aff}\,(\mathrm{T}B)$ as inductive limits.

If the sets X_i and Y_i were finite, the reformulated problem would be trivial; one would have an exact intertwining (of subsequences), just as for K_0 (by the same argument, with $\mathbf{R}$ in place of $\mathbf{Z}$). The present case, with X_i and Y_i finite unions of closed intervals, can be reduced to this case, (by an application of Theorems 2.2 and 2.1 of [8] — recalled below), by factorising the identity map on each $\mathrm{C}(X_i)$ and $\mathrm{C}(Y_i)$ through a finite direct sum of copies of $\mathrm{C}(\{0\})$, approximately, by unital positive maps. This can be done for $\mathrm{C}(X)$ for any compact space X. (For any given finite subset F of $\mathrm{C}(X)$, and $\epsilon > 0$, choose a finite open cover of X over each set of which an element of F varies by at most ϵ, choose a partition of unity (g_p) subordinate to that cover, and choose x_p in the open set of the cover supporting g_p. The map

$$\mathrm{C}(X) \ni f \mapsto (f(x_p)) \mapsto \Sigma f(x_p) g_p \in \mathrm{C}(X)$$

approximately factorises the identity on $\mathrm{C}(X)$ through $\bigoplus_p \mathrm{C}(\{0\})$ within ϵ on the finite set F.)

It follows from the compatibility of ϕ_0 and ϕ_T that the intertwinings of K_0-groups and trace simplices constructed above are at least approximately compatible, — in a suitable sense (after possibly passing to subsequences).

Rather than formulate what this means, let us just alter the simplex intertwining (at the same time passing to subsequences) to make this compatibility exact. Recall that approximate commutativity of the diagram above in the sense of Theorem 2.1 of [8] just means that starting from any fixed element x, any two paths in the diagram which coincide for sufficiently many steps to begin with, and end in the same place, approximately agree on x. It is easy to see (and this is Theorem 2.2 of [8]) that this holds if each triangle approximately commutes, on a sufficiently large finite set, to within a sufficiently fine tolerance. Here, sufficiently large means that the finite set at each stage contains all images of earlier sets, and the union of all images at infinity, in both $\mathrm{Aff}\,(\mathrm{T}A)$ and $\mathrm{Aff}\,(\mathrm{T}B)$, is dense. Sufficiently fine means that the sum of the tolerances at all stages is finite. It follows that, on changing the vertical arrows in a way that tends to zero at infinity, on each fixed element, after passing to subsequences of (X_i) and (Y_i) (to ensure summability of the new tolerances) the diagram is still approximately commutative. We shall now construct a canonical perturbation of each vertical map, asymptotically negligible in this sense, which makes it exactly compatible with the corresponding vertical maps of K_0. (More precisely, in order to do this, we must first pass to subsequences, twice, as described below.)

First, since A and B are by hypothesis simple, so that no element of $\mathrm{T}(A)$ or $\mathrm{T}(B)$ is zero on any nonzero positive element of A or B, after passing to direct summands of each A_i and B_i, so that no nonzero positive element of K_0A_i or K_0B_i is zero in K_0A or K_0B, we may pass to subsequences of (A_i) and (B_i) and suppose that the positive maps $\mathrm{C}(X_i) \to \mathrm{C}(Y_i)$ and $\mathrm{C}(Y_i) \to \mathrm{C}(X_{i+1})$ are strictly faithful on the images of K_0A_i and K_0B_i, in the sense that they map nonzero positive elements into strictly positive elements. The canonical approximation to $\mathrm{C}(X_i) \to \mathrm{C}(Y_i)$ (or $\mathrm{C}(Y_i) \to \mathrm{C}(X_{i+1})$) alluded to above then consists in considering the restriction to each direct summand of $\mathrm{C}(X_i)$ (or $\mathrm{C}(Y_i)$) isomorphic to $\mathrm{C}([0,1])$, and renormalising the restricted map, by first dividing by the image of 1, and then multiplying by the element of $\mathrm{C}(Y_i)$ (or $\mathrm{C}(X_{i+1})$) determined by the image in K_0B_i (or K_0A_{i+1}) of the class in K_0A_i (or K_0B_i) of the unit of the direct summand of A_i (or B_i) under consideration. If, at each stage, we replace Y_i by Y_j (or X_{i+1} by X_{j+1}) for a suitable $j > i$, then this approximation is arbitrarily good.

By Theorem 5, below, each compatible pair of vertical maps (beginning with the pair $\mathrm{K}_0A_1 \to \mathrm{K}_0B_1$, $\mathrm{T}B_1 \to \mathrm{T}A_1$) arises approximately (to an arbitrarily close degree of approximation) from a finite family of unital C*-algebra homomorphisms (at the first stage, from A_1 to B_1), in the sense that each gives rise precisely to the map of K_0-groups, and the average, a positive map, gives rise, approximately, to the map of simplices of tracial states. Again changing the vertical maps $\mathrm{T}B_i \to \mathrm{T}A_i$ and $\mathrm{T}A_{i+1} \to \mathrm{T}B_i$ by small amounts, we may suppose that this genesis takes place exactly.

Using again that A and B are simple, we shall now show how to pass to subsequences of (A_i) and (B_i) in such a way that the finite averages of homomorphisms $A_i \to B_i$ and $B_i \to A_{i+1}$ obtained above, giving rise to the pairs of maps $\mathrm{K}_0 A_i \to \mathrm{K}_0 B_i$, $\mathrm{T}B_i \to \mathrm{T}A_i$ and $\mathrm{K}_0 B_i \to \mathrm{K}_0 A_{i+1}$, $\mathrm{T}A_{i+1} \to \mathrm{T}B_i$, may be replaced by single homomorphisms.

What we shall use is that for each i, for each $\epsilon > 0$, and for each finite set $F \subseteq A_i$ (resp. $F' \subseteq B_i$) there exist $j \geq i$ and a homomorphism $A_i \to A_j$ (resp. $B_i \to B_j$) agreeing with the given one within ϵ on F (resp. F') and with image commuting with a finite-dimensional unital sub-C*-algebra of A_j (resp. B_j) each direct summand of which is of arbitrarily large order. (This is equivalent to saying that A and B are approximately divisible C*-algebras in the sense of [3], but the fact that A and B have this property is not contained among the results of [3].) This fact (say for A) is proved as follows. It is enough to prove the assertion made above just for the algebra A_1, and it is easy to reduce the statement to the case $A_1 = \mathrm{C}([0,1])$. As we shall show below, simplicity of the inductive limit requires that for each $\delta > 0$ the eigenvalues of the image of the canonical selfadjoint generator h $(= t \mapsto t)$ in each primitive quotient of A_j for suitably large j are δ-dense. At the same time, by Theorem 4 of [6] we may suppose, perturbing the image of h by an arbitrarily small amount, that these eigenvalues are distinct. For any given integer $n > 0$, perturbing the eigenvalues of the image of h now to have multiplicity between n and $2n$ (for small δ), — for example, with all eigenvalues of multiplicity n except the largest which has multiplicity between n and $2n-1$, — so that the eigenvalues may now only be $2n\delta$-dense, we have a perturbation of the image of h by at most $2n\delta$ which commutes with a finite-dimensional sub-C*-algebra each minimal direct summand of which has order at least n. Since n may be chosen arbitrarily large, and for fixed n, δ may be chosen so that $2n\delta$ is arbitrarily small, for sufficiently large j, this proves the assertion.

The statement that simplicity requires approximate density in $[0,1]$ of the eigenvalues of the image of h in each primitive quotient of A_j for large j is made with the understanding that the maps in the sequence $A_1 \to A_2 \to \cdots$ are injective. A given sequence representing A (or B) may be replaced with one with this property by Theorem 4.3 of [8] — or, rather, by a slight modification of this theorem with the circle replaced by the interval (the proof of which is the same).

Given that the sequence $A_1 \to A_2 \to \cdots$ consists of injective maps, and that the inductive limit, A, is simple, let us prove the statement made above concerning the approximate density of the eigenvalues of h in the interval $[0,1]$. By functional calculus, we see that it is sufficient to show that each nonzero element of A_1 has nonzero image in each primitive quotient of A_j for all sufficiently large j. If an element x of A_1 maps into zero in some primitive quotient of A_j for every j, then its image generates a proper closed two-sided ideal in each A_j, and therefore (as A is unital) also in A. The image of such an element x in A is then zero, since A is simple, and by injectivity x is zero in A_1. This proves the assertion.

It follows by Theorem 2.1 of [8] that we may pass to subsequences and replace the given homomorphisms $A_i \to A_{i+1}$ and $B_i \to B_{i+1}$ by ones constructed as above in such a way that the image of A_i in A_{i+1} (resp. of B_i in B_{i+1}) commutes with a finite-dimensional unital sub-C*-algebra with all direct summands of arbitrarily large order, — in particular, of order substantially larger than the number of unital homomorphisms from B_{i-1} to A_i (resp. from A_i to B_i), each of which gives the map on K_0 and the average of which gives the map on traces. Hence as in Lemma 3.5 of [10], we may replace each finite family of homomorphisms by a single one, a block diagonal direct sum of cutdowns of these. This requires a new approximation, since the orders of the simple finite-dimensional subalgebras may not be exactly multiples of the number of homomorphisms at each stage, but we may change the vertical simplex maps as before so that they are exactly equal to the ones arising from the single vertical homomorphisms.

To summarise, we now have a diagram

$$\begin{array}{ccccccc} A_1 & \to & A_2 & \to & \cdots & \to & A = \varinjlim A_i \\ \downarrow & \nearrow & \downarrow & \nearrow & & & \\ B_1 & \to & B_2 & \to & \cdots & \to & B = \varinjlim B_i \end{array}$$

of unital C*-algebra homomorphisms, such that the induced diagram at the level of K_0 commutes and gives rise to ϕ_0 in the limit, such that the induced diagram at the level of tracial simplices approximately commutes (in the sense of Theorem 2.1 of [8], in terms of the normed affine function spaces) and gives rise to ϕ_{T} in the limit, and such that the K_0 and trace mappings at each stage are compatible.

We may suppose that all the homomorphisms are injective. (For the horizontal ones, this was shown above. For the vertical ones, we may achieve this by adding a copy of the identity in one diagonal block at the appropriate point in the construction of each such map.)

Using again that A and B are simple, we may pass to subsequences of (A_i) and (B_i) such that the hypotheses of Theorem 6 are fulfilled, with respect to each pair of maps from A_i to A_{i+1} or from B_i to B_{i+1} (either directly across, or around the triangle), and with respect to an integer n_i or m_i (respectively) for each i, which may be set at an arbitrarily large value, — after the choice of A_i or B_i (respectively). Simplicity is used, at each stage, after choosing a nonzero continuous function between 0 and 1 supported in each of the intervals of length $1/n$ in question, to proceed to a later stage at which all traces on the images of the corresponding central elements are at least half of the smallest value in the limit (i.e., achieved by traces on A or B), which is nonzero because of injectivity of the maps, and simplicity of the limit. (What is actually desired is that these relations also hold at all stages beyond this stage; this could be established in the same way (using that Aff TA and Aff TB are the inductive limits of the sequences of normed spaces $(C(X_i))$ and $(C(Y_i))$, but, in fact, as it happens, the persistence of these relations is automatic: the convergence of the smallest value of the trace is monotone (increasing).) This makes it possible to attain the desired approximation on the various central elements specified, between the two corresponding maps on traces. More explicitly, when at the stage A_i, with A_k and B_k chosen for $k < i$, one must replace B_i by an algebra farther along the sequence, to be able to make use of the approximate commutativity of the induced diagram at the level of tracial simplices. Similarly, at the stage B_i, one must replace A_{i+1} by an algebra farther out. A more detailed explanation follows.

The passage to subsequences of (A_i) and (B_i), described in outline in the preceding paragraph, which is needed to establish the hypotheses of Theorem 6, will now be described somewhat more fully. Keep the algebra A_1 as given. With n_1 arbitrary (to be specified in the paragraph which follows), divide the interval $[0, 1]$ into n_1 equal parts (subintervals), and choose a nonzero continuous function, with values between 0 and 1, supported in each of these n_1 parts. Consider the n_1 central elements of each minimal direct summand of A_1 corresponding to these n_1 functions, and note that the images in A of all these positive elements of A_1 are nonzero (by injectivity of the maps), and hence (by simplicity of A) are nonzero on any tracial state of A. It follows that there exists $i'_1 \geq 1$ such the images of all these elements of A_1 in $A_{i'_1}$ are nonzero on any tracial state of $A_{i'_1}$. Thus, with δ_1 the smallest value of a tracial state of $A_{i'_1}$ on one of these images, $\delta_1 > 0$. Clearly, for any $j > i'_1$, and for any unital map from $A_{i'_1}$ to A_j (not just the given one), also any tracial state of A_j is greater than or equal to δ_1 on the image of any of these elements in A_j (since such a trace defines also a trace on $A_{i'_1}$). By approximate commutativity of the diagram of affine function spaces, there exists $i_1 \geq i'_1$ such that, for any $i_2 > i_1$, for any two maps from A_{i_1} to A_{i_2} obtained by following the arrows in the diagram (including also vertical arrows), the two images in A_{i_2} of any one of the above specified elements of A_1, mapped first horizontally into A_{i_1}, and then into A_{i_2} by each of the two maps under consideration, agree to strictly within δ_1 on every tracial state of A_{i_2}. The two maps from A_{i_1} to A_{i_2} which will be of interest are the horizontal one, and the vertical one from A_{i_1} to B_{i_1}, followed by the horizontal one to B_{i_2-1}, followed finally by the vertical one from B_{i_2-1} to A_{i_2}. The index i_2, which is arbitrary, will now be specified. First, fix $m_1 > 0$, to be specified in the following paragraph. Just as for A_1, above, choose m_1 central elements of each direct summand of B_{i_1} (corresponding to m_1 nonzero positive continuous functions supported in each of the m_1 consecutive subintervals of $[0,1]$ of length $1/m_1$), and choose $i_2' > i_1$ such that the images of all these elements in $B_{i_2'}$ are nonzero on any tracial state of $B_{i_2'}$ — and hence are bounded below on tracial states by some $\delta_2 > 0$. Using approximate commutativity, choose $i_2 \geq i_2'$ such that, for any $i_3 > i_2 - 1$, for any two maps from B_{i_2-1} to B_{i_3} obtained by following the arrows in the diagram, the two images in B_{i_3} of any one of the above specified elements of B_{i_1}, mapped first horizontally into B_{i_2-1}, and then into B_{i_3} by the two maps under consideration, agree to strictly within δ_2 on every tracial state of B_{i_3}. The two maps from B_{i_2-1} to B_{i_3} which will be of interest are the horizontal one, and the composition of the vertical one from B_{i_2-1} to A_{i_2}, the horizontal one from A_{i_2} to A_{i_3}, and the vertical one from A_{i_3} to B_{i_3}. The index i_3, which is arbitrary, is to be specified in the next step. Continue in the same way (i.e., as in the two steps described above). The subsequences of $A_1 \to A_2 \to \cdots$ and $B_1 \to B_2 \to \cdots$ to be considered are $A_1 \to A_{i_2} \to A_{i_4} \to \cdots$, and $B_{i_1} \to B_{i_3} \to B_{i_5} \to \cdots$.

Hence by Theorem 6, below (applied in somewhat the same way as Theorem 5.3 of [8] in the proof of Theorem 7.1 of [8] — but the present situation is much simpler), we may pass to subsequences of (A_i) and (B_i) and change the vertical maps of the (resulting) diagram, one by one, in successive triangles, composing each one with an inner automorphism of its codomain algebra, in such a way that the diagram with new vertical homomorphisms is approximately commutative in the sense of Theorem 2.1 of [8]. (This is ensured by estimates on the individual triangles by Theorem 2.2 of [8].) Then by Theorem 2.1 of [8] the downward vertical maps converge to an isomorphism $\phi:\ A \to B$, which necessarily gives rise to the given maps ϕ_0 and ϕ_{T}.

5. **Theorem** (cf. Theorem 2.1 of [10]). *Let A_1 and A_2 be finite C*-algebra direct sums of matrix algebras over* $\mathrm{C}([0,1])$. *Let*

$$\phi_0:\ \mathrm{K}_0(A_1) \to \mathrm{K}_0(A_2)$$

be a homomorphism of ordered groups preserving the class of the unit. Let

$$\phi_{\mathrm{T}}:\ \mathrm{T}(A_2) \to \mathrm{T}(A_1)$$

be a continuous affine homomorphism of the simplex of tracial states of A_2 into that of A_1. Suppose that these two maps are compatible, with respect to the pairing between traces and projections (as in Theorem 2).

It follows that for every finite subset F of A_1 and every $\epsilon > 0$ there exists a finite family of unital homomorphisms from A_1 to A_2, each giving rise to the map ϕ_0 on K_0, *such that the average of the corresponding affine maps from $\mathrm{T}A_2$ to $\mathrm{T}A_1$, say ψ, agrees with ϕ_{T} up to ϵ when measured on F:*

$$|\,\langle a,\ \psi(\tau) - \phi_T(\tau)\rangle\,| \leq \epsilon, \qquad a \in F,\ \tau \in T(A_2)\,.$$

Proof. The theorem is easily reduced, as follows, to the case that A_1 and A_2 are both just $\mathrm{C}([0,1])$, in which case the statement is Theorem 2.1 of [10].

The normed spaces $\mathrm{Aff}\,(\mathrm{T}A_1)$ and $\mathrm{Aff}\,(\mathrm{T}A_2)$ are each the direct sum of finitely many copies of $\mathrm{C}([0,1])$, each one corresponding to a minimal direct summand of A_1 or A_2. The unit of each fixed minimal direct summand of A_1 corresponds on the one hand to $1 \in \mathrm{C}([0,1])$ and on the other hand to a positive element g of $\mathrm{K}_0(A_1)$, majorised by $[1]$. Under the two given maps, the element $1 \in \mathrm{C}([0,1])$ goes into a direct sum of elements of $\mathrm{C}([0,1])$, and the element g goes into the K_0-class of a projection e in A_2, and by assumption the various functions give the traces of e in the various components of A_2. The various projections e, corresponding to different minimal direct summands of A_1, may be chosen to be orthogonal with sum 1. To construct a unital homomorphism $A_1 \to A_2$, it is sufficient to construct one from each fixed minimal direct summand of A_1 to the corresponding subalgebra eA_2e. The map which takes a tracial state τ of eA_2e, first to the unique tracial state of A_2 which is proportional to τ on each minimal direct summand of A_2, then by the given map $\mathrm{T}A_2 \to \mathrm{T}A_1$ to a tracial state of A_1, and, finally, to the normalised restriction of this to the minimal direct summand of A_1 under consideration, is compatible with the restriction to this direct summand of the given map $\mathrm{K}_0A_1 \to \mathrm{K}_0A_2$. (It is sufficient to check this on the unit of this direct summand, which maps into e.) One therefore has the hypotheses of the theorem with this minimal direct summand in place of A_1. In a similar way we can reduce the theorem to the case that A_2 has a single minimal direct summand.

The reduction from the case that A_1 is a matrix algebra of order n over $\mathrm{C}([0,1])$ to the case that A_1 is equal to $\mathrm{C}([0,1])$ is clear. (Just take any unital homomorphism on the subalgebra $\mathrm{M}_n(\mathbf{C})$, and then pass to the relative commutants in A_1 and A_2; the hypotheses are still fulfilled — with the same maps ϕ_0 and ϕ_T.)

One can now, finally, reduce the theorem to the case that A_2 is $\mathrm{C}([0,1])$. (If A_2 is a matrix algebra of order m over $\mathrm{C}([0,1])$, and A_1 is just $\mathrm{C}([0,1])$, consider the identity map

$$\mathrm{K}_0A_1 \;=\; \mathbf{Z} \;\to\; \mathrm{K}_0\mathrm{C}([0,1]) \;=\; \mathbf{Z}\,,$$

which is equal to $1/m$ times the homomorphism $\phi_0:\ \mathrm{K}_0A_1 \to \mathrm{K}_0A_2$, together with the unital positive map

$$\mathrm{Aff}\,\mathrm{T}A_1 \;=\; \mathrm{C}([0,1]) \;\to\; \mathrm{Aff}\,\mathrm{T}A_2 \;=\; \mathrm{C}([0,1]) \;=\; \mathrm{Aff}\,\mathrm{T}(\mathrm{C}([0,1]))$$

dual to $\phi_\mathrm{T}:\ \mathrm{T}A_2 \to \mathrm{T}A_1$. These maps fulfil the hypotheses of the theorem with $\mathrm{C}([0,1])$ in place of A_2. If the conclusion of the theorem is known to hold in this case, for a given finite subset F of A_1 and $\epsilon > 0$, then the composition of the resulting unital homomorphisms $A_1 \to \mathrm{C}([0,1])$ with the canonical isomorphism from $\mathrm{C}([0,1])$ onto the centre of $A_2 = \mathrm{M}_m(\mathrm{C}([0,1]))$ verify the conclusion of the theorem in the case desired.)

6. **Theorem.** *Let A_1 and A_2 be finite C^*-algebra direct sums of matrix algebras over $\mathrm{C}([0,1])$. Let ϕ and ψ be unital homomorphisms from A_1 to A_2, giving rise to the same map from K_0A_1 to K_0A_2. Let $n > 0$ be an integer, and suppose that for some $\delta > 0$ the image under each of ϕ and ψ of the canonical selfadjoint central generator of each minimal direct summand of A_1 (i.e., the element corresponding to the function $t \mapsto t$ on $[0,1]$) in each primitive quotient of A_2 has at least the fraction δ of its eigenvalues in each of the n consecutive subintervals of $]0,1]$ of length $1/n$. Suppose that the maps from $\mathrm{T}A_2$ to $\mathrm{T}A_1$ arising from ϕ and ψ agree to strictly within δ on the n central elements of each minimal direct summand of A_1 corresponding to the functions which are equal to zero from 0 to r/n, equal to one from $(r+1)/n$ to 1, and linear in between, $r = 0, 1, \ldots, n-1$.*

It follows that there exists a unitary $u \in A_2$ such that ϕ and $(\mathrm{Ad}\,u)\psi$ agree to within $3/n$ on the canonical generators of A_1 (i.e., the matrices with constant entries 0 or 1, and the central selfadjoint elements referred to above).

Proof. It is enough to consider the case that $A_1 = \mathrm{C}([0,1])$ and $A_2 = \mathrm{M}_q(\mathrm{C}([0,1]))$ for some q. By Theorem 4 of [6], we may suppose that the eigenvalues of $\phi(h)$ and $\psi(h)$ at each point are distinct, where h denotes the function $t \mapsto t$ in A_1. Hence, modifying ϕ and ψ by composing them with inner automorphisms, we may suppose that $\phi(h)$ and $\psi(h)$ are diagonal, with eigenvalues at each point in increasing order from upper left to lower right.

Let us verify that, with this change of notation, ϕ and ψ themselves are within $3/n$ on the element h (which is what the specified set of generators has reduced to). In other words, corresponding eigenvalues of $\phi(h)$ and $\psi(h)$ at each point should be within $3/n$.

Denote by k_r the characteristic function of the interval $[r/n, 1]$, $r = 1, \ldots, n-1$, so that

$$h_r k_r = k_r,\; k_r h_{r+1} = h_{r+1}, \qquad r = 1, \ldots, n-1\,,$$

where $h_1, \ldots, h_n$ denote the n functions specified in the statement of the theorem. For each $r = 1, \ldots, n-2$, at each point of the spectrum of A_2, the (normalised) trace of $\phi(k_r)$ is less than that of $\phi(h_r)$, which is within δ of that of $\psi(h_r)$, which is less than that of $\psi(k_{r-1})$. In particular, at each point of $[0,1]$,

$$\operatorname{rank} \phi(k_r) \leq \operatorname{rank} \psi(k_{r-1}) + q\delta\,.$$

Similarly,

$$\operatorname{rank} \psi(k_r) \leq \operatorname{rank} \phi(k_{r-1}) + q\delta\,.$$

In other words, the number of eigenvalues of each of $\phi(h)$ and $\psi(h)$ which are at least r/n is no more than $q\delta$ more than the number of eigenvalues of the other which are at least $(r-1)/n$ $(r = 1, \ldots, n-1)$. Since by hypothesis there are at least $q\delta$ eigenvalues of each of $\phi(h)$ and $\psi(h)$ in each interval $[(r-1)/n, r/n]$ (at each point of the spectrum of A_2), the number of eigenvalues of $\phi(h)$ greater than or equal to r/n is no more than the number of eigenvalues of $\psi(h)$ greater than or equal to $(r-2)/n$, $r = 2, \ldots, n-1$, and similarly with ϕ and ψ interchanged. It follows by a simple combinatorial analysis that each eigenvalue of $\phi(h)$ is within $3/n$ of the corresponding eigenvalue of $\psi(h)$, as desired.

7. **Remark.** The hypothesis of simplicity in Theorem 2 may not be omitted. For example, the K-theory and tracial data of the non simple C*-algebra $C([0,1]) \otimes M_{2^\infty}$ are, by Theorem 3.9 of Thomsen in [10], exhibited also by a simple C*-algebra in the class considered in Theorem 2.

8. **Problem.** It would be interesting to determine the most general pairing of a simple unital dimension group with a simplex arising in Theorem 2. It is clearly a necessary condition that every positive unital map from the dimension group into **R** arise from an element of the simplex.

9. **Example.** The simple C*-algebra in the class dealt with in Theorem 2, with K_0-group equal to **Q** (or to a dense subgroup of **Q**), with distinguished K_0-class $1 \in \mathbf{Q}$, and with simplex of tracial states of dimension one (i.e., having two extreme points), is, as we shall see below, the crossed product of the UHF C*-algebra with the K_0-group **Q** by an automorphism of order two.

In particular, this shows that such a crossed product need not have real rank zero (in the sense of [5]). This is a refinement of the observation of Blackadar in [1] that such a crossed product need not be AF. (Blackadar's crossed product did have real rank zero.)

To see this, let us review Thomsen's construction of this C*-algebra in [10]. We shall see that it is possible to make the arbitrary choices permitted by Thomsen in a way which reveals a two-fold symmetry, interchanging the two extreme traces. It turns out that the associated crossed product C*-algebra is UHF. (The assertion above then follows by Takesaki-Takai duality.)

Consider the unital embedding of $C([0,1])$ into $M_2(C([0,1]))$ which takes the function $h\colon t \mapsto t$ into the diagonal direct sum of the two functions h_0 and h_1 described in the proof of Theorem 6 (in the case $n = 2$). Iterating this gives an inductive limit C*-algebra with two extremal traces (coming from evaluation at 0 and evaluation at 1 at each stage). This C*-algebra is not simple but a minor perturbation to the construction, as performed in [10] — which consists in passing to a subsequence, and adding at each stage an identity map and a pair of one-dimensional representations, corresponding to the pair $\{q_n, 1-q_n\}$ where (q_n) is a sequence of points in $[0,1]$ which is dense — makes the algebra simple, and does not change the tracial state space.

This construction is clearly compatible with flipping the interval. More precisely, there exists an automorphism α_n at the nth stage which is the composition of the interval flip and an inner automorphism, such that α_n and α_{n+1} are intertwined by the nth embedding.

The crossed product of the inductive limit is the inductive limit of the crossed
products. Each of these is a matrix algebra over the C*-algebra of continuous 2×2 matrix valued functions on the interval $[0,1/2]$ which are diagonal at $1/2$. It is easy to verify that this inductive limit has a unique tracial state. Of course, it is simple, and it follows by Theorem 1.3 of [3] (or rather, by the proof of this theorem, modified to take account of the multiple right hand endpoint) that this algebra has real rank zero. By the classification theorem of Su for real rank zero inductive limits of matrix algebras over finite graphs with multiple vertices, given in [11], it follows that this inductive limit is AF. Since K_0 of the algebra at each stage is the ordered group direct sum $\mathbf{Z} \oplus \mathbf{Z}$, to determine K_0 of the inductive limit it is necessary to compute. A simple computation (along the lines of [4]) shows that this group is of rank one. So the crossed product algebra is in fact UHF.

Acknowledgement. During the course of this work the author was Chercheur Invité at the Centre de Recherches Mathématiques of the Université de Montréal.

References

[1] B. Blackadar, Symmetries of the CAR algebra, *Ann. of Math.* **131** (1990), 589–623.
[2] B. Blackadar, O. Bratteli, G.A. Elliott, and A. Kumjian, Reduction of real rank in inductive limits of C*-algebras, *Math. Ann.* **292** (1992), 111–126.
[3] B. Blackadar, A. Kumjian, and M. Rørdam, Approximately central matrix units and the structure of noncommutative tori, preprint.
[4] O. Bratteli, G.A. Elliott, D.E. Evans, and A. Kishimoto, Actions of finite groups on AF algebras obtained by folding the interval, *K-Theory*, to appear.
[5] L.G. Brown and G.K. Pedersen, C*-algebras of real rank zero, *J. Funct. Anal.* **99** (1991), 130–149.
[6] M.-D. Choi and G.A. Elliott, Density of the self-adjoint elements with finite spectrum in an irrational rotation C*-algebra, *Math. Scand.* **67** (1990), 73–86.
[7] G.A. Elliott, On the classification of inductive limits of semisimple finite-dimensional algebras, *J. Algebra* **38** (1976), 29–44.
[8] G.A. Elliott, On the classification of C*-algebras of real rank zero, preprint.
[9] G.A. Elliott and G. Gong, On the classification of C*-algebras of real rank zero, II, in preparation.

[10] K. Thomsen, Inductive limits of interval algebras: The tracial state space, *Amer. J. Math.*, to appear.
[11] H. Su, On the classification of C*-algebras of real rank zero: inductive limits of matrix algebras over graphs, *C. R. Math. Rep. Acad. Sci. Canada* **13** (1991), 223–228.

K-theoretic Classification for Certain Real Rank Zero C*-algebras

HONGBING SU
University of Toronto, Toronto, M5S 1A1, CANADA

Dedicated to Professor Huzihiro Araki on his sixtieth birthday.

Abstract. In this paper a K-theoretic classification is given of the real rank zero C*-algebras that can be expressed as inductive limits of sequences of finite direct sums of matrix algebras over finite connected graphs (possibly with dimension drop). This class of C*-algebras provides torsion K_1. The special case that the graphs are intervals and the connecting maps are of the same homotopy type is due to Elliott.

Key words: K-theory, real rank zero C*-algebras, classification

1. Statement of the Result

1.1. Introduction

The problem we are interested in is to classify certain C*-algebras by their K-theory data. Given a C*-algebra A there are two abelian groups associated with it. Namely, $K_0(A)$, which is often an ordered group, and $K_1(A)$ [1]. The complete invariant in the present classification will be the graded group $K_* = K_0 \oplus K_1$, together with the distinguished subset consisting of the pairs $([e],[u])$, where e is a projection in the algebra and u is a partial unitary with support (and range) e [9]. Following [9], we shall call this set the graded dimension range.

The first result of this kind was the classification of inductive limits of sequences of finite direct sums of matrix algebras (called AF algebras), in which case K_1 is trivial [8]. The group K_0, together with the dimension range, is a complete invariant. In [9], this classification was extended. It was shown that if one replaces the matrix algebras by matrix algebras over the unit circle and restricts the limit C*-algebras to have real rank zero (see Definition 2), then K_*, as a graded group, together with the graded dimension range, is a complete invariant. Recently, it was shown in [10] that this class of C*-algebras contains all the irrational rotation C*-algebras (also see [7]). In [12], this classification theorem was extended further to the real rank zero C*-algebras that can be expressed as inductive limits of finite

H. Araki et al. (eds.), Quantum and Non-Commutative Analysis, 387–397.

direct sums of matrix algebras over finite non-Hausdorff graphs. The C*-algebras of this type are important in studying group actions [2][3][5][11].

The K-groups for the C*-algebras considered above are torsion free. In [9], it was shown that K_*, together with the dimension range, continues to classify the real rank zero C*-algebras arising as inductive limits of finite direct sums of matrix algebras over the unit interval with dimension drops (see Definition 1), with a certain restriction on the homotopy type of the connecting maps. This class of C*-algebras provides torsion groups for K_1.

In this paper, the classification theorem is extended to the real rank zero C*-algebras that can be expressed as inductive limits of finite direct sums of matrix algebras over finite graphs with dimension drops at some of their vertices (see Definition 1). As a consequence, the restriction imposed in [9] on the homotopy classes of maps is shown to be unnecessary.

1.2. Basic building blocks

We shall consider real rank zero C*-algebras that can be expressed as inductive limits of sequences of finite direct sums of basic building blocks defined as follows:

Definition Let X be a finite connected graph and denote by $C(X)$ the C*-algebra of complex-valued continuous functions on X. A C*-algebra A will be said to be a basic building block if A is isomorphic to a sub-C*-algebra of $C(X) \otimes M_n$ of the following form:

$$\{f \in C(X) \otimes M_n \mid f \in M_{n/k} \text{ at some of the vertices of } X\}$$

for some positive integer n and k such that k divides n.

A basic building block as in the definition will be said to have spectrum X with dimension drop k, or simply spectrum X. It is easy to see that $K_1(A)$ may have torsion component Z_k.

Example The following is an example of basic building block:

$$A = \{f \in M_2(C[0,1]) \mid f(0) \text{ is scalar}\}.$$

Here the graph is the unit interval with dimension drop 2 at 0.

1.3. Main theorem

Definition A unital C*-algebra is said to be of real rank zero if the set of invertible self-adjoint elements is dense in the set of all self-adjoint elements [6].

In this paper, we will only consider the case that all the C^*-algebras and all the $*$-homomorphisms are unital.

The main result of this paper is the following:

Theorem *Let A and B be two C^*-algebras of real rank zero. Suppose that A and B are the inductive limits of sequences (A_i) and (B_i) of finite direct sums of basic building blocks. Suppose that the graded groups $K_*(A)$ and $K_*(B)$ are isomorphic, in a way preserving the graded dimension range. Then A and B are isomorphic.*

2. Perturbation

The following Theorem 2.1 is a perturbation result for certain $*$- homomorphisms. The maps we are interested in are those from finite direct sum of basic building blocks to finite direct sum of basic building blocks. For simplicity, we will state the result in the case that each algebra is a single basic building block.

Theorem *Let A and B be two basic building blocks with spectra X and Y and with generic fibres M_n and M_m, respectively. For any unital $*$-homomorphism ϕ from A to B, any finite subset $F \subset A$ and any $\epsilon > 0$, there exists a unital $*$-homomorphism ϕ' from A to B such that*
(1) $\|\phi(f) - \phi'(f)\| < \epsilon$ for $f \in F$,
(2) On each edge of Y, identified with $I = [0,1]$,

$$\phi'(g)(t) = W(t) \begin{pmatrix} g(S_1(t)) & & \\ & \ddots & \\ & & g(S_p(t)) \end{pmatrix} W^*(t)$$

for all $g \in A$ and all $t \in I$, where $W(\cdot)$ is a unitary in $M_m(C(I))$ and $\{S_i(\cdot)\}_{i=1}^p$, called eigenvalue maps, are in $C(I,X)$.

3. Approximate intertwining

This section is taken from [9]

3.1. Approximate intertwining

Let $A_1 \to A_2 \to \cdots$ and $B_1 \to B_2 \to \cdots$ be sequences of C*-algebras. Let us say that a sequence of C*-algebra homomorphisms $A_1 \to B_1 \to A_2 \to B_2 \to \cdots$ is an approximate intertwining of the diagram

$$\begin{array}{ccccc} A_1 & \longrightarrow & A_2 & \longrightarrow & \cdots \\ \downarrow & \nearrow & \downarrow & \nearrow & \\ B_1 & \longrightarrow & B_2 & \longrightarrow & \cdots \end{array}$$

if the diagram is approximately commutative, in the sense that for any fixed element of any A_i (or B_i), the difference of the images of this element along two different paths in the diagram, starting at A_i (or B_i) and ending at the same place, converges to zero as the number of steps for which the two paths coincide, starting at the beginning, becomes large [9].

3.2. Isomorphism theorem

Theorem *Suppose that there exists an approximate intertwining of the sequences of C^*-algebras $A_1 \to A_2 \to \cdots$ and $B_1 \to B_2 \to \cdots$. Then the inductive limit C^*-algebras $A = \varinjlim A_i$ and $B = \varinjlim B_i$ are isomorphic.*

3.3. Separable case

Theorem *Let $A_1 \to A_2 \to \cdots$ and $B_1 \to B_2 \to \cdots$ be sequences of separable C^*-algebras. Let $a_1 = (a_{1,n})$, $a_2 = (a_{2,n}), \ldots$ and $b_1 = (b_{1,n})$, $b_2 = (b_{2,n}), \ldots$ be dense sequences in $A_1, A_2, \ldots$ and $B_1, B_2, \ldots$, respectively. Let $A_1 \to B_1 \to A_2 \to B_2 \to \cdots$ be a sequence of C^*-algebra homomorphisms. For each $i = 1, 2, \ldots$ denote by S_i the finite subset of A_i consisting of the images in A_i of the first i terms of the sequences $a_1, a_2, \ldots, a_{i-1}$ and $b_1, b_2, \ldots, b_{i-1}$, along all possible paths in the diagram*

$$\begin{array}{ccccc} A_1 & \longrightarrow & A_2 & \longrightarrow & \cdots \\ \downarrow & \nearrow & \downarrow & \nearrow & \\ B_1 & \longrightarrow & B_2 & \longrightarrow & \cdots \end{array}$$

Similarly, denote by T_i the finite subset of B_i consisting of the images in B_i of the first i terms of the sequences $a_1, a_2, \ldots, a_i$ and $b_1, b_2, \ldots, b_{i-1}$, along all possible paths in the diagram. Suppose that for each i and for each $a \in S_i$, the difference between the two images of a in A_{i+1} (along the two paths going straight across and down up) is of norm at most 2^{-i}. Suppose, similarly, that for each i and for each $b \in T_i$, the difference between the two images of b in B_{i+1} (along the two paths going straight across and up down) is of norm at most 2^{-i}. Then the sequence $A_1 \to B_1 \to A_2 \to B_2 \to \cdots$ is an approximate intertwining.

4. Special graphs and special sequences

In this section, we shall show that one can replace arbitrary graphs in Theorem 1.3 by graphs of a certain special form described in 4.1 and one may require the inductive limit sequences to have special property described in 4.2.

4.1. special graphs

Let A be a basic building block with spectrum X. We can get a graph with special form from X by identifying some of its points. More precisely, pick finitely many points in the interior of each edge of X, and identify all these points as well as those vertices at which the fibres of A have no dimension drop. The new graph we get, say $\tilde{X}$, can be written as $\tilde{X} = \tilde{X}_1 \vee \tilde{X}_2 \vee \cdots \vee \tilde{X}_m$, i.e., m simple graphs $\{\tilde{X}_i\}$ joined at a single point x_0. Each $\tilde{X}_i$ is a circle or several edges connecting x_0 to another vertex x_i of X (with dimension drop). We shall call any graph of this form a special graph. Notice that $\tilde{X}$ is the spectrum of a sub-C*-algebra B of A which contains all the elements of A whose values are the same at those identified points of X. B has dimension drop at all the vertices of $\tilde{X}$ except x_0.

The Theorem 4.1 below together with Theorem 3.2 and Theorem 3.3 will allow us to replace arbitrary graphs in Theorem 1.3 by special graphs. This reduction is crucial in this work.

Suppose that ϕ is a unital $*$-homomorphism between two basic building blocks A and B with generic fibres M_n and M_m, respectively. For each y in the spectrum of B, the evaluation map τ_y from B to the fibre of B at y, say D, given by

$$\tau_y(g) = g(y) \quad \text{for} \quad g \in B$$

is a $*$-homomorphism and hence gives rise to a representation $\tau_y\phi \colon A \to D$. There are then k points $(x_i)_{i=1}^k$ in the spectrum of A and a unitary $u \in D$ such that

$$(\tau_y\phi)(f) = \mathrm{Ad}(u)\begin{pmatrix} f(x_1) & & \\ & \ddots & \\ & & f(x_k) \end{pmatrix} \quad \text{for} \quad f \in A.$$

It is clear that these points may change when y changes.

Lemma *Let A be a C*-algebra of real rank zero, arising as the inductive limit of a sequence (A_i) of basic building blocks. For any i and any $\epsilon > 0$, there exists $j > i$ such that for any two points in the spectrum of A_j, the corresponding two families of points in the spectrum of A_i can be matched up to be within ϵ point by point.*

With a slight modification, both of the statement and of the proof, the lemma still holds if each A_i is a finite direct sum of basic building blocks. This is the version used in proving Theorem 4.1 below. In fact, the lemma is still true if instead of basic building blocks one has matrix algebras over arbitrary compact metric spaces (see [4], [12]).

Before stating Theorem 4.1, we need the following:

Definition Let X be a finite connected graph with k edges, $L_1, L_2, \cdots, L_k$. Identify each of L_i with $[0,1]$. For $\epsilon > 0$, a finite subset F of a basic building block A with spectrum X and with generic fibre M_n is said to be approximately constant, to within ϵ, if there exists a unitary u in $\mathrm{C}(I, \bigoplus_{i=1}^k M_n(\mathbf{C}))$ such that the image of each element of F, under the natural embedding of A into $\mathrm{C}(I, \bigoplus_{i=1}^k M_n(\mathbf{C}))$ combined with $\mathrm{Ad}u$, when considered as a matrix-valued function on $[0,1]$, is approximately the same at each point, to within ϵ. F is said to be approximately constant to within ϵ in the strong sense if, furthermore the image of each element of F at 0, under this embedding, is in $M_n(\mathbf{C})$.

Using Theorem 2.1 together with Lemma 4.1, we can prove the following Theorem 4.1:

Theorem *Let $A_1 \to A_2 \to A_3 \to \cdots$ be a sequence of finite direct sums of basic building blocks, and suppose that the C*-algebra inductive limit is of real rank zero. For each i, each finite subset $F \subset A_i$, and each $\epsilon > 0$, there exist $j > i$, a sub-C*-algebra $A_j^0 \subset A_j$ isomorphic to a finite direct sum of basic building blocks, and a map ϕ from A_i to A_j^0 such that the component of $\phi(F)$ in each basic building block of A_j^0 is approximately constant, to within ϵ, and such that the following diagram approximately commutes, to within ϵ, on F:*

$$\begin{array}{ccc} A_i & \to & A_j \\ & \searrow & \uparrow \\ & & A_j^0 \end{array}$$

Furthermore, the spectrum of each basic building block of A_j^0 may be required to have special form.

4.2. Special sequences

Let $C_1 \to C_2 \to \cdots$ be a sequence of finite direct sums of basic building blocks with special spectra, and suppose that the C*-algebra inductive limit is of real rank zero. We shall say that the sequence is special if for each i, for each finite subset F of C_i, and for each $\epsilon > 0$, there exists $j > i$ such that the image of F in each basic building block of C_j is approximately constant in the strong sense, to within ϵ. (This property is stronger than that guaranteed by Theorem 4.1)

A proof similar to Proposition 5.4 of [12] shows that the inductive limit of a special sequence will be of real rank zero.

As a consequence of Theorem 4.1 and Theorem 3.3, we have the following:

Corollary *Let $A_1 \to A_2 \to A_3 \to \cdots$ be a sequence of finite direct sums of basic building blocks, and suppose that the C*-algebra inductive limit is of real rank zero. Then A can be expressed as the inductive limit of a special sequence.*

So we may assume at the very beginning that the sequences in Theorem 1.3 are special.

5. Standard forms and four-term exact sequences

In this section, we will introduce standard forms for $*$-homomorphisms between finite direct sums of basic building blocks (see 5.1) and we will discuss the commutative diagram of four-term exact sequences obtained from a $*$-homomorphism with the standard form (see 5.2). These are needed for proving the uniqueness Theorem 6.2 below. For the sake of simplicity, we will discuss the case that each algebra is a single basic building block. The arguements are similar for the general case.

5.1. STANDARD FORMS

Let A and B be two basic building blocks with spectra X and Y, respectively. Identify each edge of X and Y with $[0,1]$. Let ϕ be a unital $*$-homomorphism from A to B with the property ensured by the Perturbation theorem. We shall say that ϕ has standard form if each of the eigenvalue map $S_i(t)$ (see Theorem 2.1) takes one of the following forms:

$$\{t, 1-t, a\ vertex\}$$

5.2. FOUR-TERM EXACT SEQUENCES

For each basic building block D, there is a canonical ideal E associated with it. Namely, the ideal of all the elements of D that vanish at all the vertices of the spectrum of D. By Bott periodicity, one has the following four-term exact sequence:

$$0 \to K_0(D) \to K_0(D/E) \to K_1(E) \to K_1(D) \to 0$$

We will denote this four-term exact sequence by K^*.

Let $I \subset A$ and $J \subset B$ be the corresponding canonical ideals and let ϕ be a $*$-homomorphism from A to B. Suppose that ϕ has standard form. It is easy to see that ϕ induces a map ϕ_I from I to J and a map ϕ_Q from A/I to B/J. Hence, one has the following commutative diagram:

$$\begin{array}{ccccccccc} 0 & \to & K_0(A) & \to & K_0(A/I) & \to & K_1(I) & \to & K_1(A) & \to 0 \\ & & \downarrow & & \downarrow & & \downarrow & & \downarrow & \\ 0 & \to & K_0(B) & \to & K_0(B/J) & \to & K_1(J) & \to & K_1(B) & \to 0 \end{array}$$

Let ψ be another $*$-homomorphism with standard form from A to B. If ϕ and ψ induce the same commutative diagram, then one can show that for any $\epsilon > 0$ and for any finite subset $F \subset A$, there exists a unitary $U \in B$ such that

$$\|\phi(f) - AdU \circ \psi(f)\| < \epsilon$$

for all $f \in F$.

We would like to point out that $K_*\phi = K_*\psi$ is not enough to induce the same commutative diagram.

6. Equivalent relation and uniqueness

We shall introduce an equivalent relation among $*$-homomorphisms between two basic building blocks. Theorem 6.2 below shows that two maps that induce the same map on $K_0 \oplus K_1$ will be approximately unitary equivalent, as described in 5.2, provided that they are equivalent.

6.1. Equivalent class

Let A and B be two basic building blocks with dimension drop a and b, respectively. For two $*$-homomorphisms ϕ and ϕ' from A to B, we shall say that ϕ is equivalent to ϕ' if ϕ is homotopic to ϕ'.

For ϕ, one can associate a $q \times p$ matrix T_ϕ to it. More precisely, let t_{ij} be the multiplicity of x_j in a representation of ϕ at y_i (with multiplicity one), where $i, j > 0$. The matrix $T_\phi = (t_{ij})$ has non-negative entries. If ϕ' is equivalent to ϕ, $T_{\phi'}$ will be different from T_ϕ in general. However, the following relation holds:

$$T_\phi - T_{\phi'} = 0 \ mod(a)$$

This is an equivalent relation. We will denote the class of T_ϕ by $[T_\phi]$.

6.2. Uniqueness

Theorem *Let A and B be two basic building blocks with special spectra and let ϕ and ψ be two unital $*$-homomorphisms from A to B. Suppose that ϕ and ψ have the property ensured by the Perturbation theorem. For $\epsilon > 0$ and for any finite subset $F \subset A$ of approximately constant elements to within ϵ, assume that*
(i) $K_*\phi = K_*\psi$,
(ii) $[T_\phi] = [T_\psi]$,
(iii) The variation for each eigenvalue map of ϕ and of ψ over each edge of the spectrum of B is to within $\delta < 1/16$.

It follows that there exists a unitary $U \in B$ such that for all $f \in F$

$$\|\phi(f) - AdU \circ \psi(f)\| < 27\epsilon$$

We would like to point out that condition (ii) was not needed in what were called uniqueness theorems if A and B were basic building blocks in the sense of [9] and [12]. However, we do need this condition here.

Clearly, this section can be generalized to the case that the algebras are of finite direct sums of basic building blocks.

7. Classification for a special case

In this section, we shall prove the classification theorem for a special case (see Theorem 7.2). In section 8, we will show that the general case can be reduced to the special case.

7.1. Intertwining of K^*

Lemma *Let $A = \varinjlim(A_i, \xi_i)$ and $B = \varinjlim(B_i, \eta_i)$ be two inductive limit C^*-algebras. Suppose that*
(1) $A_i = B_i$ is a finite direct sum of basic building blocks for each i,
(2) Each ξ_i and each η_i is standard,
(3) $K_*\xi_i = K_*\eta_i$, and ξ_i and η_i induce the same K_1-map on the canonical ideals.

It follows that one has the following commutative diagram:

$$\begin{array}{ccccccc} K^*(A_{n_1}) & \rightarrow & K^*(A_{n_2}) & \rightarrow & \dots & \rightarrow & K^*(A) \\ \downarrow & \nearrow & \downarrow & \nearrow & & & \\ K^*(B_{m_1}) & \rightarrow & K^*(B_{m_2}) & \rightarrow & \dots & \rightarrow & K^*(B) \end{array}$$

where $m_i > n_i$ and $n_{i+1} > m_i$.

We would like to point out that A and B can not be real rank zero.

7.2. Classification for a special case

Let $A = \varinjlim(A_n, \phi_{n,n+1})$ and $B = \varinjlim(A_n, \psi_{n,n+1})$ be two special sequences in the sense of section 4.2. Assume that $K_*(\phi_{n,n+1}) = K_*(\psi_{n,n+1})$ for each n. Under these conditions, one can show that A and B are isomorphic.

First, let us deform $\phi_{n,n+1}$ and $\psi_{n,n+1}$ into standard form. Denote the new maps by $\{\phi'_{n,n+1}\}$ and $\{\psi'_{n,n+1}\}$. Since $K_*(\phi_{n,n+1}) = K_*(\psi_{n,n+1})$, it can be shown that we may require $\phi'_{n,n+1}$ and $\psi'_{n,n+1}$ to induce same K_1-map on the K_1-group of the corresponding canonical ideal. By section 7.1, we can intertwin K^* of these two new sequences:

$$\begin{array}{ccccccc} K^*(A_{n_1}) & \rightarrow & K^*(A_{n_2}) & \rightarrow & \dots & \rightarrow & K^*(A) \\ \downarrow & \nearrow & \downarrow & \nearrow & & & \\ K^*(A_{m_1}) & \rightarrow & K^*(A_{m_2}) & \rightarrow & \dots & \rightarrow & K^*(B) \end{array}$$

where $m_i > n_i$ and $n_{i+1} > m_i$. In the original sequence, deform each ϕ_{n_1,m_1} and each $\phi_{m_i,n_{i+1}}$ into standard form. By changing vertex point evalution of the new maps, one can lift each down and up map to the algebra level to get the following not necessarily commutative diagram:

$$\begin{array}{ccccccc} A_{n_1} & \rightarrow & A_{n_2} & \rightarrow & \dots & \rightarrow & A \\ \downarrow & \nearrow & \downarrow & \nearrow & & & \\ A_{m_1} & \rightarrow & A_{m_2} & \rightarrow & \dots & \rightarrow & B \end{array}$$

Although the diagram is not commutative, the two matrices defined in section 6 associated to the two maps in each triangular have the same equivalent class. By passing to subsequences and by using Theorem 6.2, one can obtain an approximate intertwining sequence in the sence of section 3. As a consequence of Theorem 3.3, A and B are isomorphic.

8. Proof of Theorem 1

8.1. Intertwining of K_*

Let $A = \varinjlim A_n$ and $B = \varinjlim B_n$ be two special sequences with same K_*. The isomorphism $\mathrm{K}_*(A) \to \mathrm{K}_*(B)$ can be lifted to an intertwining of subsequences of the sequences $\mathrm{K}_*(A_i)$ and $\mathrm{K}_*(B_i)$.

More precisely, after passing to subsequences of A_i and B_i and changing notation, one has a commutative diagram:

$$\begin{array}{ccccccc} \mathrm{K}_*A_1 & \to & \mathrm{K}_*A_2 & \to & \cdots & \to & \mathrm{K}_*A \\ \downarrow & \nearrow & \downarrow & \nearrow & & & \\ \mathrm{K}_*B_1 & \to & \mathrm{K}_*B_2 & \to & \cdots & \to & \mathrm{K}_*B \end{array}$$

where each map preserves both the graded group structure and the graded dimension range. This is due to the special nature of A_i and B_i. For example, an application of Bott periodicity shows that K_*A_i and K_*B_i, as groups, are finite direct sums of cyclic groups.

8.2. Existance

After passing to subsequences and changing notations, the map intertwining the sequences (K_*A_i) and (K_*B_i) can be lifted, individually, to C*-algebra homomorphisms. In other words, one has a not necessarily commutative diagram

$$\begin{array}{ccccccc} A_1 & \to & A_2 & \to & \cdots & \to & A \\ \downarrow & \nearrow & \downarrow & \nearrow & & & \\ B_1 & \to & B_2 & \to & \cdots & \to & B \end{array}$$

This follows from:

Theorem *Let A be a basic building block and let $B = \varinjlim B_n$ be the inductive limit of a sequence of finite direct sums of basic building blocks. Suppose that B is of real rank zero and suppose that the spectrum of A and the spectrum of each basic building block of B_n has the special form considered in 4.1. Let $\psi\colon K_*(A) \to K_*(B)$ be a group homomorphism preserving the graded group structure and graded dimension range. Then there exists an integer n and there exists a $*$-homomorphism ϕ from A to B_n such that ϕ composed with the map from B_n to B induces ψ on $K_*(A)$.*

8.3. The classification

To show that A is isomorphic to B, we shall construct two new isomorphic inductive limit C^*-algebras $A^{(1)}$ and $B^{(1)}$ and prove that A is isomorphic to $A^{(1)}$ and B is isomorphic to $B^{(1)}$.

Specificly, consider the diagram obtained in 8.2:

$$\begin{array}{ccccccc} A_1 & \to & A_2 & \to & \cdots & \to & A \\ \downarrow & \nearrow & \downarrow & \nearrow & & & \\ B_1 & \to & B_2 & \to & \cdots & \to & B \end{array}$$

Using the down-up maps, one can obtain a new sequence:

$$A_1 \to A_2 \to \cdots \to A^{(1)}$$

Similarly, by using the up-down maps, one has

$$B_1 \to B_2 \to \cdots \to B^{(1)}$$

It is easy to see that $A^{(1)}$ is isomorphic to $B^{(1)}$.

By passing to subsequences of (A_i) and (B_i) and changing notation at the beginning, we may assume that these two new sequences are special in the sense of 4.2. By section 8, A and $A^{(1)}$ are isomorphic and B and $B^{(1)}$ are isomorphic. Finally, A and B are isomorphic.

Acknowledgement

The author wishes to thank Professor George A. Elliott for suggesting the topic of this paper and for many helpful discussions.

References

1. B. Blackadar, *K-Theory for Operator Algebras*, Mathematical Sciences Research Institute Publications 5, Springer-Verlag, New York, 1986.
2. B. Blackadar, Symmetries of the CAR algebra, *Ann. of Math.* **131** (1990) 589–623.
3. O. Bratteli, G. A. Elliott, D. E. Evans, and A. Kishimoto, Finite group actions on AF algebras obtained by folding the interval, *K-Theory*, to appear.
4. O. Bratteli, G.A. Elliott, A. Kishimoto, H. Su, Small eigenvalue variation and real rank zero, in preparation.
5. O. Bratteli, A. Kishimoto, Non-commutative spheres III: Irrational rotations, preprint.
6. L. G. Brown and G. K. Pedersen, C*-algebras of real rank zero, *J. Funct. Anal.* **99** (1991), 131–149.
7. M.-D. Choi and G. A. Elliott, Density of the self-adjoint elements with finite spectrum in an irration rotation C*-algebra, *Math. Scand.* **67** (1990) 73–86
8. G. A. Elliott, On the classification of inductive limits of sequences of semisimple finite-dimensional algebras, *J. Algebra* **38** (1976), 29–44.
9. G. A. Elliott, On the classification of C*-algebras of real rank zero, preprint.
10. G. A. Elliott and D. E. Evans, The structure of the irrational rotation C*-algebra, *Ann. of Math,* to appear.
11. D. E. Evans and A. Kishimoto, Compact group actions on UHF algebras obtained by folding the interval, *J. Funct. Anal.* **98** (1991), 346–360.
12. H. Su, On the classification of C*-algebras of real rank zero: inductive limits of matrix algebras over non-Hausdorff graphs, preprint.

Inductive Limits of Interval Algebras: The Simple Case

KLAUS THOMSEN
Institute of Mathematics, University of Aarhus, Ny Munkegade, 8000 Aarhus C, Denmark

Abstract. We generalize Elliott's classification of simple inductive limits of finite direct sums of interval algebras to the non-unital case and we show that a *-homomorphism between two such C^*-algebras is determined up to approximate inner equivalence by its action on K_o and on the space of densely defined lower semi-continuous traces.

This is a successor of the author's previous papers on the C^*-algebras that are built as inductive limits of finite direct sums of interval algebras, cf. [5],[6]. It is motivated and inspired by the remarkable result of Elliott, [4], which classifies the simple unital C^*-algebras of this type. Thanks to this result it is now possible in many cases to decide whether or not specific examples of sequences of finite direct sums of interval algebras are *-isomorphic or not. Furthermore, the analysis of [4] enables one to construct homomorphisms and automorphisms of such C^*-algebras with given action on the tracial state space. Thus we are rapidly coming to the point where such C^*-algebras can be analysed and understood to the same extent as AF-algebras, at least as long as they are simple.

The following lemma is a slight variation of a result of Elliott, Theorem 6 of [4]. The proof is essentially the same as Elliott's so we omit the details.

Lemma 1. Let $a, b \in C[0,1] \otimes M_n$ be two selfadjoints with spectral contained in $\{-1\} \cup [0,1]$. Let $N \in \mathbf{N}$ and define for each $i = 0, 1, 2, \ldots, N-1$ continuous functions $f_i, g_i : \mathbf{R} \to [0,1]$ such that $g_i(t) = 0, t \notin]\frac{i}{N}, \frac{i+1}{N}[$ and

$$f_i(t) = \begin{cases} 0 & , \quad t \leq \frac{i}{N}, \\ Nt - i & , \quad \frac{i}{N} < t < \frac{i+1}{N}, \\ 1 & , \quad t \geq \frac{i+1}{N}, \end{cases}$$

and let $g : \mathbf{R} \to \mathbf{R}$ be any continuous function vanishing on $[0,1]$ and taking the value 1 at -1. Assume that $g(a) = g(b)$ and that there is $\delta > 0$ such that

H. Araki et al. (eds.), Quantum and Non-Commutative Analysis, 399–404.

$$| \tau(f_i(a)) - \tau(f_i(b)) | < \delta \tag{1}$$

and

$$\tau(g_i(a)) > \delta, \tag{2}$$

for all $i = 0, 1, 2, 3, \ldots, N-1$, and for all tracial states τ of $C[0,1] \otimes M_n$. Then there is a unitary $u \in C[0,1] \otimes M_n$ such that $\| uau^* - b \| \leq \frac{3}{N}$. □

In the following we let $T(A)$ denote the tracial state space of a unital C^*-algebra A.

Theorem 2. Let A and B be simple unital C^*-algebras that are inductive limits of sequences of finite direct sums of interval algebras. Let $\alpha, \beta : A \to B$ be unital $*$-homomorphisms such that $\alpha_* = \beta_*$ on $K_0(A)$ and $\tau \circ \alpha = \tau \circ \beta$ for all $\tau \in T(B)$.

It follows that α and β are approximately inner equivalent.

Proof. By [6] it suffices to show that α and β induce the same map on S_A. Furthermore, since $\mathbf{N}\Sigma_A$ is dense in S_A it suffices to show that α and β induces the same map om Σ_A. To this end it suffices to show that $\alpha(a)$ and $\beta(a)$ are approximately unitarily conjugate in B for all $a = a^* \in A$. In a finite direct sum of interval algebras the selfadjoint elements with a spectrum equal to a finite union of closed intervals with non-empty interior form a dense subset of the set of all selfadjoint elements. Since, by [4], A can be realized as the inductive limit of finite direct sums of interval algebras with injective connecting *-homomorphisms, this conclusion extends to A. It follows that we can assume that $\hat{a} = \alpha(a)$ and $\hat{b} = \beta(a)$ have spectrum equal to a finite union of closed intervals with non-empty interior. By [6], if x_1, y_1, x_2, y_2 are selfadjoints in B such that $x_1 y_1 = x_2 y_2 = 0$ and x_1 is approximately unitarily conjugate to x_2 and y_1 to y_2, then $x_1 + y_1$ is approximately unitarily conjugate to $x_2 + y_2$. An obvious application of spectral theory therefore reduces the proof to the case where a, and hence also $\hat{a}$ and $\hat{b}$, have spectrum $[s,t] \cup \{0\}$ for some $s < t$ in $\mathbf{R}$. For simplicity of exposition we assume that $0 \notin [s,t]$. (The case where $0 \in [s,t]$ is handled similarly, and is in fact slightly simpler.) After applying a homeomorphism of $\mathbf{R}$, we can assume that the spectrum of a is contained in $\{-1\} \cup [0,1]$. Let $N \in \mathbf{N}$ be arbitrary. It suffices to show that there is a unitary u in B such that $\| u\hat{a}u^* - \hat{b} \| < \frac{5}{N}$. Let g and $f_i, g_i, i = 0, 1, 2, \ldots, N-1$, be the functions from Lemma 1. Then $[g(\hat{a}] = \alpha_*[g(a)] = \beta_*[g(a)] = [g(\hat{b})]$ in $K_o(B)$ bacause $\alpha_* = \beta_*$, so there is a unitary $w \in B$ such that $wg(\hat{a})w^* = g(\hat{b})$. For our present purpose we can therefore assume that $g(\hat{a}) = g(\hat{b})$. Now let $B_1 \xrightarrow{\phi_1} B_2 \xrightarrow{\phi_2} B_3 \xrightarrow{\phi_3} \cdots$ be a sequence of finite direct sums of interval algebras with unital ϕ_i's such that $B = \lim_{\rightarrow}(B_i, \phi_i)$. By [4], we can assume that the ϕ_i's are injective, and it follows that so are the canonical $*$-homomorphisms $\rho_i : B_i \rightarrow B$. A familiar approximation shows that after conjugation with another unitary, we can assume that $g(\hat{a}) = g(\hat{b}) = \phi_L(e)$ for some $L \in \mathbb{N}$ and a projection $e \in B_L$. Since $g_i(\hat{a}) \notin 0$ and B is simple, there is a $\delta > 0$ so that

$$\omega(g_i(\hat{a})) > 2\delta, \omega \in T(B), i = 0, 1, 2, \ldots, N-1. \tag{3}$$

Choose $K \geq L$ and $a_k, b_k \in B_k$ such that

$$\| \rho_k(a_k) - \hat{a} \| < \frac{1}{N}, \| \rho_k(b_k) - \hat{b} \| < \frac{1}{N}, \tag{4}$$

$$\| \rho_k(g_i(a_k)) - g_i(\hat{a}) \| < \delta/2, \| \rho_k(g_i(b_k)) - g_i(\hat{b}) \| < \delta/2, i = 0, 1, 2, 3, \ldots, N-1, \tag{5}$$

$$\| \rho_k(g_i(a_k)) - g_i(\hat{a}) \| < \delta, i = 0, 1, 2, \ldots, N-1, \tag{6}$$

and

$$g(a_k) = g(b_k) = \phi_{k,L}(e). \tag{7}$$

By a simple argument based on spectral theory, we can assume that both a_k and b_k have spectrum contained in $\{-1\} \cup [0,1]$. From (3) and (6) we conclude that $\omega(\rho_k(g_i(a_k))) > \delta$ for all i and all $\omega \in T(B)$. It follows that for some sufficiently large m_o, we have that $\omega(\phi_{m,k}(g_i(a_k)) > \delta$ for all i and all $\omega \in T(B_m)$ when $m \geq m_o$. By [5]

$$\lim_{m \rightarrow \infty} \sup\{| \omega(\phi_{m,k}(f_i(a_k)) - \phi_{m,k}(f_i(b_k))) | : \omega \in T(B_m)\} =$$
$$\sup\{| \omega(\rho_k(f_i(a_k)) - \rho_k(f_i(b_k))) | : \omega \in T(B)\}$$

for all i. By assumption $\omega(f_i(\hat{a})) = \omega \circ \alpha(f_i(a)) = \omega \circ \beta(f_i(a)) = \omega(f_i(\hat{b}))$ for all $\omega \in T(B)$ and all i, so from (5) we get that

$$\lim_{m \rightarrow \infty} \sup\{| \omega(\phi_{m,k}(f_i(a_k)) - \phi_{m,k}(f_i(b_k))) | : \omega \in T(B_m)\} < \delta.$$

Hence we can assume that m_o is so large that $|\omega(\phi_{m,k}(f_i(a_k)) - \omega(\phi_{m,k}(f_i(b_k))| < \delta$ for all $\omega \in T(B_m)$ and all i when $m \geq m_o$. From Lemma 1 we conclude that for arbitrary $m \geq m_o$, there is a unitary $v \in B_m$ such that $\| v\phi_{m,k}(a_k)v^* - \phi_{m,k}(b_k) \| < \frac{3}{N}$. Set $u = \rho_m(v)$. Using (4) we find that $\| u\hat{a}u^* - \hat{b} \| < \frac{5}{N}$, as desired. □

Theorem 2 will be generalized to the non-unital case below.

Let now A be a simple C^*-algebra, not necessarily unital, but obtained as the inductive limit of a sequence of finite direct sums of interval algebras. If $e \in A$ is a non-zero projection, then because A is simple, there is an isometry

V in the multiplier algebra of $A \otimes \mathcal{K}$, where $\mathcal{K}$ denotes the compact operators on an infinite dimensional separable Hilbert space, such that $VV^* = e \otimes 1$, cf. [1]. The $*$-isomorphism $AdV^* : eAe \otimes \mathcal{K} \to A \otimes \mathcal{K}$ induces an ordered group isomorphism $K_o(eAe) \to K_o(A)$ which agrees with the map I_{e*} induced by the inclusion $I_e : eAe \to A$. In particular, this shows that $[e] \in K_o[A]$ is an order unit.

Let $\mathcal{T}_A$ denote the set of lower-semicontinuous densely defined traces on A. Since the Pedersen ideal of A contains all projections, every $\tau \in \mathcal{T}_A$ takes finite values on projections. We endow $\mathcal{T}_A$ with the weakest topology such that the functional $\tau \to \tau(a)$ is continuous for any $a \in A_+$ dominated by a projection. Since A has an approximate unit consisting of projections, this is a Hausdorff topology.

Lemma 3. Let $e \in A$ be a non-zero projection and set $\mathcal{T}_A^e = \{\tau \in \mathcal{T}_A : \tau(e) = 1\}$. Then $\mathcal{T}_A^e$ is compact relative to the topology of $\mathcal{T}_A$ and the restriction map $R^e : \mathcal{T}_A^e \to T(eAe)$ is a homeomorphism.

Proof. By [2], 4.7, R^e is an affine bijection. Since it is obviously continuous we only have to show that the inverse is also continuous. So let $\{\omega_i\}$ be a net in $T(eAe)$ converging to $\omega \in T(eAe)$. Take $a \in A_+$ such that a is dominated by the non-zero projection $p \in A$. As observed above, $[e]$ is an order unit in $K_o(A)$, so there is an integer $n \in \mathbf{N}$ such that $[p] \leq n[e]$ in $K_o(A)$. Since $a \leq p$ it follows that there is an element $b \leq ne$ such that $\tau(a) = \tau(b)$ for all $\tau \in \mathcal{T}_A$. In particular, we conclude that $(R^e)^{-1}(\omega_i)(a) = \omega_i(b) \to \omega(b) = (R^e)^{-1}(\omega)(a)$. □

Since every element of $\mathcal{T}_A$ is finite on projections and extends naturally to a trace on matrix algebras over A with the same property, there is a natural pairing $(\tau, x) \to < \tau, x > \in \mathbf{R}, \tau \in \mathcal{T}_A, x \in K_o(A)$. Note that for any non-zero projection e in A,

$$< R^e(\tau), x >=< \tau, I_{e*}(x) >, \tau \in \mathcal{T}_A, x \in K_o(eAe). \tag{8}$$

Theorem 4. Let A and B be simple C^*-algebras that are inductive limits of sequences of finite direct sums of interval algebras. Let $\phi : K_o(A) \to K_o(B)$ be an isomorphism of scaled ordered groups and $\phi_T : \mathcal{T}_B \to \mathcal{T}_A$ an affine homeomorphism such that $< \phi_T(\tau), x >=< \tau, \phi(x) >, \tau \in \mathcal{T}_B, x \in K_o(A)$.

Then there is a $*$-isomorphism $\psi : A \to B$ such that $\psi_* = \phi$ and $\tau \circ \psi = \phi_T(\tau), \tau \in \mathcal{T}_B$.

Proof. Let $e_1 \leq e_2 \leq e_3 \leq \ldots$ be a approximate unit for A consisting of projections and $f_1 \leq f_2 \leq f_3 \leq \ldots$ an approximate unit for B consisting of projections. We will construct strictly increasing sequences $a_1 < a_2 < a_3 < \ldots$ and $b_1 < b_2 < b_3 < \ldots$ in $\mathbf{N}$, projections $p_1 \leq p_2 \leq p_3 \leq \ldots$ in A and $q_1 \leq q_2 \leq q_3 \leq \ldots$ in B such that $p_{2i-1} = e_{a_i}, q_{2i} = f_{b_i}, \phi[p_i] = [q_i]$ for all i, isomorphisms $\psi_i : p_iAp_i \to q_iBq_i$ such that $\psi_i(p_{i-1}) = q_{i-1}$, for i odd, and $\xi_i : q_iBq_i \to p_iAp_i$ such that $\xi_i(q_{i-1}) = p_{i-1}$, for i even. Furthermore, we want these isomorphisms to be compatible with ϕ and ϕ_T in the sense that $\psi_{i*} = (I_{q_i^*})^{-1} \circ \phi \circ I_{p_i^*}, \tau \circ \psi_i = R_{p_i} \circ \phi_T \circ R_{q_i}^{-1}(\tau), \tau \in T(q_iBq_i)$ and $\xi_{i*} = (I_{p_{i*}})^{-1} \circ \phi^{-1} \circ I_{q_i^*}, \tau \circ \xi_i = R_{q_i} \circ \phi_T^{-1} \circ (R_{p_i})^{-1}(\tau), \tau \in T(p_iAp_i)$.

Assume that this has been achieved. Then $(\xi_{i+1} \circ \psi_i)_*$ is the identity map on $K_o(p_iAp_i)$ and $\tau \circ \xi_{i+1} \circ \psi_i = \tau, \tau \in T(p_iBp_i)$, so by Theorem 2 $\xi_{i+1} \circ \psi_i$ is approximately inner on p_iAp_i. Similarly, $\psi_{i+1} \circ \xi_i$ will be approximately inner on q_iBq_i. So by conjugating the ξ_i's and the ψ_i's by inductively (and carefully) chosen unitaries from p_iAp_i and q_iBq_i, respectively, we can achieve that the diagram

$$\begin{array}{ccccccccc} p_1Ap_1 & \subseteq & p_2Ap_2 & \subseteq & p_3Ap_3 & \subseteq & p_4Ap_4 & \subseteq & \cdots \\ \psi_1 \downarrow & & \xi_2 \uparrow & & \psi_3 \downarrow & & \xi_4 \uparrow & & \\ q_1Bq_1 & \subseteq & q_2Bq_2 & \subseteq & q_3Bq_3 & \subseteq & p_4Bq_4 & \subseteq & \cdots \end{array}$$

satisfies the assumption of Theorem 2.2 of [3]. Since $\cup_i p_iAp_i$ is dense in A (this is why we wanted $p_{2i-1} = e_{a_i}$) and $\cup_i q_iBq_i$ in B, this result of Elliott gives us a $*$-isomorphism $\psi : A \to B$ such that $\psi(x) = \lim_{k\to\infty} \psi_k(x), x \in p_iAp_i$. It is then straightforward to check that $\psi_* = \phi$ and $\tau \circ \psi = \phi_T(\tau), \tau \in T_B$.

Now to the construction. To start the induction, set $a_1 = 1$ and $p_1 = e_1$. Choose a projection $q_1 \in B$ such that $\phi([p_1]) = [q_1]$. We can assume that $q_1 \leq f_{b_1}$ for some $b_1 \in \mathbf{N}$. Set $\phi_1 = (I_{q_1^*})^{-1} \circ \phi \circ I_{p_1^*} : K_o(p_1Ap_1) \to K_o(q_1Bq_1)$ and $\phi_T^1 = R_{p_1} \circ \phi_T \circ (R_{q_1})^{-1} : T(q_1Bq_1) \to T(p_1Ap_1)$. Then by the assumptions on ϕ and ϕ_T and (11), $< \phi_T^1(\tau), x >=< \tau, \phi_1(x) >, \tau \in T(q_1Bq_1), x \in K_o(p_1Ap_1)$. Thus by [4], there is an isomorphism $\psi_1 : p_1Ap_1 \to q_1Bq_1$ such that $\psi_{1*} = \phi_1$ and $\phi_T^1(\tau) = \tau \circ \psi_1, \tau \in T(q_1Bq_1)$. This starts the induction. Let $k \in \mathbf{N}$ and $i = 2k-1$ and assume that everything has been constructed up to i. As seen from the preceding, we may assume that $q_i \leq f_{b_k}$ for some $b_k > b_{k-1}$. We must then just make sure that the corresponding $i+2$ condition is fulfilled when ψ_{i+2} and q_{i+2} are constructed. Set $q_{i+1} = f_{b_k}$. Then choose $a_{k+1} > a_k$ and a projection $p_{i+1} \leq e_{a_{k+1}}$ such that $\phi^{-1}[q_{i+1}] = [p_{i+1}]$. Since $p_i = e_{a_{k+1}}$ and $[p_i] = \phi^{-1}[q_i] \leq \phi^{-1}[q_{i+1}] = [p_{i+1}]$, we can assume that $p_i \leq p_{i+1} \leq e_{a_{k+1}}$. As above we can use Elliott's result from [4] to construct a $*$-isomorphism $\xi_{i+1} : q_{i+1}Bq_{i+1} \to p_{i+1}Ap_{i+1}$ such that $\xi_{i+1*} = (I_{p_{i+1*}})^{-1} \circ \phi^{-1} \circ I_{q_{i+1*}}$ on $K_o(q_{i+1}Bq_{i+1})$ and $\tau \circ \xi_{i+1} = R_{q_{i+1}} \circ \phi_T^{-1} \circ (R_{p_{i+1}})^{-1}(\tau), \tau \in T(p_{i+1}Ap_{i+1})$. Since $\xi_{i+1*}[q_i] = [p_i]$ in $K_o(p_{i+1}Ap_{i+1})$, we can assume after conjugation with a unitary in $p_{i+1}Ap_{i+1}$, that $\xi_{i+1}(q_i) = p_i$. Set $p_{i+2} = e_{a_{k+1}}$. Then $\phi[p_{i+2}] = [q_{i+2}]$ for some projection q_{i+2} dominated by an $f_{b_{k+1}}$ for some $b_{k+1} > b_k$. An argument completely symmetric to the preceding now gives us the $*$-isomorphism $\psi_{i+2} : p_{i+2}Ap_{i+2} \to q_{i+2}Bq_{i+2}$ with the desired properties. This completes the induction step and hence the proof. □

As shown by the next theorem the $*$-isomorphism of Theorem 4 is unique up to approximately inner equivalence.

Theorem 5. Let A and B be inductive limit C^*-algebras of sequences of finite direct sums of interval algebras. Assume that A and B are simple. Let $\alpha, \beta : A \to B$ be *-homomorphisms such that $\alpha_* = \beta_*$ on $K_o(A)$ and $\tau \circ \alpha = \tau \circ \beta$ for all $\tau \in \mathcal{T}_B$.

It follows that α and β are approximate inner equivalent, i.e. there is a sequence $\{u_n\}$ of unitaries in the C^*-algebra obtained by adjoining a unit to B such that $\lim_{k\to\infty} u_k \alpha(a) u_k^* = \beta(a), a \in A$.

Proof. By Theorem 2.2 of [6] it suffices to show that α and β induce the same map on S_A. To this end it suffices to take a projection $p \in A$ and an element $0 \leq a \leq p$ in A and show that there is a sequence $\{w_n\}$ of partial isometries in B such that $\lim_{k\to\infty} w_n \alpha(a) w_n^* = \beta(a)$. But $\alpha_*[p] = \beta_*[p]$ in $K_o(B)$, so there is a partial isometry $v \in B$ such that $vv^* = \beta(p)$ and $v^*v = \alpha(p)$. Set $\alpha' = v\alpha(\cdot)v^*|_{pAp}$ and $\beta' = \beta|_{pAp}$. Then α' and β are unital $*$-homomorphisms mapping from pAp to $\beta(p)B\beta(p)$. Since $\alpha'_* = \beta'_*$ on $K_o(pAp)$ and $\tau\circ\alpha' = \tau\circ\beta', \tau \in T(pAp)$ by Lemma 3, we can use Theorem 2 to conclude that there is a sequence $\{v_n\}$ of unitaries in $\beta(p)B\beta(p)$ such that $\lim_{n\to\infty} v_n \alpha'(a) v_n^* = \beta(a)$. Set $w_n = v_n v, n \in \mathbf{N}$. □

References

[1] L. Brown: Stable isomorphisms of hereditary subalgebras of C^*-algebras, Pacific J.Math.71(1977), 335-348.
[2] J.Cuntz and G.K.Pedersen: Equivalence and traces on C^*-algebras, J.Func.Anal. 33(1979), 135-164.
[3] G.A. Elliott: On the classification of C^*-algebras of real rank zero, Preprint.
[4] G.A. Elliott: A Classification of Certain Simple C^*-algebras, Preprint.
[5] K. Thomsen: Inductive limits of interval algebras: The Tracial State Space, Amer.J.Math., to appear.
[6] K.Thomsen: Inductive Limits of Interval algebras: Unitary Orbits of Positive Elements, Math.Ann.293(1992), 47-63.

Operator Algebras and Abstract Duals: Progress and Problems

S. DOPLICHER
Dipartimento di Matematica
Università di Roma "La Sapienza"
00185 Roma, Italy

Dedicated to Professor Huzihiro Araki on the occasion of his "Kanreki". Mount Fuji and the Ocean are immune to aging; so may Huzihiro long continue being volcanic and deep.

Key words: operator algebras, endomorphisms, group duals.

1. Introduction

The theory of Tannaka and Krein [1] characterized the category of (continuous unitary finite dimensional) representations of a compact group G (a *concrete* G–dual) within the category $\mathrm{Vect}_{\mathbb{C}}$ of finite dimensional vector (Hilbert) spaces.

A genuine application of this beautiful theory should produce the group G out of such a dual arising *per se* in some other context.

A challenge in this direction came from Algebraic Quantum Field Theory in the early seventies [2,3] where a candidate for a G–dual, the *superselection category* $\mathcal{T}$, arises from the local observables. However, $\mathcal{T}$ is only given as an *abstract category* acting on the C^*–algebra $\mathfrak{A}$ generated by the local observable. More specifically $\mathcal{T}$ is a full subcategory of $\mathrm{End}\,\mathfrak{A}$ ([4]; cf. below).

In order to apply the Tannaka–Krein theory one would need the further assignement of a representation F of $\mathcal{T}$ in $\mathrm{Vect}_{\mathbb{C}}$; this functor would then factorize

$$\begin{array}{ccc} \mathcal{T} & \xrightarrow{F} & Vect_{\mathbb{C}} \\ \downarrow_H & & \nearrow \\ R_{ep}G & & \end{array} \tag{1}$$

where H is an isomorphism with a representation category of G and the third arrow in the commutative diagram is the "forgetful functor" which assigns to each representation the underlying vector space and imbeds the

H. Araki et al. (eds.), Quantum and Non-Commutative Analysis, 405–418.

intertwining operators in the linear operators between the corresponding spaces. In other words, if *the pair* $(\mathcal{T}, F)$ were given fulfilling suitable axioms we could construct G.

In this way a compelling motivation was given to the following natural problem: can we characterize compact group duals as *abstract* categories? In other words, for which categories $\mathcal{T}$ can we construct a pair (F, G) as in (1), where G is a uniquely determined compact group? More specifically, can we characterize such categories among the full subcategories of End$\mathfrak{A}$, where $\mathfrak{A}$ is a C^*–algebra?

The solution of the second problem [4] was the central step leading to the answer to the first question as well [5]. The core result was the construction, for suitable $\mathcal{T}$, of a crossed product C^*–algebra $\mathfrak{A} \times \mathcal{T}$ containing $\mathfrak{A}$ as a subalgebra so that G appears as the group of all automorphisms of $\mathfrak{A} \times \mathcal{T}$ leaving $\mathfrak{A}$ pointwise fixed, and $\mathfrak{A}$ is the set of all G–fixed points in $\mathfrak{A} \times \mathcal{T}$, i.e. G is a Galois–closed subgroup of Aut $\mathfrak{A} \times \mathcal{T}$ (cf. [6] for a survey).

In order to cover the duality theory of compact *Lie* groups, it suffices to deal with *minimal* duals $\mathcal{T}$, where the objects are all powers of a given endomorphism ρ of $\mathfrak{A}$. If $\mathfrak{A}$ is minimal too, in the sense that the arrows of $\mathcal{T}$ generate $\mathfrak{A}$ as a C^*–algebra, $\mathfrak{A} \times \mathcal{T}$ is just the Cuntz algebra $\mathcal{O}_d$ [7], where d is the dimension of ρ. The action of G is a canonical action and ρ is induced by the fundamental multiplet, so that $\{\mathfrak{A}, \rho\}$ is identified with the *universal model* $\{\mathcal{O}_G, \sigma_G\}$. Here $\mathcal{O}_G$ is the fixed point subalgebra of $\mathcal{O}_d$ under the canonical action of G and σ_G is the restriction to $\mathcal{O}_G$ of the canonical endomorphism of $\mathcal{O}_d$.

Therefore, in order to generalize the abstract duality theory beyond compact groups, one should first study appropriate model actions. Some preliminary results in this directions will be outlined in Section 3, while in Section 2 we discuss few open problems and motivations, after reviewing the key result on the crossed product $\mathfrak{A} \times \mathcal{T}$ and a recent corollary on extensions of automorphisms [8].

2. Crossed products and extensions of automorphisms. Abstract duals and open problems

Let $\mathfrak{A} \subset \mathcal{B}$ be C^*–algebras with unit I such that the relative commutant is trivial:

$$\mathfrak{A}' \cap \mathcal{B} = \mathbb{C} \cdot I \; . \tag{2}$$

If ρ is a unital endomorphism of $\mathfrak{A}$ define

$$H = \{\psi \in \mathcal{B} \mid \psi A = \rho(A)\psi \; , \quad A \in \mathfrak{A}\} \tag{3}$$

Then H is a *Hilbert space* in $\mathcal{B}$ [2,9]: it is a norm closed vector subspace and

$$\psi^*\psi' \in \mathfrak{A}' \cap \mathcal{B} = \mathbb{C} \cdot I \; ; \quad \psi, \; \psi' \in H \tag{4}$$

so that $\psi^*\psi' = (\psi, \psi') \cdot I$ defines a scalar product on H whose associated norm coincides with the $\mathfrak{A}$ norm by the C^*–identity. If a subgroup G of Aut$\mathcal{B}$ leaves $\mathfrak{A}$ pointwise fixed, it leaves also H globally stable by (3) and, by (4), H induces a *unitary representation* u_ρ of G:

$$u_\rho(g)\psi = g(\psi) \; ; \quad g \in G \, , \quad \psi \in H \, . \tag{5}$$

In general H might be $\{0\}$; say that ρ is *inner in* $\mathcal{B}$ if the associated Hilbert space H has zero left annihilator. If H is finite dimensional (or if $\mathfrak{A}$ is a von Neumann algebra) the range projections of the $\psi \in H$ have a l.u.b. $I_H \in \mathfrak{A}$, the *support* of H, and ρ inner in $\mathcal{B}$ means $I_H = I$.

Let now $\mathfrak{A}$ be the G–fixed points:

$$\mathfrak{A} = \mathcal{B}^G \; ; \tag{6}$$

the correspondence $\rho \to u_\rho$, for those unital endomorphisms whose associated Hilbert space H_ρ (eq. (3)) is *finite dimensional with support* I, can be viewed as an isomorphism of categories [9].

Let us first recall the definition of the monoidal category End$\mathfrak{A}$. The objects of End$\mathfrak{A}$ are the unital endomorphisms $\rho, \sigma, \ldots$, of $\mathfrak{A}$. The arrows from ρ to σ are given by

$$(\rho, \sigma) = \{T \in \mathfrak{A} \mid T\rho(A) = \sigma(A)T \, , \quad A \in \mathfrak{A}\} \, . \tag{7}$$

The monoidal operations are: composition of maps on the objects, and on the arrows:

$$R \in (\rho, \rho'), S \in (\sigma, \sigma') \to R \times S \in (\rho\sigma, \rho'\sigma') \, , \quad R \times S = R\rho(S) \, . \tag{8}$$

These operations are *strictly* associative and, in a natural way, compatible with the composition of maps $\circ$, norm, and the adjoint $*$, so that End$\mathfrak{A}$ is a *strict monoidal C^*–category* [4,5]; the identity automorphism is the monoidal unit.

We turn now again to the inclusion $\mathfrak{A} \subset \mathcal{B}$ fulfilling (2) and (6). Let $S(\mathcal{B})$ denote the full[1] subcategory of End$\mathfrak{A}$ with objects those endomorphisms ρ for which H_ρ is *finite dimensional of support* I. If $T \in (\rho, \sigma)$, by (2), (6), (7) we have for each $g \in G$, $\psi \in H_\rho$,

$$TH_\rho \subset H_\sigma \; ; \quad g(T\psi) = Tg(\psi) \, , \tag{9}$$

so that T intertwines the corresponding unitary representations, i.e. in a natural way $T \in (u_\rho, u_\sigma)$.

Hence $\rho \to u_\rho$, $T \to T$ is a functor of $S(\mathcal{B})$ onto a subcategory of Rep$G$. The linear operators (H_ρ, H_σ) from H_ρ to H_σ are naturally identified with

[1] Namely, if ρ, σ are objects of the subcategory, the arrows (ρ, σ) are all the arrows in the ambient category.

(the linear span of) $H_\sigma H_\rho^*$ whose G–fixed points are, by (5) and (7), in (ρ, σ); one concludes that we have a faithful functor onto a full subcategory of Rep$G$. (The linear span of) $H_\rho H_\sigma$ provides a *strictly associative* realization of the tensor product; it is again a Hilbert space with support I and by (5) it coincides with $H_{\rho\sigma}$. One checks easily that our functor carries $T \times S$ onto the intertwiner $T \otimes S \in (u_\rho \otimes u_\sigma, u_{\rho'} \otimes u_{\sigma'})$, if $T \in (\rho, \rho')$, $S \in (\sigma, \sigma')$. Thus we have a *strict monoidal* functor [9].

The next piece of structure to be mentioned is *symmetry*. Since HK spans a tensor product of H, K, for any finite dimensional Hilbert spaces with support I in $\mathcal{B}$ we have a unitary

$$\vartheta_{H,K} \in (HK, KH) \subset \mathcal{B} ,$$
$$\vartheta_{H,K}\varphi\psi = \psi\varphi , \quad \varphi \in H, \ \psi \in K . \tag{10}$$

If ρ, σ are objects of $S(\mathcal{B})$, obviously $\vartheta_{H_\rho, H_\sigma}$ intertwines $u_\rho \otimes u_\sigma$ and $u_\sigma \otimes u_\rho$ hence is in $\mathfrak{A}$:

$$\vartheta(\rho, \sigma) \equiv \vartheta_{H_\rho, H_\sigma} \in (\rho\sigma, \sigma\rho) . \tag{11}$$

Then $(S(\mathcal{B}), \vartheta)$ is a *strict symmetric* monoidal C^*–category, in the sense that ϑ assignes to objects ρ, σ a unitary fulfilling, with (11), the following easily checked relations:

$$\vartheta(\rho', \sigma') \circ R \times S = S \times R \circ \vartheta(\rho, \sigma) , \quad R \in (\rho, \rho') , \quad S \in (\sigma, \sigma') ; \tag{12}$$

$$\vartheta(\rho\sigma, \tau) = \vartheta(\rho, \tau) \times I_\sigma \circ I_\rho \times \vartheta(\sigma, \tau) ; \tag{13}$$

$$\vartheta(\rho, \sigma) \circ \vartheta(\sigma, \rho) = I_{\rho\sigma} . \tag{14}$$

Here I_τ denotes the identity element, when viewed as a selfintertwiner of the object τ.

If $\mathcal{T}$ is a full monoidal subcategory of End$\mathfrak{A}$, a *symmetry* ε for $\mathcal{T}$ assignes to each pair of objects a unitary $\varepsilon(\rho, \sigma)$ fulfilling the relations (11–14) obeyed by ϑ.

We will say that $(\mathcal{T}, \varepsilon)$ is a *symmetric monoidal subcategory* of $(S(\mathcal{B}), \vartheta)$ if the objects of $\mathcal{T}$ are objects of $S(\mathcal{B})$, i.e. they are endomorphisms *inner* in $\mathcal{B}$, induced by a finite dimensional Hilbert space, and if

$$\varepsilon(\rho, \sigma) = \vartheta(\rho, \sigma)$$

for any objects ρ, σ of $\mathcal{T}$.

We will denote by $\mathcal{H}(\mathcal{T})$ the collection of Hilbert spaces in $\mathcal{B}$ inducing the objects of $\mathcal{T} : \mathcal{H}(\mathcal{T}) = \{H_\rho; \rho \in |\mathcal{T}|\}$.

We can now define our crossed products as follows

Definition 1. *Let* $\mathfrak{A}$ *be a* C^**–algebra with centre* $\mathbb{C} \cdot I$*;* $\mathcal{T}$ *a full monoidal subcategory of* End$\mathfrak{A}$ *and* ε *a symmetry for* $\mathcal{T}$.

A C^**–algebra* $\mathcal{B} \supset \mathfrak{A}$ *is a crossed product* $\mathfrak{A} \times \mathcal{T}$ *if there is a group* G *of automorphisms of* $\mathcal{B}$ *s.t.*

a) $\mathfrak{A} = \mathcal{B}^G$;

b) $\mathfrak{A}' \cap \mathcal{B} = \mathbb{C} \cdot I$;

c) $(\mathcal{T}, \varepsilon)$ *is a symmetric monoidal subcategory of* $(S(\mathcal{B}), \vartheta)$;

d) $\mathfrak{A}$ *and* $\mathcal{H}(\mathcal{T})$ *generate* $\mathcal{B}$ *as a* C^**–algebra.*

Remark. Given such a system $(\mathcal{B}, G)$, the strong (point–norm) closure $\overline{G}$ of G provides another system $(\mathcal{B}, \overline{G})$ fulfilling the same condition where now $\overline{G}$ is *compact* [9]. We will hereafter take $G = \overline{G}$ by definition.

The last piece of structure of $\mathcal{T}$, needed to state the existence of a crossed product, requires that there are sufficient *special* objects in $\mathcal{T}$, i.e. objects with "determinant" one, in the following sense.

The symmetry ε determines, for each object ρ, a unitary representation ε_ρ of the rational permutations $\mathbf{P}_\infty$ of the integers, such that $\varepsilon_\rho(i, i+1) = I_{\rho^{i-1}} \times \varepsilon(\rho, \rho) = \rho^{i-1}(\varepsilon(\rho, \rho))$ [5, Appendix]. Obviously $\varepsilon_\rho(p) \in (\rho^n, \rho^n)$ if $p \in \mathbf{P}_n$.

With A_d the total antisymmetric projection in the group algebra of $\mathbf{P}_d$, say that ρ is *special* if there is an integer d and an isometry $R \in (\iota, \rho^d)$ such that

$$\begin{aligned} RR^* &= \varepsilon_\rho(A_d) \ , \\ R^* \rho(R) &= \frac{1}{d}(-1)^{d-1} \cdot I \ . \end{aligned} \tag{15}$$

Definition 2. $(\mathcal{T}, \varepsilon)$ *is specially directed if every finite set of objects is dominated by a special object.*

Recall that σ is dominated by ρ if there are partially isometric intertwiners from σ to powers of ρ with initial projections summing up to I_σ.

The main result of [4] can now be stated.

Theorem 3. *Let* $\mathfrak{A}$ *be a* C^**–algebra with centre* $\mathbb{C} \cdot I$, $(\mathcal{T}, \varepsilon)$ *a full symmetric monoidal subcategory of* End$\mathfrak{A}$ *which is specially directed. There exists a unique crossed product* $\mathfrak{A} \times \mathcal{T}$. *Furthermore the compact group* G *is given by*

$$G = \mathrm{Aut}_{\mathfrak{A}} \mathfrak{A} \times \mathcal{T} \ , \tag{16}$$

the (pointwise) stabilizer of $\mathfrak{A}$ *in* Aut $\mathfrak{A} \times \mathcal{T}$, *and is accordingly Galois–closed and uniquely associated to* $(\mathcal{T}, \varepsilon)$.

Remark. The functor H of diagram (1) is given here, together with F, as a result of the construction, by the maps discussed earlier on:

$$\begin{aligned} &\rho \in |\mathcal{T}| \to u_\rho \ ; \\ &T \in (\rho, \sigma) \to T \in (u_\rho, u_\sigma) \subset (H_\rho, H_\sigma) \ . \end{aligned}$$

If $\mathcal{T}$ is now an abstract C^*–category, s.t. $(\mathcal{T}, \varepsilon)$ is strict symmetric monoidal in the above sense (cf. [5] for the full definitions), say that $\mathcal{T}$ *has conjugates* if for each object ρ there is an object $\overline{\rho}$ and arrows $R, \overline{R}$ s.t.

$$\begin{aligned} &R \in (\iota, \overline{\rho}\rho)\ , \quad \overline{R} \in (\iota, \rho\overline{\rho}) \\ &\overline{R}^* \times I_\rho \circ I_\rho \times R = I_\rho \\ &R^* \times I_{\overline{\rho}} \circ I_{\overline{\rho}} \times \overline{R} = I_{\overline{\rho}} \\ &\overline{R} = \varepsilon(\overline{\rho}, \rho) \circ R\ . \end{aligned} \tag{17}$$

Then $R^* \circ R \in (\iota, \iota)$ and if the selfintertwiners of the monoidal unit reduce to $\mathbb{C}$ we get a number, which turns out to be an *integer*: the *dimension* of ρ. If $\mathcal{T}$ has subobjects and direct sums, we can define a *determinant*, and objects with determinant one are *special* in the above sense, and $(\mathcal{T}, \varepsilon)$ is *specially directed* [5].

If ρ is a special object, the full subcategory $\mathcal{T}_\rho$ with objects the powers of ρ can be identified with an action on a minimal C^*–algebra $\mathcal{O}_\rho$ [5, Section 4], so that it describes a minimal dual of a compact Lie group G_ρ. Since $\mathcal{T}$ is specially directed, the G_ρ form a projective family whose projective limit is a compact group G. Along with the construction of G the functor H of (1) is obtained [5, Sect. 6]. This leads to the main Theorem of [5].

Theorem 4. *A strict symmetric monoidal C^*-category with conjugates, subobjects and direct sums, such that the selfintertwiners of the monoidal unit reduce to $\mathbb{C}$, is the abstract dual of a unique compact group.*

The central tool in the characterization of our minimal duals is the eminently non commutative analysis in the Cuntz algebras: an example in the spirit of this meeting of how non commutative analysis might solve classical problems.

A parallel result to Theorem 4 by P. Deligne [10] then characterized the linear representations of algebraic groups as abstract categories; unitarity is not required in that theory but an integrality axiom of the dimensions has to be added.

The other chief application of Theorem 3 concerns its original motivation, Quantum Field Theory.

The local structure of the C^*-algebra $\mathfrak{A}$ of local observable determines the *superselection structure* [3] which can be described as a full monoidal subcategory $\mathcal{T}$ of End$\mathfrak{A}^d$, $\mathfrak{A}^d$ being the C^*-algebra associated to the dual net [11]. This discussion applies to localizable charges [3]; with appropriate modifications [11] it covers the case of quantum topological charges too (the most general case which might appear in massive theories [12]).

Through the basic locality postulate, the geometry of ordinary 4-dimensional space time plays a role and one can define a canonical *symmetry* $\hat{\varepsilon}$

for $\mathcal{T}$, so that the results of [3] allow one to conclude that $(\mathcal{T}, \hat{\varepsilon})$ is symmetric monoidal with conjugates. Then Theorem 3 produces the inclusions

$$\mathfrak{A} \subset \mathfrak{A}^d \subset \mathfrak{A}^d \times \mathcal{T} \hookrightarrow \mathcal{F} \tag{18}$$

where the last arrows embeds $\mathfrak{A}^d \times \mathcal{T}$ in the C^*-algebra of *field operators* obtained inducing up the vacuum representation of $\mathfrak{A}$ to $\mathfrak{A}^d \times \mathcal{T}$ *via* the conditional expectation defined by the average over G, and taking the norm closed union of the weakly closed local subalgebras. One has [11]

$$G = \mathrm{Aut}_{\mathfrak{A}^d}\mathcal{F} \, . \tag{19}$$

G is the group of *exact* gauge symmetries of first kind on $\mathcal{F}$, while

$$\Gamma \equiv \mathrm{Aut}_{\mathfrak{A}}\mathcal{F} \tag{20}$$

defines the (possibly larger) group of gauge symmetries of first kind, including the *spontaneously broken symmetries* [11,13].

Coming back to Theorem 3, note that the crossed product might be the same for different choices of $\mathcal{T}$. In particular we could let $\hat{\mathcal{T}}$ be the full subcategory of End$\mathfrak{A}$ with objects all the endomorphisms dominated by some object of $\mathcal{T}$, and extend uniquely the symmetry ε to a symmetry $\hat{\varepsilon}$ of $\hat{\mathcal{T}}$. Obviously

$$\mathfrak{A} \times \mathcal{T} = \mathfrak{A} \times \hat{\mathcal{T}} \equiv \mathcal{B} \, ,$$

but moreover one easily gets from [9, Lemma 6.2] that

$$(\mathcal{S}(\mathcal{B}), \vartheta) = (\hat{\mathcal{T}}, \hat{\varepsilon}) \, . \tag{21}$$

As another example consider an automorphism β of the inclusion $\mathfrak{A} \subset \mathcal{B}$ i.e.

$$\beta \in \mathrm{Aut}(\mathfrak{A}, \mathcal{B})$$

and define

$$(\mathcal{T}', \varepsilon') = \beta(\mathcal{T}, \varepsilon)$$

where the objects and the arrows of $\mathcal{T}'$ are the coniugates and the immages by β of the objects and arrows of $\mathcal{T}$ respectively, and

$$\varepsilon'(\beta\rho\beta^{-1}, \beta\sigma\beta^{-1}) \equiv \beta(\varepsilon(\rho, \sigma)) \, .$$

Again we have:

$$\mathfrak{A} \times \mathcal{T} = \mathfrak{A} \times \mathcal{T}' \, .$$

Note also that by definition and by (21)

$$\beta(\hat{\mathcal{T}}, \hat{\varepsilon}) = (\hat{\mathcal{T}}, \hat{\varepsilon}) \, . \tag{22}$$

Now we can state a converse [8].

Theorem 5. *Let $\mathfrak{A}$ and $(\mathcal{T}, \varepsilon)$ be as in Theorem 3. An automorphism β of $\mathfrak{A}$ extends to an automorphism of $\mathfrak{A} \times \mathcal{T}$ if and only if*

$$\beta(\hat{\mathcal{T}}, \hat{\varepsilon}) = (\hat{\mathcal{T}}, \hat{\varepsilon}) \ .$$

If $M \subset N$ is an inclusion of infinite factors such that

$$M' \cap N = \mathbb{C} \cdot I$$

$$M = N^G$$

for a compact group G of automorphisms of N, then

$$N = M \overset{w^*}{\times} \mathcal{T}$$

where $\overset{w^*}{\times}$ is a w^*-version of the crossed product of Theorem. 3 (cf. [4, Sect. 7]) and

$$(\mathcal{T}, \varepsilon) = (S(M), \vartheta) \ . \tag{23}$$

As a Corollary [8] we then obtain that an automorphism β of M extends to an automorphism of N if and only if

$$\beta(\mathcal{T}, \varepsilon) = (\mathcal{T}, \varepsilon) \ . \tag{24}$$

Such a result ought to extend to more general inclusions of factors and to the associated invariants recently extensively studied after the pioneering works of V.F.R. Jones.

We close this section turning to discuss some open problems. They concern first a specialization of Theorem 4, then its possible generalizations.

Let Rep$G$ denote the category of all finite dimensional continuous unitary representations of a compact group G. If we were to disregard the arrows we would consider only the unitary equivalence classes of those representations, equipped with the operations of direct sum, tensor product and complex conjugation. Would this "reduced dual" suffice to describe the group? The answer is well known to be negative in general, a counter example being the two groups, of order 8, D_4 and Q_2, where even the dimensions coincide [1].

However if G is at the opposite extreme to finite groups, namely is compact and *connected*, a recent result of D. Handelman [14] shows that the "reduced dual" determines G uniquely. In this case the reduced dual is "torsion free": no nontrivial finite subset of irreducible classes is stable under (taking components of) tensor products and conjugates. This suggest the following

Problem. *Characterize abstractly torsion free reduced duals and provide the construction of the associated connected compact groups from the abstract data.*

Turning to discuss generalizations of Theorem 3, we first point out that there is a rich history of generalizations of the Tannaka Krein Theory up to date. Generalizations to locally compact groups have been studied by Ernest, Stinespring, Tatsuuma, and Takesaki [15] where duality takes a symmetric form in the context of Hopf-von Neumann algebras and references to earlier work can be found. Generalizations to compact quantum groups have been recently given by Woronowicz [16] and in the setting of category theory by Yetter [17].

A unified treatment covering all those cases has been given by Baaj and Skandalis based only on multiplicative unitaries [18].

All these developments deal with *concrete duals* in the spirit of Tannaka-Krein theory where in the diagram (1) the *pair* $(\mathcal{T}, F)$ has to be assigned. Again Quantum Field Theory poses a challenge to this theory: model theories on a 2-dimensional space time, where the geometry is essentially different from ordinary Minkonski space, give rise to superselection structures described, as shown by Fredenhagen, Rehren and Schroer (cf. [19], and references quoted there) by a *braided* monoidal subcategory [20] of End$\mathfrak{A}$, where equation (14) no longer holds[2]. The exchange operators $I_{\rho^{i-1}} \times \varepsilon(\rho, \rho) = \varepsilon_\rho(i, i+1)$ now generate representations of the *braid groups* which do not factor through the permutation groups in general. This leads to the following question [11, Sect. 7]:

Problem. *Extend the duality theory of Theorem 4 to strict braided monoidal C^*-categories with conjugates and $(\iota, \iota) = \mathbb{C}$, and describe the appropriate class of "Quantum Groups" dual to them.*

As in the theory of group duals, the central step can be expected to be a construction of a crossed product $\mathfrak{A} \times \mathcal{T}$ of C^*-algebras by the action of such a category, parallel to Theorem 3. Minimal systems $(\mathfrak{A}, \mathcal{T})$ should again be identified with model actions. Therefore the study of model actions ought to be the first step.

Model actions of locally compact groups and their duals will be discussed in the next section, based on joint work with T. Ceccherini, C. Pinzari and J.E. Roberts ([21]; cf. the announcements [22,23]). Model actions of finite dimensional Hopf algebras have been recently investigated by Cuntz [24]. Related to the work of Cuntz and based on the theory of inclusions of factors, a different approach to the construction of a crossed product has been very recently developed by R. Longo ([29], cf. these Proceedings).

[2] In presence of quantum topological charges, the superselection category cannot be embedded in End$\mathfrak{A}$ [12,11]; it might be a braided category even in three space time dimensions [19].

3. Model actions on the Extended Cuntz Algebra

If G is a group of unitaries on the Hilbert space H, the canonical action of G on the Cuntz algebra $\mathcal{O}_H$ is *prime* (i.e. $\mathcal{O}'_G \cap \mathcal{O}_H = \mathbb{C} \cdot I$) if $\dim H < \infty$ [25]. If $\dim H = \infty$, the conclusion remains true if G has compact strong closure; however if G has trivial compactification the action is ergodic, i.e. $\mathcal{O}_G = \mathbb{C} \cdot I$ [26]. (For further results on $\mathcal{O}_G$ cf. [27] and C. Pinzari, these Proceedings).

To get nontrivial actions of a G-dual one has therefore to investigate other model actions of G. Let H be a Hilbert space and realize it as a Hilbert space with support I in $\mathcal{B}(\mathcal{H})$ for some other Hilbert space $\mathcal{H}$ (cf. e.g. [23]). The (norm closed linear span of) H^n realize the tensor powers $H \otimes \ldots \otimes H = H^{\otimes n}$ and the compact, respectively bounded, linear operators between these spaces are realized as

$$\mathcal{K}(H^n, H^m) \equiv \text{ norm closed l.s. } H^m H^{*n} \cong \mathcal{K}(H^{\otimes n}, H^{\otimes m}) \ , \qquad (25)$$

$$(H^n, H^m) \equiv \ uw \text{ closed l.s. } H^m H^{*n} \cong \mathcal{B}(H^{\otimes n}, H^{\otimes m}) \ . \qquad (26)$$

The C^*-algebra generated by the subspaces (25) as $n, m = 0, 1, 2, \ldots$, is the Cuntz algebra $\mathcal{O}_d$, $d = \dim H = 2, 3, \ldots, \infty$. The C^*-algebra generated by the subspaces (26) differs only if $d = \infty$ in which case it is the *extended Cuntz algebra* $\tilde{\mathcal{O}}_H$ [21]. It is a simple C^*-algebra covariantly associated with H (for H separable infinite dimensional there is a unique algebra $\tilde{\mathcal{O}}$). A representation π of $\tilde{\mathcal{O}}_H$ is ultraweakly continuous on each subspace (26) if and only if it is normal on each (H^n, H^n), $n = 1, 2, \ldots$, and if and only if $\pi(H)$ has support $\pi(I)$.

The unitary group $\mathcal{U}(H)$ is embedded in $\mathrm{Aut}\tilde{\mathcal{O}}_H$ as the set of all automorphisms leaving H globally stable; the embedding is the *canonical action*

$$\alpha_U(\psi) = U\psi \ ; \quad \psi \in H \ , \qquad U \in \mathcal{U}(H) \ . \qquad (27)$$

As in the case of $\mathcal{O}_d$, $d < \infty$, there is a *canonical endomorphism* $\tilde{\sigma}$ of $\tilde{\mathcal{O}}_H$ s.t.

$$\psi B = \tilde{\sigma}(B)\psi \ , \quad \psi \in H \ , \quad B \in \tilde{\mathcal{O}}_H \ , \qquad (28)$$

which commutes with the canonical action, hence induces an endomorphism $\tilde{\sigma}_G$ of $\tilde{\mathcal{O}}_G$, the fixed points in $\tilde{\mathcal{O}}_H$ under the canonical action of $G \subset \mathcal{U}(H)$. Now $\tilde{\mathcal{O}}_G \supset \tilde{\mathcal{O}}_{\mathcal{U}(H)}$ which coincides with the C^*-algebra generated by the permutations $\vartheta(p)$, $p \in \mathbb{P}_\infty$ acting on the monomials in elements of H by permuting the factors with the permutation p. This subalgebra is proved, as in the case $\dim H < \infty$, to have *trivial relative commutant*, hence every canonical action on $\tilde{\mathcal{O}}$ is *prime*.

One can then conclude that the full subcategory of unitary representations of G with objects the tensor powers of the defining representation u on H, say $(\mathcal{T}_G, \vartheta)$ with ϑ the symmetry

$$\vartheta(u^{\otimes n}, u^{\otimes s})x \otimes y = y \otimes x \,, \quad x \in H^{\otimes n}, y \in H^{\otimes s} \,,$$

is *isomorphic* as a symmetric monoidal category to $(\tilde{\mathcal{T}}_G, \vartheta)$, the full subcategory of End$\tilde{\mathcal{O}}_G$ with objects the powers of $\tilde{\sigma}_G$ and the symmetry, also denoted by ϑ, given by $\vartheta(\tilde{\sigma}_G^n, \tilde{\sigma}_G^s) = \vartheta(n, s)$ (the flip of the first r factors with the next s factors in monomials of the elements of H).

These categories do not determine G in general (they depend only upon its strong closure in $\mathcal{U}(H)$). However if G is *locally compact* and acts on H *via* its *(right) regular representation*, the duality Theorem of Tatsuuma [28] allows us to reconstruct G from $\mathcal{T}_G$ and yields to the following

Theorem 6. [21] *Every (second countable infinite) locally compact group G, embedded in* Aut$\tilde{\mathcal{O}}$ *as the canonical action of its right regular representation, is a Galois closed subgroup of* Aut$\tilde{\mathcal{O}}$:

$$G = \mathrm{Aut}_{\tilde{\mathcal{O}}_G} \tilde{\mathcal{O}} \tag{29}$$

and the symmetric monoidal subcategory $(\tilde{\mathcal{T}}_G, \vartheta)$ of End$\tilde{\mathcal{O}}_G$ *determines G.*

Furthermore the following result holds

Theorem 7. (C. Pinzari) *If G is embedded in* Aut$\tilde{\mathcal{O}}$ *as in Theorem 6 and G is amenable, then $\tilde{\mathcal{O}}_G$ is isomorphic to $\tilde{\mathcal{O}}$. If G is of finite order d, then $\mathcal{O}_G$ is isomorphic to the Cuntz algebra $\mathcal{O}_d$.*

The analysis of $\tilde{\mathcal{O}}_G$ can be carried over to the case of regular multiplicative unitaries which have on associated braiding, and Theorem 7 extends too [21].

In view of Theorem 6, we can state the abstract duality problem for locally compact groups, as already formulated in the Nara Conference:

Problem. [21,23] *Given a C^*-algebra $\mathfrak{A}$ with centre $\mathbb{C} \cdot I$, and a unital endomorphism ρ of $\mathfrak{A}$, when is*

$$\{\mathfrak{A}, \rho\} \sim \{\tilde{\mathcal{O}}_G, \tilde{\sigma}_G\} \tag{30}$$

for some locally compact group G? Or, when is $\mathfrak{A}$ ultraweakly dense in $\tilde{\mathcal{O}}_G$ in some faithful representation?

The properties of $\{\tilde{\mathcal{O}}_G, \tilde{\sigma}_G\}$ reflecting properties of the regular representation of G should be a guide. Stated for $\{\mathfrak{A}, \rho\}$ they are *necessary* conditions for (30) to hold. To spell them out, denote by $\mathcal{T}_\rho$ the full subcategory of End$\mathfrak{A}$ with objects the powers of ρ. Our necessary conditions can then be formulated as follows [21,23]:

1. $(\mathcal{T}_\rho, \varepsilon)$ *is a symmetric monoidal subcategory of* End$\mathfrak{A}$.

2. *There is a Hilbert space K in $\mathfrak{A}$, $K \subset (\rho, \rho^2)$ with support I in some faithful representation of $\mathfrak{A}$ and s.t., for $r, s = 1, 2, \ldots$,*

$$(K^r, K^s) \subset \mathfrak{A} ;$$

3. $(\rho, \rho) \subset (K, K)$;
4. (ρ^r, ρ^s), $r, s = 1, 2, \ldots$, *generate $\mathfrak{A}$ as a C^*-algebra.*

Property 4 applies to $\tilde{\mathcal{O}}_G$ when G is *amenable.*

The Hilbert space K in properties 2,3 is given by $K = WH$, where H is the generating Hilbert space of $\tilde{\mathcal{O}}$ and W the image in $\tilde{\mathcal{O}}$ of the fundamental unitary on $H \otimes H = L^2(G \times G)$ given by $(Wf)(s,t) = f(st^{-1}, t)$.

Theorem 7 follows from conditions 2,3,4: the endomorphism of $\tilde{\mathcal{O}}$ induced by W, mapping $\psi \in H$ to $W\psi \in K$ (as in the Cuntz theory, cf. [24]) maps $\tilde{\mathcal{O}}$ onto $\tilde{\mathcal{O}}_G$. Denoting by τ its restriction to $\tilde{\mathcal{O}}_G$ we have, combining with Theorem 6, that

$$G = \mathrm{Aut}_{\tau(\mathfrak{A})} \mathfrak{A} . \tag{31}$$

Then our problem takes the more precise form: given a system $\{\mathfrak{A}, \rho\}$ fulfilling conditions 1.-4., is there an endomorphism τ s.t. (30), (31) hold?

References

[1] E. Hewitt, K.A. Ross, Abstract harmonic analysis II, Springer-Verlag, New York (1970).

[2] S. Doplicher, J.E. Roberts, Fields, statistics and non-abelian gauge groups, *Commun. Math. Phys.*, **28**, 331–348 (1972).

[3] S. Doplicher, R. Haag, J.E. Roberts, Local observables and particle statistics I, *Commun. Math. Phys.*, **23**, 199–230 (1971); II, *Commun. Math. Phys.*, **35**, 49–95 (1974).

[4] S. Doplicher, J.E. Roberts, Endomorphisms of C^*-algebras, cross products and duality for compact groups, *Ann. Math.*, **130**, 75–119 (1989).

[5] S. Doplicher, J.E. Roberts, A new duality theory for compact groups, *Inventions Math.*, **89**, 157–218 (1989).

[6] S. Doplicher, Abstract compact group duals, operator algebras and Quantum Field Theory, Proceedings of the ICM–90, Kyoto; Springer, 1991.

[7] J. Cuntz, Simple C^*-algebras generated by isometries, *Commun. Math. Phys.*, **57**, 173–185 (1977).

[8] D. Buchholz, S. Doplicher, R. Longo, J.E. Roberts, *Extension of Automorphisms and Gauge symmetries*, *Commun. Math. Phys.*, to appear.

[9] S. Doplicher, J.E. Roberts, Compact group actions on C^*-algebras, *Journ. Operator Theory*, **19**, 293–305 (1988).

[10] P. Deligne, Categories tannakiennes, Grothendieck Festschrift, Birkhäuser (1990).

[11] S. Doplicher, J.E. Roberts, Why there is a field algebra with a compact gauge group describing the superselection structure in particle physics, *Commun. Math. Phys.*, **131**, 51–107 (1990).

[12] D. Buchholz, K. Fredenhagen, Locality and the structure of particle states, *Commun. Math. Phys.*, **84**, 1–54 (1982).

[13] D. Buchholz, S. Doplicher, R. Longo, J.E. Roberts, A new look at Goldstone's Theorem, Reviews in Mathematical Physics, to appear.

[14] D. Handelman, Representation rings as invariants for compact groups and limit ratio Theorems for them, University of Ottawa preprint, 1991.

[15] M. Takesaki, A characterization of group algebras as a converse of Tannaka–Stinespring–Tatsuuma duality Theorem, *American Journ. Math.*, **91**, 529–564 (1969).

[16] S.L. Woronowicz, Tannaka–Krein duality for compact matrix pseudogroups. Twisted $SU(N)$ groups, *Inventiones Math.*, **93**, 35–76 (1988).

[17] D. Yetter, *Math. Proceedings Camb. Phil. Soc.*, **108**, 261–290 (1990).

[18] S. Baaj, G. Skandalis, Unitaires multiplicatifs et dualité pour les produits croises de C^*-algebres, preprint.

[19] K. Fredenhagen, K.H. Rehren, B. Schroer, Superselection sectors with braid group statistics and exchange algebras I, *Commun. Math. Phys.*, **125**, 201–226 (1989); II preprint.

[20] A. Joyal, R. Street, Braided monoidal categories, Mcquarie Mathematics Reports n. 860081 (1986).

[21] T. Ceccherini, S. Doplicher, C. Pinzari, J.E. Roberts, A generalization of the Cuntz Algebras and Model Actions, preprint.

[22] S. Doplicher, Operator algebra and group duality in Current topics in operator algebras, H. Araki, H. Choda, Y. Nakagami, K. Saitô and J. Tomiyama editors, World Scientific, 1991.

[23] S. Doplicher, Operator Algebras, group actions and abstract duals, to appear in the proceedings of the Istambul conference on Operator Algebras, 1991.

[24] J. Cuntz, Regular actions of Hopf algebras on the C^*-algebra generated by a Hilbert space, preprint.

[25] S. Doplicher, J.E. Roberts, Duals of compact Lie groups realized in the Cuntz algebras and their action on C^*-algebras, *Journ. Functional Analysis*, **73**, 96–120 (1987).

[26] T. Ceccherini, C. Pinzari, Canonical actions on $\mathcal{O}_\infty$, *Journ. Functional Analysis*, **103**, 26–39 (1992).
[27] C. Pinzari, Semigroups of non–unital endomorphisms of C^*–algebras and compact group duality, *Journ. Functional Analysis*, (to appear).
[28] N. Tatsuuma, A duality Theorem for locally compact groups, *J. Math. Kyoto University*, **6**, 197–293 (1967).
[29] R. Longo, A duality for Hopf algebras and for subfactors I, preprint.

Regular Actions of Compact Groups on Cuntz Algebras

C. PINZARI
Dipartimento di Matematica
Università di Roma "Tor Vergata"
via della Ricerca Scientifica
I-00133 Roma, Italy

Key words: endomorphism semigroup, group duality, C*-category,K-theory

1. Introduction

Let G be a compact and metrizable group and $\mathcal{O}_{L^2(G)}$ the Cuntz algebra canonically generated by $L^2(G)$ [3]. The regular action α of G on $\mathcal{O}_{L^2(G)}$ is defined by extending the (left) regular representation λ of G on $L^2(G)$ to automorphisms of $\mathcal{O}_{L^2(G)}$.

The basic subject of this report will be the fixed point algebra $\mathcal{O}_{\lambda(G)}$ which is a simple, nuclear, separable and purely infinite C^*–algebra canonically associated with G.

In section 2 we shall report from [6], [2] and [12] some general properties of canonical actions of topological groups G on Cuntz algebras $\mathcal{O}_H$ induced by (s.o. continuous unitary) representations u on H of the underlying group. The case of compact group action turns out to be very interesting since the fixed point algebra $\mathcal{O}_{u(G)}$ has always trivial relative commutant in $\mathcal{O}_H$, and is often simple (it suffices either G finite or u selfconjugate). Gift $\mathcal{O}_{u(G)}$ with a suitable semigroup Δ of endomorphisms induced by finite–dimensional G–invariant Hilbert spaces of $\mathcal{O}_H$. Following Doplicher–Roberts [7] the full monoidal C^*–subcategory $\mathcal{T}_\Delta$ of End$\mathcal{O}_{u(G)}$, the category of (not–necessarily unital) endomorphisms of $\mathcal{O}_{u(G)}$, defined by Δ is naturally permutation symmetric and identifies with a category of (unitary finite–dimensional) representations of G.

If $u = \lambda$ a distinguished semigroup Δ_λ is generated by those endomorphisms induced by irreducible G–submodules of $H \equiv L^2(G)$. Section 3 is dedicated to the problem of characterizing the categories which arise in this way. Our theorem could be considered a variant of the results of Doplicher

H. Araki et al. (eds.), Quantum and Non-Commutative Analysis, 419–426.

and Roberts about their duality theory for compact groups [8], [9]. In fact the core of its proof is in part a generalization of some arguments of them to certain categories of non–unital endomorphisms.

In the last section we report from [13] some results about the classification of the C^*–algebras $\mathcal{O}_{\lambda(G)}$'s and its K–theory.

2. Main properties of canonical group actions on Cuntz algebras

Let G be a topological group and u a strongly continuous unitary representation of G on a separable Hilbert space H. Define a canonical action of G by automorphisms of the Cuntz algebra $\mathcal{O}_H$ generated by H by:

$$\alpha_g \psi = u(g)\psi \ , \quad \psi \in H \ , \quad g \in G \ .$$

The map α is continuous if Aut$\mathcal{O}_H$ is equipped with the pointwise norm convergence topology.

We call α quasi–special if $u(G)$ is precompact in the strong operator topology and u is direct sum of quasi–special representations $\{u_i\}_{i \in N}$ of G, i.e. (determinant $u_i(G))^{r_i} = 1$ for some $r_i \in \mathbb{N}$ and all $g \in G$, $i \in N$. Here N denotes a countable set. The case of special action (i.e. dim $H < +\infty$ and $u(G) \subset SU(H)$) dealt in [6] can be generalized to get the following

Theorem 1. *Assume $u(G)$ s.o. precompact. Then*

i) *$\mathcal{O}_{u(G)}$ has trivial relative commutant in $\mathcal{O}_H$,*

ii) *the stabilizer of $\mathcal{O}_{u(G)}$ in $\mathcal{O}_H$ coincides with α_G,*

iii) *if α is quasi–special, the fixed point algebra $\mathcal{O}_{u(G)}$ is simple.*

Let $H = \bigoplus_{i \in N} H_i$ be an orthogonal decomposition of H into finite dimensional subspaces $\{H_i\}_{i \in N}$ (not necessarily quasi–special) and define a natural unitary action of $U = \prod_{i \in N} U(H_i)$ on H. If $u(G)$ is strongly precompact then $u(G)$ is contained in some U, hence $\mathcal{O}_{u(G)} \supseteq \mathcal{O}_U$. Now $\mathcal{O}_U$ turns out to be the smallest Banach subspace of $\mathcal{O}_H$ containing the permutation operators $\Theta = \{\vartheta_{i_1 \ldots i_r}(p), i_1, \ldots, i_r \in \mathbb{N}, p \in \mathbb{P}_r, r \in \mathbb{N}\}$, where

$$\vartheta_{i_1 \ldots i_r}(p) = \sum_{h_1, \ldots, h_r} \psi_{h_1}^{(i_1)} \ldots \psi_{h_r}^{(i_r)} (\psi_{h_{p(1)}}^{i_{p(1)}} \ldots \psi_{h_{p(r)}}^{i_{p(r)}})^*$$

with $\{\psi_h^{(i)}\}_h$ an orthonormal basis of H_i. The proof of Theorem 1, i) is thus reduced to show that the relative commutant of Θ in $\mathcal{O}_H$ is trivial. Theorem 1, iii) is essentially due to the abundance of isometries in $\mathcal{O}_{u(G)}$: if a G–invariant subspace $K \subset H$ carries a quasi–special representation of

G then a suitable tensor power of a totally antisymmetric vector S_K of K^d, $d = \dim K$, belongs to $\mathcal{O}_{u(G)}$.

Fixed another sequence of positive integers $\{r_i\}_{i\in N}$ let G_0 be the subgroup of all $g \in U$ such that (determinant $g|_{H_i})^{r_i} = 1$, $i \in N$. The simple C^*–algebra $\mathcal{O}_{G_0}$ is generated, as a C^*–algebra, by Θ and the isometries $S_{H_1}^{r_1}$, $S_{H_2}^{r_2}, \ldots$.

The non–comapct group action case is dealt in the following. Let E be the orthogonal projector of H onto the closed subspace generated by its finite–dimensional G–submodules. The C^*–algebra $C^*(EH)$ generated by EH is clearly globally G–invariant, and it actually determines all the G–invariant elements of $\mathcal{O}_H$:

Theorem 2.

i) $C^*(EH)_{u(G)} = \mathcal{O}_{u(G)}$, *so* α *is ergodic iff* u *has no non–zero finite–dimensional subrepresentation,*

ii) *if* $0 < E < I$, *then* $\mathcal{O}_{u(G)}$ *has trivial relative commutant in* $\mathcal{O}_H$ *iff some tensor power* $u^{\otimes n}$ *of* u *contains the trivial one–dimensional representation.*

Theorem 2, ii) provides an obstruction to realize a model action of any full subcategory of Rep(G) (the category of unitary so–continuous representations of (G) on $\mathcal{O}_{u(G)}$, as described in [7] and shortly outlined in the following (cf..also S. Doplicher, these Proceedings).

Let $\alpha : G \to \mathrm{Aut}B$ be a faithful pointwise norm–continuous homomorphism of G into the automorphism group of a unital C^*–algebra B and set $\mathfrak{A} = B^\alpha$. Any G–invariant finite dimensional Hilbert space K of B induces an endomorphism σ_K of B which restricts to ρ_K on $\mathfrak{A}$.

Define $u_K = \alpha|_K$. If $\mathfrak{A}$ has trivial relative commutant in B the space of intertwiners $(\rho_K, \rho_{K'}) = \{T \in \mathfrak{A} | T\rho_K(A) = \rho_{K'}(A)T,\ A \in \mathfrak{A},\ T\rho_K(I) = T\}$ in End$\mathfrak{A}$ identifies naturally with the space $(u_K, u_{K'})$ of intertwiners in Rep(G). The map F which assignes u_K to ρ_K and is the identity from $(\rho_K, \rho_{K'})$ to $(u_K, u_{K'})$ is a faithful functor from a full subcategory of Rep(G) onto the full subcategory $S(B)$ of End$\mathfrak{A}$ with objects the endomorphisms of the form ρ_K. The image $S(B)$ is a strictly monoidal C^*–category with permutation symmetry $\theta(\rho_K, \rho_{K'})\psi \otimes \psi' = \psi' \otimes \psi$, $\psi \in K$, $\psi' \in K'$.

Theorem 2 shows that $\mathcal{O}_\infty$ is too thin to define model actions of duals of non–compact groups on $\mathcal{O}_{u(G)}$. One is naturally led to define an extended Cuntz algebra $\tilde{\mathcal{O}}_H$ canonically associated to a Hilbert space H and to consider canonical (locally compact) group actions by automorphisms on it. The first results about these topics are contained in [1].

3. Duality for compact and metrizable groups

Let us restrict ourselves to the case G a compact group and $u = \lambda$ its (left) regular representation. The simple C^*–algebra $\mathcal{O}_{\lambda(G)}$ canonically associated with G has a distinguished semigroup Δ_λ of non–unital endomorphisms: the one generated through compositions, direct sums and subobjects by the endomorphisms induced by an irreducible decomposition of λ.

Recall that if $\mathfrak{A}$ is a unital C^*–algebra and ρ, σ are objects of $\mathrm{End}\mathfrak{A}$ the C^*–category of not–necessarily unital endomorphisms of $\mathfrak{A}$, ρ is called a subobject of σ if it is of the form $\rho(\cdot) = E\sigma(\cdot)$ for some projection $E \in (\sigma, \sigma)$ (here (ρ, σ) denotes the usual space of intertwiners from ρ to σ in $\mathrm{End}\mathfrak{A}$).

If $\rho(I)$ and $\sigma(I)$ are orthogonal projections the direct sum $\rho \oplus \sigma$ of ρ and σ is defined as the pointwise sum of ρ and σ.

Note that our notions of subobjects and direct sums slightly differ from those of [9], which are defined only up to unitary equivalence.

Let ϑ be the permutation simmetry of the full subcategory $\mathcal{T}_\lambda$ of $\mathrm{End}\mathcal{O}_{\lambda(G)}$ generated by Δ_λ as defined in the above section. The triple $(\mathcal{O}_{\lambda(G)}, \mathcal{T}_\lambda, \vartheta)$ provides, then, an abstract G–dual which contains any irreducible. The following theorem characterizes such triples.

Let $\mathfrak{A}$ be a unital C^*–algebra and Δ a semigroup with identity automorphism $\imath$ of endomorphisms of $\mathfrak{A}$ with permutation symmetry ε (cf. S. Doplicher, these Proceedings); Doplicher–Roberts in [9] defined a (necessarily unique up to unitary equivalence) conjugate endomorphism $\overline{\rho}$ of ρ by the property that there is $V \in (\imath, \overline{\rho}\rho)$, such that:

$$\overline{V}^* \rho(V) = \rho(I) \tag{3.1}$$

$$V^* \overline{\rho}(V) = \overline{\rho}(I)$$

with $\overline{V} = \varepsilon(\overline{\rho}, \rho)V$. If $\mathfrak{A}$ has trivial centre they pr oved that V^*V in an integer multiple of I, and defines the dimension of ρ denoted $\dim \rho$.

We call Δ orthogonally generated if it is generated as a semigroup with subobjects and direct sums by a countable set P of endomorphisms with pairwise orthogonal supports fulfilling $\sum_{\rho \in P} \rho(I) = I$ if finite. Such a subset P is called regular if it is selfconjugate, each $\rho \in P$ is irreducible: $(\rho, \rho) = \mathbb{C}\rho(I)$, occurs $\dim \rho$ times in P up to unitary equivalence and for each $\rho \in \Delta$ there are $\rho_1, \ldots, \rho_n \in P$ and isometries $S_i \in (\rho_i, \rho)$ whose range projections sum up to $\rho(I)$.

Theorem 3. *Let $\mathfrak{A}$ be a C^*–algebra with trivial centre and Δ a semigroup of endomorphisms of $\mathfrak{A}$ with permutation symmetry ε orthogonally generated by a regular subset P If $\mathfrak{A}$ is generated as a Banach space by the arrows of Δ then the triple $(\mathfrak{A}, \Delta, \varepsilon)$ is of the form $(\mathcal{O}_{\lambda(G)}, \Delta_\lambda, \vartheta)$. This G is unique up to isomorphisms and does not depend on P.*

Theorem 3 is actually a consequence of a more general theorem which characterizes quasi–special actions of compact groups on C^*–algebras. This theorem generalizes the Doplicher–Roberts construction of the cross product $\mathfrak{A} \times \mathcal{T}_\rho$ of $\mathfrak{A}$ by the full monoidal subcategory $\mathcal{T}_\rho$ of End$\mathfrak{A}$ generated by a single special unital endomorphism ρ.

Given in general $(\mathfrak{A}, \Delta, \varepsilon)$ and $r \in \mathbb{N}$ an element $\rho \in \Delta$ is called r–special (or quasi–special) if there is a positive integer d and an isometry $R \in (i, \rho^{rd})$ such that

$$RR^* = \varepsilon^{(d)}(A_d) \times \ldots \times \varepsilon^{(d)}(A_d) \quad r\text{–times} \tag{3.2}$$

$$R^* \varepsilon^{(rd+1)}(1\ rd+1)R = \frac{\rho(I)}{d} \tag{3.3}$$

where for each $s \in \mathbb{N}$ $\varepsilon^{(s)} : C^*(\mathbb{P}_s) \to (\rho^s, \rho^s)$ denotes the canonical representation of the group algebra $C^*(\mathbb{P}_s)$ of the permutation group $\mathbb{P}_s$ defined by assigning $\rho^{t-1}(\varepsilon(\rho, \rho))\rho^s(I)$ to the exchange $(t\, t+1)$, $\times$ denotes the monoidal product in the full monoidal subcategory $\mathcal{T}_\Delta$ of End$\mathfrak{A}$ defined by Δ [9] and A_d is the totally antisymmetric projection of $\mathbb{P}_d$. If in particular $r = 1$, this definition is equivalent to the usual definition of special endomorphism of [8]: there is an isometry $R \in (i, \rho^d)$ which satisfies (3.2) with $r = 1$ and

$$R^*\rho(R) = \frac{(-1)^{d-1}}{d}\rho(I) \ ,$$

cf. [12]. If ρ is r–special with associated integer d and $\mathfrak{A}$ has trivial centre ρ^{rd-1} contains a conjugate of ρ as a subobject and d is uniquely determined as the dimension of ρ.

Theorem 4. *Let $\mathfrak{A}$ be a C^*–algebra with centre $\mathbb{C}I$ and (Δ, ε) a permutation symmetric semigroup of endomorphisms of $\mathfrak{A}$ orthogonally generated by a subset P constituted by quasi–special endomorphisms. Then there is a unique cross-product C^*–algebra $\mathfrak{A} \times \mathcal{T}_\Delta$ of $\mathfrak{A}$ by $\mathcal{T}_\Delta$, which does not depend on P.*

We refer the reader to [S. Doplicher, these Proceedings] for the definition of cross product $\mathfrak{A} \times \mathcal{T}_\Delta$.

4. On the classification of $\mathcal{O}_{\lambda(G)}$ and K–theory

The C^*–algebra $\mathcal{O}_{\lambda(G)}$ is gifted with an automorphic action γ of the circle group $\mathbb{T}$: $\gamma_z\psi = z\psi$, $\psi \in L^2(G)$, $z \in \mathbb{T}$. Let $\mathcal{O}^0_{\lambda(G)}$ denote the γ–fixed points, also called the zero–grade. A natural question is: when is $\mathcal{O}_{\lambda(G)}$ isomorphic to $\mathcal{O}_{\lambda(G')}$ according to a zero–grades preserving isomorphism? The general answer is given in the following:

Theorem 5.

i) *if G is finite there is a canonical isomorphism*

$$\tau : \mathcal{O}_{L^2(G)} \to \mathcal{O}_{\lambda(G)}$$

such that $\tau \circ \gamma_z = \gamma_z \circ \tau$, $z \in \mathbb{T}$;

ii) *if G and G' are compact and infinite there is an isomorphism $\beta : \mathcal{O}_{\lambda(G)} \to \mathcal{O}_{\lambda(G')}$ with $\beta(\mathcal{O}^0_{\lambda(G)}) = \mathcal{O}^0_{\lambda(G')}$ if and only if the group C^*–algebras $C^*(G)$ and $C^*(G')$ are isomorphic.*

Part i) of Theorem 5 is proved in [1] in the more general case of the regular action of a locally compact amenable group G on the extended Cuntz algebra $\tilde{\mathcal{O}}_H$. We shall give a short outline of this.

Let H be a separable Hilbert space contained in some $B(\mathcal{H})$ with support I (the support of H is defined as the least upper bound of the range projection of any orthonormal basis of H). For any pair of non–negative integers r,s define (H^r, H^s) as the ultraweakly closed subspace of $B(\mathcal{H})$ generated by $H^s(H^r)^*$. Here $H^0 = \mathbb{C}I$.

Let $\tilde{\mathcal{O}}_H$ be the C^*–algebra generated by all the subspaces (H^r, H^s). If $d = \dim H < +\infty$ $\tilde{\mathcal{O}}_H$ is the Cuntz algebra $\mathcal{O}_H$. The non–degenerate Hilbert space representations π of $\tilde{\mathcal{O}}_H$ ultraweakly continuous on each (H^r, H^s) (shortly locally ultraweakly continuous) are characterized by support $\pi(H) = I$. $\tilde{\mathcal{O}}_H$ is a simple C^*–algebra covariantly associated with H (cf. [1]).

Let a locally compact second countable group G act on $\tilde{\mathcal{O}}_H$ via its left regular representation λ:

$$\alpha_g \psi = \lambda(g)\psi \ , \quad g \in G \ , \quad \psi \in L^2(G) \equiv H$$

and define $(H^r, H^s)_{\lambda(G)}$ as the intersection of (H^r, H^s) with the fixed point algebra $\tilde{\mathcal{O}}_{\lambda(G)}$.

Let i denote the trivial representation of G on $L^2(G)$. Any unitary intertwiner U from $i \otimes \lambda$ to $\lambda \otimes \lambda$ defines a Hilbert space $\tau_U(H) \equiv K_U$ with support I in $(H, H^2)_{\lambda(G)} \subset \tilde{\mathcal{O}}_{\lambda(G)}$ by:

$$\tau_u(\psi) = Ul(\psi) \ , \quad \psi \in H$$

and $\tau_U : H \to K_U$ is unitary. Here $l(\psi) \in (H, H^2)$ the intertwiner from λ to $i \otimes \lambda$ tensorizing on the left by ψ. Now $K_U \subset (H, H^2)_{\lambda(G)}$ implies $(K_U^r, K_U^s) \subset (H^{r+1}, H^{s+1})_{\lambda(G)} \subset \tilde{\mathcal{O}}_{\lambda(G)}$ hence, by the covariance property τ_U extends uniquely to a locally u.w. continuous monomorphism $\tau_U : \tilde{\mathcal{O}}_H \to \tilde{\mathcal{O}}_{\lambda(G)}$. Choose the fundamental unitary:

$$U\psi(h,g) = \psi(g^{-1}h, g) \ , \quad \psi \in H \otimes H = L^2(G \times G)$$

and denote by τ and K the corresponding Hilbert space and monomorphism. We have that $(H^r, H^s)_{\lambda(G)} \subset (K^r, K^s)$, $r, s = 0, 1, 2, \ldots$, if moreover G is amenable there is a conditional expectation $m : \tilde{\mathcal{O}}_H \to \tilde{\mathcal{O}}_{\lambda(G)}$, hence τ is surjective and provides the desired isomorphism.

If $\dim H < +\infty$ more in general R. Longo and M. Izumi proved independently that the fixed point algebra under the regular action of a finite–dimensional Hopf algebra considered by Cuntz in [5] is again isomorphic to a Cuntz algebra, although this isomorphism is not canonical [11], [14].

If G and G' are two compact infinite groups with $C^*(G) \cong C^*(G')$ it is not difficult to exhibit a zero–grades preserving isomorphism $\beta : \mathcal{O}_{\lambda(G)} \to \mathcal{O}_{\lambda(G')}$ (cf. [13]). The converse is due to the fact that the dimensions of all the irreducibles of $C^*(G)$ are the square roots of the dimensions of GNS representations arising from the minimal trace states of $\mathcal{O}^0_{\lambda(G)}$. In [13] all the traces of $\mathcal{O}^0_{\lambda(G)}$ are determined. If t is such a trace, then the composition of t with the conditional expectation $m_0 : \mathcal{O}_{\lambda(G)} \to \mathcal{O}^0_{\lambda(G)}$ obtained overaging over the $\mathbb{T}$–action defines a canonical state φ_t on $\mathcal{O}_{\lambda(G)}$ which can be used to induce Hilbert space representations π_{φ_t} of $\mathcal{O}_{\lambda(G)}$ via the GNS construction. This suggests the following:

Problem. Define $N_t(G) = \pi_{\varphi_t}(\mathcal{O}^0_{\lambda(G)})$ and $M_t(G) = \pi_{\varphi_t}(\mathcal{O}_{\lambda(G)})$. Is the inclusion $N_t(G) \subset M_t(G)$ irreducible (i.e. $N_t(G)' \cap M_t(G) = \mathbb{C}I$)? In case of positive answer, which is the Murray–von Neumann type of $M_t(G)$ and $N_t(G)$?

Turning to the classification problem of $\mathcal{O}_{\lambda(G)}$, we do not know whether $\mathcal{O}_{\lambda(G)} \simeq \mathcal{O}_{\lambda(G')}$ implies $C^*(G) \cong C^*(G')$ in general.

Nevertheless $C^*(G)$ is determined by certain C^*–subalgebras of $\mathcal{O}_{\lambda(G)}$: $\mathcal{O}_{\lambda(G)}$ is the C^*–inductive limit of certain canonical C^*–subalgebras which are extensions of Cuntz algebras of finite orders often by finite direct sums of the compacts. The arising orders are all the dimensions of finite–dimensional subrepresentations of λ (for more details we refer to [13]).

We conclude this section with the following result about the K–theory of $\mathcal{O}_{\lambda(G)}$. Let $K_{\lambda(G)}(H)$ denote the C^*–subalgebra of $\mathcal{O}_{\lambda(G)}$ of compact intertwiners of λ. It is isomorphic to the C^*–subalgebra of $\bigoplus_{u \in \hat{G}} M_{d_u}(\mathbb{C})$ of functions f on $\hat{G}$, a complete set of irreducibles of G, with $f(u) \in M_{d_u}(\mathbb{C})$ vanishing at infinity. For each $u \in \hat{G}$ by d_u we mean its dimension.

Theorem 6. *If G is a compact infinite and metrizable group, then $K_1(\mathcal{O}_{\lambda(G)}) = \{0\}$ and the inclusion*

$$K_{\lambda(G)}(H) \to \mathcal{O}_{\lambda(G)}$$

induces an isomorphism at K_0–groups level, hence

$$K_0(\mathcal{O}_{\lambda(G)}) = \bigoplus_{\mathbb{N}} \mathbb{Z} \, .$$

We refer the reader again to [13] for the computation of the K_0–ordered group of the canonical C^*–subalgebra $\mathcal{O}^0_{\lambda(G)}$ of $\mathcal{O}_{\lambda(G)}$.

References

[1] T. Ceccherini, S. Doplicher, C. Pinzari, J.E. Roberts, A generalization of Cuntz algebras and model actions, In preparation.

[2] T. Ceccherini, C. Pinzari, Canonical actions on $\mathcal{O}_\infty$, *J. Funct. Anal.* **103**, 26–39 (1992).

[3] J. Cuntz, Simple C^*–algebras generated by isometries, *Commun. Math. Phys.* **57**, 173–185 (1977).

[4] J. Cuntz, K–theory for certain C^*–algebras, *Ann. Math.* **113**, 181–197 (1981).

[5] J. Cuntz, Regular actions of Hopf algebras on the C^*–algebra generated by a Hilbert space, preprint (1991).

[6] S. Doplicher, J.E. Roberts, Duals of compact Lie groups realized in the Cuntz algebras and their actions on C^*–algebras, *J. Funct. Anal.* **74**, 96–120 (1987).

[7] S. Doplicher, J.E. Roberts, Compact group actions on C^*–algebras, *J. Op. Th.* **19**, 283–305 (1988).

[8] S. Doplicher, J.E. Roberts, Endomorphisms of C^*–algebras, cross products and duality for compact groups, *Ann. Math.* **130**, 75–119 (1989).

[9] S. Doplicher, J.E. Roberts, A new duality theory for compact groups, *Inv. Math.* **89**, 157–218 (1989).

[10] S. Baaj, G. Skandalis, Unitaries multiplicatifs et dualité pour le produits croises de C^*–algebras, preprint (1991).

[11] R. Longo, A duality for Hopf algebras and for subfactors. I, preprint (1992).

[12] C. Pinzari, Semigroups of non–unital endomorphisms of C^*–algebras and compact group duality, *J. Funct. Anal.*, to appear.

[13] C. Pinzari, Simple C^*–algebras associated with compact groups and K–theory, preprint (1992).

[14] M. Izumi, On subalgebras of non–AF algebras with finite Watatani index. I, preprint (1992).

Almost Shift Invariant Projections in Infinite Tensor Products

OLA BRATTELI
Department of Mathematics, University of Oslo, P.O.Box 1053 Blindern, N-0316 Oslo, Norway

DAVID E.EVANS
Department of Mathematics and Computer Science, University College of Swansea, Singleton Park, Swansea SA2 8PP, Wales, U.K.

and

AKITAKA KISHIMOTO
Department of Mathematics, Hokkaido University, Sapporo, 060 Japan

Abstract. The Rohlin property for an automorphism σ of a UHF algebra is a key property to show that σ is almost a product type action and was established for the shift on a UHF algebra of Glimm's type n^∞ with n even. We shall try to prove this property in the case n is odd and show an *approximate* Rohlin property.

Key words: Rohlin property, shift, UHF algebra, CAR algebra

1. The Rohlin property

Let σ be an automorphism of a unital C^*-algebra A. Following [Con], [HO], [BEK] we will say that the dynamical system (A, σ) has the *Rohlin property* if there exists a strictly increasing sequence $(n_k)_k$ of positive integers such that for each $m = n_k$ there is a sequence

$$(e_i^{(j)} | i = 0, 1, \dots, m)_j$$

of families of projections in A with

$$e_m^{(j)} = e_0^{(j)},$$

$$\sum_{i=0}^{m-1} e_i^{(j)} = 1,$$

$$\lim_{j \to \infty} \|\sigma(e_i^{(j)}) - e_{i+1}^{(j)}\| = 0$$

for $i = 0, 1, \dots, m-1$, and

$$\lim_{j \to \infty} \| [x, e_i^{(j)}] \| = 0$$

H. Araki et al. (eds.), Quantum and Non-Commutative Analysis, 427–434.

for all $x \in A$, $i = 0, 1, \dots, m-1$.

It was established in [Voi, Lemma 3.3] (see also [BKRS], Lemma 2.1]) that if A is a UHF algebra (or more generally, if A is a unital AF algebra and σ is an approximately inner automorphism) and (A, σ) has the Rohlin property, then σ is an almost inductive limit automorphism. This means that for any $B \in \mathcal{F}(A) =$ the set of finite dimensional *-subalgebras of A containing the unit of A, and any $\varepsilon > 0$, there is a $C \in \mathcal{F}(A)$ such that $B \subseteq^{\varepsilon} C$ and $d(\alpha(C), C) < \varepsilon$. Here we use the terminilogy $C_1 \subseteq^{\varepsilon} C_2$ iff

$$\sup\{\inf\{\|x-y\| \big| y \in C_2, \|y\| \le 1\} \big| x \in C_1, \|x\| \le 1\} < \varepsilon$$

and

$$d(C_1, C_2) = \inf\{\varepsilon > 0 | C_1 \subseteq^{\varepsilon} C_2 \text{ and } C_2 \subseteq^{\varepsilon} C_1\}$$

whenever C_1, C_2 are subspaces of A. If σ is an almost inductive limit automorphism and $\varepsilon > 0$, then there exists a unitary $v \in A$ such that $Adv \circ \sigma$ is an inductive limit automorphism and $\|v - 1\| < \varepsilon$, [Voi, Proposition 2.3]. That $\beta = Adv \circ \sigma$ is an inductive limit automorphism means that there is an increasing sequence $(A_n) \subseteq \mathcal{F}(A)$ such that $\bigcup_n A_n$ is dense in A and $\beta(A_n) = A_n$ for all n.

It is an open problem whether or not any automorphism of a UHF algebra is an almost inductive limit automorphism. If in particular $A = M_{2^\infty} = \bigotimes^{\infty} M_2$, i.e. A is the CAR algebra, and σ is the (Bernoulli) shift on A, it was established in [BKRS] that σ has the Rohlin property, and hence σ is an almost inductive limit automorphism. Let σ be a quasifree automorphism of A, i.e.

$$\sigma(a(f)) = a(Uf)$$

where $f \in \mathcal{H} =$ the one particle space, $a(f)$ is the annihilator corresponding to f, and U is a unitary operator on $\mathcal{H}$. It was established in [BEK] that σ has the Rohlin property if and only if $U^n - 1$ is not of Hilbert-Schmidt class for all $n \in \mathbf{Z} \setminus \{0\}$, i.e. if and only if σ^n is outer in the trace representation for all $n \in \mathbf{Z} \setminus \{0\}$. Thus σ is an almost inductive limit automorphism in this case. If on the other hand $U^n - 1$ is a Hilbert-Schmidt operator for some $n \neq 0$, then U has purely discrete spectrum and σ is a product type automorphism, and hence σ is a (strict) inductive limit automorphism.

The present paper came out of an attempt to establish that the shift σ on $A = M_{n^\infty} = \bigotimes^{\infty} M_n$ has the Rohlin property also for $n = 3, 5, \dots$ (when n is even, the Rohlin property follows from the Rohlin property of the shift on M_{2^∞}). We have, however, only been able to establish the following approximate version of the Rohlin property.

Theorem 1. *Let σ be the one-sided shift on M_{2^∞}, and let τ be the trace state on M_{n^∞}. For any finite subset $\{x_1, \dots, x_k\} \subseteq M_{n^\infty}$, any $m \in \mathbf{N}$ and* *any* $\varepsilon > 0$ *there exists projections* $e_0, e_1, \ldots, e_{2^m}$ *in* M_{n^∞} *such that*

$$e_{2^m} = e_0\,,$$
$$e = \sum_{i=0}^{2^m-1} e_i \quad \text{is a projection,}$$
$$\tau(e) > 1 - \varepsilon\,,$$
$$\|\sigma(e_i) - e_{i+1}\| < \varepsilon$$

for $i = 0, 1, \ldots, 2^m - 1$, *and*

$$\|[x_j, e_i]\| < \varepsilon$$

for $i = 0, 1, \ldots, 2^m - 1$, $j = 1, \ldots, k$.

Combining Theorem 1 with the techniques of [Voi], [BKRS] we then establish that the shift is approximately an almost inductive limit automorphism in the following sense:

Corollary 2. *Let* σ *be the one-sided shift on* M_{n^∞}. *For any finite dimensional sub-*algebra* D *of* M_{n^∞} *and any* $\varepsilon > 0$, *there exists a projection* $e \in D' \cap M_{n^\infty}$ *such that*

$$\|\sigma(e) - e\| < \varepsilon$$

and

$$\tau(e) > 1 - \varepsilon\,,$$

*and there exists a finite-dimensional *-subalgebra* $E \subseteq eM_{n^\infty}e$ *such that*

$$De \subseteq^{\varepsilon} E$$

and

$$d(E, \sigma(E)) < \varepsilon\,.$$

2. The embedding of $GICAR(\mathcal{H})$ into M_{n^∞}

In this section we prove Theorem 1 and Corollary 2. Let σ be the one-sided shift on M_{n^∞}, $n = 2, 3, 4, \ldots$. Let $GICAR(\mathcal{H})$ be the gauge-invariant part of the CAR algebra $CAR(\mathcal{H})$ over the Hilbert space $\mathcal{H}$, i.e., $GICAR(\mathcal{H})$ consists of the elements in $CAR(\mathcal{H})$ which are invariant under the quasi-free action of the circle group T given by

$$a(f) \to a(zf)$$

for $z \in \mathbf{T} \subseteq \mathbf{C}$, $f \in \mathcal{H}$. Following [CE] (see also [E]), $GICAR(\mathcal{H})$ can be embedded into M_{n^∞} as follows: Let $(f_{ij})_{i,j=1}^n$ be a complete set of matrix units for M_n, and define projections $e_1, e_2, \ldots$ in $M_{n^\infty} = \bigotimes_1^\infty M_n$ as follows

$$\begin{aligned} e_1 &= \sum_{ij} \frac{1}{n} f_{ij} \otimes 1 \otimes 1 \otimes \cdots, \\ e_2 &= \sum_i f_{ii} \otimes f_{ii} \otimes 1 \otimes \cdots, \\ e_3 &= \sum_{ij} \frac{1}{n} 1 \otimes f_{ij} \otimes 1 \otimes \cdots, \\ e_4 &= \sum_i 1 \otimes f_{ii} \otimes f_{ii} \otimes \cdots, \end{aligned}$$

etc. The e_i's satisfy the Temperley-Lieb relations

$$\begin{aligned} e_i e_j = e_j e_i \qquad &\text{if} \quad |i-j| \geq 2, \\ e_k e_{k\pm 1} e_k = \frac{1}{n} e_k \,. \end{aligned}$$

Furthermore,

$$\sigma(e_k) = e_{k+2}$$

for $k = 1, 2, \ldots$

Let $f_1, f_2, \ldots$ be a basis for $\mathcal{H}$. The one-sided quasi-free shift is the injective morphism of $CAR(\mathcal{H})$ defined by

$$\beta(a_k) = a_{k+1}$$

where $a_k = a(f_k)$, $k = 1, 2, \ldots$. The shift β restricts to an injective morphism of $GICAR(\mathcal{H})$ which we also denote by β.

The embedding γ of $GICAR(\mathcal{H})$ into M_{n^∞} is given by

$$\begin{aligned} &\gamma(a_i^* a_i) = e_{2i} \\ &\gamma(a_i^* a_{i+1}) = e_{2i}(1 - n e_{2i+1}) e_{2i+2} \,, \end{aligned}$$

[CE]. We outline the argument from [CE]: First note that if $gl(\mathbf{N})$ denotes the Lie algebra of complex $\infty \times \infty$ matrices such that only finitely many matrix elements are non-zero, then

$$H = [H_{ij}] \in gl(\mathbf{N}) \to Q(H) = \sum_{ij} H_{ij} a_i^* a_j$$

is a Lie algebra morphism, i.e.

$$Q([H, H']) = [Q(H), Q(H')] \,.$$

Since

$$[e_{i,i+1}, e_{i+1,i+2}] = e_{i,i+2}$$

etc., it follows that

$$[a_i^* a_{i+1}, a_{i+1}^* a_{i+2}] = a_i^* a_{i+2}$$

etc. It follows that the *-algebra generated by $a_i^* a_i, a_i^* a_{i+1}$ contains all elements of the form $a_i^* a_j$, $i, j = 1, 2, \ldots$ Since any gauge invariant polynomial in a_i, a_j^*, $i, j = 1, \ldots$ may be written as a polynomial in $a_i^* a_j$ it follows that $f_i \equiv a_i^* a_i$ and $u_i \equiv a_i^* a_{i+1}$, $i = 1, 2, \ldots$ generate $GICAR(\mathcal{H})$ as a C^*-algebra. Furthermore, $GICAR(\mathcal{H})$ can be characterized as the universal C^*-algebra generated by elements f_i, u_i, $i = 1, 2, \ldots$ such that the f_i are projections and the map

$$\begin{aligned} e_{ii} &\to f_i \\ e_{i,i+1} &\to u_i \end{aligned}$$

extends to a Lie algebra morphism $gl((\mathbf{N}) \to GICAR(\mathcal{H})$. More explicitely, $GICAR(\mathcal{H})$ is the universal C^*-algebra generated by elements u_i, f_i, $i \in \mathbf{N}$ satisfying

(a) f_i is a projection.

(b) $[f_i, f_j] = 0$.

(c) $[u_i, f_j] = 0$ if $j \neq i, i+1$.

(d) $[u_i, u_j^{\#}] = 0$ if $|i - j| \geq 2, \# = \cdot, *$.

(e) $u_i^* u_i = f_{i+1}(1 - f_i)$,
$u_i u_i^* = f_i(1 - f_{i+1})$.

Thus, to verify that $\gamma : GICAR(\mathcal{H}) \to M_{n^\infty}$ is a morphism, one merely have to check that if f_i, u_i are defined by

$$\begin{aligned} f_i &= e_{2i} \\ u_i &= e_{2i}(1 - n e_{2i+1}) e_{2i+2} \end{aligned}$$

then f_i, u_i satisfy the relations (a)–(e). This is immediate.

Now, since $\sigma(e_k) = e_{k+2}$ and $\beta(a_i^* a_j) = a_{i+1}^* a_{j+1}$ it follows from the definition that γ interwines σ and β:

$$\gamma\beta(x) = \sigma\gamma(x)$$

for all $x \in GICAR(\mathcal{H})$.

Now, let τ be the unique normalized trace state on M_{n^∞}. Then $\tau \circ \gamma$ identifies with the quasifree state defined by the operator $\frac{1}{n}1$, restricted to $GICAR(\mathcal{H})$, by the remark before Lemma 5.2 in [CE]. (See [Ren, Appendix] for a description of this extremal trace state on $GICAR(\mathcal{H})$.) This follows from the computations:

$$\begin{aligned}
&\tau \circ \gamma(a_i^* a_i) = \tau(e_{2i}) = \frac{1}{n}, \\
&\tau \circ \gamma(a_i^* a_{i+1}) = \tau(e_{2i+2} e_{2i}(1 - n e_{2i+1}) e_{2i+2}) \\
&= \tau(e_{2i}(e_{2i+2}(1 - n e_{2i+1}) e_{2i+2})) \\
&= \tau(e_{2i}(e_{2i+2} - e_{2i+2})) \\
&= \tau(e_{2i} 0) = 0 .
\end{aligned}$$

We are now ready to construct the approximate Rohlin towers of Theorem 2. For economy of notation we identify $GICAR(\mathcal{H})$ with its image in M_{n_∞}.

If $\lambda, \mu \in \mathbf{T} \subseteq \mathbf{C}$, we may choose unit vectors $f, g \in \mathcal{H}$ such that f, g are orthogonal, and

$$Uf \approx \lambda f , \qquad Ug \approx \mu g$$

where U is the isometry implementing the one-sided shift on $\mathcal{H}$. Then

$$v = a^*(f) a(g)$$

is a partial isometry with

$$\sigma(v) \approx \lambda \bar{\mu} v .$$

Furthermore

$$\begin{aligned}
v^* v &= a(g)^* a(f) a(f)^* a(g) \\
&= a(f) a(f)^* a(g)^* a(g) .
\end{aligned}$$

Here $a(f)a(f)^*$ and $a(g)^*a(g)$ are commuting projections in different tensor factors (if f, g are chosen suitably) of trace $\frac{n-1}{n}$ and $\frac{1}{n}$, respectively, so

$$\tau(v^* v) = \frac{n-1}{n^2}$$

Also, as

$$\begin{aligned}
v^* v &= a(f)^* a(g) a(g)^* a(f) \\
&= a(f)^* a(f) a(g) a(g)^* .
\end{aligned}$$

and the projections $a(f)^*a(f)$, $a(g)a(g)^*$ are orthogonal to $a(f)a(f)^*$, $a(g)^*a(g)$, respectively, the projection $v v^*$ is orthogonal to $v^* v$,

$$v^* v v v^* = 0 = v v^* v^* v$$

Hence the C^*-algebra $C^*(v)$ generated by v is isomorphic to M_2, and if $1_v = v^* v + v v^*$ is the identity of this C^*-algebra, then

$$\tau(1_v) = \frac{2(n-1)}{n^2}$$

Also

$$\sigma\Big|_{C^*(v)} \approx \mathrm{Ad}\begin{pmatrix} 1 & 0 \\ 0 & \lambda\bar{\mu} \end{pmatrix}$$

Now, repeat the construction with $f_2, g_2 \perp f, g$ and (f_2, g_2) orthonormal, and cut down by $1 - 1_v = 1 - vv^* - v^*v$ to obtain a partial isometry v_2 with

$$\tau(v_2v_2^*) = \tau(v_2^*v_2) = \frac{n-1}{n^2}\left(1 - 2\frac{(n-1)}{n^2}\right)$$

and

$$\sigma(v_2) \approx \lambda\bar{\mu}v_2$$

so

$$\tau(1 - vv^* - v^*v - v_2v_2^* - v_2^*v_2) = \left(1 - 2\frac{(n-1)}{n^2}\right)^2$$

Continuing in this manner, we may construct partial isometries $v_1 = v, \ldots, v_m$ such that the projections $v_1v_1^*, v_1^*v_1, v_2v_2^*, v_2^*v_2, \ldots, v_mv_m^*, v_m^*v_m$ are all orthogonal, and

$$\sigma(v_k) \approx \lambda\bar{\mu}v_k$$

for all $k = 1, \ldots, m$ and

$$\tau\left(1 - \sum_{k=1}^{m}(v_kv_k^* + v_k^*v_k)\right) = \left(1 - 2\frac{n-1}{n^2}\right)^m$$

Now, put

$$u_m = v_1 + v_2 + \cdots + v_m$$

then u_m is a partial isometry such that $u_mu_m^* \perp u_m^*u_m, \sigma(u_m) \approx \lambda\bar{\mu}u_m$ and

$$\tau(1 - u_mu_m^* - u_m^*u_m) = \left(1 - 2\frac{n-1}{n^2}\right)^m$$

Now repeating this construction, going further and further out in $\mathcal{H} \simeq l^2(\mathrm{N})$ to find f, g's and using $\frac{1}{2\pi i}\log(\lambda\bar{\mu}) = 2^{-1}, 2^{-2}, 2^{-3}, \ldots, 2^{-m}$ (with the last m being that of Theorem 1), one finds a set of matrix units e_{ij}, $i, j = 1, \ldots, 2^m$, which may be taken to approximately commute with any finite set in M_{n^∞}, such that $\tau\left(\sum_{i=1}^{2^m} e_{ii}\right) > 1 - \varepsilon$ and

$$\sigma(e_{ij}) \approx e_{i+1,j+1}$$

where the addition of indices is modulo 2^m (see [BEK] for details). Putting $e_i = e_{ii}$, Theorem 1 follows.

The proof of Corollary 2 from Theorem 1 now follows the lines of the proofs of [Voi, Lemma 3.3] or [BKRS, Lemma 2.1]. By [Voi, Lemma 3.1], for $D \in \mathcal{F}(M_{n^\infty})$ and $m \in \mathbf{N}$ there are $B_j \in \mathcal{F}(M_{n^\infty})$, $j = 0, 1, \ldots, m$ with

$$\begin{aligned} &B_0 = B_m\,, \\ &d(\sigma(B_i), B_{i+1}) < \frac{5\pi}{m}\,, \qquad i = 0, \ldots, m-1\,, \\ &D \subseteq B_i\,, \qquad i = 0, 1, \ldots, m\,. \end{aligned}$$

If $e_0, \ldots, e_m$ is a Rohlin tower approximately commuting with $B_0, \ldots, B_m$ we may assume that $e = \sum_{k=0}^{m-1} e_k$ commute with D by a small perturbation of e, and putting

$$E \approx \sum_{i=0}^{m-1} B_i e_i\,,$$

E has the desired properties.

References

[1] O. Bratteli, D. E. Evans and A. Kishimoto, The Rohlin property for quasifree automorphisms of the fermion algebra, under typing.

[2] O. Bratteli, A. Kishimoto, M. Rørdam and E. Størmer, The crossed product of a UHF algebra by a shift, *Preprint, July'92*.

[3] A. Connes and D. E. Evans, Embedding of $U(1)$-current algebras in noncommutative algebras of classical statistical mechanics, *Commun. Math. Phys.* **12** (1989), 507–525.

[4] A. Connes, Outer conjugacy classes of automorphisms of factors, *Ann. Scient. Ec. Norm. Sup., 4e série*, **8** (1975), 383–420.

[5] D. E. Evans, C^*-algebraic methods in statistical mechanics and field theory, *Inter. J. Modern Physics* **4** (1990), 1069–1118.

[6] R. H. Herman and A. Ocneanu, Stability for integer actions on UHF C^*-algebras, *J. Funct. Anal.* **59** (1984), 132–144.

[7] J. Renault, A Groupoid Approach to C^*-algebras, *LNM 793*, Springer Verlag 1980.

[8] D. Voiculescu, Almost inductive limit automorphisms and embeddings into AF-algebras, *Ergod. Th. & Dyman. Sys.* **6** (1986), 475–484.

Strongly Elliptic and Subelliptic Operators on Lie Groups

DEREK W. ROBINSON
Centre for Mathematics and its Applications, School of Mathematical Sciences, Australian National University, GPO Box 4, Canberra, ACT 2601, Australia

1. Introduction

Partial differential operators, and in particular elliptic operators, on a Riemannian manifold provide a rich source of problems of non-commutative analysis. In the simplest situation the manifold is a Lie group and one can examine elliptic operators constructed from a vector space basis of the Lie algebra or subelliptic operators constructed from a Lie algebraic basis. In the latter case the choice of algebraic basis affects the characteristic geometric features of the subelliptic operators and their description involves an interesting interplay between analysis and geometry.

The theory of strongly elliptic operators, with either constant or variable coefficients, is described in [8] together with the basic features of second-order subelliptic operators with real constant coefficients. Subsequently, in a paper with Bratteli [1] and a series of papers with ter Elst [4] [5] [6], the general theory of subelliptic operators has been developed. In this review we summarize some of the main features of these latter developments. We concentrate on the description of the principal results and an explanation of one of the key methods of analysis, the parametrix method. For simplicity and brevity we restrict attention to second-order operators with constant coefficients throughout most of the article but in the conclusion we discuss some of the additional features of the theory of higher order operators and operators with variable coefficients.

Background material can be found in [8], in the book by Davies [3], the review article by Jerison and Sánchez-Calle [7] or in the forthcoming book by Coulhon, Saloff-Coste and Varopoulos [2]. These sources contain extensive references to the literature and hence we only give references for a few specific points which receive particular emphasis in the course of the discussion.

H. Araki et al. (eds.), Quantum and Non-Commutative Analysis, 435–453.

2. General formulation

Let G be a connected Lie group, with identity e, represented by bounded linear operators $g \in G \mapsto U(g)$ acting on a Banach space $\mathcal{X}$. We assume the representation is weakly continuous or, if $\mathcal{X}$ has a predual $\mathcal{X}_*$, weakly* continuous. If $a_1, a_2, \ldots$ are elements of the Lie algebra $\mathfrak{g}$ of G then we let $A_1, A_2, \ldots$ denote the generators of the continuous one-parameter subgroups $t \in \mathbf{R} \mapsto U(e^{-ta_1})$, $t \in \mathbf{R} \mapsto U(e^{-ta_2}), \ldots$. Thus the A_i are closed densely-defined operators which can be viewed as abstract partial differential operators. If, for example, $G = \mathbf{R}^d$ and U is the regular representation of G on one of the standard function spaces then the generators corresponding to the usual Cartesian basis $x_1, \ldots, x_d$ are the partial differential operators $\partial/\partial x_1, \ldots, \partial/\partial x_d$.

Let d denote the dimension of G and $a_1, \ldots, a_d$ a vector space basis of the Lie algebra $\mathfrak{g}$. Then a subbasis $a_1, \ldots, a_{d'}$ is defined to be an algebraic basis of $\mathfrak{g}$ if the Lie algebra generated by $a_1, \ldots, a_{d'}$ is equal to $\mathfrak{g}$. It follows automatically from this definition that there is an integer r such that the $a_1, \ldots, a_{d'}$ and their commutators of order r or less span $\mathfrak{g}$. The minimum such r is called the rank of the basis. In particular a vector space basis has rank one.

Next let

$$\mathcal{X}_2' = \bigcap_{i,j=1}^{d'} D(A_i A_j)$$

denote the common domain of products of the representatives of the algebraic basis. Now define the second-order operator H by setting $D(H) = \mathcal{X}_2'$ and

$$H = -\sum_{i,j=1}^{d'} c_{ij} A_i A_j + \sum_{i=1}^{d'} c_i A_i \qquad (1)$$

where the coefficients c_{ij}, c_i are complex numbers. Then H is said to be *subelliptic* whenever the real part $\Re C = (C + C^*)/2$ of the matrix $C = (c_{ij})$ of principal coefficients is strictly positive-definite. In the special case that $a_1, \ldots, a_{d'}$ is a vector space basis of the Lie algebra H is said to be *strongly elliptic*. In both cases, however, the minimum eigenvalue λ_c of $\Re C$ is called the ellipticity constant.

Subelliptic operators have many remarkable properties which we describe in the sequel. But we conclude this section with two simple examples which are useful in illustrating features of the general scheme.

Example 2.1 a. The Lie algebra of the three-dimensional group of rotations has a vector space basis a_1, a_2, a_3 satisfying the commutation relations

$$[a_1, a_2] = a_3 \quad , \quad [a_2, a_3] = a_1 \quad , \quad [a_3, a_1] = a_2 \quad .$$

Therefore a_1, a_2 is an algebraic basis of rank two.
b. The Lie algebra of the Heisenberg group has a vector space basis a_1, a_2, a_3 satisfying the commutation relations

$$[a_1, a_2] = a_3 \quad , \quad [a_2, a_3] = 0 \quad , \quad [a_3, a_1] = 0 \quad .$$

Therefore a_1, a_2 is again an algebraic basis of rank two.

3. Heat semigroups

The simplest strongly elliptic operator is the Laplacian

$$\Delta = -\sum_{i=1}^{d} A_i^2$$

corresponding to the vector space basis $a_1, \ldots, a_d$ of $\mathfrak{g}$. It is well known that in each of the regular representations of G the closure $\overline{\Delta}$ of Δ generates a holomorphic semigroup analogous to the usual heat semigroup. The first theorem on subellipticity gives a general statement of this nature.

Theorem 3.1 *Let H be a subelliptic operator in a general representation $(\mathcal{X}, G, U)$.*

I. *The operator H is closable.*

II. *The closure $\overline{H}$ of H generates a continuous semigroup S.*

III. *The semigroup S is holomorphic and if the matrix C of principal coefficients is hermitian then S is holomorphic in the open right half-plane.*

IV. *If $\mathcal{X}_\infty$ denotes the C^∞-elements of the representation then $S_t\mathcal{X} \subseteq \mathcal{X}_\infty$ for all $t > 0$.*

Although this theorem is simply stated it is long and difficult to prove. Hence we restrict ourselves to a general commentary on some of the key arguments used in the proof.

First, H is closable because it has a densely defined adjoint. Indeed if $\mathcal{F}$ denotes the dual $\mathcal{X}^*$ of $\mathcal{X}$, or, in the weakly* continuous case the predual $\mathcal{X}_*$, then there is a weakly* continuous, or weakly continuous, dual representation U_* on $\mathcal{F}$ such that

$$U_*(g) = U(g^{-1})^* \quad .$$

Now if A_{*i} denotes the representative of a_i with respect to the dual representation then H extends the formal adjoint $H^\dagger$ defined on $\mathcal{F}_2'$ by

$$H^\dagger = -\sum_{i,j=1}^{d'} \overline{c}_{ji} A_{*i} A_{*j} - \sum_{i=1}^{d'} \overline{c}_i A_{*i} \quad .$$

In fact one can also establish that $\overline{H} = (H^\dagger)^*$.

Secondly, the most difficult part of the proof is to establish that $\overline{H}$ generates a holomorphic semigroup. If, however, one can prove that each subelliptic operator generates a continuous semigroup then these semigroups are automatically holomorphic. This follows by observing that if S is the semigroup generated by $\overline{H}$ then for small θ the semigroup $t \geq 0 \mapsto S_{e^{i\theta}t}$ corresponds to the subelliptic operator with coefficients $e^{i\theta}c_{ij}$ and $e^{i\theta}c_i$. This operator is indeed subelliptic because

$$\begin{aligned}\Re e^{i\theta}C &= \cos\theta\,\Re C - \sin\theta\,\Im C\\ &\geq \Big(\lambda_c\cos\theta - |\sin\theta|\cdot\|\Im C\|\Big)I \quad ,\end{aligned}$$

where $\Im C = (C - C^*)/2i$, and hence $\Re e^{i\theta}C$ is strictly positive-definite whenever $|\tan\theta| < \lambda_c/\|\Im C\|$. Note that this argument gives a representation independent estimate on the sector of holomorphy and establishes holomorphy in the open right half-plane whenever the matrix C is hermitian.

Thirdly, the last statement of the theorem concerning the smoothness of the action of S relative to the group representation is a consequence of the holomorphy of S, which implies

$$S_t\mathcal{X} \subseteq \bigcap_{n\geq 1} D(\overline{H}^n)$$

for all $t > 0$, and an elliptic regularity result. Let $\mathcal{X}'_n$ denote the common domain of all monomials of order n in the generators $A_1, \ldots, A_{d'}$. Then clearly one has

$$\mathcal{X}'_{2n} \subseteq D(\overline{H}^n) \quad .$$

But conversely one can establish the regularity result

$$D(\overline{H}^n) \subseteq \mathcal{X}'_m$$

for all integers n, m satisfying $mr < 2n$. Consequently,

$$\bigcap_{n\geq 1} D(\overline{H}^n) = \bigcap_{n\geq 1} \mathcal{X}'_n = \mathcal{X}_\infty$$

and hence $S_t\mathcal{X} \subseteq \mathcal{X}_\infty$.

Therefore the real difficulty in the proof of the theorem lies in establishing the generator property of $\overline{H}$. This is achieved by adaptation of the parametrix method of partial differential equations on $\mathbf{R}^d$ to the Lie group setting.

4. Parametrices and generators

We now simplify notation and use H to denote the closed subelliptic operator. In order to establish that it generates a holomorphic semigroup S it suffices to prove that there is a $\varphi \in \langle 0, \pi/2]$ and an $\omega \geq 0$ such that

1. $(\lambda I + H)^{-1}$ exists as a bounded operator,

and

2. $\|(\lambda I + H)^{-1}\| \leq c|\lambda|^{-1}$ for some $c > 0$,

for all $\lambda \in \Gamma_\varphi$ where $\Gamma_\varphi = \{\lambda \in \mathbf{C}; |\arg \lambda| < \pi/2 + \varphi \text{ and } |\lambda| > \omega\}$. Then it follows from standard semigroup theory that S is holomorphic in the sector $\Delta(\varphi) = \{z \in \mathbf{C}; |\arg z| < \varphi\}$ and satisfies bounds $\|S_t\| \leq Me^{\omega t}$ for some $M \geq 1$ and all $t \geq 0$. The parametrix method achieves this result by a scheme of successive approximation which is reminiscent of perturbation theory. It does not require, however, a single unperturbed system and the zero-order approximation on which the method is based can vary continuously over the group. Nevertheless it is illuminating to compare the method with perturbation theory.

The main idea of both approaches is to investigate properties of the inverse of $(\lambda I + H)$ by solving the equation

$$(\lambda I + H)R_\lambda = I \tag{2}$$

iteratively. In perturbation theory one assumes that $H = H_0 + H_1$ and that the corresponding equation for H_0 can be solved, i.e., one has an appropriate R^0_λ satisfying

$$(\lambda I + H_0)R^0_\lambda = I \quad .$$

But then

$$(\lambda I + H)R^0_\lambda = I + Q_\lambda \tag{3}$$

with $Q_\lambda = H_1 R^0_\lambda$. Therefore if $\|Q_\lambda\| < 1$ one can invert $(I + Q_\lambda)$ by a power series expansion and hence

$$(\lambda I + H)R^0_\lambda(I + Q_\lambda)^{-1} = I \quad .$$

Now R^0_λ corresponds to the inverse $(\lambda I + H_0)^{-1}$ and one has the perturbation expansion

$$\begin{aligned} R_\lambda &= R^0_\lambda(I + Q_\lambda)^{-1} \\ &= \sum_{n \geq 0} R^0_\lambda(-H_1 R^0_\lambda)^n = \sum_{n \geq 0} (\lambda I + H_0)^{-1}(-H_1(\lambda I + H_0)^{-1})^n \end{aligned}$$

for R_λ. The condition $\|Q_\lambda\| < 1$ corresponds to the constraint $\|H_1(\lambda I + H_0)^{-1}\| < 1$ which is equivalent to the usual statement that H_1 is relatively bounded by H_0 with relative bound $b < 1$.

The parametrix method is similar insofar it solves (2) starting from an approximate solution P_λ, the parametrix, which satisfies the required estimate $\|P_\lambda\| < c|\lambda|^{-1}$ for $\lambda \in \Gamma_\varphi$ and an equation

$$(\lambda I + H)P_\lambda = I + Q_\lambda$$

analogous to (3). Again P_λ is chosen such that $\|Q_\lambda\| < 1$ uniformly for $\lambda \in \Gamma_\varphi$ and then

$$R_\lambda = P_\lambda(I + Q_\lambda)^{-1} = \sum_{n\geq 0} P_\lambda(-Q_\lambda)^n \quad .$$

The difference between the methods lies in the construction of the approximate solution. The P_λ is not necessarily the solution corresponding to an approximating operator but its properties depend critically on the nature of H as a partial differential operator. The construction of P_λ is best illustrated by two examples which describe complementary aspects of the parametrix method. The first covers the original problem for which the method was developed.

Example 4.1 Let H be a second-order strongly elliptic operator on $\mathbf{R}^d$ with smooth coefficients, i.e.,

$$H = -\sum_{i,j=1}^{d} c_{ij}\partial_{x_i}\partial_{x_j}$$

where $\partial_x = \partial/\partial x$, $c_{ij} \in C^\infty(\mathbf{R}^d)$ and the real part of the matrix $C = (c_{ij})$ is strictly positive-definite uniformly over $\mathbf{R}^d$. If $R_\lambda = (\lambda I + \overline{H})^{-1}$ exists then it is to be expected that its action is given by an integral kernel, i.e.,

$$(R_\lambda\varphi)(y) = \int_{\mathbf{R}^d} dx\, r_\lambda(y;\, x)\varphi(x)$$

for a suitable function r_λ, the Greens function. The value of $r_\lambda(y;\, x)$ represents the effect at the point y of the action of the operator at the point x. Moreover, if the coefficients c_{ij} are constant then the corresponding kernel should be invariant, i.e., $r_\lambda(y;\, x) = r_\lambda(y - x)$.

Now consider the operator H_y obtained by fixing the coefficients of H at the point y,

$$H_y = -\sum_{i,j=1}^{d} c_{ij}(y)\partial_{x_i}\partial_{x_j} \quad ,$$

and let r_λ^y denote the corresponding kernel. The existence of the kernel is easily established because H_y is a strongly elliptic operator with constant coefficients. Since H and H_y have a 'similar' action near the point y it is then reasonable to presume that the effect $r_\lambda(y;\, x)$ of x at y should be well approximated by $r^y_\lambda(y-x)$, i.e., by the value corresponding to the operator H_y. Hence the operator P_λ defined by

$$(P_\lambda\varphi)(y) = \int_{\mathbf{R}^d} dx\, r^y_\lambda(y-x)\varphi(x)$$

should be a suitable candidate for the parametrix. Now to verify this hypothesis let us compute $((\lambda I + H)P_\lambda\varphi)(y)$. The derivatives in H act on $y \mapsto r^y(y-x)$. Let ∂^i denote the action on the variable in the superscript and ∂_i the action on the argument. Then

$$\begin{aligned} \partial_{y_i} r^y_\lambda(y-x) &= \partial_i r^y_\lambda(y-x) + \partial^i r^y_\lambda(y-x) \\ &= -\partial_{x_i} r^y_\lambda(y-x)\partial^i r^y_\lambda(y-x) \quad . \end{aligned}$$

Therefore

$$\begin{aligned} (\lambda I + H)r^y_\lambda(y-x) &= ((\lambda I + H_y)r^y_\lambda)(y-x) + K r^y_\lambda(y-x) \\ &= \delta(y-x) + K r^y_\lambda(y-x) \end{aligned}$$

where K is a second-order operator acting on the y-variable but with at most one derivative on the variable in the argument and we have used the observation that r^y_λ is the Greens function of $(\lambda I + H_y)$. Hence in operator form

$$(\lambda I + H)P_\lambda = I + Q_\lambda$$

with Q_λ given by

$$(Q_\lambda\varphi)(y) = \int_{\mathbf{R}^d} dx\, K r^y_\lambda(y-x)\varphi(x) \quad .$$

Now if H acts on $C_0(\mathbf{R}^d)$ one readily computes that

$$\|P_\lambda\| = \sup_y \int_{\mathbf{R}^d} dx\, |r^y_\lambda(x)| = \sup_y \|(\lambda I + H_y)^{-1}\| \le c|\lambda|^{-1}$$

for all appropriate λ. At this point the ellipticity condition on the coefficients of H is essential. Using this condition, and the special form of K, it is also not difficult to establish that $\|Q_\lambda\| \le c'|\lambda|^{-1/2}$. Hence the parametrix equation can be solved whenever $|\lambda|$ is sufficiently large and one has

$$R_\lambda = P_\lambda(I + Q_\lambda)^{-1} = \sum_{n\ge 0} P_\lambda(-Q_\lambda)^n \quad .$$

Therefore by this construction one obtains information concerning the operator with variable coefficients from the operators with constant coefficients.

Note that in this example $H - H_y$ is in no way small with respect to H. The difference is a second-order operator which is not comparable with H on $C_0(\mathbf{R}^d)$, or $L_\infty(\mathbf{R}^d; dx)$. It is comparable on the L_p-spaces with $p \in \langle 1, \infty \rangle$ but then it is not small except locally, i.e., except in a small neighbourhood of y. Therefore this example of the parametrix method is not directly related to a perturbation calculation. Nevertheless a closer analogy with perturbation theory can be drawn by emphasizing that the parametrix method is inherently local.

If $\psi \in C_c^\infty(\mathbf{R}^d)$ has $\psi(0) = 1$ one can define a local parametrix by

$$(P_\lambda^\psi \varphi)(y) = \int_{\mathbf{R}^d} dx\, \psi(y-x) r_\lambda^y(y-x)\varphi(x)$$

and one then obtains an equation

$$(\lambda I + H)P_\lambda^\psi = I + Q_\lambda^\psi$$

with Q_λ^ψ a localized version of the previous Q_λ. This can again be solved by iteration if $|\lambda|$ is sufficiently large. The corresponding convergent series,

$$R_\lambda = \sum_{n\geq 0} P_\lambda^\psi (-Q_\lambda^\psi)^n \quad ,$$

then has the interesting property that the n-th term is an operator with kernel supported in W^n where W is the support of ψ. Thus successive terms have larger and larger localization regions but are smaller and smaller in size.

The localization properties of the parametrix are of fundamental importance for applications to the non-commutative Lie group theory. Each d-dimensional Lie group G is locally isomorphic to $\mathbf{R}^d$. Hence a local parametrix can be constructed for each strongly elliptic operator on G from a solution of the corresponding problem on $\mathbf{R}^d$; since the construction is local the argument can be expressed in terms of the corresponding strongly elliptic operator on $\mathbf{R}^d$.

Example 4.2 Let H be a second-order strongly elliptic operator associated with the left regular representation of the d-dimensional Lie group G on $C_0(G)$,

$$H = -\sum_{i,j=1}^{d} c_{ij} A_i A_j + \sum_{i=1}^{d'} c_i A_i \quad ,$$

where the coefficients c_{ij}, c_i are complex numbers. The exponential map gives a C^∞-diffeomorphism $\Lambda: W \to \Omega$ from an open neighbourhood W of the origin of $\mathbf{R}^d$ into an open neighbourhood Ω of the identity in G,

$$\Lambda(x) = \exp\{\sum_{i=1}^{d} x_i a_i\} \quad .$$

This can then be used to identify the generators A_i of left translations on $C_0(G)$ with C^∞-vector fields X_i on $\mathbf{R}^d$,

$$X_i = (\Lambda^{-1})_*(A_i) \quad .$$

The C^∞-vector fields X_i differ from the partial derivatives ∂_{x_i} on $\mathbf{R}^d$ by vector fields Z_i with smooth coefficients which vanish at the origin, i.e.,

$$X_i = \partial_{x_i} + Z_i$$

with $|(Z_i\varphi)(x)| = O(x)$ as $x \to 0$ for each $\varphi \in C_c^\infty(W)$. Thus the image H_* of H under the inverse exponential map has the form

$$H_* = (\Lambda^{-1})_*(H) = H_\partial + H_1$$

where H_∂ is the strongly elliptic partial differential operator on $C_0(W)$ constructed with the coefficients of H and the partial derivatives ∂_{x_i} and H_1 is a second-order operator with smooth coefficients a_{ij}, a_i such that $|a_{ij}(x)| = O(x)$ as $x \to 0$. Now let r_λ be the Greens function of $(\lambda I + H_\partial)$ with H_∂ viewed as an operator on $C_0(\mathbf{R}^d)$, i.e.,

$$((\lambda I + H_\partial)r_\lambda)(x) = \delta(x)$$

for all $x \in \mathbf{R}^d$. Further let $\psi \in C_c^\infty(W)$ satisfy $\psi(0) = 1$ and set $p_\lambda = r_\lambda\psi$. Then one has

$$(\lambda I + H_*)p_\lambda = \delta + q_\lambda \quad .$$

Again $\|p_\lambda\|_1 \leq c|\lambda|^{-1}$ and $\|q_\lambda\|_1 \leq c|\lambda|^{-1/2}$ for all suitable large λ. The second estimates are not immediately evident because q_λ has terms of the form $a_{ij}(\partial_{x_i}\partial_{x_j}r_\lambda)\psi$ and the second derivatives of r_λ are not locally integrable. But the fact that $|a_{ij}(x)| = O(x)$ overcomes this difficulty and leads to the required estimate.

Next let $\varphi_\lambda, \psi_\lambda \in C_0(\Omega)$ denote the images of p_λ, q_λ under the exponential map Λ. Then

$$(\lambda I + H)\varphi_\lambda = \delta + \psi_\lambda$$

and φ is a parametrix for H which can be used to calculate the resolvent $(\lambda I + H)^{-1}$ as an operator of convolution on $C_0(G)$. Alternatively it can be used to construct the resolvent for the operator H corresponding to the general representation $(\mathcal{X}, G, U)$. Setting

$$U(\varphi_\lambda) = \int_\Omega dg\, U(g)\varphi_\lambda(g) \quad , \quad U(\psi_\lambda) = \int_\Omega dg\, U(g)\psi_\lambda(g)$$

one verifies that

$$(\lambda I + H)U(\varphi_\lambda) = I + U(\psi_\lambda) \quad .$$

Thus $U(\varphi_\lambda)$ is a parametrix for H and

$$(\lambda I + H)^{-1} = U(\varphi_\lambda)(I + U(\psi_\lambda))^{-1}$$

for all appropriate λ.

Therefore the resolvent of the strongly elliptic operator H associated with a general representation of the Lie group is finally calculated in terms of the resolvent of the corresponding operator associated with the left regular representation of $\mathbf{R}^d$, i.e., the solution to the Lie group problem is obtained by the parametrix method from the solution to the comparable problem for the Euclidean group $\mathbf{R}^d$.

One could apply the foregoing reasoning to the subelliptic operator H but then the corresponding H_∂ on $\mathbf{R}^d$ is no longer strongly elliptic and the basic estimates on the parametrix fail. Therefore one must adopt a different tactic. The idea is to embed G locally in a group $\widetilde{G}$ which is more complicated than $\mathbf{R}^d$ insofar it is non-abelian but which nevertheless retains some of the simplifying features of the abelian group. Then the Lie group problem is solved in terms of the comparable problem on $\widetilde{G}$.

Let $\tilde{\mathfrak{g}} = \mathfrak{g}(d', r)$ be the nilpotent Lie algebra free of step r with d' generators $\tilde{a}_1, \ldots, \tilde{a}_{d'}$ and $\widetilde{G} = G(d', r)$ the connected simply connected Lie group with Lie algebra $\tilde{\mathfrak{g}}$. For example, $G(2,2)$ is the Heisenberg group mentioned in Example 2.1 b. Next let $\tilde{d}$ denote the dimension of $\tilde{\mathfrak{g}}$ and extend the algebraic basis $\tilde{a}_1, \ldots, \tilde{a}_{d'}$ to a vector space basis $\tilde{a}_1, \ldots, \tilde{a}_{\tilde{d}}$ of $\tilde{\mathfrak{g}}$. Now G can be identified locally with $\mathbf{R}^d$ by the inverse exponential map and the group $\widetilde{G}$ can be identified globally with $\mathbf{R}^{\tilde{d}}$. But $\tilde{d} \geq d$ and one can parametrize $\mathbf{R}^{\tilde{d}}$ with a pair of co-ordinates $\xi \in \mathbf{R}^d$, $\eta \in \mathbf{R}^{\tilde{d}-d}$ such that the vector fields X_i introduced above act on $C_c^\infty(W \times \mathbf{R}^{\tilde{d}-d})$ whilst the image $\widetilde{X}_i$ of the generators $\widetilde{A}_i$ of left translations on $C_0(\widetilde{G})$ act on $C_c^\infty(\mathbf{R}^{\tilde{d}})$. It is then a basic observation of Rothschild and Stein [9] that the X_i and $\widetilde{X}_i$ are 'very similar'. In particular there exist bounded open neighbourhoods $V, \widetilde{W}$ of the origin in $\mathbf{R}^{\tilde{d}}$ and a C^∞-diffeomorphism Θ from V onto $\widetilde{W}$ such that $V \subset W \times \mathbf{R}^{\tilde{d}-d}$, $\Theta(0) = 0$ and $\Theta_*(X_i) = \widetilde{X}_i + Y_i$ where the correction terms Y_i have particularly manageable properties which we will not discuss in detail (see [5], Section 5). The important feature is that if H_X denotes the operator constructed from the coefficients of H and the vector fields X_i and if $H_{\widetilde{X}}$ is the comparable operator constructed from the $\widetilde{X}_i$ then $\Theta_*(H_X) = H_{\widetilde{X}} + H_1$ where H_1 is essentially a perturbation of $H_{\widetilde{X}}$. (A similar situation occurred in Example 4.2 where H_∂ plays the role of $H_{\widetilde{X}}$.) Therefore the construction of a parametrix for H, or for H_X, is effectively reduced to the construction of a parametrix for the corresponding subelliptic operator $H_{\widetilde{X}}$ associated with left translations on $C_0(\widetilde{G})$. Thus we have reduced the calculation to a local problem with G replaced by $\tilde{G} = G(d', r)$ in much the same way that G was replaced by $\mathbf{R}^d = G(d, 1)$ in the strongly elliptic case.

It might seem that the replacement of G by $\tilde{G}$ merely replaces one intractable problem with another. But the group $\tilde{G}$ has significant simplifying feature. If $u \in \mathbf{R}_+ \mapsto \gamma_u$ denotes the map of $\tilde{\mathfrak{g}} \to \tilde{\mathfrak{g}}$ such that $\gamma_u \tilde{a}_i = u\tilde{a}_i$ for $i = 1, \ldots, d'$ then γ extends via the exponential map to a family of dilations of $\tilde{G}$. It is the existence of these dilations, and the relative freedom of the group, which simplifies analysis on $\tilde{G}$. The dilations reflect a homogeneity of $\tilde{G}$ analogous to the homogeneity of the Euclidean group. Therefore the groups $G(d', r)$ have the key features of $\mathbf{R}^d$ and correspond to non-commutative versions of the Euclidean spaces.

The further analysis of the subelliptic operators on $\tilde{G}$ combines L_2-techniques, hypoellipticity arguments, Sobolev estimates and scaling. It is considerably more complicated than the analogous analysis of constant coefficient operators on $\mathbf{R}^d$ which is the vital element in the discussion of the strongly elliptic operators. Nevertheless it is feasible and finally information about the subelliptic operators on G is obtained by consideration of the corresponding operators on the particular group $\tilde{G} = G(d', r)$ appropriate to the choice of algebraic basis in the definition of the operators. The details, together with references to the earlier literature, are given in [5].

5. Semigroups and kernels

The universal properties of the subelliptic semigroups discussed in the previous section are really a consequence of the existence of a heat kernel satisfying Gaussian bounds. These latter bounds, which are stated in the next theorem, involve a distance $|\cdot|'$ and a dimension D' characteristic of the algebraic basis. These will be defined and discussed below.

Theorem 5.1 *Let S denote the semigroup generated by the subelliptic operator H.*

I. *The semigroup has an integral kernel, i.e., there exists a unique family of appropriately integrable functions K_t over G such that*

$$S_t = \int_G dg\, K_t(g) U(g) \quad .$$

II. *$t > 0 \mapsto K_t$ extends to a function analytic in a sector $\Delta(\theta) = \{z \in \mathbf{C} : |\arg z| < \theta\}$ and jointly C^∞ on $G \times \langle 0, \infty \rangle$.*

III. *For each multi-index α there exist $a, b > 0$ and $\omega \geq 0$ such that*

$$|(A^\alpha K_z)(g)| \leq a|z|^{-(D'+|\alpha|)/2} e^{\omega|z|} e^{-b(|g|')^2/|z|}$$

for all $g \in G$ and $z \in \Delta(\varphi)$ with $\varphi < \theta$.

IV. *$K_t \geq 0$ if and only if $c_i, c_{ij} \in \mathbf{R}$ and then there exist $a', b' > 0$ and $\omega' \geq 0$ such that*

$$K_t(g) \geq a' t^{-D'/2} e^{-\omega' t} e^{-b'(|g|')^2/t}$$

for all $g \in G$ and $t > 0$.

In the third statement we have used multi-index notation. The multi-index α is an element of $I_n(d') = \{1, \ldots, d'\}^n$ for some $n = |\alpha|$ and $A^\alpha = A_{i_1} \ldots A_{i_n}$ if $\alpha = (i_1, \ldots, i_n)$. Moreover, the distance $|g|'$ associated with the algebraic basis $a_1, \ldots, a_{d'}$ is defined by considering absolutely continuous paths $\gamma: [0,1] \to G$ from the identity $e \in G$ to $g \in G$ such that the tangents are almost everywhere in the span of $a_1, \ldots, a_{d'}$. Then if ψ is a smooth function over G and $A_1, \ldots, A_{d'}$ are the generators of left translations there are tangential coordinates such that

$$\frac{d\psi(\gamma(t))}{dt} = \sum_{i=1}^{d'} \gamma_i(t) \, (A_i \psi)(\gamma(t)) \quad .$$

The distance $|g|'$ of g is defined in terms of these coordinates by

$$|g|' = \inf_{\gamma(0)=e,\, \gamma(1)=g} \int_0^1 dt \left(\sum_{i=1}^{d'} \gamma_i(t)^2 \right)^{1/2} \quad .$$

It can be established that if

$$B'_\rho = \{g \in G : |g|' < \rho\}$$

there is an integer $D' \geq d$ and $C, c > 0$ such that

$$c\rho^{D'} \leq |B'_\rho| \leq C\rho^{D'}$$

for $\rho \in [0,1]$ where $|B'_\rho|$ is the volume of B'_ρ with respect to left, or right, Haar measure. The integer D' is referred to as the *local dimension* of the subbasis $a_1, \ldots, a_{d'}$. If the subbasis $a_1, \ldots, a_{d'}$ is completed to a full vector space basis $a_1, \ldots, a_{d'}, \ldots, a_d$ then one can repeat the above definitions and obtain a modulus $|\cdot|$ which automatically satisfies

$$|g| \leq |g|'$$

for all $g \in G$ and hence $D' \geq d$. The last two statements of the theorem show that D' and $|\cdot|'$ are the appropriate parameters for the semigroup kernel.

It should be stressed that positivity of the kernel only occurs for second-order operators with real coefficients. Therefore in the analysis of operators with complex coefficients, or operators of higher order, it is not possible to exploit positivity techniques. In particular one cannot use the maximum principle or related results. Nevertheless one can use the parametrix method in its second 'time-dependent' form.

In Section 4 we emphasized that the parametrix construction for the resolvent $(\lambda I + H)^{-1}$ is very similar to 'time-independent' perturbation theory. In the analysis of the semigroup kernel it is important that there is an analogous construction of a parametrix for the semigroup e^{-tH}. This latter construction is similar to 'time-dependent' perturbation theory and as an illustration we briefly discuss it in the framework of Examples 4.1 and 4.2.

Example 5.2 Adopt the assumptions of Example 4.1. The operator H_y obtained from H by fixing the coefficients at the point y generates a semigroup S^y on $C_0(\mathbf{R}^d)$ with a Gaussian kernel K_t^y which vanishes for all $t \leq 0$ and satisfies the heat equation

$$((\partial_t + H_y)K_t^y)(x) = \delta(t)\delta(x)$$

for all $t \in \mathbf{R}$ and all $x \in \mathbf{R}^d$. Then defining $K_t^{(0)}$ by

$$K_t^{(0)}(y;\, x) = K_t^y(y - x)$$

one verifies that the function $y \mapsto K_t^{(0)}(y;\, x)$ also vanishes for all $t \leq 0$ and satisfies a perturbed heat equation

$$((\partial_t + H)K_t^{(0)})(y;\, x) = \delta(t)\delta(y - x) + L_t(y;\, x)$$

for all $t \in \mathbf{R}$ and all $x, y \in \mathbf{R}^d$. Therefore the solution of

$$((\partial_t + H)K_t)(y;\, x) = \delta(t)\delta(y - x) \quad ,$$

the heat equation expected for the kernel of the semigroup S, is formally given by the series expansion

$$K_t = \sum_{n \geq 0} K_t^{(n)} \tag{4}$$

with the $K^{(n)}$ defined recursively by

$$K_t^{(n)} = -(K^{(n-1)} \hat{*} L)_t \tag{5}$$

and the convolution product $\hat{*}$ by

$$(\varphi \hat{*} \psi)_t(y;\, x) = \int_{\mathbf{R}} ds \int_{\mathbf{R}^d} dz\, \varphi_{t-s}(y;\, z)\psi_s(z;\, x) \quad .$$

But one can establish that the series expansion, which is reminiscent of the 'time-dependent' perturbation expansion, converges for all $t > 0$ and $x \in \mathbf{R}^d$. This universal convergence is somewhat unexpected. It is, however, a straightforward consequence of the Gaussian form of $K_t^{(0)}$ and Gaussian bounds on L_t with an additional factor $t^{-1/2}$ for small t. (The additional $t^{-1/2}$ arises because L_t contains derivatives of $K_t^{(0)}$.) Since the series then defines the required semigroup kernel for all $t > 0$ and $x \in \mathbf{R}^d$ one can exploit it to obtain the Gaussian bounds of the kernel corresponding to the operator with variable coefficients from analogous bounds on the kernels K_t^y of the operators with fixed coefficients.

In Example 4.1 the resolvent of the strongly elliptic operator with variable coefficients on $\mathbf{R}^d$ was constructed from the analogous resolvent for operators with constant coefficients. The above example gives a similar construction for the semigroup kernel. Again the approximation technique is local and one could have chosen the alternative approximant $K_t^{\psi}(y;x) = K_t^y(y-x)\psi(y-x)$ where $\psi \in C_c^\infty(W)$ and $\psi(0) = 1$. Then the corresponding L_t^{ψ} also has compact support in W. This local version of the parametrix for the semigroup kernel can be used to analyze the Lie group operators.

Example 5.3 Adopt the assumptions of Example 4.2 and again let H_∂ denote the strongly elliptic operator on $\mathbf{R}^d$ constructed with the coefficients of the strongly elliptic operator H on G and the partial derivatives ∂_{x_i}. Then the image H_* of H under the inverse exponential map can be expressed in the form $H_* = H_\partial + H_1$ given in Example 4.2. Now if k_t is the heat kernel of H_∂, i.e., the unique solution of the heat equation

$$((\partial_t + H_\partial)k_t)(x) = \delta(t)\delta(x) \quad ,$$

and $\psi \in C_c^\infty(W)$ with $\psi(0) = 1$, define $k_t^{(0)}$ by $k_t^{(0)} = k_t\psi$. Then

$$((\partial_t + H_*)k_t^{(0)})(x) = \delta(t)\delta(x) + l_t(x)$$

where l_t is readily calculable from $k_t^{(0)}$. This equation can then be solved by iteration to give a solution analogous to (4) and (5). Since this solution is constructed from functions with support in the bounded neighbourhood W of the origin in $\mathbf{R}^d$ it can be transformed by the exponential map into the desired semigroup kernel for H on G. Thus the resulting kernel K_t is expressed in terms of the kernel k_t corresponding to H_∂ and properties of K_t such as Gaussian bounds, analyticity, etc. can be inferred from similar properties of k_t.

The foregoing example illustrates the basic tactic used to prove Theorem 5.1. But the analysis of the subelliptic operators cannot be reduced to the

$\mathbf{R}^d$-case. It can, however, be reduced, in the manner briefly described at the end of Section 4, to the discussion of operators on the groups $G(d', r)$. Details can again be found in [5].

If one specializes to the consideration of subelliptic operators with real coefficients one can obtain improved estimates on the semigroup kernel. Then the upper and lower bounds of Theorem 5.1 establish that $t \mapsto K_t(e)$ has a singularity $t^{-D'/2}$ for small t and this is optimal. The large t behaviour can also be estimated but this depends on the structure of the group. There are three types of group, compact groups, polynomial groups for which there in an integer D such that

$$c\rho^D \leq |B'_\rho| \leq C\rho^D$$

as $\rho \to \infty$, and exponential group for which the volume $|B'_\rho|$ of the ball B'_ρ grows exponentially fast with ρ as $\rho \to \infty$. Thus in each case one can chose an integer D such that

$$|B'_\rho| \geq c\rho^D$$

as $\rho \to \infty$. For compact groups $D = 0$ and for exponential groups it can be chosen arbitrarily large.

Theorem 5.4 *If G is unimodular, $c_{ij} = c_{ji}$, $c_i \in \mathbf{R}$ and*

$$|B'_\rho| \geq c\rho^D$$

for all $\rho \geq 1$ then there is an $a > 0$ such that

$$0 < K_t(g) \leq a(1 \wedge \varepsilon t)^{-D'/2}(1 \vee \varepsilon t)^{-D/2} \inf_{\rho \geq 0} e^{\rho^2(1+\varepsilon)\|C\|t - \rho(|g|' - |c|t)}$$

for all $t > 0$, $g \in G$ and $\varepsilon \in \langle 0, 1]$ where $|c| = \sum_{i=1}^{d'} |c_i|$ and $\|C\|$ is the norm of the matrix $C = (c_{ij})$.

If $|g|' \leq |c|t$ then the theorem gives uniform bounds

$$0 < K_t(g) \leq a(1 \wedge t)^{-D'/2}(1 \vee t)^{-D/2} .$$

but if $|g|' > |c|t$ then one has

$$0 < K_t(g) \leq a(1 \wedge \varepsilon t)^{-D'/2}(1 \vee \varepsilon t)^{-D/2} e^{-(|g|' - |c|t)^2/4(1+\varepsilon)\|C\|t}$$

with $\varepsilon \in \langle 0, 1]$. In both cases one sees that $\|K_t\|_\infty$ is bounded by a multiple of $t^{-D/2}$ in the limit $t \to \infty$.

These bounds are most interesting for polynomial groups because D can then be chosen to have its optimal value, the 'dimension at infinity' of the group. Then one has bounds $\|K_t\|_\infty \leq c|B_t'|^{-1/2}$ for all $t > 0$. Saloff-Coste [10] has improved these bounds for operators with no first-order terms.

Theorem 5.5 *If G is polynomial, $c_{ij} = c_{ji} \in \mathbf{R}$ and $c_i = 0$ then there are $a, a', b, b' > 0$ such that*

$$a|B_t'|^{-1/2}e^{-b(|g|')^2/t} \geq K_t(g) \geq a'|B_t'|^{-1/2}e^{-b'(|g|')^2/t}$$

for all $t > 0$ and $g \in G$.

The theorem establishes that the kernel, which in this case is a probability distribution, is asymptotically uniform over the group and its rate of decrease is governed by the volume available, e.g., $\|K_t\|_\infty$ behaves like $|B_t'|^{-1/2}$. If the group is compact one can obtain stronger ergodicity properties [8] by exploitation of spectral properties of the subelliptic operator but if the group is exponential the asymptotic behaviour is less well understood. In the latter case the best estimate on the kernel [8] appears to be an upper bound $\|K_t\|_\infty \leq ce^{-bt^{1/3}}$ for large t. Finally if G is non-unimodular the large t behaviour cannot be well controlled but otherwise the bounds of Theorem 5.4 still apply.

6. Conclusion

The theory we have sketched can be extended in two different directions. One can analyze operators with variable coefficients associated with the regular representations of G on one of the standard function spaces. Alternatively one can consider higher order subelliptic operators. We conclude with some brief comments on these two options.

Let A_i denote the generators of left translations in the left regular representation of G on $C_0(G)$, or on one of the L_p-spaces over G. Then if $c_i, c_{ij} \in C_b(G)$ are interpreted as multiplication operators the operator (1) corresponds to a partial differential operator with variable coefficients. The operator is then defined to be subelliptic if the real part of the matrix $C = (c_{ij})$ of principal coefficients is strictly positive-definite uniformly over G. The theory of such operators is similar to the theory of constant coefficient operators. One can pass from the special case of constant coefficients to the general case of variable coefficients by the parametrix method illustrated for $\mathbf{R}^d$ in Examples 4.1 and 5.2. The replacement of $\mathbf{R}^d$ by the general group G has little effect on the arguments which require, however, some smoothness of the coefficients; twice differentiable coefficients suffice to reestablish the analogues of Theorems 3.1 and 5.1 [6]. Subsequently these smoothness hypotheses can be weakened by supplementary approximation arguments [1].

The extension of the notion of subellipticity to higher order operators is less straightforward. The definition (1) of a second-order partial differential operator has a straightforward extension to m-th order,

$$H_m = \sum_{\alpha:|\alpha|\leq m} c_\alpha A^\alpha \quad , \tag{6}$$

with $c_\alpha \in \mathbf{C}$ where we have again used multi-index notation. The domain $D(H_m)$ of H_m is taken to be the subspace $\mathcal{X}'_m$ which can now be identified as

$$\mathcal{X}'_m = \bigcap_{\alpha\in I_m(d')} D(A^\alpha)$$

with $I_m(d') = \{1, \ldots, d'\}^m$.

If $d' = d$, i.e., if one has a full vector space basis, the natural definition of strong ellipticity is the restriction

$$\mathrm{Re}\Big((-1)^{m/2} \sum_{\alpha\in I_m(d)} c_\alpha \xi^\alpha\Big) \geq \lambda_c |\xi|^m \tag{7}$$

for some $\lambda_c > 0$ and all $\xi \in \mathbf{R}^d$. This corresponds exactly to the definition adopted in the classical theory of elliptic partial differential operators. The appropriate definition of subellipticity in the non-commutative theory is not so evident but the suggestion given in [5] again involves the groups $G(d', r)$. In order to motivate and explain this suggestion we first reformulate the foregoing expression of strong ellipticity.

Note that the condition (7) only involves the principal coefficients of the operator (6) and the group $\mathbf{R}^d$. (The group G enters peripherally through the dimension d.) Therefore the definition is unchanged if one replaces the representation $(\mathcal{X}, G, U)$ by the left regular representation of $\mathbf{R}^d$ on $L_2(\mathbf{R}^d; dx)$. Then $A_i = \partial_{x_i}$ and it follows by Fourier transformation and a scaling argument that the strong ellipticity condition (7) is equivalent to the *Gårding inequality*

$$\mathrm{Re}(\varphi, P_m\varphi) \geq \lambda_c \Big(\|\varphi\|_{2;m/2}\Big)^2 \tag{8}$$

for all $\varphi \in L_{2;m} = \bigcap_{\alpha\in I_m(d)} D(\partial_x^\alpha)$ where P_m is the principal part of H_m, i.e.,

$$P_m = \sum_{\alpha\in I_m(d)} c_\alpha \partial_x^\alpha$$

and

$$\|\varphi\|_{2;n} = \sup_{\alpha\in I_n(d)} \|\partial_x^\alpha \varphi\|_2$$

is the natural C^n-seminorm on $L_{2;n}$. A further scaling argument then establishes that (7) is equivalent to the Gårding inequality (8) for all $\varphi \in L_{2;m}$

with compact support in some bounded open neighbourhood W of the origin in $\mathbf{R}^d$.

The moral of the foregoing discussion is that strong ellipticity is equivalent to a local $\mathbf{R}^d$-Gårding inequality. But in the previous discussion of subellipticity the role of the group $\mathbf{R}^d$ was assumed by the group $G(d',r)$ appropriate to the algebraic basis. Therefore in [5] it was proposed that the general definition of subellipticity should be in terms of a local $G(d',r)$-Gårding inequality. Specifically, if $a_1, \ldots, a_{d'}$ is an algebraic basis of rank r then the m-th order operator H_m given by (6) is defined to be *subelliptic* if its principal part

$$P_m = \sum_{\alpha \in I_m(d')} c_\alpha A^\alpha$$

acting in the left regular representation of the group $\widetilde{G} = G(d',r)$ on the space $L_2(\widetilde{G};\, dg)$ satisfies the Gårding inequality

$$\mathrm{Re}(\varphi, P_m \varphi) \geq \lambda_c \Big(\|\varphi\|'_{2;m/2} \Big)^2$$

for some $\lambda_c > 0$ and all $\varphi \in \bigcap_{\alpha \in I_m(d')} D(A^\alpha)$ with compact support in some open neighbourhood Ω of the identity e of $\widetilde{G}$. In this definition we have used the seminorm

$$\|\varphi\|'_{2;n} = \sup_{\alpha \in I_n(d')} \|A^\alpha \varphi\|_2 \quad .$$

Although the definition is somewhat indirect it does have several advantages.

First, it incorporates the case of strong ellipticity if $d' = d$ and $r = 1$. Secondly, it is equivalent to the earlier definition of subellipticity for second-order operators. If $r = 1$ then $d' = d$ and the equivalence is a corollary of the foregoing dicussion of strong ellipticity. But if $r \geq 2$ then the equivalence is established in Proposition 6.7 of [5]. Thirdly, the definition covers such obviously interesting operators as the higher order Laplacians

$$H_{2m} = \sum_{i=1}^{d'} (-A_i^2)^m$$

or the operators

$$H_{2m} = (-1)^m \sum_{\alpha,\beta \in I_m(d')} c_{\alpha\beta} A^{\alpha_*} A^\beta$$

with the the real part of the matrix $(c_{\alpha\beta})_{\alpha,\beta \in I_m(d')}$ strictly positive-definite and with $\alpha_* = (i_n, \ldots, i_1)$ whenever $\alpha = (i_1, \ldots, i_n)$. Finally, all the principal features of Theorems 3.1 and 5.1 can be reestablished with this general notion of subellipticity. Again the details can be found in [5].

References

[1] Bratteli, O., and Robinson, D.W., Subelliptic operators on Lie groups: variable coefficients. Research Report CMA-MR3-92, The Australian National University, Canberra, Australia, 1992.

[2] Coulhon, T., Saloff-Coste, L. and Varopoulos, N. Th., *Analysis on Lie groups*. Cambridge Tracts in Mathematics. Cambridge University Press, Cambridge etc.. To appear.

[3] Davies, E. B., *Heat kernels and spectral theory*. Cambridge Tracts in Mathematics. Cambridge University Press, Cambridge etc., 1989 .

[4] Elst, A.F.M. ter, and Robinson, D.W., Subelliptic operators on Lie groups: regularity. *Journal of the Australian Mathematical Society* (1992). To appear.

[5] ———, Subcoercivity and subelliptic operators on Lie groups. Research Report CMA-MR4A-92, The Australian National University, Canberra, Australia, 1992.

[6] ———, Subcoercive and subelliptic operators on Lie groups: variable coefficients. Research Report CMA-MR16-92, The Australian National University, Canberra, Australia, 1992.

[7] D. Jerison and A. Sánchez-Calle, Subelliptic, second order differential operators. In C. A. Berenstein, eds., *Complex analysis III*, Lecture Notes in Mathematics 1277. Springer-Verlag, Berlin etc., 46-77 (1987).

[8] Robinson, D.W., *Elliptic operators and Lie groups*. Oxford Mathematical Monographs. Oxford University Press, Oxford etc., 1991.

[9] Rothschild, L.P., and Stein, E.M., Hypoelliptic differential operators and nilpotent groups. *Acta Math.* **137** , 247–320 (1976).

[10] Saloff-Coste, L., Analyse sur les groupes de Lie à croissance polynômiale *Arkiv för Mat.* **28** , 315–331 (1990).

Postscript

FORMER STUDENTS

Professor Huzihiro Araki was born on the 28th of July, 1932 in Tokyo as the eldest son of the physicist, Gentaro Araki. He went to the Primary School Attached to the Tokyo Higher Normal School, where an advanced method of teaching was being implemented and he already showed his interest and talent in arithmetic. Incidentally, his classmates there include S. T. Kuroda (for 4 years), a grandson of the famous mathematician Teiji Takagi, and Kojiro Nishina (for 5 years), the second son of the famous physicist Yoshio Nishina. His family moved to Kyoto in March, 1944 when his father was appointed as a professor in the Faculty of Engineering of Kyoto Imperial University.

The government of Japan, which was fighting abroad, established special classes for science education for upper primary school pupils and junior high school students at several different specific schools in order to bring up scientific talents who would shoulder future science in Japan. In Kyoto in May, 1945, one such class for the first, second and third year students each was established at the First Kyoto Prefectual Middle School (the Alma Mater of Yukawa and Tomonaga) by the initiative of eminent professors of the Faculty of Science of Kyoto University such as Professor Yukawa. Students of the special classes, about 30 each, were selected from among middle school students in the Kinki area by several steps of examinations (including interviews by Yukawa and other professors as well as an interview at the psychology department). Araki was admitted in this special class and he seems to have enjoyed it very much. Some of the lessons such as physics and chemistry were taught by young faculty members of Kyoto University (which included Chushiro Hayashi, later winner of the Eddington Medal). His house was full of his father's books on mathematics and physics when there was nothing to enjoy or even to eat outside. Quite naturally he used to read these books as if they were comics.

The special classes for science education in Kyoto were dissolved in March, 1948 at the occasion of the transition to the new educational system after the war. But Araki's interest in science grew. Since the time at the special class, he self-studied various subjects of mathematics at the university and postgraduate levels as well as quantum mechanics. He led the activities of the mathematics club of his high school and gave constantly lectures on

H. Araki et al. (eds.), Quantum and Non-Commutative Analysis, 455–460.

advanced mathematics at that club.

It is said that the young Araki could not ride a bicycle (probably cannot now). The responsibility for this should be shared with his mother Kiyoko, who was afraid of his having an accident. His life seems to have been filled with stories. Among them, what makes his character stand out is that he is very quick, incisive and well organized in doing anything — aware or not. This is, according to Professor Araki himself, due to the hardships most people had to go through after the war; he was no exception. Well, never mind his theory does not explain why most of his generation are not so quick and incisive.

He entered Kyoto university in 1951, and started his academic career with a joint paper with his father in 1953 on the spectrum of cyanine dyes, in which he contributed Fourier series calculations. He was then twenty-one years old. He went on to postgraduate studies in particle physics ; his supervisor was the late Professor Hideki Yukawa.

In 1957 Araki went to Princeton as a Fulbright Fellow, where he studied axiomatic aspects of quantum field theory, under the supervision of Professor Rudolf Haag who was visiting there for two years and completed the Ph.D thesis on Hamiltonian formalism, showing that a representation of the canonical commutation relations more or less determines the Hamiltonian. (An appendix of the thesis, which was not included in the published version, dealt with some concrete constructions of examples which led to a later work on free Bose gas with E.J.Woods and to another later work on the non-commutative extension of the Levy–Kinchin formula.)

While in Princeton, Araki took advantage of various research activities in theoretical physics then going on there to enlarge the area of his interests and produced seveen other works on topics such as positive mass problem in general relativity (then in its infancy), quantum mechanical measurement theory, clustering property of Wightman functions, generalization of retarded functions, non-existence of a local field whose current is a Wick polynomial of free fields and the connection bewteen spin and statistics.

Araki met the late Res Jost at the Boulder Summer Gathering in 1960 and the last work mentioned above drew Jost's attention. After going back to Japan and being married with Chieko Kawaguchi, Araki went to ETH Zürich in the fall of 1961 by the invitation of Jost, and gave a series of lectures on axiomatic quantum field theory, which is known today as the ETH lecture notes by Araki.

During the one year stay in Zürich, Professor Y. Akizuki visited Araki twice and told him about the planned Mathematical Science Institute to be set up in Kyoto. After this one year stay in Zürich and another one and a half year stay at the University of Illinois with Haag, Araki came back to Kyoto in the spring of 1964 and joined the Research Institute for Mathematical Sciences at Kyoto University, which had started one year before, as associate professor. He was promoted to full professor in January 1966.

During his stay in Princeton, Araki was advised by Haag to study the theory of operator algebras with a book of Naimark (German version which Haag was carrying) and the original papers of von Neumann. The thesis of the fellow Ph.D. student J. S. Lew on the structure of representations of the canonical commutation relations with an introduction to some aspects of C*-algebra theory as an appendix was also another material which drew him to operator algebra theory.

Thus Araki's interest gradually became focused on the theory of operator algebras on the pure mathematics side and its application to quantum field theory and statistical mechanics on the mathematical physics side.

His interests are so wide that none of his students can cover all of them. Some on the physics side studied his ETH lecture notes — whose updated version in Japanese was recently published; some could be totally ignorant of this, studying instead his papers on entropy; others studied only his CCR and CAR papers.

Araki has worked hard not only for his fields of interest but also in promoting mathematical sciences in this country. He has also made a tremendous effort to promote international cooperation in mathematical sciences. He was instrumental in founding of the International Association of Mathematical Physics in 1977. Many students learned what scientific research should be, not only directly from Araki but also through his efforts in the community of mathematical sciences.

He has been associated with Communications in Mathematical Physics since 1965 when it first appeared, first in its Advisory Board and then in its Editorial Board since 1973. He is also the chief editor of Publications of the Research Institute for Mathematical Sciences and Reviews in Mathematical Physics, in the editorial board of Letters in Mathematical Physics, Reports on Mathematical Physics, Nuovo Cimento B and Open & Information Dynamics, a series editor of Springer Lecture Notes in Physics, and a series associate editor of Birkhauser Monographs in Mathematics.

He is at present the director of the Research Institute of Mathematical Sciences. There is no doubt that Professor H. Araki continues to be one of the most originative and productive members of the Japanese scientific community, despite occupying this time-consuming position.

List of Selected Papers of Professor Huzihiro Araki in 1950-76 [1]

1. Interaction Between Electrons in One-Dimensional Free Electron Model with Application to Absorption Spectra of Cyanine Dyes (with G. Araki), Prog. Theor. Phys. **11** (1954), 20—24.
2. On Weak Time–Symmetric Gravitational Waves, Ann.Physics **7** (1959),456–465.
3. Measurement of Quantum Mechanical Operators (with M. Yanase), Phys. Rev. **120** (1960), 622–626.
4. On Asymptotic Behavior of Vacuum Expectation Valuess at Large Space–like Separation, Ann. Physics **11** (1960), 260–274.
5. Properties of the Momentum Space Analytic Function (with N. Burgoyne), Nuovo Cimento Ser. X **18** (1960), 342–346.
6. Hamiltonian Formalism and the Canonical Commutation Relations in Quantum Field Theory, J. Mathematical Phys. **1** (1960),492–504.
7. Generalized Retarded Functions and Analytic Function in Momentum Space in Quantum Field Theory, J. Mathematical Physics **2** (1961),163–177.
8. The Determination of a Local or Almost Local Field from a Given Current (with R. Haag and B. Schroer), Nuovo Cimento Ser. X **19** (1961), 90–102.
9. On the Connection of Spin and Commutation Relations Between Different Fields, J. Mathematical Physics **2** (1961), 267–270.
10. Wightman Functions, Retarded Functions and Their Analytic Continuations, Progr. Theoret. Phys. , Suppl. No. **18** (1961), 83–125.
11. On the Asymptotic Behaviour of Wightman Functions in Space–like Directions (with K. Hepp and D. Ruelle), Helv. Phys. Acta **35** (1962), 164–174.
12. A Generalization of Borchers' Theorem, Helv. Phys. Acta **36** (1963), 132–139.
13. Representations of the Canonical Commutation Relations Describing a Nonrelativstic Infinite Free Bose Gas (with E. J. Woods), J. Mathematical Phys. **4** (1963), 637–662.
14. A Lattice of von Neumann Algebras Associated with the Quantum Theory of a Free Bose Field, J. Mathematical Phys. **4** (1963),1343–1362.
15. Type of von Neumann Algebra Associated with Free Field, Progr. Theor. Phys. **32**, (1964) 956–965.
16. Einführung in die Axiomatische Quantenfeldtheorie, I und II, Lecture Notes at ETH, Zürich, 1961–62. Notes taken by K. Hepp, F. Riahi and W. Wyss.

[1] These are a part of his publications which are related to the postscript.

17. Complete Boolean Algebras of Type I factors (with E. J. Woods),Publ. RIMS, Kyoto Univ. , Ser. A, **2** (1966), 157–242.
18. A Remark on Piron's Paper (with I. Amemiya), Publ. RIMS, Kyoto Univ. Ser. A, **2** (1967), 423–427.
19. Collision Cross Sections in Terms of Local Observables (with R. Haag), Commun. Math. Phys. **4** (1967), 77–91.
20. A Classification of Factors (with E. J. Woods), Publ. RIMS, Kyoto Univ., Ser. A, **4** (1968), 51–130.
21. A Classification of Factors (II), Publ. RIMS, Kyoto Univ., Ser. A, **4** (1969), 585–593.
22. Gibbs States of a One–Dimensional Quantum Lattice, Commun. Math. Phys. **14** (1969), 120–157.
23. Factorizable Representation of Current Algerbra, Non–Commutative Extension of the Levy–Kinchin Formula and Cohomology of a Solvable Group with Values in a Hilbert Space, Publ. RIMS, Kyoto Univ., Ser. A, **5** (1970), 361–422.
24. Local Quantum Theory, I. Proceedings of the International School of Physics "Enrico Fermi", Course 45, ed. R. Jost. (Academic Press, N. Y. and London, 1969), 65–96.
25. Entropy Inequalities (with E. H. Lieb), Commun. Math. Phys. **18** (1971), 160–170.
26. On Quasifree States of CAR and Bogolyubov Automorphisms, Publ. RIMS, Kyoto Univ. **6** (1971), 385–442.
27. On the Homotopical Significance of the Type of von Neumann Algebra factors (with Mi–Soo Bae Smith and Larry Smith), Commun. Math. Phys. **2** (1971), 711–88.
28. Some Properties of Modular Conjugation Operator of von Neumann Algebra and a Non–Commutative Radon–Nikodym Theorem with a Chain Rule, Pacific J. Math. **50** (1973), 309–354.
29. On the Definition of C*–algebras (with G. A. Elliott), Publ. RIMS, Kyoto Univ. **9** (1973), 93–112.
30. Relative Hamiltonian for Fiathful Normal States of a von Neumann Algebra, Publ. RIMS, Kyoto Univ. **9** (1973), 165–209.
31. Structure of Some von Neumann Algebras with Isolated Discrete Modular Spectrum, Publ. RIMS, Kyoto Univ. **9** (1973), 1–44.
32. On the Equivalence of the KMS condition and the Variational Principle for Quantum Lattice Systems, Commun. Math. Phys. **38** (1974), 1–10.
33. On Uniqueness of KMS States of One–Dimensional Quantum Lattice Systems, Commun. Math. Phys. **44** (1975), 1–7.
34. Relative Entropy of States of von Neumann Algebras, Publ. RIMS, Kyoto Univ. **11** (1975–76), 809–833.
35. On Clustering Property (with A. Kishimoto), Rep. Mathematical Phys. **10** (1976), 119–208.

36. Extension of KMS State and Chemical Potential (with D. Kastler, M. Takesaki and R. Haag), Commun. Math. Phys. **53** (1977), 97–134.
37. Relative Entropy for States of von Neumann Algebras, II. Publ. RIMS, Kyoto Univ. **13** (1977), 173–192.

former students

Yoshiomi Nakagami	Akitaka Kishimoto	Keiichi R. Ito
Izumi Ojima	Tetsuya Masuda	Shigeru Yamagami
Taku Matsui	Masaki Izumi	

List of Speakers

Asao Arai
Department of Mathmatics
Hokkaido University
Sapporo 060, JAPAN
arai@math.hokudai.ac.jp

Huzihiro Araki
RIMS
Kyoto Univesity
Sakyo-ku, Kyoto 606-01
JAPAN
araki@kurims.kyoto-u.ac.jp

Jean Bellissard
Laboratoire de Physique Quantique
Université Paul Sabatier
118 Route de Narbonne
F-31062, Toulouse Cedex
FRANCE
bellissa@frpqt51.bitnet

Hans-J. Borchers
Institut für Theoretische Physik
der Universität Göttingen
Bunsenstrasse 9
W-3400 Göttingen
GERMANY
hborche@dgogwdg1.bitnet

Ola Bratteli
Matematisk Institutt
Universitetet i Oslo
PB1053, Blindern, N-0316
Oslo 3, NORWAY
bratteli@math.uio.no

Marie Choda
Department of Mathematics
Osaka Kyoiku University
Minamikawahori-cho,Tennoji
Osaka 543, JAPAN
f00527@sinet.ad.jp

Man-Duen Choi
Department of Mathematics
University of Toronto
Toronto, Ontario, M5S 1A1
CANADA
choi@math.toronto.edu

Sergio Doplicher
Dipartimento di Matematica
Universita di Roma
La Sapienza
I-00185 Roma, ITALY
serdopli@itcaspur.caspur.it

George A. Elliott
Department of Mathematics
University of Toronto
Toronto, Ontario, M5S 1A1
CANADA
elliott@math.toronto.edu

David E. Evans
Department of Mathematics
University College of Swansea
Singleton Park
Swansea SA2 8PP, Wales
U.K.
madai@pyramid.swansea.ac.uk

Moshe Flato
20 rue de Tournon
F-75006, Paris
FRANCE
flato@frccub11.bitnet

Petr P. Kulish
Steklov Mathematical Institute
St. Petersbrug
Fontanka 27, St. Petersbrug 191011
RUSSIA

Antti J. Kupiainen
Department of Mathematics
Helsinki University
Hallituskatu 15, 00100 Helsinki
FINLAND
akupiai@cc.helsinki.fi

Lawrence J. Landau
Department of Mathematics
King's College London
Strand, London WC2R 2LS
U.K.
udah194@oak.cc.kcl.ac.uk

Elliott Lieb
Jadwin Hall
Princeton University
Princeton, PO Box 708
NJ 08540, U.S.A.
lieb@math.princeton.edu

Roberto Longo
Dipartimento di Matematica
Universita di Roma
Tor Vergata, Via Fontanile
di Caracaricola
I-00133 Roma, ITALY
longo@mat.utovrm.it

Tetsuya Masuda
Institute of Mathematics
University of Tsukuba
Tsukuba, Ibaraki 305
JAPAN
tetsuya@math.tsukuba.ac.jp

Taku Matsui
Department of Mathematics
Tokyo Metropolitan University
1-1 Minamiosawa, Hachioji
Tokyo 192-03, JAPAN
matsui@math.metro-u.ac.jp

Tetsuji Miwa
RIMS, Kyoto University
Sakyo-ku, Kyoto 606-01
JAPAN
miwa@kurims.kyoto-u.ac.jp

Yoshiomi Nakagami
Department of Mathematics
Faculty of Litrature and Science
Yokohama City University
22-2 Seto, Kanazawa-ku
Yokohama, Kanagawa 236
JAPAN
e00551@sinet.ad.jp

Toshikazu Natsume
Department of Mathematics
State University of New York
Buffalo, NY 14214
U.S.A.
mthnatsu@ubvms.bitnet

Yong Moon Park
Department of Mathematics
Yonsei University
Seoul 120-749
KOREA

Claudia Pinzari
Dipartimento di Matematica
Universita di Roma
La Sapienza, I-00185 Roma
ITALY

John Roberts
Dipartimento di Matematica
Universita di Roma
"Tor Verhata" via della
Ricera Scientitica
I-00133 Roma, ITALY
roberts@mat.utovrm.it

Derek W. Robinson
Department of Mathematics
Australian National University
Canberra, AUSTRALIA
derek@pell.anu.edu.au

Daniel Sternheimer
c/o Prof. M. Flato
20 rue de Tournon
F-75006, Paris
FRANCE
sternhei@frcirp81.bitnet

Raymond F. Streater
Department of Mathematics
King's College London
Strand, London WC2R 2LS
U.K.
udah110@oak.cc.kcl.ac.uk

Hongbing Su
Department of Mathematics
University of Toronto
Toronto, Ontario, M5S 1A1
CANADA
hongbing@math.toronto.edu

Masamichi Takesaki
Department of Mathematics
U.C. Los Angeles
CA 90024-1555
U.S.A.
mt@math.ucla.edu

Klaus Thomsen
Mathematisk Institut
Ny Munkegade
DK-8000 Aarhus C
DENMARK
matkt@mi.aau.dk

Yasuo Watatani
Department of Mathematics
Faculty of Science
Hokkaido University
Sapporo 060, JAPAN
watatani@math.hokudai.ac.jp

Shigeru Yamagami
Department of Mathematics
College of General Education
Tôhoku University
Aoba-ku, Sendai 980, JAPAN
e22236@cctu.ee.tohoku.ac.jp

GPSR Compliance

The European Union's (EU) General Product Safety Regulation (GPSR) is a set of rules that requires consumer products to be safe and our obligations to ensure this.

If you have any concerns about our products, you can contact us on ProductSafety@springernature.com

In case Publisher is established outside the EU, the EU authorized representative is:

Springer Nature Customer Service Center GmbH
Europaplatz 3
69115 Heidelberg, Germany

Batch number: 10614741

Printed by Printforce, the Netherlands